TEST BANK

To Accompany

ORGANIC CHEMISTRY

SIXTH EDITION

T. W. GRAHAM SOLOMONS
University of South Florida

ROBERT G. JOHNSON
Xavier University
Cincinnati, Ohio

JOHN WILEY & SONS, INC.
New York • Chichester • Brisbane • Toronto • Singapore

Copyright © 1996 by John Wiley & Sons, Inc.

This material may be reproduced for testing or
instructional purposes by people using the text.

ISBN 0-471-12091-X

Printed in the United States of America

10 9 8 7 6 5 4 3 2 1

CONTENTS

Chapter 1: Carbon Compounds and Chemical Bonds

1. Listed below are electron dot formulas for several simple molecules and ions. All valence electrons are shown; however, electrical charges have been omitted deliberately.

```
               H          H          H          H
               .          .          .          .
   H:Be:H    H:B:H      H:N:       H:N:H      H:O:H
               .          .          .          .
               H          H          H

     I         II        III         IV          V
```

Which of the structures actually bear(s) a positive charge?

A) I
B) II
C) III
D) III and V
* E) IV and V

2. In quantum mechanics a node (nodal surface or plane) is:

A) a place where X is negative.
B) a place where X is positive.
* C) a place where X = 0.
D) a place where X^2 is large.
E) a place where X^2 is negative.

3. Which of the structures below would be trigonal planar (a planar triangle)? (Electrical charges have been deliberately omitted.)

```
    .                                            .
   :F:                           H              :F:
    . .            .             .              . .
    B:F:         H:O:H          H:C:          :F:N:
    . .            .             .              . .
   :F:             H             H              :F:
    .                                            .

    I              II            III             IV
```

* A) I B) II C) III D) IV E) I and IV

4. What is the formal charge on oxygen in the following structure?

```
        .
   CH3DODCH3
        3
       CH3
```

A) +2 * B) +1 C) 0 D) -1 E) -2

5. In which structure(s) below does the oxygen have a formal charge of +1?

```
                                              H
        .                .                3  .               .
      HDODH            HDO:            HDCMODH          CH3DODCH3
        3                .                                  3
        H                                                  CH3

        I               II                III               IV
```

A) I only
B) II only
C) I and III
D) I and IV
* E) I, III, and IV

6. Which of the following would have a trigonal planar (or triangular) structure?

```
     -              +                                            +
   :CH3            CH3           :NH3          BF3             :OH3

     I              II            III           IV               V
```

A) I, II, and IV
* B) II and IV
C) IV
D) II, IV, and V
E) All of these

7. Which structure(s) contain(s) an oxygen that bears a formal charge of +1?

```
     .                                                         .
    :O                                                         O:
     :            .    .           .             .           . : .
    HDC        :OMCMO:        CH3DODH       CH3DODCH3      :ODCDO:
       \                          3              3           .    .
       :O:                        H             CH3
        .

     I            II             III             IV            V
```

A) I and II
* B) III and IV
C) V
D) II
E) I and V

8. Which of the following molecules or ions has a nitrogen with a formal charge of -1? (Charges on ions have been omitted.)

* A)
```
  .
:NDH
 3
 H
```

B)
```
   .
HDNDH
  3
  H
```

C)
```
   .
HDNDCH3
  3
  H
```

D)
```
       .
CH3DNDCH3
       3
       H
```

E) CH3CpN:

9. When the 1s orbitals of two hydrogen atoms combine to form a hydrogen molecule, how many molecular orbitals are formed?

A) 1 * B) 2 C) 3 D) 4 E) 5

10. Which molecule has a zero dipole moment?

A) SO_2
* B) CO_2
C) CO
D) $CHCl_3$
E) None of these

11. In which structure(s) below does nitrogen have a formal charge of +1?

```
      H                          H
      3                .          \. .        .         .
 CH3DNMCH2         CH3DNDH        NDOH     CH3NH2    CH3DNDCH3
                       .          /  .                   3
                                 H                      CH3

      I               II           III       IV          V
```

* A) I B) II C) III D) IV E) V

12. Select the most electronegative element.

A) H * B) O C) N D) B E) C

13. Which molecule has a zero dipole moment?

A) CH_3Cl
B) CH_2Cl_2
C) $CHCl_3$
* D) CCl_4
E) None of these

14. Which principle(s) or rule must be used to determine the correct electronic configuration for carbon in its ground state?

A) Aufbau Principle
B) Hund's Rule
C) Pauli Exclusion Principle
D) (A) and (B) only
* E) All three

15. What point on the potential energy diagram below represents the most stable state for the hydrogen molecule?

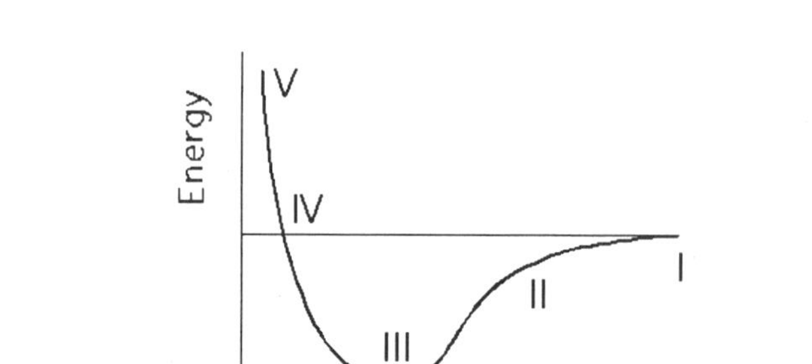

A) I B) II * C) III D) IV E) V

16. According to molecular orbital theory, which molecule could not exist?

A) H_2 * B) He_2 C) Li_2 D) F_2 E) N_2

17. Select the hybridized atomic orbital.

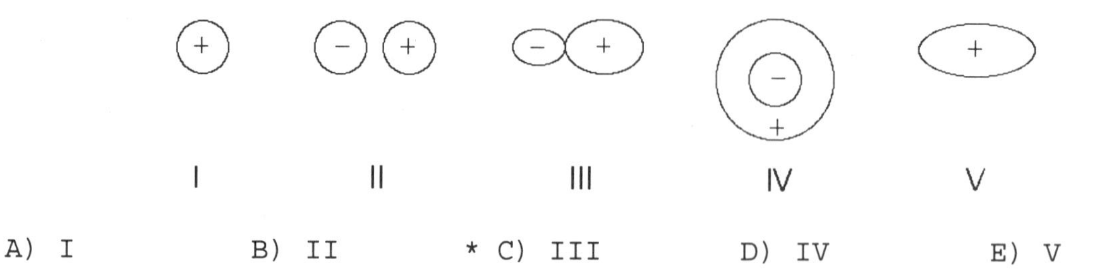

A) I B) II * C) III D) IV E) V

18. What bond angle is associated with a tetrahedral molecule?

A) 120x * B) 109.5x C) 180x D) 90x E) 45x

19. What would be the spatial arrangement of the atoms of the methyl anion, $:CH_3^-$?

A) Octahedral
B) Tetrahedral
C) Trigonal planar
D) Linear
* E) Trigonal pyramidal

20. Which of the following is an ion with a single negative charge?

A)
```
       .
   CH3DO:
       .
```

B)
```
     . . .
   :ODNMO:
     .
```

C)
```
        .
       :O:
    .  /
   :OMN
       \
       :O:
        .
```

* D) All of these

E) None of these

21. How many 2p atomic orbitals from boron must be mixed with a 2s atomic orbital to yield the bonding hybrid atomic orbitals in BF_3?

A) 1   * B) 2   C) 3   D) 4   E) 5

22. The formal charge on sulfur in sulfuric acid is:

```
       O
       :
   HDODSDODH
       :
       O
```

* A) 0   B) -1   C) +1   D) +2   E) -2

23. Which of the following could not be a resonance structure of CH_3NO_2?

A)
```
            -
     H :Ȯ:
     3  3
   HDDCDDNMMȮ
     3  +   ·
     H
```

B)
```
     H  Ȯ:
     3  :
   HDDCDDNDDȮ:
     3      ·
     H  +   -
```

C)
```
            -
     H :Ȯ:
     3  3     -
   HDDCDDNDDȮ:
     3  +2  ·
     H
```

* D)
```
            -
     H :Ȯ:
     3  3
   HDDCMMNDDȮDDH
        +   ·
```

E) Both C and D

24. Which molecule would be linear? (In each case you should write a Lewis structure before deciding.)

A) SO_2 * B) HCN   C) H_2O_2 D) H_2S E) OF_2

25. Which molecule would have a dipole moment greater than zero?

A) $BeCl_2$ B) BCl_3 C) CO_2 * D) H_2O E) CCl_4

26. Which is NOT a correct Lewis structure?

* A)

```
  .
HDNMF:
  3 .
  H
```

B)

```
  H
  3 .
HDCDF:
  3 .
  H
```

C)

```
  . .
HDODODH
  . .
```

D)

```
          .
         ODH
     .  /.
HDODB
     .  \.
         ODH
          .
```

E) None of these

27. In which molecule is the central atom sp^3 hybridized?

A) CH_4

B) $:NH_3$

C)

```
  .
H2O:
```

* D) All of these

E) None of these

28. Which of the following pairs are NOT resonance structures?

A)

```
                                  +
     . . .                     . . .  -
CH3DODNMO:      and      CH3DOMNDO:
     .                          .
```

B)

```
 .   .                  . + .  -
:OMCMO:         and    :OMCDO:
                             .
```

* C)

```
                              :O:
     . . .                    +N
CH3DODNMO:      and     CH3DN
     .                        \ -
                              :O:
                               .
```

D) Each of these pairs represents resonance structures.

E) None of these pairs represents resonance structures.

29. Listed below are electron dot formulas for several simple molecules and ions. All valence electrons are shown; however, electrical charges have been omitted deliberately.

```
                 H              H              H              H
                 .              .              .              .
   H:Be:H      H:B:H          H:N:          H:N:H          H:O:H
                 .              .              .              .
                 H              H              H

     I           II            III            IV              V
```

Which of the structures is negatively charged?

A) I
* B) II
C) III
D) III and V
E) IV and V

30. How many resonance structures can be written for the NO_3^- ion in which the nitrogen atom bears a formal charge of +1?

A) 1
B) 2
* C) 3
D) 4
E) 5

31. In which of the following would you expect the central atom to be sp^3 hybridized (or approximately sp^3 hybridized)?

A) BH_4^-
B) NH_4^+
C) CCl_4
D) $CH_3:^-$
* E) All of these

32. What shape does the methyl cation, CH_3^+, have?

A) Octahedral
B) Tetrahedral
* C) Trigonal planar
D) Linear
E) Trigonal pyramidal

33. When the 1s orbitals of two hydrogen atoms combine to form a hydrogen molecule, how many molecular orbitals are formed?

A) One bonding molecular orbital only
B) Two bonding molecular orbitals
* C) One bonding molecular orbital and one antibonding molecular orbital
D) Two antibonding molecular orbitals
E) Three bonding molecular orbitals

34. The electron configuration, $\underset{1s}{\underline{tu}}\ \underset{2sp^3}{\underline{tu}}\ \underset{2sp^3}{\underline{t}}\ \underset{2sp^3}{\underline{t}}\ \underset{2sp^3}{\underline{t}}$ represents:

A) the ground state of boron.
B) the sp^3 hybridized state of carbon.
* C) the sp^3 hybridized state of nitrogen.
D) the ground state of carbon.
E) an excited state of carbon.

35. What is the empirical formula for cyclohexane? (Its molecular formula is C_6H_{12}.)

A) CH * B) CH_2 C) C_2H_4 D) C_6H_6 E) C_2H_2

36. A compound has the empirical formula, CCl. Its molecular weight is 285 +/- 5. What is the molecular formula for the compound?

A) C_2Cl_2 B) C_3Cl_3 C) C_4Cl_4 D) C_5Cl_5 * E) C_6Cl_6

37. A compound consists only of carbon, hydrogen and oxygen. Elemental analysis gave: C, 70.5%, H, 13.8%. The molecular weight of the compound was found to be 103 +/- 3. What is the molecular formula for the compound?

A) $C_6H_{12}O$ B) $C_5H_{12}O_2$ C) $C_3H_2O_4$ D) $C_3H_6O_3$ * E) $C_6H_{14}O$

38. Which of the following is a set of constitutional isomers?

I: Br (skeletal structure)

II: $CH_3CH(CH_3)CH_2Br$ — drawn as CH_3 above, CH_3CHCH_2Br

III: CH_2Br above, $CH_3CHCH_2CH_3$

IV: CH_3 above, CH_3CCH_2Br, CH_3 below

A) I and II
B) II and III
C) I, II, and III
D) II, III, and IV
* E) I, III, and IV

39. Credit for the first synthesis of an organic compound from an inorganic precursor is usually given to:

A) Berzelius.
B) Arrhenius.
C) Kekule.
* D) Wohler.
E) Lewis.

40. The greatest degree of ionic character is anticipated for the bond between:

A) H and C * B) H and Cl C) C and Cl D) H and Br E) Br and Cl

41. Expansion of the valence shell to accomodate more than eight electrons is possible with:

A) Fluorine.
B) Nitrogen.
C) Carbon.
* D) Sulfur.
E) Beryllium.

42. $CH_3CH_2OCH_2CH_3$ and $CH_3CH_2CH_2CH_2OH$ are examples of what are now termed:
A) Structural isomers.
B) Resonance structures.
C) Functional isomers.
D) Empirical isomers.
* E) Constitutional isomers.

43. For a molecule to possess a dipole moment, the following condition is necessary but not sufficient.
A) Three or more atoms in the molecule
* B) Presence of one or more polar bonds
C) A non-linear structure
D) Presence of oxygen or fluorine
E) Absence of a carbon-carbon double or triple bond

44. Which set contains non-equivalent members?
A) Enthalpy and heat content
B) Endothermic reaction and +(H
C) Exothermic reaction and -(H
D) Kinetic energy and energy of motion
* E) High energy and high stability

45. VSEPR theory predicts an identical shape for all of the following, except:
A) NH_3
B) H_3O^+
* C) BH_3
D) CH_3^-
E) All have the same geometry.

46. A non-zero dipole moment is exhibited by:
* A) SO_2
B) CO_2
C) CCl_4
D) BF_3
E)
```
H     H
 \   /
  CMC
 /   \
H     H
```

47. Which of these substances contains both covalent and ionic bonds?
* A) NH_4Cl B) H_2O_2 C) CH_4 D) HCN E) H_2S

48. The bond angles in PH_3 would be expected to be approximately:
A) 60x B) 90x C) 105x * D) 109x E) 120x

49. In which case do the members of the following pairs of structures represent different compounds?

A)
```
CH3DCH2DCH2        and        ClCH2DCH2
        3                             3
        CH2Cl                   CH3DCH2
```

* B)
```
CH2DCH2DCH3            and            H
3                                     3
CH3                               CH3DCDCH3
                                      3
                                      CH3
```

C)
```
   H              and              H
   3                               3
BrDCDBr                        BrDCDH
   3                               3
   H                               Br
```

D)
```
  Br                            and                     H
  3                                                     3
HDCDCH2DCH2DCH3                              CH3DCH2DCH2DCDBr
  3                                                     3
  Br                                                    Br
```

E)
```
     CH3            and             H
     3                              3
CH3DCDH                        CH3DCDCH3
     3                              3
     CH3                            CH3
```

50. Which of these is a correct electron-dot representation of the nitrite ion, NO_2^-?

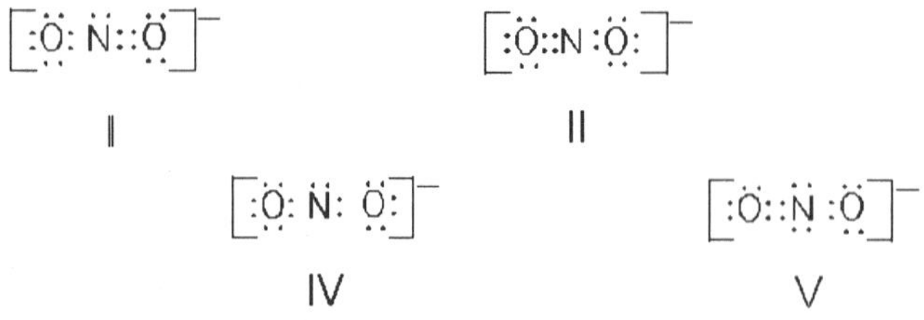

[:Ö: N::Ö]⁻ I

[:Ö::N :Ö:]⁻ II

[:Ö::N::Ö:]⁻ III

[:Ö: N: Ö:]⁻ IV

[:Ö::N :Ö]⁻ V

* A) I B) II C) III D) IV E) V

51. What is the approximate hybridization state of the nitrogen atom in trimethylamine?

A) sp B) sp^2 * C) sp^3 D) p^3 E) d^2sp^3

52. Which molecule has a non-linear structure (i.e., for which molecule are the nuclei <u>not</u> in a straight line)?

A) OMCMO * B) HDODH C) HDCl D) HDCpN E) HDCpCDH

53. Based on VSEPR theory, which of the following would have a trigonal planar shape?

A) $(CH_3)_3N$ B) HCN C) NH_4^+ D) CH_3^- * E) CH_3^+

54. Which of the following structures represent compounds that are constitutional isomers of each other?

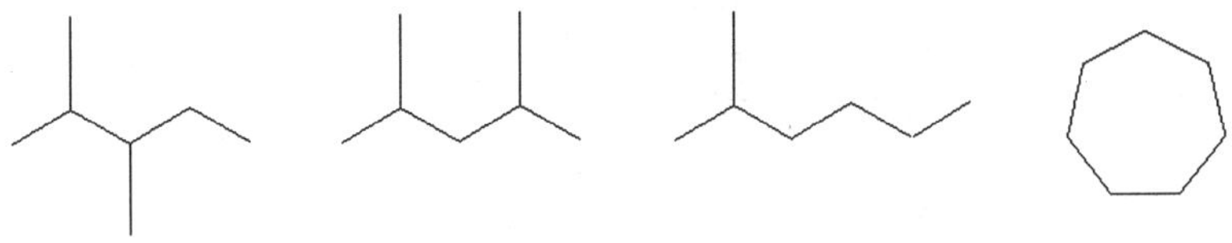

I II III IV

A) I and II
B) I and III
* C) I, II, and III
D) I, II, III, and IV
E) II and III

55. Which compound contains a nitrogen atom with a formal positive charge?

CH_3NHCH_3 (I) $CH_3CH_2N(CH_3)_3$ (II) (III: pyrrolidine N bearing CH_3 and CH_2CH_3)

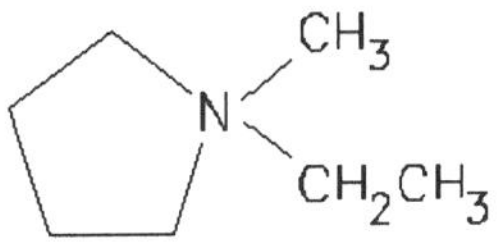

A) I
B) II
C) III
* D) More than one of the above
E) None of the above

56. Which compound is not an isomer of the others?

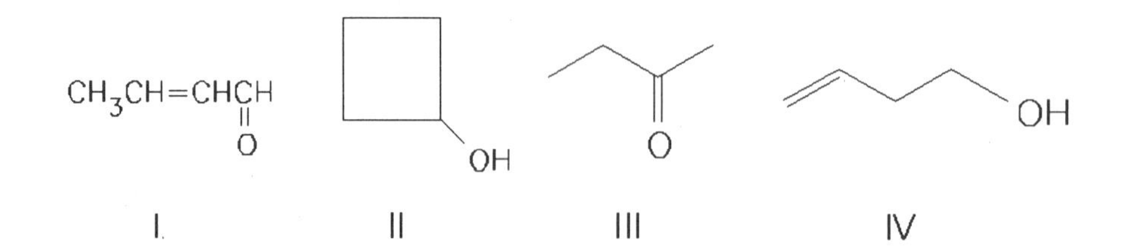

* A) I
B) II
C) III
D) IV
E) All of the above are isomers of each other.

57. Considering Lewis structures, which of these compounds possesses a single unpaired electron?

A) N_2 B) N_2O * C) NO D) N_2O_4 E) O_2

58.
```
  .
  Z
 /·\
Y   Y
```
is a generalized structural representation which can be used for all of the following, except:

A) OF_2
B) NH_2^D
C) H_2S
* D) $BeBr_2$
E) There is no exception.

59. In which of these cases, does the central atom have a zero formal charge?

A) HFH

B)
```
   H
   3
CH3OCH3
```

C)
```
 F
 3
FBF
 3
 F
```

D)
```
H   CH3
 \ /
  N
 / \
H   CH3
```

* E)
```
  CH3
   3
CH3CCH3
   3
  CH3
```

60. Which of the structures below is not expected to contribute to the CO_2 resonance hybrid?

A)
```
 + -
OMCDO
```

B)
```
- +
ODCMO
```

C) OMCMO

* D)
```
- ++ -
ODCDO
```

E)
```
-    +
ODCpO
```

61. Which of these structures would be a perfectly regular tetrahedron?

A) CH_3Br
B) CH_2Br_2
C) $CHBr_3$
* D) CBr_4
E) More than one of these

1. Which compound would you expect to have the lowest boiling point?

A) $CH_3CH_2CH_2NH_2$

B) CH_3CHCH_3
3
NH_2

C) CH_3CH_2NH
3
CH_3

* D) CH_3DNDCH_3
3
CH_3

E) CH_3
3
CH_3DCDNH_2
3
CH_3

2. Which compound listed below is a secondary alcohol?

* A) $CH_3CH_2CHCH_3$
3
OH

B) CH_3CHCH_2OH
3
CH_3

C) CH_3
3
CH_3DCDOH
3
CH_3

D) $CH_3CH_2CH_2CH_2OH$

E) $CH_3CH_2CH_2OCH_3$

3. Which compound is a secondary amine?

A) $CH_3CH_2CH_2NH_2$

B) CH_3CHCH_3
3
NH_2

* C) CH_3CH_2NH
3
CH_3

D) CH_3DNDCH_3
3
CH_3

E) $CH_3CH_2CHNH_2$
3
CH_3

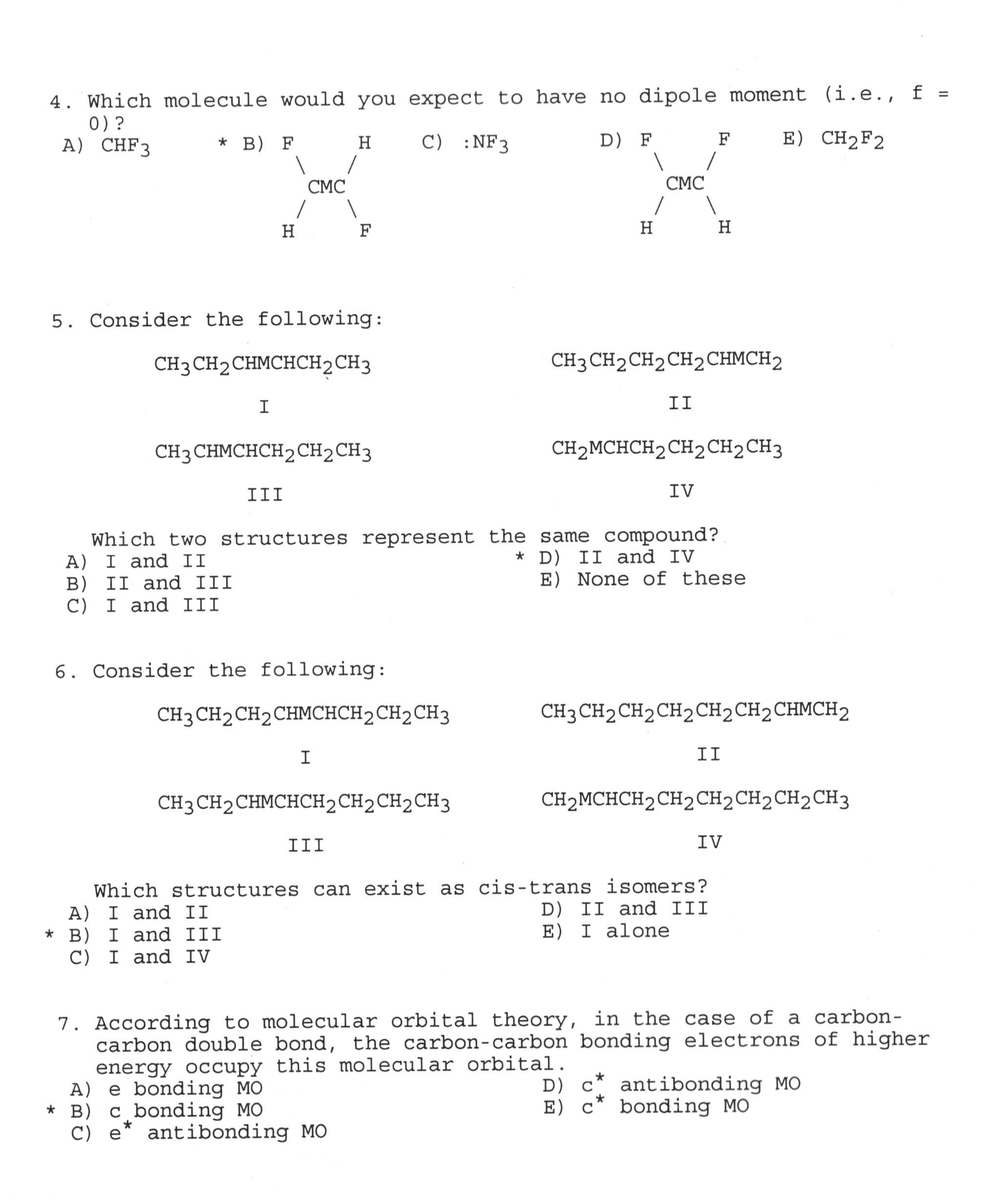

4. Which molecule would you expect to have no dipole moment (i.e., f = 0)?

A) CHF_3

* B)
```
F       H
 \     /
  CMC
 /     \
H       F
```

C) $:NF_3$

D)
```
F       F
 \     /
  CMC
 /     \
H       H
```

E) CH_2F_2

5. Consider the following:

$CH_3CH_2CHMCHCH_2CH_3$ (I)

$CH_3CH_2CH_2CH_2CHMCH_2$ (II)

$CH_3CHMCHCH_2CH_2CH_3$ (III)

$CH_2MCHCH_2CH_2CH_2CH_3$ (IV)

Which two structures represent the same compound?

A) I and II
B) II and III
C) I and III
* D) II and IV
E) None of these

6. Consider the following:

$CH_3CH_2CH_2CHMCHCH_2CH_2CH_3$ (I)

$CH_3CH_2CH_2CH_2CH_2CH_2CHMCH_2$ (II)

$CH_3CH_2CHMCHCH_2CH_2CH_2CH_3$ (III)

$CH_2MCHCH_2CH_2CH_2CH_2CH_2CH_3$ (IV)

Which structures can exist as cis-trans isomers?

A) I and II
* B) I and III
C) I and IV
D) II and III
E) I alone

7. According to molecular orbital theory, in the case of a carbon-carbon double bond, the carbon-carbon bonding electrons of higher energy occupy this molecular orbital.

A) e bonding MO
* B) c bonding MO
C) e^* antibonding MO
D) c^* antibonding MO
E) c^* bonding MO

8. Which compound is an aldehyde?

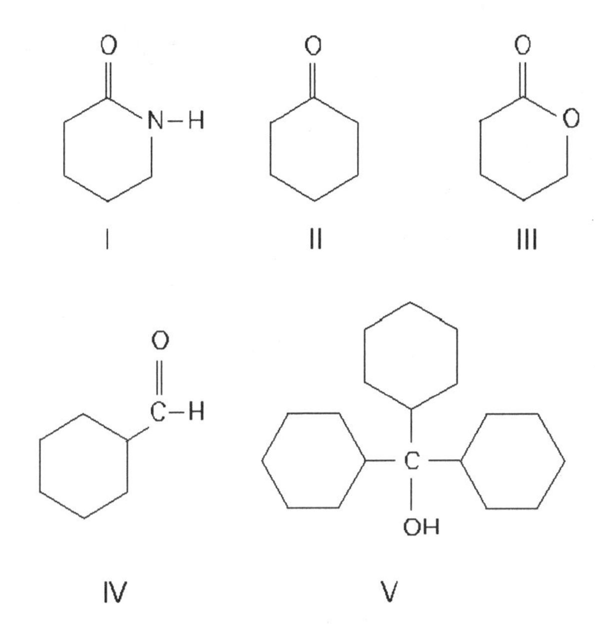

A) I B) II C) III * D) IV E) V

9. Which compound is a ketone?

A)
```
 O
 :
HCOH
```

* B)
```
CH3CCH2CH3
    :
    O
```

C)
```
 O
 :
HCOCH3
```

D)
```
H
 \
  CMO
 /
H
```

E)
```
CH3
  \
   CHOH
  /
CH3
```

10. Which compound is an ester?

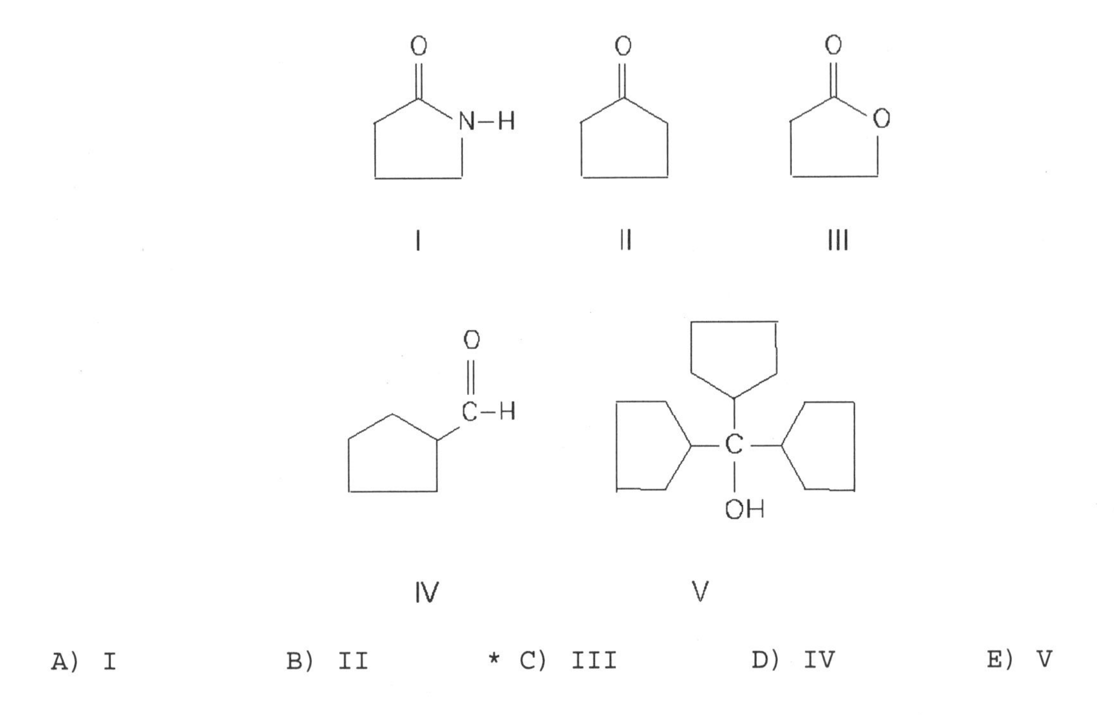

A) I B) II * C) III D) IV E) V

11. The compound shown below is a synthetic estrogen. It is marketed as an oral contraceptive under the name Enovid.

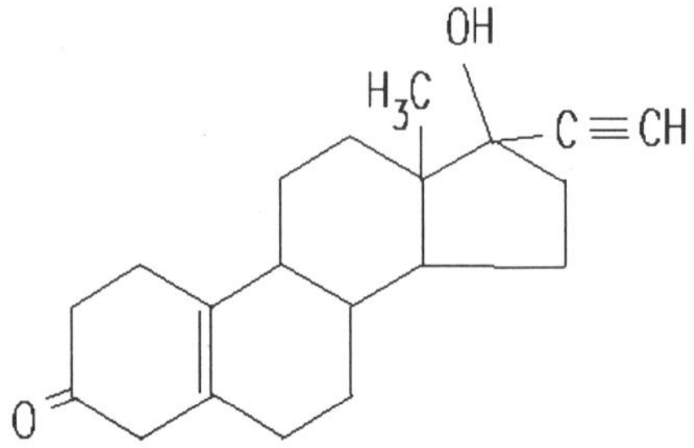

In addition to an alkane (actually cycloalkane) skeleton, the Enovid molecule also contains the following functional groups:

A) Ether, alcohol, alkyne
B) Aldehyde, alkene, alkyne, alcohol
C) Alcohol, carboxylic acid, alkene, alkyne
* D) Ketone, alkene, alcohol, alkyne
E) Amine, alkene, ether, alkyne

12. Which of these compounds would have the highest boiling point?

A) $CH_3OCH_2CH_2CH_2OCH_3$

B) $CH_3CH_2OCH_2CH_2OCH_3$

C) $CH_3CH_2OCH_2OCH_2CH_3$

D) $CH_3OCH_2CHOCH_3$
```
        3
        CH3
```

* E) $HOCH_2CH_2CH_2CH_2CH_2OH$

13. Which of these would you expect to have the lowest boiling point?

A) $CH_3CH_2CH_2OH$

B) CH_3CHCH_3
```
    3
    OH
```

* C) $CH_3OCH_2CH_3$

D) $CH_3CH_2CH_2CH_2OH$

E) $CH_3CH_2OCH_2CH_3$

14. How many sigma (1s-2sp^3) bonds are there in ethane?

A) 7 * B) 6 C) 5 D) 3 E) 1

15. Which of the following represent pairs of constitutional isomers?

A)
```
        O                O
        :                :
  CH3CH2COH    and   CH3COCH3
```

B)
```
  Cl     H           H     Cl
    \   /              \   /
     CMC       and      CMC
    /   \              /   \
  H      Cl          H      Cl
```

C)
```
  Cl     H           Cl     Cl
    \   /              \   /
     CMC       and      CMC
    /   \              /   \
  H      Cl          H      H
```

* D) More than one of these pairs

E) All of these pairs

16. Which is a 3x alkyl halide?

* A)
```
      CH3
      3
   CH3CDCl
      3
      CH3
```

B)
```
      CH3
      3
   CH3CCH2Cl
      3
      CH3
```

C) $CH_3CH_2CHCH_3$
```
          3
          Cl
```

D) $CH_3CH_2CH_2CH_2Cl$

E) $ClCHCH_2CH_3$
```
    3
    CH3
```

17. Which compound would have the highest boiling point?

A) $CH_3CH_2CH_2CH_2CH_2CH_3$

B) $CH_3CH_2OCH_2CH_2CH_3$

* C) $CH_3CH_2CH_2CH_2CH_2OH$

D) $CH_3CH_2OCH(CH_3)_2$

E) $CH_3OCH_2CH_2CH_2CH_3$

18. Which is not an intermolecular attractive force?

A) Ion-ion

B) van der Waals

C) Dipole-dipole

* D) Resonance

E) Hydrogen bonding

19. Which is a 3^x amine?

A)
```
      CH3
      3
CH3DCDNH2
      3
      CH3
```

B) $CH_3CH_2CH_2NH_2$

* C) $CH_3CH_2N(CH_3)_2$

D)
```
CH3CHNHCH3
    3
    CH3
```

E) $(CH_3)_3CNHCH_3$

20. Which molecule contains an sp-hybridized carbon?

* A) HCN

B) CH_2MCH_2

C) CH_3Cl

D)
```
H
 \
  CMO
 /
H
```

E) CH_3CH_3

21. Which functional group is not contained in prostaglandin E_1?

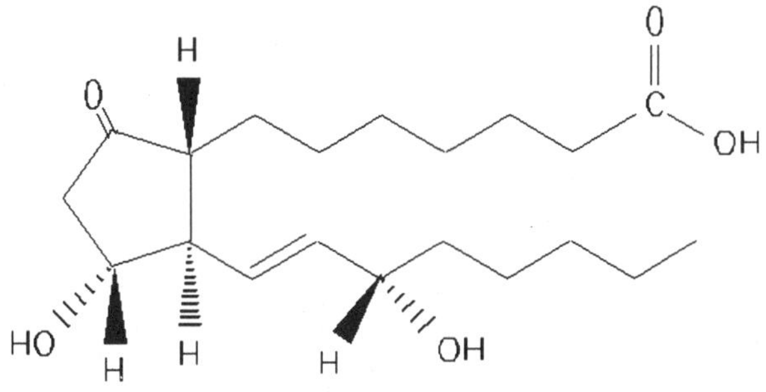

Prostaglandin E_1

A) Ketone
B) 2^x alcohol
* C) 3^x alcohol
D) Carboxylic acid
E) Alkene

22. Which compound would you expect to have the lowest boiling point?

A)
```
          O
          :
CH3CH2CH2CNH2
```

B)
```
      O
      :
CH3CH2CNHCH3
```

* C)
```
   O
   :
CH3CN(CH3)2
```

D)
```
      O
      :
CH3CHCNH2
    3
    CH3
```

E)
```
O
:
HCNHCH2CH2CH3
```

23. Which of the following pairs of formulas represent constitutional isomers?

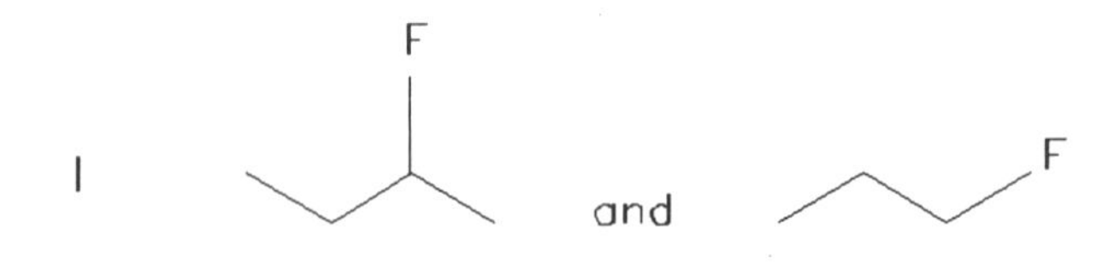

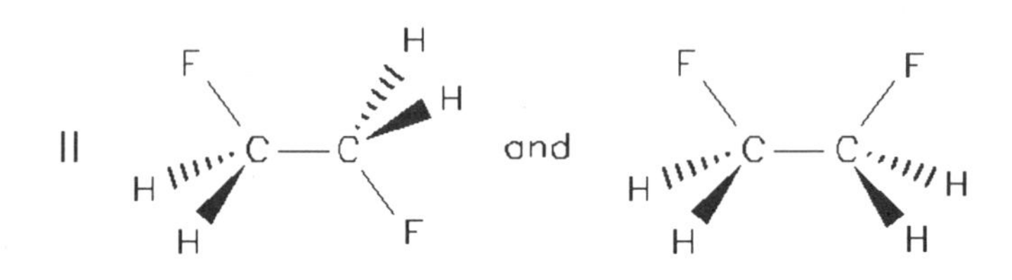

III $CH_3CH_2CH_2NH_2$ and CH_3NCH_3 (with CH_3 on N)

A) I
B) II
* C) III
D) More than one of these pairs
E) All of these pairs

24. Which compound would you expect to have the highest boiling point?

A) $CH_3OCH_2CH_2OCH_3$
B) $CH_3OCH_2OCH_2CH_3$
* C) $HOCH_2CH_2CH_2CH_2OH$
D) $CH_3OCH_2CH_2CH_2OH$
E) $(CH_3O)_2CHCH_3$

25. Which of the following represent a pair of constitutional isomers?

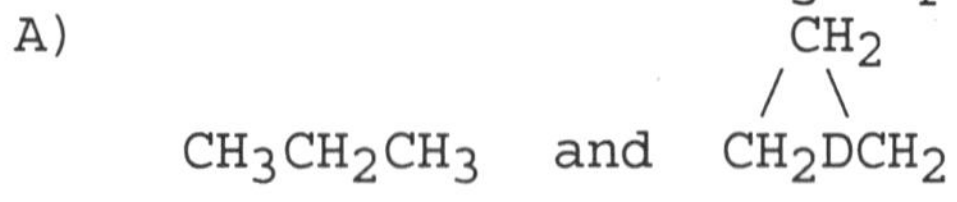

C)
```
Br      Br         H      Br
  \    /            \    /
   CMC      and      CMC
  /    \            /    \
H       H         Br      H
```

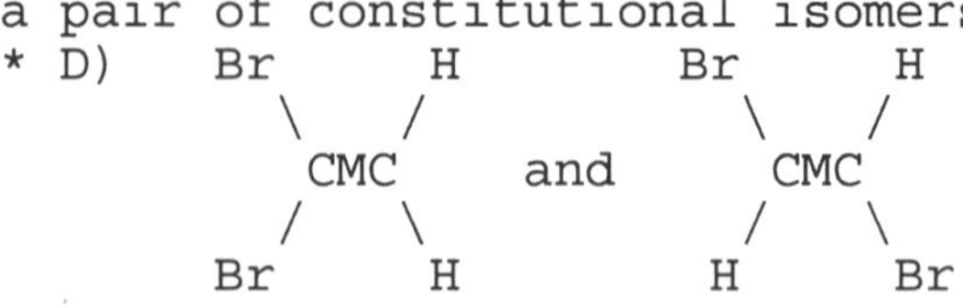

E) More than one of these

26. The compound below is an adrenocortical hormone called cortisone. Which functional group is not present in cortisone?

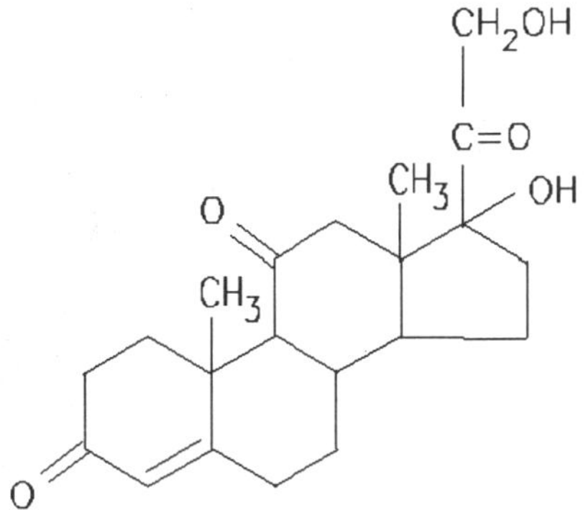

A) 1^x alcohol
* B) 2x alcohol
C) 3x alcohol
D) Ketone
E) Alkene

27. Which of the following would have no net dipole moment (f = 0)?

A) CBr_4
B) <u>cis</u>-1,2-Dibromoethene
C) <u>trans</u>-1,2-Dibromoethene
D) 1,1-Dibromoethene
* E) More than one of these

28. Which compound has the shortest carbon-carbon bond(s)?

A) CH_3CH_3

B) CH_2MCH_2

* C) HCpCH

D) $CH_3CH_2CH_3$

E) All carbon-carbon bonds are the same length.

29. The compound shown below is a steroid called progesterone.

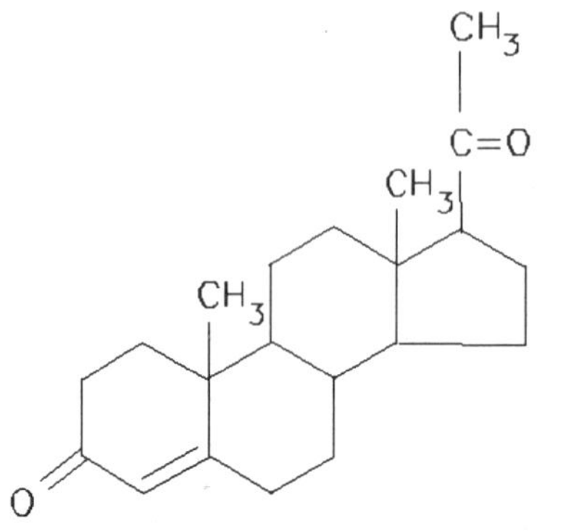

In addition to the cycloalkane skeleton, the progesterone molecule contains which other functional groups?

* A) Alkene and ketone
B) Alkene, ketone, and ether
C) Aldehyde, alkene, and ketone
D) Ester, alkene, and ketone
E) Ether, ester, and alkyne

30. Drawn below is atropine.

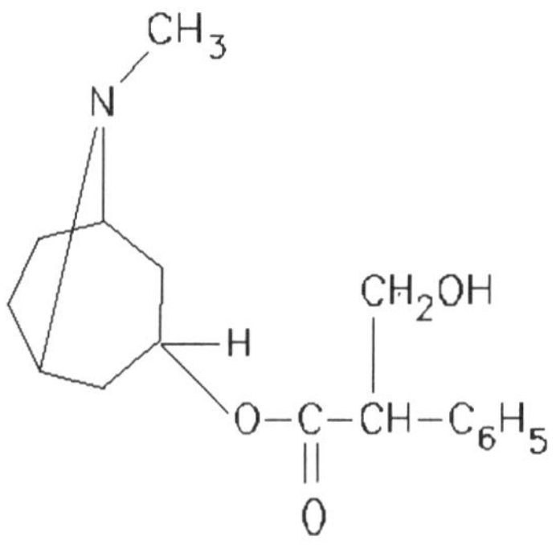

Which of the following functional groups is NOT in atropine?

A) Amine
B) Ester
C) Alcohol
D) Benzene ring
* E) Ketone

31. The compound shown below is the male sex hormone, testosterone.

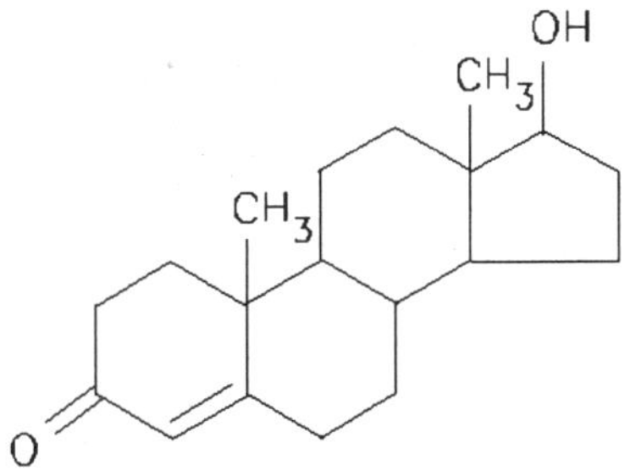

In addition to a cycloalkane skeleton, testosterone also contains the following functional groups:

A) Alkene, ester, tertiary alcohol
B) Alkene, ether, secondary alcohol
* C) Alkene, ketone, secondary alcohol
D) Alkyne, ketone, secondary alcohol
E) Alkene, ketone, tertiary alcohol

32. Which formula represents a compound that is different from all the rest?

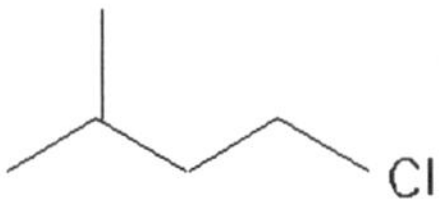

I

$ClCH_2CH_2CH(CH_3)_2$

II

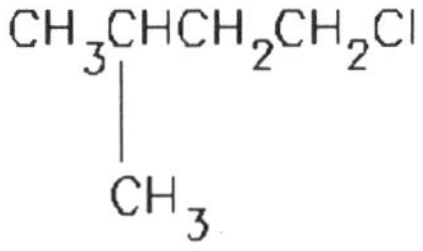

III

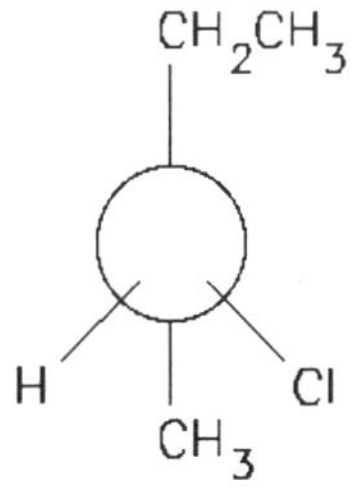

IV

A) I
B) II
C) III
* D) IV
E) None of these

33. Which is a carboxylic acid?

A) O
   :
   $HCCH_2CH_3$

B) $HOCH_2CH_2CH_3$

C) O
   :
   $CH_3CH_2OCCH_3$

D) $CH_3CH_2CCH_2CH_3$
   :
   O

* E) O
   :
   $HOCCH(CH_3)_2$

34. Which compound is a tertiary alcohol?

CH_3 | CH_3-O-H | CH_3

I

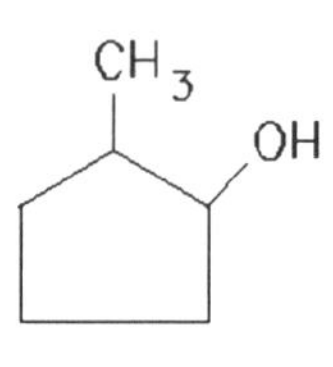

II

CH_2OH

III

OCH_3

IV

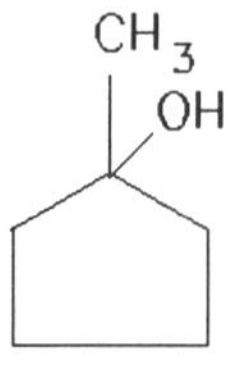

V

A) I B) II C) III D) IV * E) V

35. Which molecule has dipole moment greater than zero?

A)

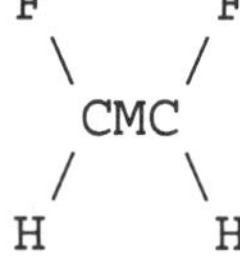

B) F H
   \ /
   CMC
   / \
   F H

C) F H
   \ /
   CMC
   / \
   H F

* D) More than one of these

E) None of these

36. Which compound is a primary amine with the formula $C_5H_{13}N$?

CH_3 N–H NH_2 NH_2

I II III

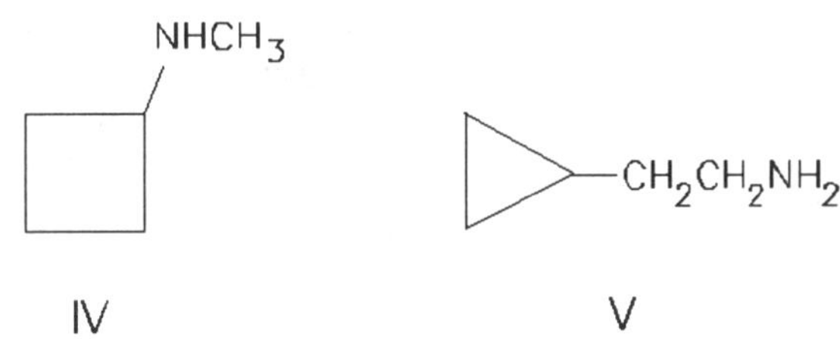

A) I B) II * C) III D) IV E) V

37. Which compound is an alcohol?

A) CH_3CONH_2

B) CH_3CCH_3 (with :O below the central C)

C) $CH_3CO_2CH_2CH_3$

D) CH_3CHO

* E) $(CH_3)_3COH$

38. Which of the following contains an sp^2-hybridized carbon?

A) CH_4 B) $CH_3:^-$ C) CH_3CH_3 * D) CH_3^+ E) HCpCH

39. Identify the atomic orbitals in the C-C sigma bond in ethyne:

A) $(2sp^2, 2sp^2)$

B) $(2sp^3, 2sp^3)$

* C) (2sp, 2sp)

D) (2p, 2p)

E) (2sp, 1s)

40. The C-O-C bond angle in diethyl ether is predicted to be approximately:

A) 90x * B) 105x C) 110x D) 120x E) 180x

41. Which compound contains a secondary carbon atom?

A) CH_4

B) CH_3CH_3

* C) $CH_3CH_2CH_3$

D) CH_3CHCH_3
   3
   CH_3

E) CH_3
   3
   CH_3CH
   3
   CH_3

42. Overlap of p-orbital lobes of opposite signs results in the formation of:

A) a bonding e orbital.
B) an antibonding e orbital.
C) a bonding c orbital.
* D) an antibonding c orbital.
E) None of the above

43. For ethene, the molecular orbitals increase in energy in the order:

* A) e < c < c* < e*
B) e < c < e* < c*
C) e < e* < c < c*
D) e* < c* < c < e
E) e* < e < c* < c

44. The strongest of attractive forces are:

A) van der Waals
B) Ion-dipole
C) Dipole-dipole
* D) Cation-anion
E) Hydrogen bonds

45. The number of unique monochloro derivatives of propene is:

A) 2 B) 3 * C) 4 D) 5 E) 6

46. Of the following compounds, the one with the highest boiling point is:

A) CH_3CH_3

B) CH_3CH_2Cl

C) H
   3
   CH_3CMO

* D) CH_3CH_2OH

E) $CH_3CH_2OCH_2CH_3$

47. This alkane is predicted to have the highest melting point of those shown:

A) $CH_3CH_2CH_2CH_3$

B) CH_3CHCH_3
      3
      CH_3

C) $CH_3CH_2CH_2CH_2CH_3$

D) $CH_3CHCH_2CH_3$
      3
      CH_3

* E) CH_3
      3
      CH_3CCH_3
      3
      CH_3

48.

H
3
The carbon-carbon bond in CH_3CMO results from the overlap of

2 1

which orbitals (in the order C_1, C_2)?

A) $sp\text{-}sp^2$ B) $sp\text{-}sp^3$ C) $sp^2\text{-}sp^2$ * D) $sp^2\text{-}sp^3$ E) $sp^3\text{-}sp^3$

49. An example of a tertiary amine is:

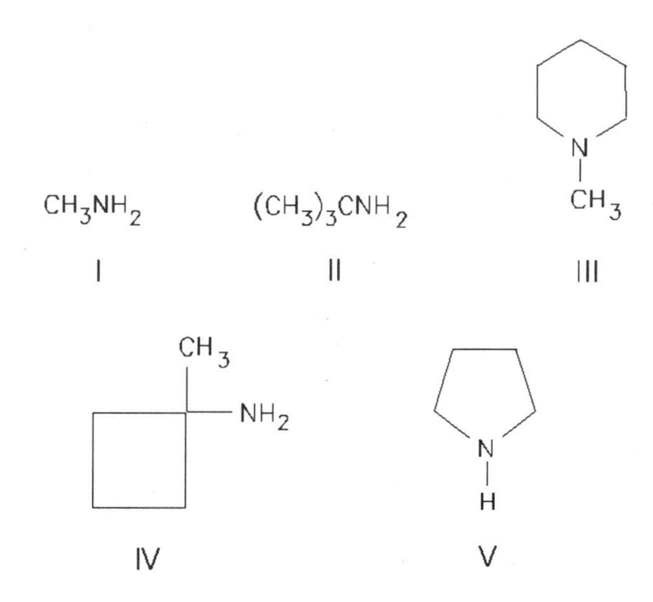

A) I   B) II   * C) III   D) IV   E) V

50. The solid alkane $CH_3(CH_2)_{18}CH_3$ is expected to exhibit the greatest solubility in which of the following solvents?

* A) CCl_4
B) CH_3OH
C) H_2O
D) CH_3NH_2
E) $HOCH_2CH_2OH$

51. <u>Cis-trans</u> isomerism is possible only in the case of:
A) CH_2MCBr_2 B) CH_2MCHBr * C) BrCHMCHBr D) $Br_2CMCHBr$ E) Br_2CMCBr_2

52. What is the shortest of the carbon-carbon single bonds indicated by arrows in the following compounds?

A) u
H_3CDCH_3

B) u
$H_3CDCpCH$

C) u
$H_3CDCHMCH_2$

* D) u
HCpCDCpCH

E) u
$H_2CMCHDCpCH$

53. Which compound would have the lowest boiling point?

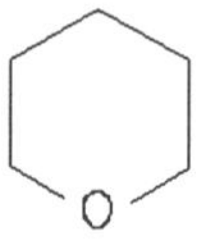

O
I

OH
II

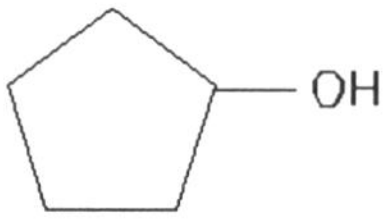

OH
III

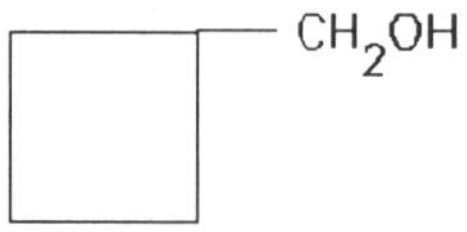

CH_2OH
IV

OH
V

* A) I B) II C) III D) IV E) V

54. Which functional groups are present in the following compound?

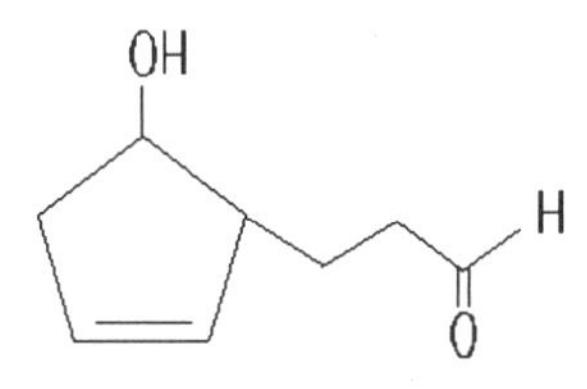

A) Alkene, 1x alcohol, ketone
* B) Alkene, 2x alcohol, aldehyde
C) Alkene, 2x alcohol, ketone
D) Alkyne, 1x alcohol, aldehyde
E) Alkyne, 2x alcohol, ketone

55. How many constitutional isomers are possible with the formula $C_4H_{10}O$?

A) 3 B) 4 C) 5 D) 6 * E) 7

56. A tertiary carbon atom is present in which of these compounds?

CH_3 | CH_3CCH_3 | CH_3	$CH_3CH_2CH_3$	(cyclopentane)–CH_3	(cyclobutane) with CH_3, CH_3
I	II	III	IV

A) I
B) II
* C) III
D) IV
E) None of these

57. Which of these chlorinated ethenes has the largest dipole moment?

```
* A)  H      H
       \    /
        C=C
       /    \
     Cl      Cl
  B)  H      Cl
       \    /
        C=C
       /    \
     Cl      H
  C) H      Cl
      \    /
       C=C
      /    \
     H      Cl
  D)  H      Cl
       \    /
        C=C
       /    \
     Cl      Cl
  E) Cl      Cl
      \    /
       C=C
      /    \
     Cl      Cl
```

58. Which of these compounds is a secondary alkyl chloride?

A) $CH_3CH_2CH_2CH_2CH_2Cl$

B)
CH_3
3
$CH_3CCH_2CH_3$
3
Cl

C) $CH_3CHCH_2CH_2CH_3$
3
Cl

D) CH_3CH_2CHCl
3
CH_2CH_3

* E) Two of these

59. How many 2x alkyl bromides, neglecting stereoisomers, exist with the formula $C_6H_{13}Br$?

A) 4 B) 5 * C) 6 D) 7 E) 8

60. Many organic compounds contain more than one functional group. Which of the following is both an alcohol and a ketone?

A)
OCH_3
/
CH_3CH
\
OH

B)
O
:
HOCH

C)
H
3
$HOCH_2C{=}O$

* D)
O
:
$HOCH_2CCH_3$

E)
O
:
$HCCCH_3$
:
O

Chapter 3: Acids and Bases

1. Adding sodium hydride to ethanol would produce:
   A) $CH_3CH_2OCH_2CH_3$ + H_2
   B) $CH_3CH_2OCH_2CH_3$ + NaOH
   * C) CH_3CH_2ONa + H_2
   D) CH_3CH_2Na + NaOH
   E) CH_3CH_3 + NaOH

2. Which acid-base reaction would <u>not</u> take place as written?
   A) CH_3Li + CH_3CH_2OH DDDDDD4 CH_4 + CH_3CH_2OLi
   * B) HCpCH + NaOH DDDDDD4 HCpCNa + H_2O
   C) HCpCNa + H_2O DDDDDD4 HCpCH + NaOH
   D) CH_3OH + NaH DDDDDD4 CH_3ONa + H_2
   E) CH_3CO_2H + CH_3ONa DDDDDD4 CH_3CO_2Na + CH_3OH

3. According to the Lewis definition, a base is a(n):
   A) Proton donor.
   * B) Electron pair donor.
   C) Hydroxide ion donor.
   D) Hydrogen ion donor.
   E) Electron pair acceptor.

4. Which of the following is <u>not</u> both a Bronsted-Lowry acid and a Bronsted-Lowry base?
   A) HSO_4^- B) $H_2PO_4^-$ C) HCO_3^- * D) OH^- E) SH^-

5. Which of the following is <u>not</u> a conjugate acid - conjugate base pair (in that order)?
   A) H_3PO_4, $H_2PO_4^-$
   B) HBF_4, BF_4^-
   C) CH_3CH_2OH, $CH_3CH_2O^-$
   D) H_3O^+, H_2O
   * E) HPO_4^{--}, $H_2PO_4^-$

6. The conjugate base of sulfuric acid is:
   A) $H_3SO_4^+$ B) SO_3 * C) HSO_4^- D) H_2SO_3 E) HSO_3^-

7. The amide ion, NH_2^-, is a base which can be used only in which of the solvents shown below:
   A) CH_3OH
   B) CH_3CH_2OH
   C) H_2O
   D) D_2O
   * E) Liquid NH_3

8. Which of the following is <u>not</u> a Lewis base?
   A) NH_3 B) H^- * C) BF_3 D) H_2O E) H_3C^-

9. Which of the following is not a Bronsted-Lowry acid?
A) H_2O * B) $(CH_3)_3N$ C) NH_4^+ D) CH_3CO_2H E) HCpCH

10. Consider the equilibrium

$$PO_4^{---} + H_2O \text{ QMMMMMO } HPO_4^{--} + OH^-$$

Which are the Bronsted-Lowry bases?
A) PO_4^{---} and HPO_4^{--}
* B) PO_4^{---} and OH^-
C) PO_4^{---} and H_2O
D) H_2O and OH^-
E) H_2O and HPO_4^{--}

11. Acetic acid dissociates most completely in:
A) CCl_4
B) $Cl_2C{=}CCl_2$
* C) H_2O
D) $(CH_3CH_2)_2O$
E) the gas phase.

12. Which of these is not a true statement?
A) All Lewis bases are also Bronsted-Lowry bases.
* B) All Lewis acids contain hydrogen.
C) All Bronsted-Lowry acids contain hydrogen.
D) All Lewis acids are electron deficient.
E) According to the Bronsted-Lowry theory, water is both an acid and a base.

13. For the equilibrium

$$CH_3NH_3^+ + H_2O \text{ QMMMMMMO } CH_3NH_2 + H_3O^+$$

the two substances which are both acids are:
A) H_2O and H_3O^+
B) $CH_3NH_3^+$ and H_2O
C) $CH_3NH_3^+$ and CH_3NH_2
* D) $CH_3NH_3^+$ and H_3O^+
E) CH_3NH_2 and H_2O

14. The compounds ethane, ethene, and ethyne exhibit this order of increasing acidity:
A) Ethyne < ethene < ethane
B) Ethene < ethyne < ethane
C) Ethane < ethyne < ethene
* D) Ethane < ethene < ethyne
E) Ethene < ethane < ethyne

15. Which is an incorrect statement?
A) RSH compounds are stronger acids than ROH compounds.
B) PH_3 is a weaker base than NH_3.
C) NH_2^- is a stronger base than OH^-.
* D) OH^- is a stronger base than OR^-.
E) H^- is a stronger base than OR^-.

16. For the reaction:

$$HCN \quad + \quad H_2O \quad QMMMO \quad H_3O^+ \quad + \quad CN^-$$

which pair of substances both are bases?

* A) H_2O and CN^-
B) H_3O^+ and H_2O
C) HCN and H_3O^+
D) HCN and CN^-
E) H_3O^+ and CN^-

17. The correct sequence of the ions shown, in order of <u>increasing</u> basicity, is:

A) $CH_3CH_2:^-$ < $CH_2MCH:^-$ < $HCpC:^-$
B) $CH_3CH_2:^-$ < $HCpC:^-$ < $CH_2MCH:^-$
C) $HCpC:^-$ < $CH_3CH_2:^-$ < $CH_2MCH:^-$
D) $CH_2MCH:^-$ < $HCpC:^-$ < $CH_3CH_2:^-$
* E) $HCpC:^-$ < $CH_2MCH:^-$ < $CH_3CH_2:^-$

18. Which is a protic solvent?

A) CCl_4
B) $HCCl_3$
* C) CH_3OH
D) $CH_3(CH_2)_4CH_3$
E) $CH_3CH_2OCH_2CH_3$

19. If a 0.01 <u>M</u> solution of a weak acid has a pH of 4.0, the pK_a of the acid is:

A) 10.0 B) 8.0 * C) 6.0 D) 4.0 E) 2.0

20. Which reaction of these potential acids and bases does <u>not</u> occur to any appreciable degree due to an unfavorable equilibrium?

A) NaOH (aq) + $CH_3CH_2CH_2CO_2H$
* B) CH_3CH_2ONa in ethanol + ethene
C) CH_3Li in hexane + ethyne
D) $NaNH_2$ in liq. NH_3 + ethanol
E) $NaC_2H_3O_2$ (aq) + HI

21. Which combination of substances below does <u>not</u> constitute a Lewis acid-Lewis base reaction?

A) PH_3 + H^+
B) Ag^+ + NH_3/H_2O
C) BF_3 + NH_3
D) $CH_3CH_2OCH_2CH_3$ + $AlCl_3$
* E) OH^- + NH_3/H_2O

22. Which one of the following is a true statement?

A) The stronger the acid, the larger is its pK_a.
B) The conjugate base of a strong acid is a strong base.
C) Acid-base reactions always favor the formation of the stronger acid and the stronger base.
* D) Strong acids can have negative pK_a values.
E) Hydrogen need not be present in the molecular formula of a Bronsted-Lowry acid.

23. The basic species are arranged in <u>decreasing</u> order of basicity in the sequence:

A) F^- > OCH_3^- > NH_2^- > $CH_3CH_2^-$
B) OCH_3^- > $CH_3CH_2^-$ > NH_2^- > F^-
* C) $CH_3CH_2^-$ > NH_2^- > OCH_3^- > F^-
D) NH_2^- > $CH_3CH_2^-$ > F^- > OCH_3^-
E) NH_2^- > OCH_3^- > $CH_3CH_2^-$ > F^-

24. A particular acid has $K_a = 2.0 \times 10^{-5}$ (in aqueous solution). The evaluation of which of these expressions gives the value for pK_a?

A) $10^{-14}/2.0 \times 10^{-5}$
B) $10^{-14}(2.0 \times 10^{-5})$
* C) $5 - \log 2.0$
D) $-5 + \log 2.0$
E) $2.0 \times 10^{-5}/10^{-14}$

25. As a consequence of the "leveling effect," the strongest acid which can exist in appreciable concentration in aqueous solution is:

* A) H_3O^+ B) H_2SO_4 C) $HClO_4$ D) HCl E) HNO_3

26. Based on the position of the central atom in the periodic chart, we predict that the strongest acid of the following is:

A) H_2O B) H_2S C) H_2Se * D) H_2Te

27. An acid, HA, has the following thermodynamic values for its dissociation in water at 27x C: (H = -2.0 kcal mol^{-1}; (S = -17 cal $K^{-1}mol^{-1}$.

The (G for the process is:

A) +7.1 kcal mol^{-1}
* B) +3.1 kcal mol^{-1}
C) -1.5 kcal mol^{-1}
D) -3.1 kcal mol^{-1}
E) -7.1 kcal mol^{-1}

28. Which of these is <u>not</u> a Lewis acid?

A) $AlCl_3$ B) H^+ C) BCl_3 D) SO_3 * E) H_2S

29. Which of these bases is the strongest one which can be used (and retains its basic character) in aqueous solution?

A) OCH_3^- B) F^- * C) OH^- D) $C_2H_3O_2^-$ E) HSO_4^-

30. The acidity constant, K_a, differs from the equilibrium constant, K_{eq}, for the dissociation of the same acid in water at the same temperature and concentration in what way?

A) K_a can be determined experimentally with less accuracy than K_{eq}.
B) The two terms are identical numerically.
C) K_a is used for strong acids only; K_{eq} for weak acids.
D) K_a is the reciprocal of K_{eq}.
* E) $K_{eq} = K_a/[H_2O]$.

31. Which of the following is an <u>untrue</u> statement?
A) The % dissociation of a weak acid increases with increasing dilution of the acid solution.
B) The stronger an acid, the weaker its conjugate base.
C) The larger the value of K_a for an acid, the smaller the value of its pK_a.
D) Comparison of the acidity of strong acids in solution requires the use of a solvent less basic than water.
* E) The stronger the acid, the more positive the value of (Gx for the dissociation.

32. When proton transfer reactions reach equilibrium, there have been formed:
* A) the weaker acid and the weaker base.
B) the weaker acid and the stronger base.
C) the stronger acid and the weaker base.
D) the stronger acid and the stronger base.
E) All proton transfers go to completion; they are not equilibrium processes.

33. This species is a carbon-based Lewis acid:
A) CH_4 B) $HCCl_3$ * C) CH_3^+ D) $:CH_3^-$ E) yCH_3

34. For the simple hydrides, MH_n, pK_a values <u>decrease</u> in the order:
* A) CH_4 > NH_3 > H_2O > H_2S > HBr
B) HBr > H_2S > H_2O > NH_3 > CH_4
C) HBr > H_2O > NH_3 > H_2S > CH_4
D) NH_3 > H_2S > CH_4 > H_2O > HBr
E) H_2S > H_2O > HBr > NH_3 > CH_4

35. Which combination of reagents is the least effective in generating sodium ethoxide, CH_3CH_2ONa?
A) CH_3CH_2OH + NaH
B) CH_3CH_2OH + $NaNH_2$
* C) CH_3CH_2OH + NaOH
D) CH_3CH_2OH + CH_3Li
E) CH_3CH_2OH + HCpCNa

36. The compound aniline, $C_6H_5NH_2$, has weakly basic properties in aqueous solution. In this other solvent, aniline would behave as a strong base.
A) CH_3OH
B) CH_3CH_2OH
* C) CF_3CO_2H
D) Liquid NH_3
E) $CH_3(CH_2)_4CH_3$

37. Which of the following organic compounds is the strongest acid?
A) C_6H_{12} pK_a = 52
B) CH_3CH_3 pK_a = 50
C) CH_3CH_2OH pK_a = 18
D) CH_3CO_2H pK_a = 5
* E) CF_3CO_2H pK_a = 0

38. What is the conjugate base of ethanol?

* A) $CH_3CH_2O^-$
B) $CH_3CH_2^-$
C) $CH_3CH_2OH_2^+$
D) CH_3CH_3
E) CH_3OCH_3

39. Which of the acids below would have the strongest conjugate base?

* A) CH_3CH_2OH pK_a = 18
B) CH_3CO_2H pK_a = 4.75
C) $ClCH_2CO_2H$ pK_a = 2.81
D) Cl_2CHCO_2H pK_a = 1.29
E) Cl_3CCO_2H pK_a = 0.66

40. In the reaction, $Na^+NH_2^-$ + CH_3OH DDDDDD4 $CH_3O^-Na^+$ + NH_3, the stronger base is:

* A) $NaNH_2$
B) CH_3OH
C) CH_3ONa
D) NH_3
E) This is not an acid-base reaction.

41. Which of the following is a Lewis acid?

A) H_3O^+ * B) BF_3 C) NF_3 D) OH^- E) NpN

42. Which sequence is the best one to use to prepare CH_3CpCD?

* A) CH_3CpCH DDDDDD4 (NaH) DDDDDD4 (D_2O)

B) CH_3CpCH DDDDDD4 (NaOH) DDDDDD4 (D_2O)

C) CH_3CpCH DDDDDD4 (CH_3ONa) DDDDDD4 (D_2O)

D) CH_3CpCH DDDDDD4 (DOH)

E) None of these

43. Adding sodium hydride, NaH, to water produces:

* A) H_2 and NaOH(aq)
B) H^-(aq) + Na^+(aq)
C) H_3O^+(aq) + Na^+(aq)
D) H_3O^-(aq) + Na^+(aq)
E) Na_2O + H_2

44. Which is the strongest acid?

A) CH_3CH_2OH * B) CH_3CO_2H C) HCpCH D) CH_2MCH_2 E) CH_3CH_3

45. Which reaction will yield CH_3CH_2-D?

A) CH_3CH_3 + D_2O
* B) CH_3CH_2Li + D_2O
C) CH_3CH_2OLi + D_2O
D) CH_3CH_2OH + D_2O
E) More than one of these

46. A product of the reaction, CH_3CH_2Li + D_2O DDDDDD4, is

A) CH_3CH_2OD
B) $CH_3CH_2CH_2CH_3$
C) CH_2MCH_2
* D) CH_3CH_2D
E) $CH_3CH_2OCH_2CH_3$

47. Which of the following correctly lists the compounds in order of <u>decreasing</u> acidity?

* A) H_2O > HCpCH > NH_3 > CH_3CH_3
B) HCpCH > H_2O > NH_3 > CH_3CH_3
C) CH_3CH_3 > HCpCH > NH_3 > H_2O
D) CH_3CH_3 > HCpCH > H_2O > NH_3
E) H_2O > NH_3 > HCpCH > CH_3CH_3

48. Select the strongest base.

A) OH^- B) $RCpC^-$ C) NH_2^- D) CH_2MCH^- * E) $CH_3CH_2^-$

49. A group of acids arranged in order of decreasing acidity is:
HNO_3 > CH_3COOH > C_6H_5OH > H_2O > HCpCH

What is the arrangement of the conjugate bases of these compounds in <u>decreasing</u> order of basicity?

A) NO_3^- > CH_3COO^- > $C_6H_5O^-$ > OH^- > $HCpC^-$
B) CH_3COO^- > $C_6H_5O^-$ > NO_3^- > OH^- > $HCpC^-$
C) $C_6H_5O^-$ > NO_3^- > $HCpC^-$ > OH^- > CH_3COO^-
* D) $HCpC^-$ > OH^- > $C_6H_5O^-$ > CH_3COO^- > NO_3^-
E) No prediction of relative base strength is possible.

50. What prediction can be made of the relative strengths of the conjugate bases of: H_2S, HCl, SiH_4, PH_3?

A) PH_2^- > SiH_3^- > HS^- > Cl^-
* B) SiH_3^- > PH_2^- > HS^- > Cl^-
C) Cl^- > HS^- > PH_2^- > SiH_3^-
D) HS^- > Cl^- > SiH_3^- > PH_2^-
E) Cl^- > PH_2^- > SiH_3^- > HS^-

51. Which of these species is <u>not</u> amphoteric?

A) $HCpC^-$ B) HS^- C) NH_3 * D) CH_3^- E) HPO_4^{--}

52. What compounds are produced when sodium nitrate is added to a mixture of water and ethanol?

A) HNO_3 + NaOH
B) HNO_3 + CH_3CH_2ONa
C) NaOH + CH_3CH_2ONa
D) $CH_3CH_2OCH_2CH_3$ + NaOH
* E) No reaction occurs.

53. Which of these phosphorus-based acids is dibasic?

```
A)   O       * B)   O       C)   O       D)   O O       E)   O O
     :              :            :            : :            : :
   HOPOH          HOPOH         HPH         HOPOPOH        HOPDPOH
     3              3            3            3 3            3 3
     OH             H            OH          HO OH          HO OH
```

54. Why cannot one determine the relative acid strengths of $HClO_4$ and HNO_3 using aqueous solutions of these acids?
A) The acids are insufficiently soluble for the measurements.
B) A more basic solvent than H_2O must be used.
* C) H_2O is too basic a solvent for the distinction to be made.
D) These oxidixing acids cause redox reactions to occur.
E) Actually, the acid strengths can be determined using aqueous solutions.

55. Which of these is not a diprotic acid?
A) H_2S B) H_2SO_4 * C) H_2O D) $(COOH)_2$ E) $H_2PO_4^-$

Chapter 4: Alkanes and Cycloalkanes

1. The least stable conformation of butane is:

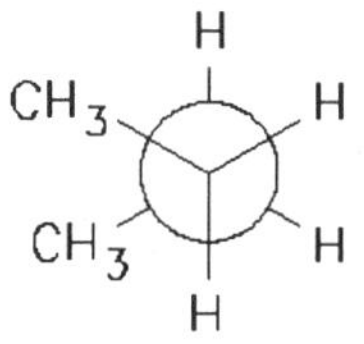

I 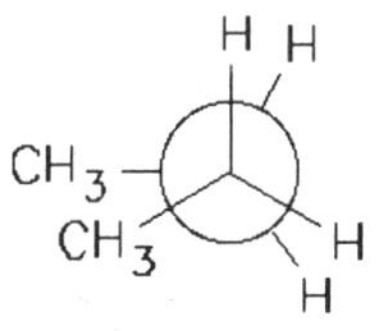 II 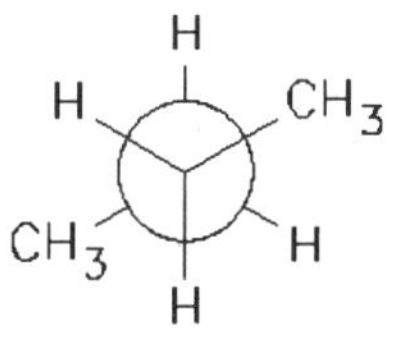 III

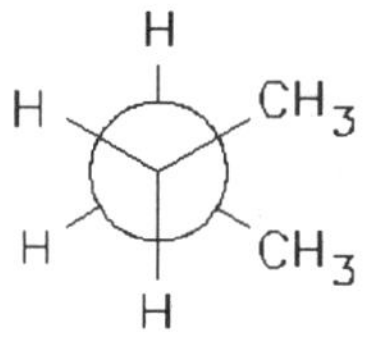

IV 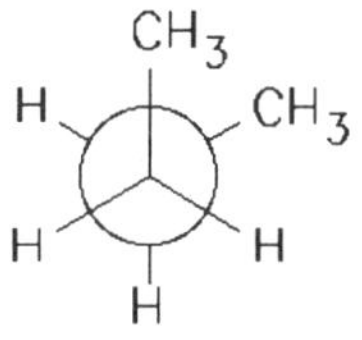 V

A) I * B) II C) III D) IV E) V

2. The IUPAC name for $CH_3CH_2CHDDCHDDCH_2CHCH_3$ is:

```
                    3    3      3
                   CH3  CH3    CH2
                                3
                               CH3
```

A) 6-Ethyl-3,4-dimethylheptane
B) 2-Ethyl-4,5-dimethylheptane
* C) 3,4,6-Trimethyloctane
D) 3,5,6-Trimethyloctane
E) 2-(1-Methylpropyl)-4-methylhexane

3. cis-1,3-Dibromocyclohexane is represented by structure(s):

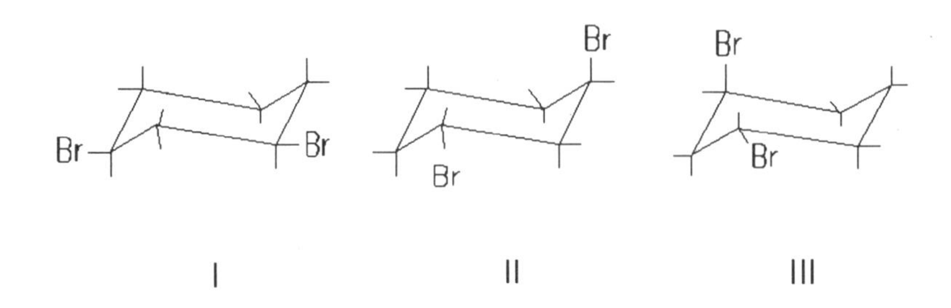

* A) I
B) II
C) III
D) II and III
E) I and II

4. The preferred conformation of cis-3-tert-butyl-1-methylcyclohexane is the one in which:

A) the tert-butyl group is axial and the methyl group is equatorial.
B) the methyl group is axial and the tert-butyl group is equatorial.
C) both groups are axial.
* D) both groups are equatorial.
E) the molecule exists in a boat conformation.

5. An IUPAC name for is:

CH_3
3
$CH_3CH_2CH_2CHCHCH_3$
3
$CHCH_3$
3
CH_2
3
CH_3

A) 5-Methyl-4-(1-methylpropyl)hexane
B) 2-Methyl-3-(1-methylpropyl)hexane
C) 2-Methyl-3-(2-methylpropyl)hexane
* D) 3-Methyl-4-(1-methylethyl)heptane
E) 5-Methyl-4-(1-methylethyl)heptane

6. What structure represents the most stable conformation of <u>cis</u>-1,3-dimethylcyclohexane?

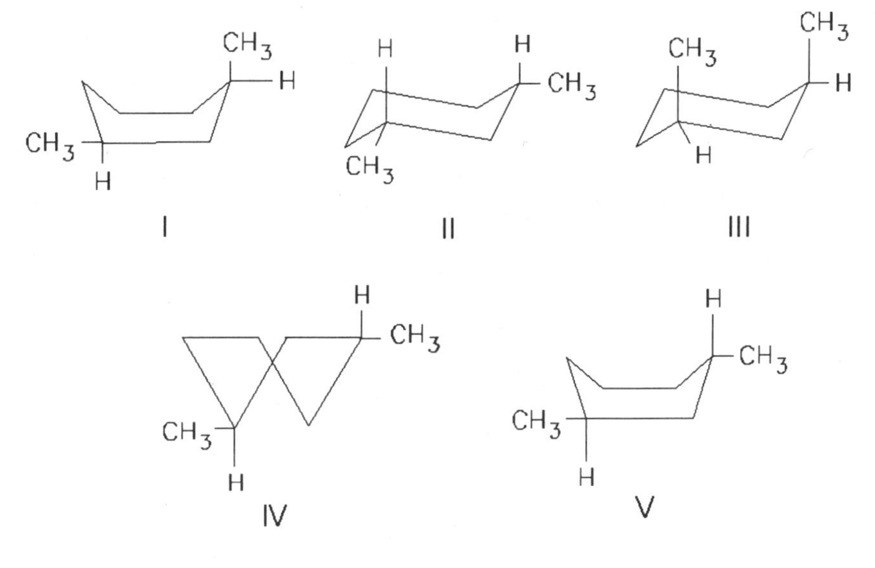

A) I * B) II C) III D) IV E) V

7. Select the reagents necessary to convert cyclopentene into cyclopentane.

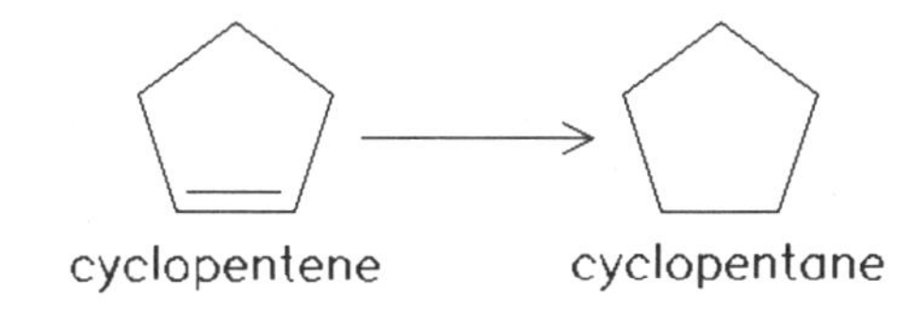

* A) H_2 and Ni B) H_2O C) Heat D) Zn, H^+ E) Light

8. A correct IUPAC name for the following compound is:

CH_3
3
$CH_3CH_2CHCH_2CHCHCH_3$
3 3
CH_3 $CH_2CH_2CH_3$

A) 2,5-Dimethyl-3-propylheptane
B) 3,6-Dimethyl-5-propylheptane
C) 6-Methyl-4-(1-methylethyl)octane
D) 2-Methyl-3-(2-methylbutyl)hexane
* E) 3-Methyl-5-(1-methylethyl)octane

9. An IUPAC name for the following compound is:

```
                  CH3
                  3
             CH3  CH2
             3    3
         CH3DCDCH2DCHDCH3
             3
             CH3
```

A) 4-Ethyl-2,2-dimethylpentane
B) 2-Ethyl-4,4-dimethylpentane
* C) 2,2,4-Trimethylhexane
D) 3,5,5-Trimethylhexane
E) 1-tert-Butyl-2-ethylpropane

10. Which of the following pairs of compounds represent pairs of constitutional isomers?
A) 2-Methylbutane and pentane
B) 2-Chlorohexane and 3-chlorohexane
C) sec-Butyl bromide and tert-butyl bromide
D) Propyl chloride and isopropyl chloride
* E) All of the above

11. Which of the following is bicyclo[3.2.2]nonane?

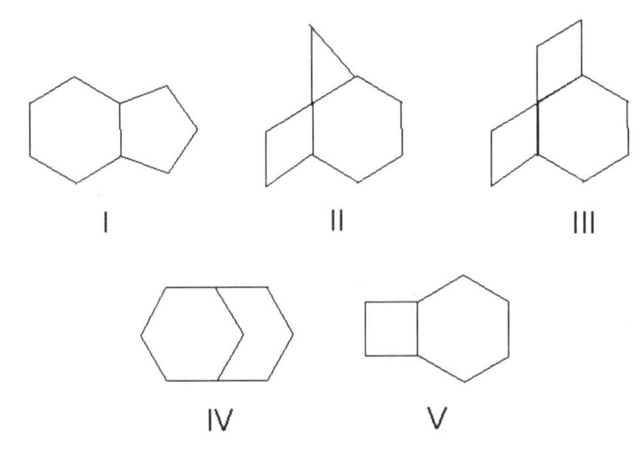

A) I B) II * C) III D) IV E) V

12. Select the reagents necessary to convert 3-bromohexane into hexane.
* A) Zn, H^+ B) CuI C) H_2O D) H_3O^+ E) OH^-

13. Which cycloalkane has the largest heat of combustion per CH_2 group?

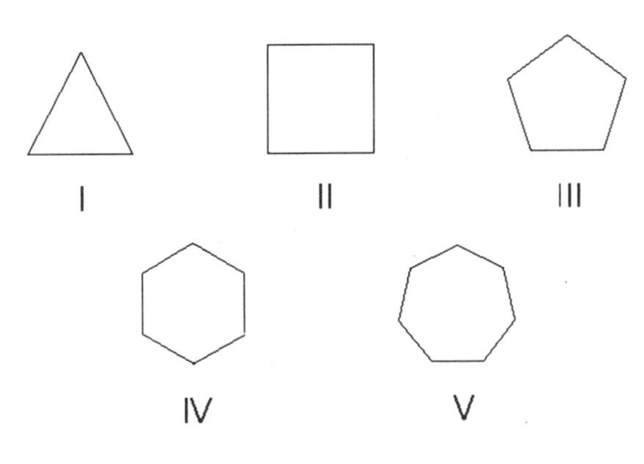

* A) I B) II C) III D) IV E) V

14. Select the systematic name for

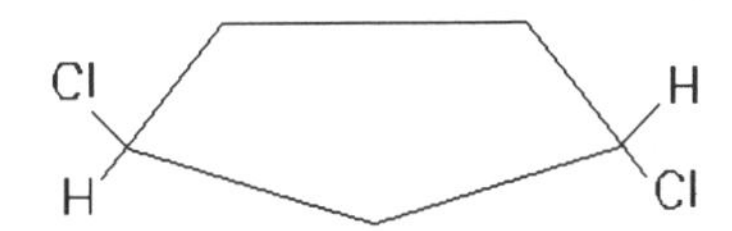

A) cis-1,3-Dichlorocyclopentane
B) trans-1,4-Dichlorocyclopentane
C) cis-1,2-Dichlorocyclopentane
* D) trans-1,3-Dichlorocyclopentane
E) 1,1-Dichlorocyclopentane

15. Which compound is a bicycloheptane?

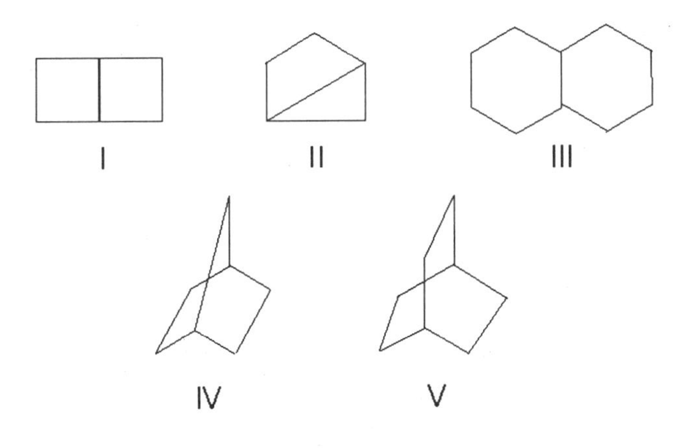

A) I B) II C) III * D) IV E) V

16. Which cycloalkane has the lowest heat of combustion per CH_2 group?

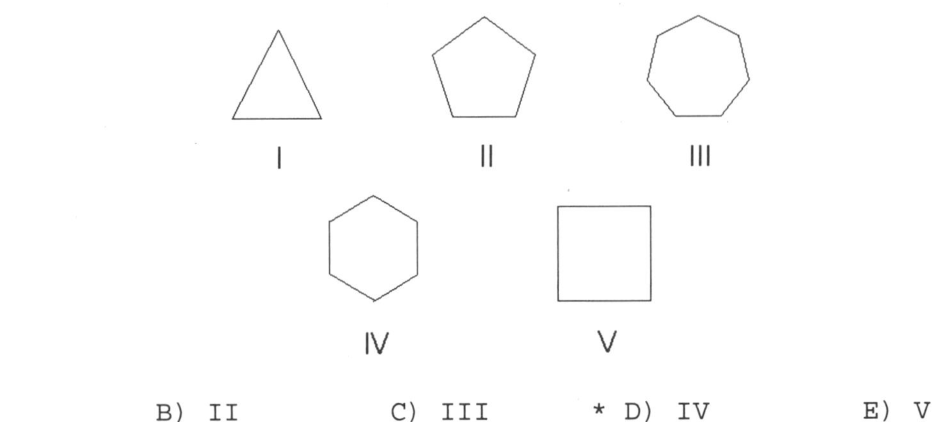

A) I   B) II   C) III   * D) IV   E) V

17. Which is the most stable conformation of cyclohexane?

* A) Chair
  B) Twist
  C) Boat
  D) One-half chair
  E) Staggered

18. Which cycloalkane has the greatest ring strain?

* A) Cyclopropane
  B) Cyclobutane
  C) Cyclopentane
  D) Cyclohexane
  E) Cycloheptane

19. When chlorinated, an alkane, C_6H_{14}, gives only two products with the formula $C_6H_{13}Cl$. The structure of the alkane is:

A) $CH_3CH_2CH_2CH_2CH_2CH_3$

B) $CH_3CHCH_2CH_2CH_3$
 3
 CH_3

* C) $CH_3CHDDCHCH_3$
 3   3
 CH_3 CH_3

D) $(CH_3)_3CCH_2CH_3$

E) $CH_3CH_2CHCH_2CH_3$
 3
 CH_3

20. How many constitutional isomers are possible for the formula C_6H_{14}?

A) 2   B) 3   C) 4   * D) 5   E) 6

21. Which formula represents a compound that is different from all the others?

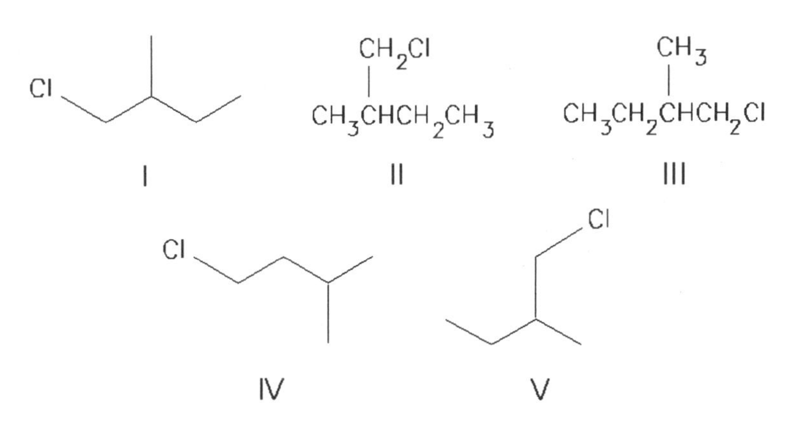

A) I B) II C) III * D) IV E) V

22. Which isomer of C_5H_{10} would you expect to have the smallest heat of combustion?

* A) Cyclopentane
  B) Methylcyclobutane
  C) 1,1-Dimethylcyclopropane
  D) cis-1,2-Dimethylcyclopropane
  E) trans-1,2-Dimethylcyclopropane

23. Which is the correct name for the compound shown below?

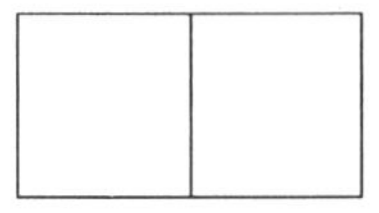

* A) Bicyclo[2.2.0]hexane
  B) Bicyclo[2.2.0]butane
  C) Bicyclo[2.2.2]hexane
  D) Bicyclo[2.2.1]hexane
  E) Disquarane

24. What is the name of this compound?

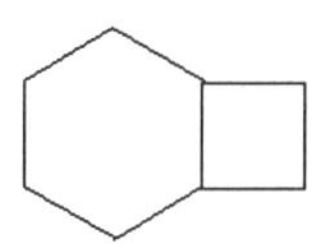

  A) Bicyclo[2.2.2]octane
  B) Bicyclo[3.2.1]octane
  C) Bicyclo[4.1.1]octane
* D) Bicyclo[4.2.0]octane
  E) Bicyclo[3.3.0]octane

25. An unknown compound with a molecular formula of C_8H_{18} reacts with chlorine in the presence of light and heat. From the reaction mixture only one monochlorination product (i.e., only one compound with the formula $C_8H_{17}Cl$) can be isolated. What is the most likely structure of the original unknown compound? (Note: dichloro-, trichloro-, and more highly chlorinated products are also isolated, but consideration of these is unnecessary.)

A) $CH_3CH_2CH_2CH_2CH_2CH_2CH_2CH_3$

B)
```
    CH3
    3
CH3CCH2CH2CH2CH3
    3
    CH3
```

C)
```
    CH3
    3
CH3CCH2CHCH3
    3    3
    CH3  CH3
```

* D)
```
 H3C CH3
   3 3
CH3CDCCH3
   3 3
 H3C CH3
```

E)
```
   CH3 CH3
   3   3
CH3CHCHCHCH3
       3
       CH3
```

26. The most stable conformation of butane is:

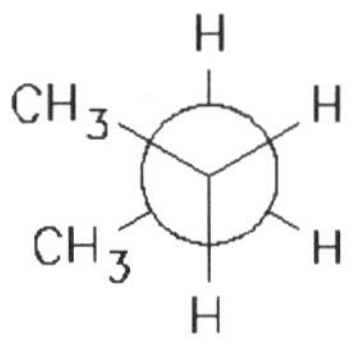

I 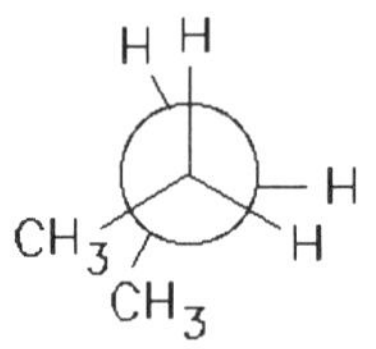 II 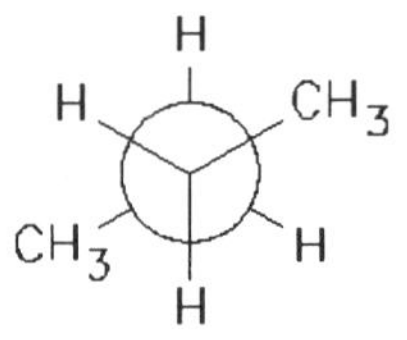 III 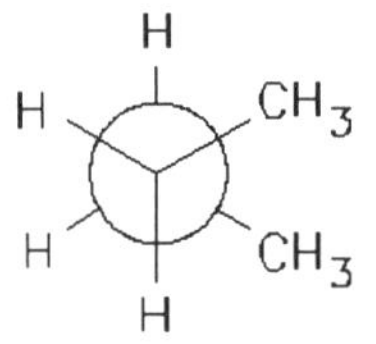 IV 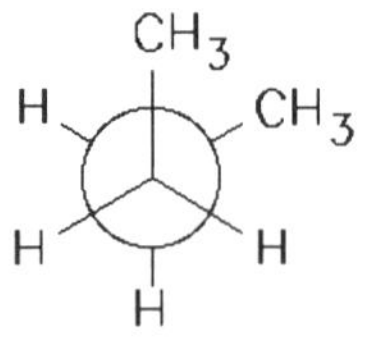 V

A) I B) II * C) III D) IV E) V

27. What is the common name for this compound?

```
     CH3
     3
CH3CHCH2Br
```

* A) Isobutyl bromide
B) tert-Butyl bromide
C) Butyl bromide
D) sec-Butyl bromide
E) Bromo-sec-butane

28. The most stable conformation of 1,2-diphenylethane is:

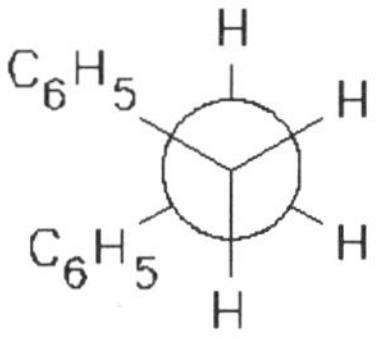

I 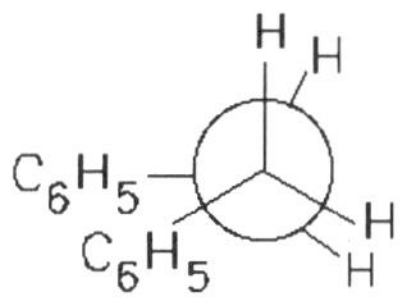   II 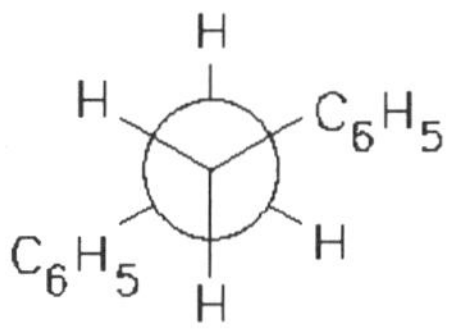   III

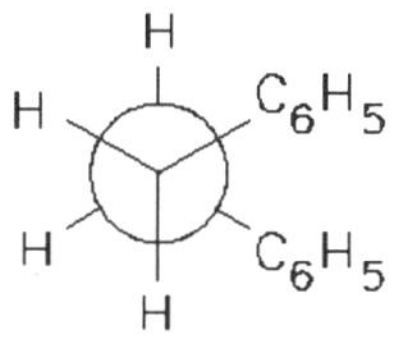

IV 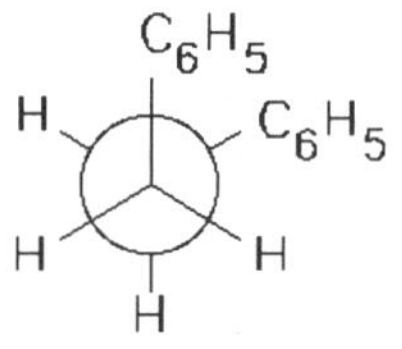   V

A) I   B) II   * C) III   D) IV   E) V

29. The IUPAC name for the following compound is:

```
         CH3
         3
CH3CHCHCH2CHCH3
    3      3
    CH3    CH2
           3
           CH3
```

A) 2-Ethyl-4,5-dimethylhexane
B) 5-Ethyl-2,3-dimethylhexane
C) 5-Ethyl-2,3-dimethylhexane
D) 3,5,6-Trimethylheptane
* E) 2,3,5-Trimethylheptane

30. Which of the following yields one monosubstituted chloroalkane upon chlorination?

A) Isobutane
* B) Cyclopentane
C) Butane
D) Propane
E) None of these

31\. An alkane, C_6H_{14}, reacts with chlorine to yield four constitutional isomers with the formula $C_6H_{13}Cl$. The structure of the alkane is:

A) $CH_3CH_2CH_2CH_2CH_2CH_3$

B) $CH_3CHCH_2CH_2CH_3$
3
CH_3

* C) $CH_3CH_2CHCH_2CH_3$
3
CH_3

D) CH_3
3
$CH_3CCH_2CH_3$
3
CH_3

E) $CH_3CHDDCHCH_3$
3 3
CH_3 CH_3

32\. A correct name for the following compound is:

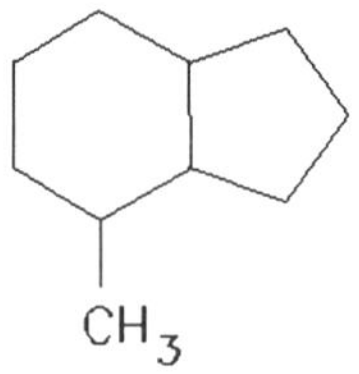

CH_3

* A) 2-Methylbicyclo[4.3.0]nonane
B) 1-Methylbicyclo[4.3.1]nonane
C) 7-Methylbicyclo[4.3.0]nonane
D) 2-Methylbicyclo[4.3.1]nonane
E) 1-Methylbicyclo[4.3.0]nonane

33\. Which cycloalkane has the least ring strain?
A) Cyclopropane
B) Cyclobutane
C) Cyclopentane
* D) Cyclohexane
E) Cycloheptane

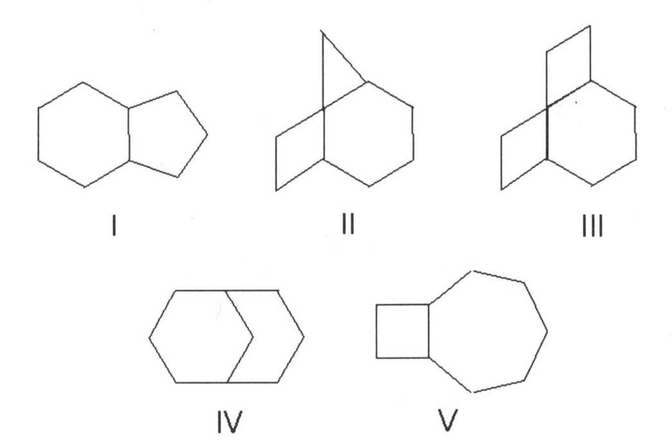

34. Which of the above is bicyclo[3.3.1]nonane?
  A) I B) II C) III * D) IV E) V

35. Which of the above is bicyclo[5.2.0]nonane?
  A) I B) II C) III D) IV * E) V

36. Which of the above is bicyclo[4.3.0]nonane?
* A) I B) II C) III D) IV E) V

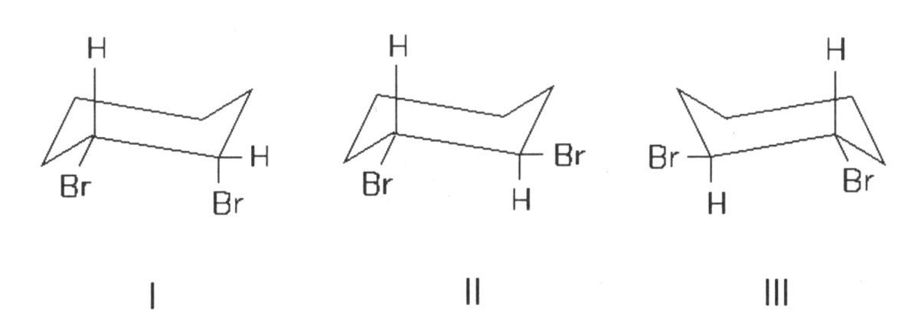

37. <u>trans</u>-1,2-Dibromocyclohexane is represented by structure(s):
  A) I
  B) II
  C) III
* D) II and III
  E) I and II

38. cis-1,2-Dibromocyclohexane is represented by structure(s):

* A) I
  B) II
  C) III
  D) II and III
  E) I and II

39. The most stable conformation of cis-1-tert-butyl-2-methylcyclohexane is the one in which:

  A) the tert-butyl group is axial and the methyl group is equatorial.
* B) the methyl group is axial and the tert-butyl group is equatorial.
  C) both groups are axial.
  D) both groups are equatorial.
  E) the twist boat conformation is adopted.

40. The most stable conformation of trans-1-tert-butyl-2-methylcyclohexane is the one in which:

  A) the tert-butyl group is axial and the methyl group is equatorial.
  B) the methyl group is axial and the tert-butyl group is equatorial.
  C) both groups are axial.
* D) both groups are equatorial.
  E) None of the above

41. The most stable conformation of trans-1-tert-butyl-3-methylcyclohexane is the one in which:

  A) the tert-butyl group is axial and the methyl group is equatorial.
* B) the methyl group is axial and the tert-butyl group is equatorial.
  C) both groups are axial.
  D) both groups are equatorial.
  E) the twist boat conformation is adopted.

42. The most stable conformation of trans-1-tert-butyl-4-methylcyclohexane is the one in which:

  A) the tert-butyl group is axial and the methyl group is equatorial.
  B) the methyl group is axial and the tert-butyl group is equatorial.
  C) both groups are axial.
* D) both groups are equatorial.
  E) None of the above

43. The most stable conformation of cis-1-tert-butyl-4-methylcyclohexane is the one in which:

  A) the tert-butyl group is axial and the methyl group is equatorial.
* B) the methyl group is axial and the tert-butyl group is equatorial.
  C) both groups are axial.
  D) both groups are equatorial.
  E) the twist boat conformation is adopted.

44. In the most stable conformation of cis-1,4-dimethylcyclohexane, the methyl groups are:

* A) one axial, one equatorial.
  B) both axial.
  C) both equatorial.
  D) alternating between being both axial and both equatorial.
  E) None of the above

45. Neopentane and neohexane are valid chemical names. However, "neoheptane" is ambiguous because this number of seven-carbon alkanes possess the structural feature which characterizes neopentane and neohexane, i.e., a quaternary carbon atom.

A) 2 * B) 3 C) 4 D) 6 E) 8

46. How many compounds with the formula C_7H_{16} (heptanes) contain a single 3x carbon atom?

A) 2 B) 3 * C) 4 D) 5 E) 6

47. An IUPAC name for the group CH_3CHCH_2- is:

```
CH3CHCH2-
   3
   CH2
   3
   CH3
```

A) Isopentyl
B) Isoamyl
C) sec-Butylmethyl
* D) 2-Methylbutyl
E) 2-Ethylpropyl

48. The neopentyl group has the alternative name:

A) 1,1-Dimethylpropyl
B) 1,2-Dimethylpropyl
* C) 2,2-Dimethylpropyl
D) 1-Methylbutyl
E) 2-Methylbutyl

49. The correct IUPAC name for the following is:

```
          CH(CH3)2
          3
CH3CHCH2CCH3
   3    3
   Br   Cl
```

A) 2-Bromo-4-chloro-4-isopropylpentane
B) 4-Bromo-2-chloro-2-isopropylpentane
* C) 5-Bromo-3-chloro-2,3-dimethylhexane
D) 2-Bromo-4-chloro-4,5-dimethylhexane
E) 2-(2-Bromopropyl)-2-chloro-3-methylbutane

50. Which of the following is a correct name which corresponds to the common name _tert_-pentyl alcohol?

A) 2,2-Dimethyl-1-propanol
B) 2-Ethyl-2-propanol
* C) 2-Methyl-2-butanol
D) 3-Methyl-1-butanol
E) Methyl _tert_-butanol

51. The correct IUPAC name for the following compound is:

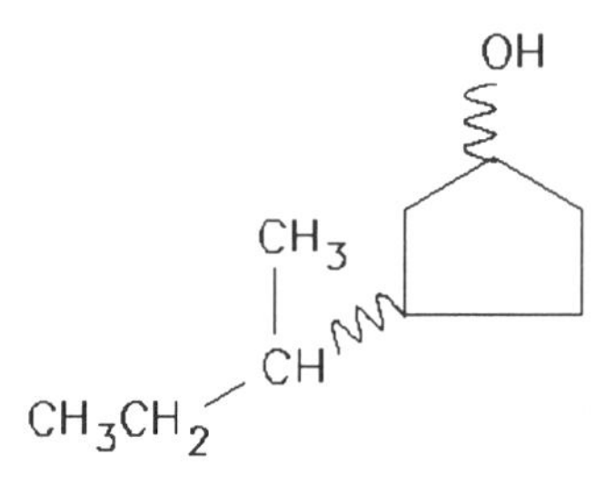

A) 1-Hydroxy-3-_sec_-butylcyclopentane
* B) 3-_sec_-Butyl-1-cyclopentanol
C) 1-_sec_-Butyl-3-cyclopentanol
D) 4-_sec_-Butyl-1-cyclopentanol
E) 3-Isobutyl-1-cyclopentanol

52. Which structure represents the most stable conformation of _cis_-1,4-di-_tert_-butylcyclohexane?

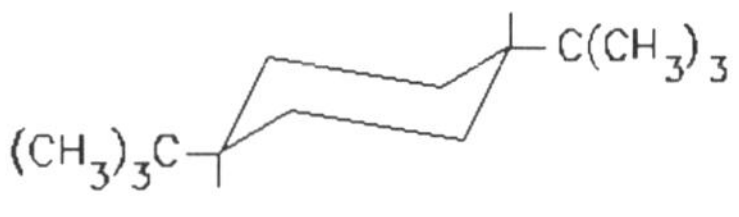

$(CH_3)_3C$ — $C(CH_3)_3$ (I) $(CH_3)_3C$ — $C(CH_3)_3$ (II) $C(CH_3)_3$ — $(CH_3)_3C$ (III)

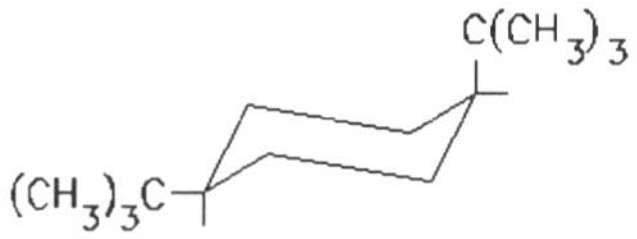

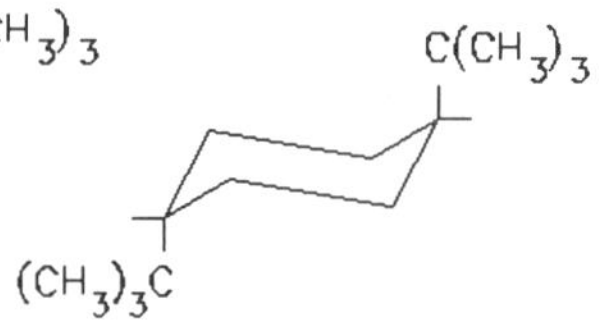

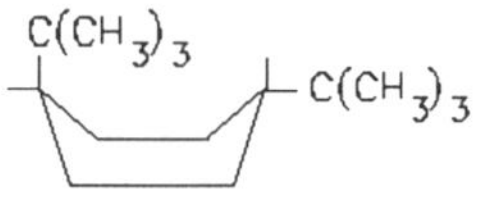

$C(CH_3)_3$ — $C(CH_3)_3$ (IV) $(CH_3)_3C$ — $C(CH_3)_3$ (V)

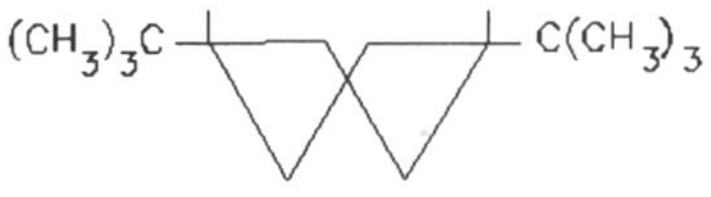

A) I B) II C) III D) IV * E) V

53. The most stable conformation of 1,2-ethanediol (ethylene glycol) is shown below. It is the most stable conformation because:

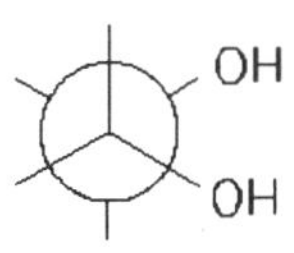

A) this corresponds to an anti conformation.
B) in general, gauche conformations possess the minimum energy.
* C) it is stabilized by hydrogen bonding.
D) it is a staggered conformation.
E) it has the highest energy of all the possibilities.

54. Catalytic hydrogenation of which of the following will yield isopentane?

A) $CH_3CH_2CCH_3$
:
CH_2

B) $CH_2MCHCH(CH_3)_2$

C) $CH_3CHMC(CH_3)_2$

D) $CH_3CH_2CMCH_2$
3
CH_3

* E) All of the above

55. Which of the following can be reduced with zinc and aqueous acid to form isohexane?
* A) 1-Bromo-2-methylpentane
B) 2-Bromo-3-methylpentane
C) 3-Bromo-3-methylpentane
D) 1-Bromo-2,2-dimethylpentane
E) 1-Bromo-3,3-dimethylpentane

56. The reaction of lithium di-<u>sec</u>-butylcuprate with isopentyl bromide yields:
* A) 2,5-Dimethylheptane
B) 2,6-Dimethylheptane
C) 3,5-Dimethylheptane
D) 3,4-Dimethylheptane
E) 3,6-Dimethylheptane

57. Which combination of reagents is to be preferred for the synthesis of 2,4-dimethylhexane by the Corey-Posner, Whitesides-House procedure?
A) Lithium diisobutylcuprate + <u>sec</u>-butyl bromide
B) Lithium dimethylcuprate + 2-bromo-4-methylhexane
C) Lithium dimethylcuprate + 4-bromo-2-methylhexane
* D) Lithium diisopropylcuprate + 1-bromo-2-methylbutane
E) Lithium di(2-methylbutyl)cuprate + isopropyl bromide

58. The graph below is a plot of the relative energies of the various conformations of:

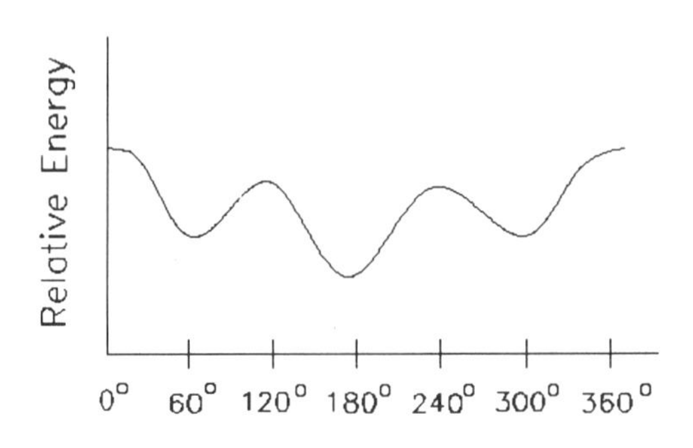

A) Ethane
B) Propane
C) Chloroethane
* D) 1-Chloropropane (C_1-C_2 rotation)
E) Butane (C_1-C_2 rotation)

59. Which conformation of <u>trans</u>-1-isopropyl-3-methylcyclohexane would be present in greatest amount at equilibrium?

A) The conformation with the methyl group equatorial and the isopropyl group axial
* B) The conformation with the methyl group axial and the isopropyl group equatorial
C) The conformation with both groups axial
D) The conformation with both groups equatorial
E) None of the above

60. What is a correct name for the following compound?

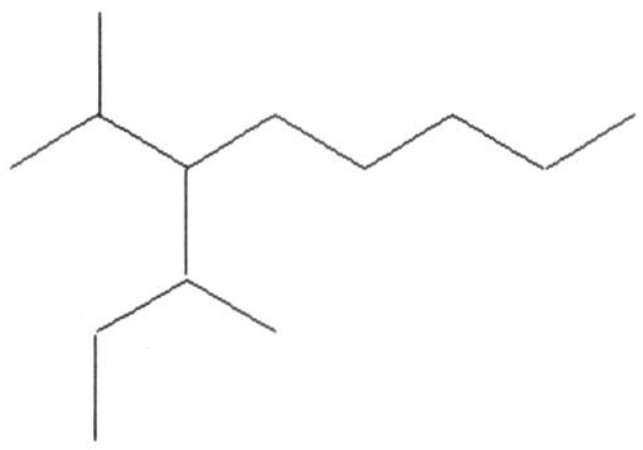

A) 3-Isobutyl-2-methylheptane
B) 3-<u>sec</u>-Butyl-2-methyloctane
C) 5-Isobutyl-6-methylheptane
D) 2-Ethyl-3-isopropyloctane
* E) 4-Isopropyl-3-methylnonane

61. What is the correct name of the following compound?

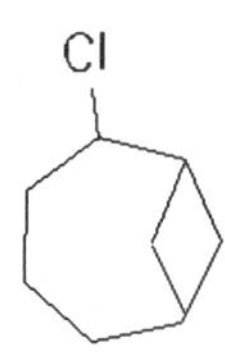

A) 1-Chlorobicyclo[4.1.1]octane
B) 2-Chlorobicyclo[4.1.0]octane
* C) 2-Chlorobicyclo[4.1.1]octane
D) 2-Chlorobicyclo[4.1.1]heptane
E) 5-Chlorobicyclo[4.1.1]octane

62. What is the correct IUPAC name for the following compound?

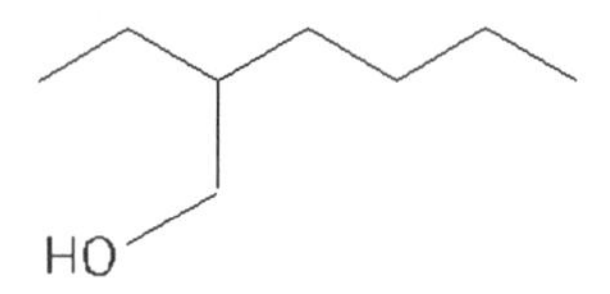

A) 3-Hydroxymethylheptane
B) 3-Hydroxymethylhexane
C) 3-Methyloxyheptane
* D) 2-Ethyl-1-hexanol
E) 2-Ethyl-1-heptanol

63. Which of the following structures represents bicyclo[3.2.1]octane?

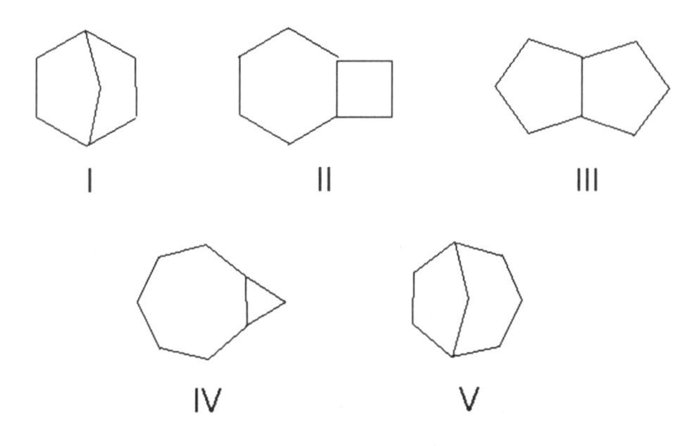

A) I
B) II
C) III
D) IV
* E) V

64. Which of these is the common name for the 1,1-dimethylpropyl group?
A) tert-Butyl
* B) tert-Pentyl
C) Isopentyl
D) Neopentyl
E) sec-Pentyl

65. Neglecting stereochemistry, which of these common group names is ambiguous, i.e., does not refer to one specific group?

A) Butyl
B) sec-Butyl
C) tert-Pentyl
D) Neopentyl
* E) sec-Pentyl

66. Which conformation represents the most stable conformation of cis-1-tert-butyl-4-methylcyclohexane?

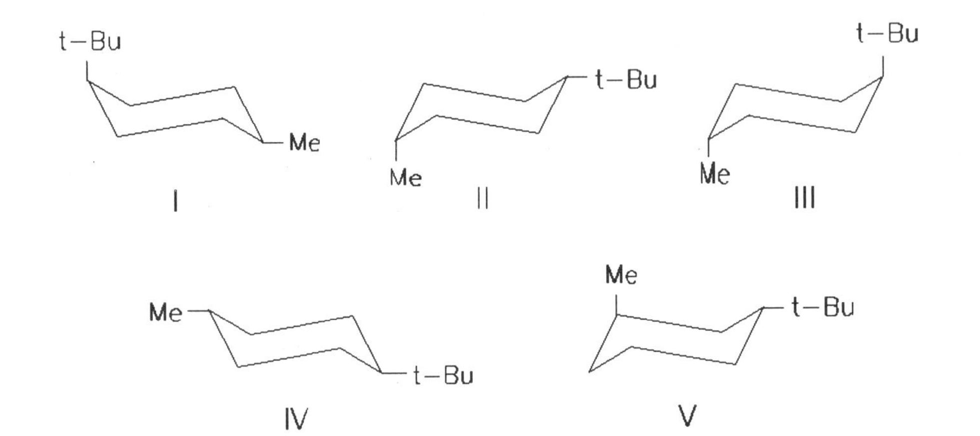

A) I   * B) II   C) III   D) IV   E) V

67. What is the correct IUPAC name for the following compound?

OH

A) 5-Ethyl-3-methylhexanol
B) 5-Ethyl-3-methyl-1-hexanol
C) 2-Ethyl-4-methyl-6-hexanol
D) 3,5-Dimethyl-7-heptanol
* E) 3,5-Dimethyl-1-heptanol

68. Isopentyl is the common name for which alkyl group?

A) $CH_3CH_2CH_2CH$-
3
CH_3

B) $CH_3CH_2CHCH_2$-
3
CH_3

* C) $CH_3CHCH_2CH_2$-
3
CH_3

D) CH_3CH_2CH-
3
CH_2CH_3

E) CH_3
3
CH_3CCH_2-
3
CH_3

69. Which isomer would have the largest heat of combustion?
* A) Propylcyclopropane
  B) Ethylcyclobutane
  C) Methylcyclopentane
  D) Cyclohexane
  E) Since they are all isomers, all would have the same heat of combustion.

70. There are unambiguous common names for all alkyl groups of a particular number of carbon atoms, up to and including this number of carbons.
  A) 3 * B) 4 C) 5 D) 6 E) 7

71. The <u>least</u> stable conformation of cyclohexane is the:
  A) boat.
  B) twist boat.
  C) chair.
* D) half-chair.
  E) twist chair.

72. What is the simplest alkane, i.e., the one with the smallest molecular weight, which possesses primary, secondary and tertiary carbon atoms?
  A) 2-Methylpropane
* B) 2-Methylbutane
  C) 2-Methylpentane
  D) 3-Methylpentane
  E) 2,2-Dimethylbutane

73. Which of these $C_{10}H_{18}$ isomers is predicted to be the most stable?

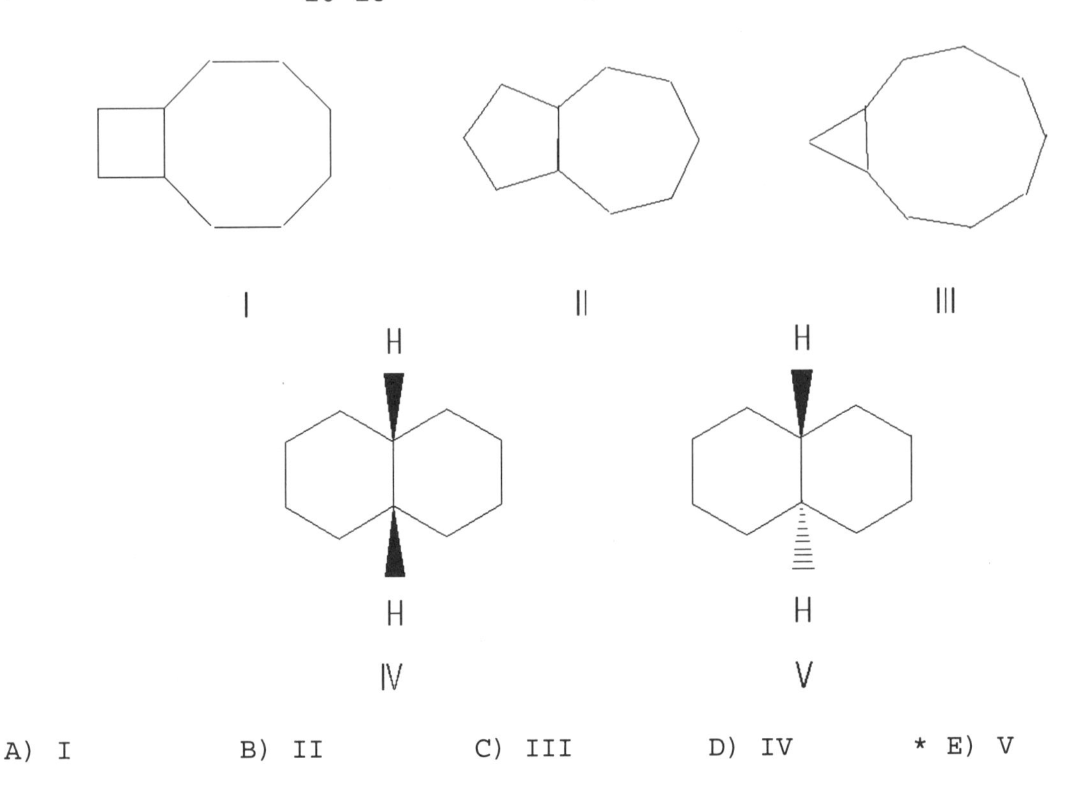

A) I B) II C) III D) IV * E) V

74. The twist boat conformation is the preferred conformation for this compound.

* A) cis-1,4-Di-tert-butylcyclohexane
B) trans-1,4-Di-tert-butylcyclohexane
C) cis-1,3-Di-tert-butylcyclohexane
D) trans-1,2-Di-tert-butylcyclohexane
E) None of these

75. Which is a correct representation of isopentane?

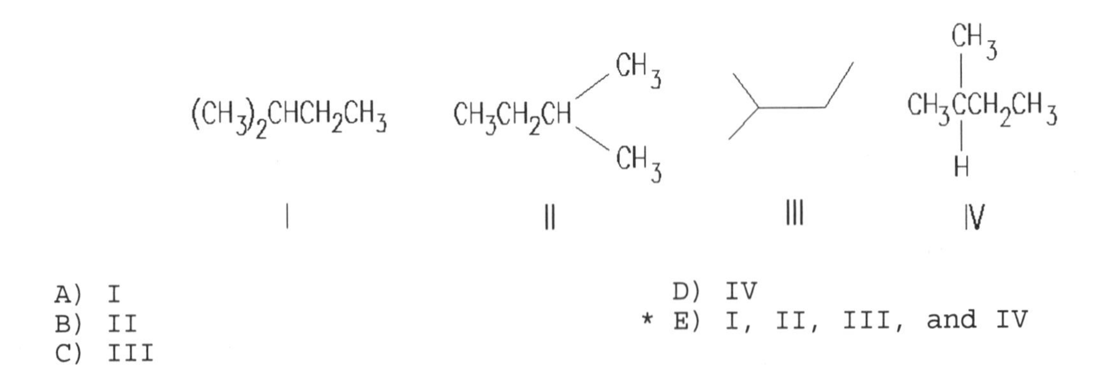

A) I
B) II
C) III
D) IV
* E) I, II, III, and IV

76. How many alkanes of formula C_7H_{16} possess a quaternary carbon atom?
A) 1 B) 2 * C) 3 D) 4 E) 5

77. Express, quantitatively, the difference in stability of the two structures shown below.

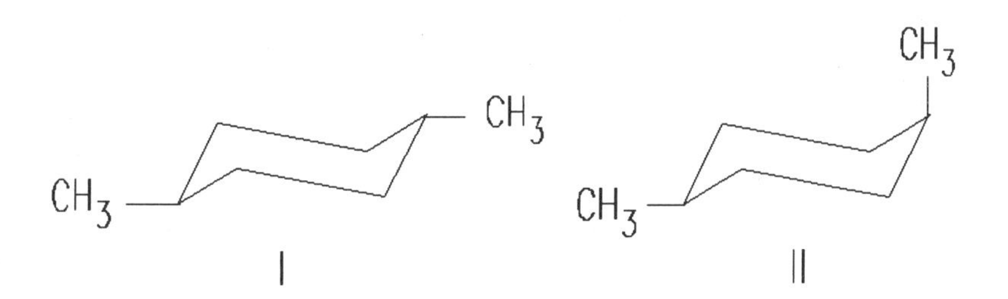

* A) I is more stable than II by 1.8 kcal mol^{-1}.
B) I is more stable than II by 3.6 kcal mol^{-1}.
C) II is more stable than I by 1.8 kcal mol^{-1}.
D) II is more stable than I by 3.6 kcal mol^{-1}.
E) The two are equal in stability.

78.

```
                                  CH3
                                  3
An IUPAC name for    CH3CHCH2CHCHCH3    is:
                        3    3
                        CH3  CHCH3
                             3
                             CH2
                             3
                             CH3
```

A) 3-Isobutyl-2,4-dimethylhexane
B) 3-*sec*-Butyl-2,5-dimethylhexane
C) 4-*sec*-Butyl-2,5-dimethylhexane
* D) 4-Isopropyl-2,5-dimethylheptane
E) 4-Isopropyl-3,6-dimethylheptane

79. An IUPAC name for the following compound is:

```
       H3C CH3
         3 3
CH3CH2CHCCH2CH2CH3
           3
           CH2CHCH3
               3
               CH3
```

A) 4-Isobutyl-3,4-dimethylheptane
B) 4-*sec*-Butyl-2,4-dimethylheptane
* C) 2,4,5-Trimethyl-4-propylheptane
D) 3,4,6-Trimethyl-4-propylheptane
E) 4-Isobutyl-4,5-dimethylheptane

Chapter 5: Stereochemistry

1. Which molecule has a plane of symmetry?

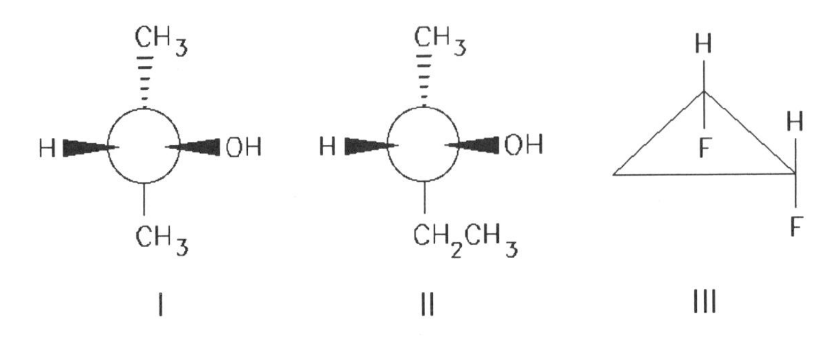

A) I
B) II
C) III
* D) More than one of these
E) None of these

2. (R)-2-Chlorobutane is represented by:

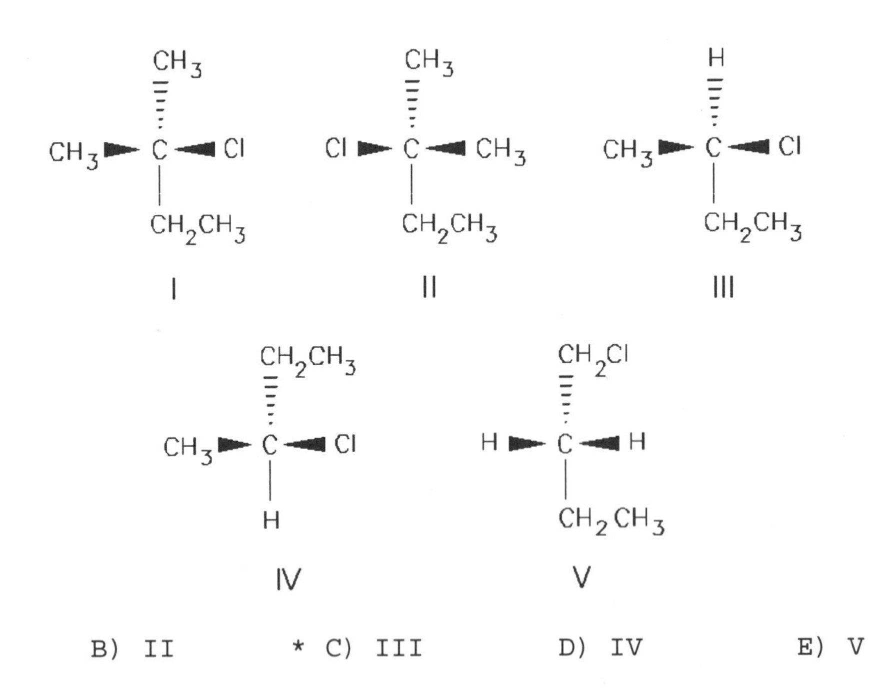

A) I B) II * C) III D) IV E) V

3. Hexane and 3-methylpentane are examples of:
A) enantiomers.
B) stereoisomers.
C) diastereomers.
* D) constitutional isomers.
E) None of these

4. I and II are:

I: Newman projection — front: CH_3 (upper left), Cl (upper right), H (bottom); back: CH_2–CH_3 (top)

II: Newman projection — front: H (upper left), H (upper right), CH_2–CH_3 (bottom); back: CH_2Cl (top)

* A) constitutional isomers.
B) enantiomers.
C) non-superposable mirror images.
D) diastereomers.
E) not isomeric.

5. Pairs of enantiomers are:

I: C with CH_3 (top, hashed), CH_3 (left, wedge), Cl (right, wedge), $CH_2CH_2CH_3$ (bottom)

II: C with CH_3 (top, hashed), Cl (left, wedge), CH_3 (right, wedge), $CH_2CH_2CH_3$ (bottom)

III: C with H (top, hashed), CH_3 (left, wedge), Cl (right, wedge), $CH_2CH_2CH_3$ (bottom)

IV: C with $CH_2CH_2CH_3$ (top, hashed), CH_3 (left, wedge), Cl (right, wedge), H (bottom)

V: C with CH_2Cl (top, hashed), H (left, wedge), H (right, wedge), $CH_2CH_2CH_3$ (bottom)

A) I, II and III, IV
B) I, II
* C) III, IV
D) IV, V
E) None of the structures

6. Chiral molecules are represented by:

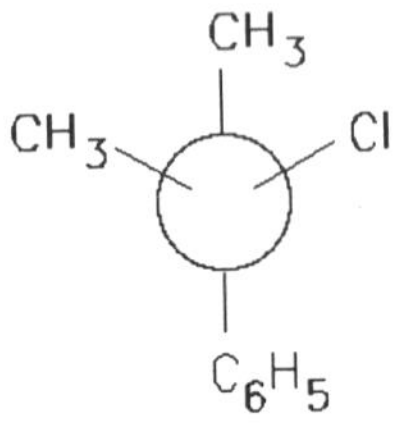

I

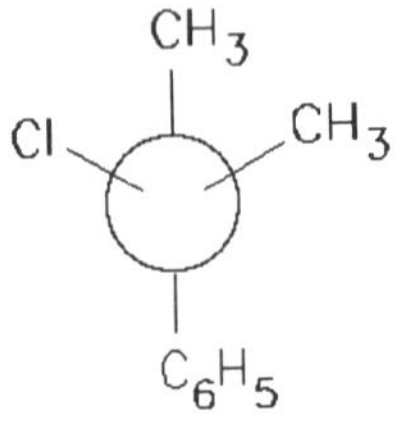

II

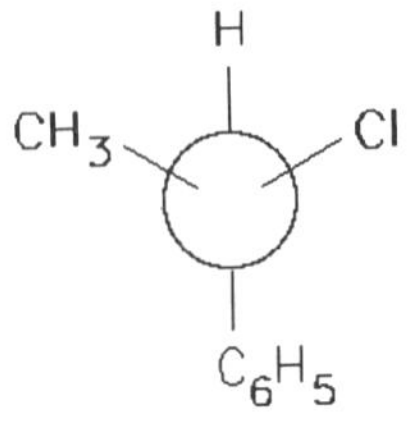

III

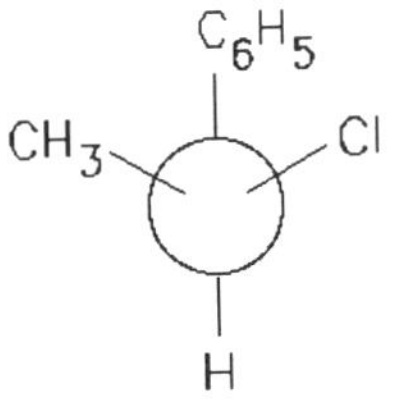

IV

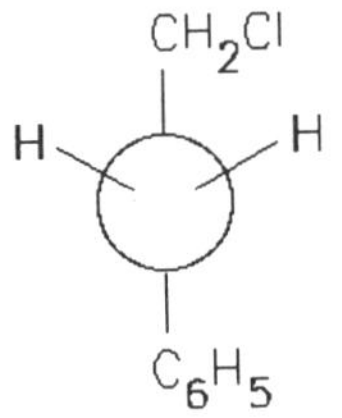

V

A) I, II, III, IV and V
B) I, II, III and IV
C) I and II
* D) III and IV
E) IV alone

7. The compounds whose molecules are shown below would have:

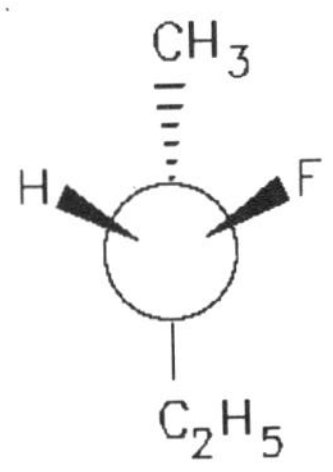

I

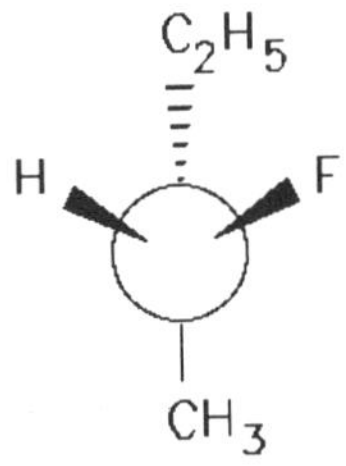

II

A) the same melting point.
B) different melting points.
C) equal but opposite optical rotations.
* D) More than one of the above
E) None of the above

8. Enantiomers are:

A) molecules that have a mirror image.
B) molecules that have stereogenic carbon atoms.
C) non-superposable molecules.
D) non-superposable constitutional isomers.
* E) non-superposable molecules that are mirror images of each other.

9. Which of the following is the enantiomer of

?

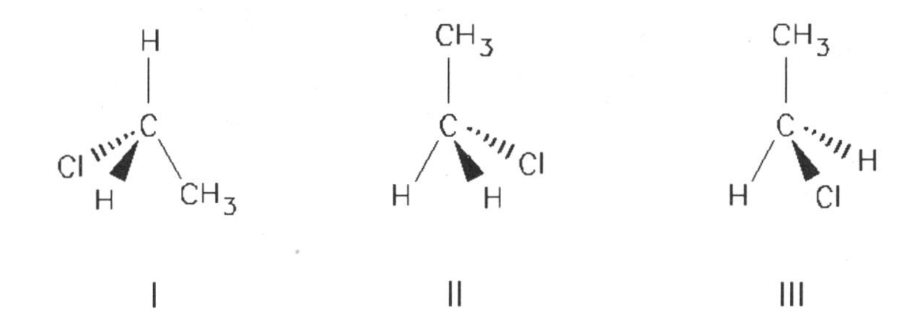

A) I
B) II
C) III
* D) It does not have a non-superposable enantiomer.
E) Both II and III

10. Which of the following molecules is achiral?

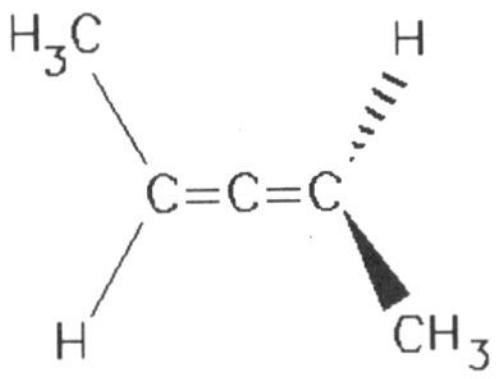

I

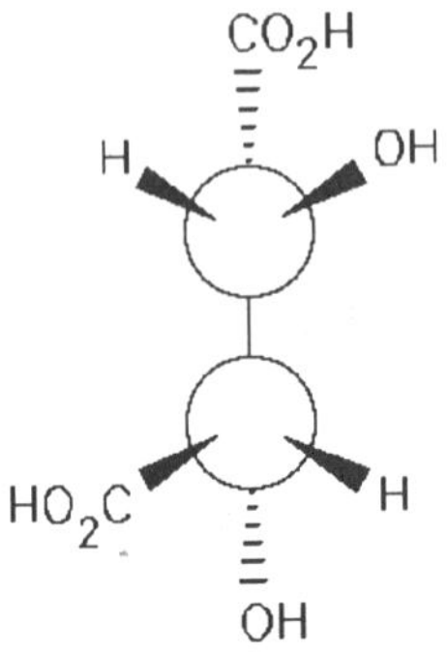

II

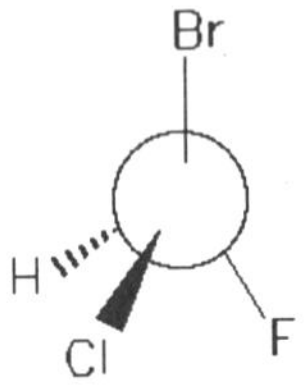

III

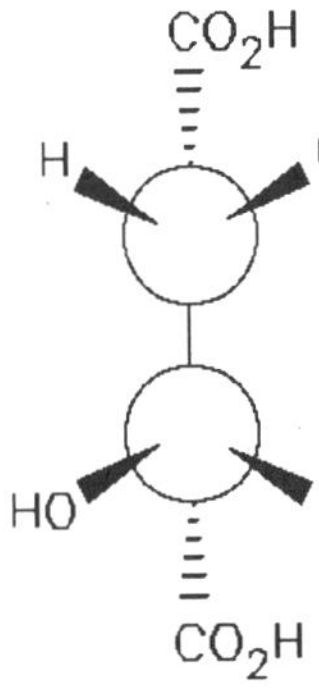

IV

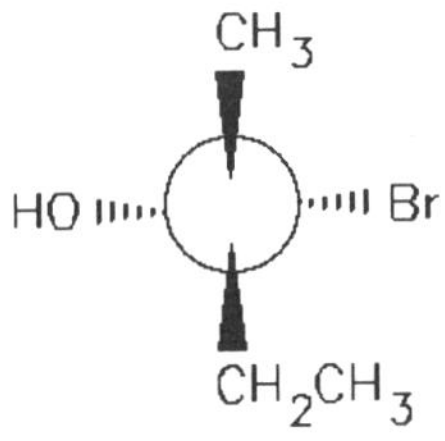

V

A) I   * B) II   C) III   D) IV   E) V

11. Which of the following is a meso compound?

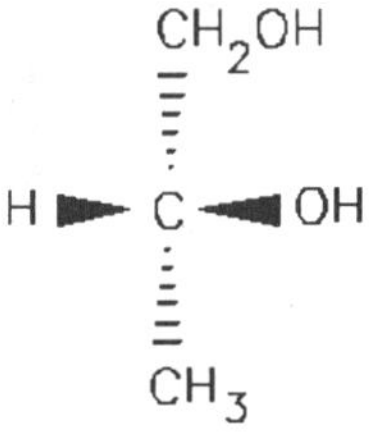

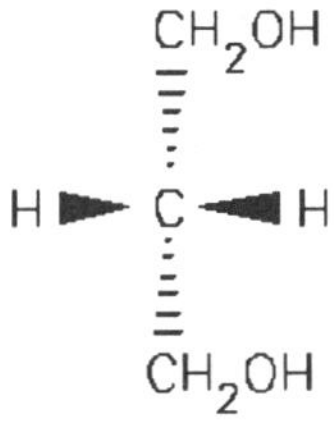

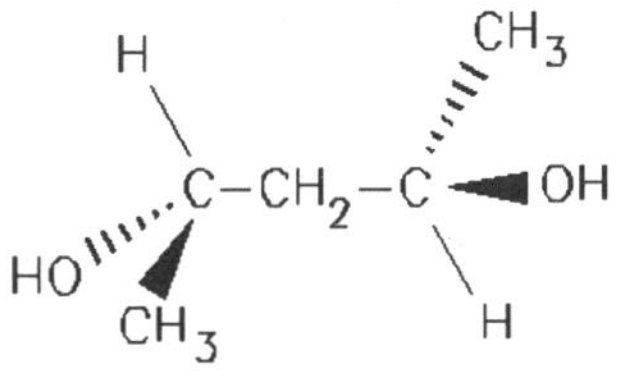

A) I B) II * C) III D) IV E) V

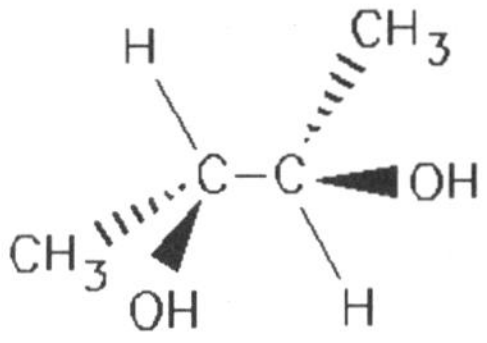

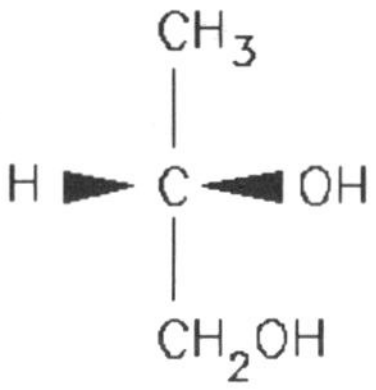

12. Which molecule has a plane of symmetry?

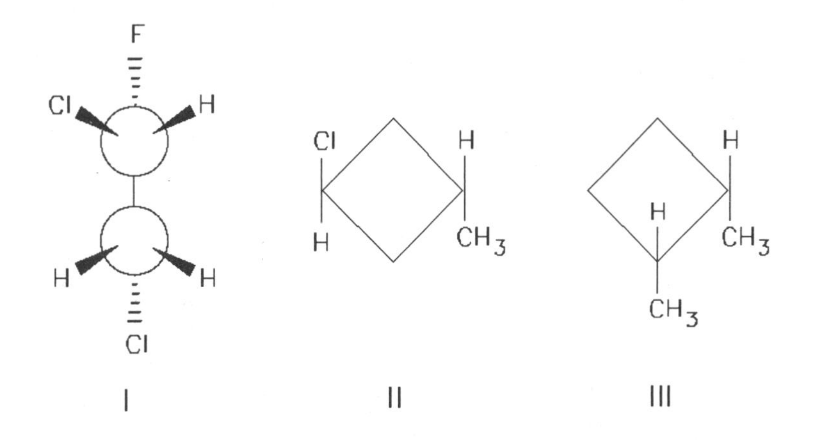

A) I
B) II
C) III
* D) More than one of these
E) None of these

13. Which of the following reactions <u>must</u> occur with retention of configuration?

A)

$$\text{(+)}\ \underset{\underset{OH}{3}}{CH_3CH_2CHCH_3}\ \ \overset{KMnO_4}{DDDDDD4}\ \ \underset{\underset{O}{:}}{CH_3CH_2CCH_3}$$

B)

$$\text{(+)}\ \underset{\underset{OH}{3}}{CH_3CH_2CHCH_3}\ \ \overset{POCl_3}{DDDDDD4}\ \ \underset{\underset{Cl}{3}}{CH_3CH_2CHCH_3}$$

C)

$$\text{(-)}\ \underset{\underset{Cl}{3}}{CH_3CHCO_2H}\ \ \overset{NH_3}{DDDDDD4}\ \ \underset{\underset{NH_2}{3}}{CH_3CHCO_2H}$$

* D)

$$\text{(+)}\ \underset{\underset{CH_3O}{3}}{CH_3CHCO_2H}\ \ \overset{PCl_5}{DDDDDD4}\ \ \underset{\underset{CH_3O}{3}}{CH_3CHCOCl}$$

E)

$$\text{(-)}\ \underset{\underset{Cl}{3}}{CH_3CHCO_2H}\ \ \overset{P}{\underset{Cl_2}{DDDDDD4}}\ \ \overset{\overset{Cl}{3}}{\underset{\underset{Cl}{3}}{CH_3CCO_2H}}$$

14. The molecules shown are:

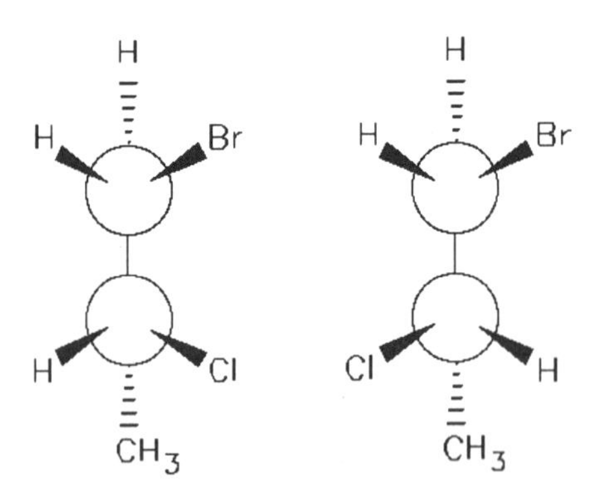

A) constitutional isomers.
* B) enantiomers.
C) diastereomers.
D) identical.
E) None of these

15. Which of the following is true about <u>any</u> (R)-enantiomer?
A) It is dextrorotatory.
B) It is levorotatory.
C) It is an equal mixture of + and -.
* D) It is the mirror image of the (S)-enantiomer.
E) (R) indicates a racemic mixture.

16. Which of the following represent (R)-2-butanol?

I: front H (top), CH_3 (upper left), OH (upper right); back CH_2CH_3 (bottom)

II: front CH_3 (top), H (upper left), OH (upper right); back CH_2CH_3 (bottom)

III: front H (top), HO (upper left), CH_3 (upper right); back CH_2CH_3 (bottom)

IV: front CH_3 (top), HO (upper left), H (upper right); back CH_2CH_3 (bottom)

V: back CH_3CH_2 (top); front H (lower left), OH (lower right), CH_3 (bottom)

A) III and V
B) I, III, IV and V
* C) I, IV and V
D) I and III
E) I, II, IV and V

17. Which of the following reactions might be safely used to relate configurations?

A)

PBr_3

(+) CH_3CHCH_2OH DDDDD4 CH_3CHCH_2Br

3 3

Br Br

B)

O

:

CH_3COH

(+) CH_3CHCH_2OH DDDDDD4 $CH_3CHCH_2OCCH_3$ (O above, : below, over the C of OCC)

H^+

3 3

Br Br

C)

H_2O

(+) CH_3CHCH_2OH DDDDD4 CH_3CHCH_2OH

3 3

Br OH

D) All of the above

* E) Answers A) and B) only

18.

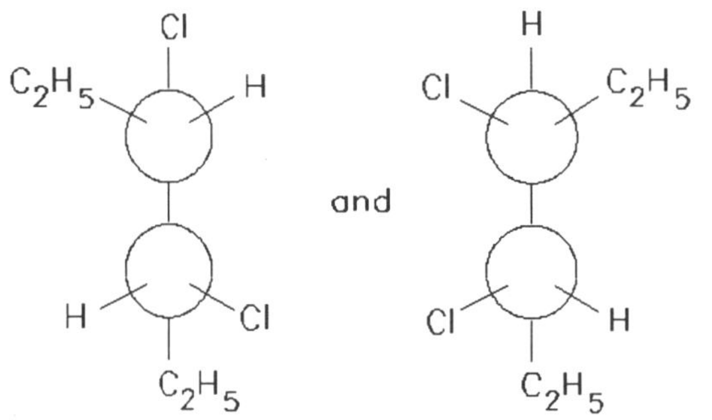

Cl

C_2H_5 H

H Cl

C_2H_5

and

H

Cl C_2H_5

Cl H

C_2H_5

are:

A) enantiomers.

* B) diastereomers.

C) constitutional isomers.

D) two conformations of the same molecule.

E) not isomeric.

19. The molecules shown are:

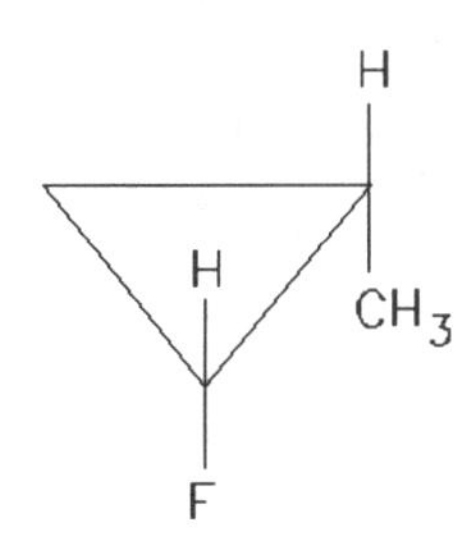

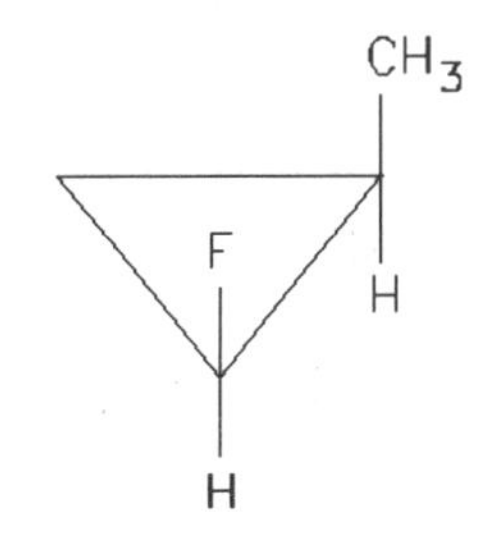

A) constitutional isomers.
* B) enantiomers.
C) diastereomers.
D) identical.
E) None of these

20. The compounds whose molecules are shown below would have:

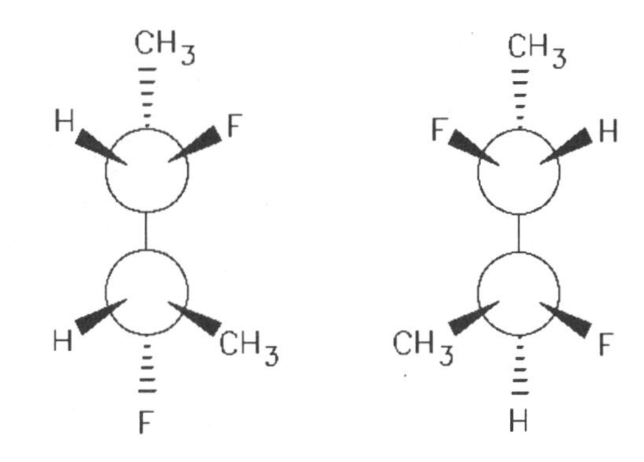

A) the same melting point.
* B) different melting points.
C) equal but opposite optical rotations.
D) More than one of the above
E) None of the above

21. The molecules below are:

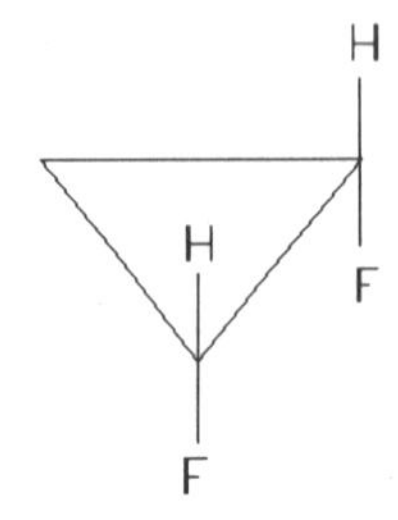

* A) constitutional isomers.
  B) enantiomers.
  C) diastereomers.
  D) identical.
  E) stereoisomers.

22.

and are:

* A) enantiomers.
  B) diastereomers.
  C) constitutional isomers.
  D) two different conformations of the same molecule.
  E) not isomeric.

23. Which compound does NOT possess a plane of symmetry?

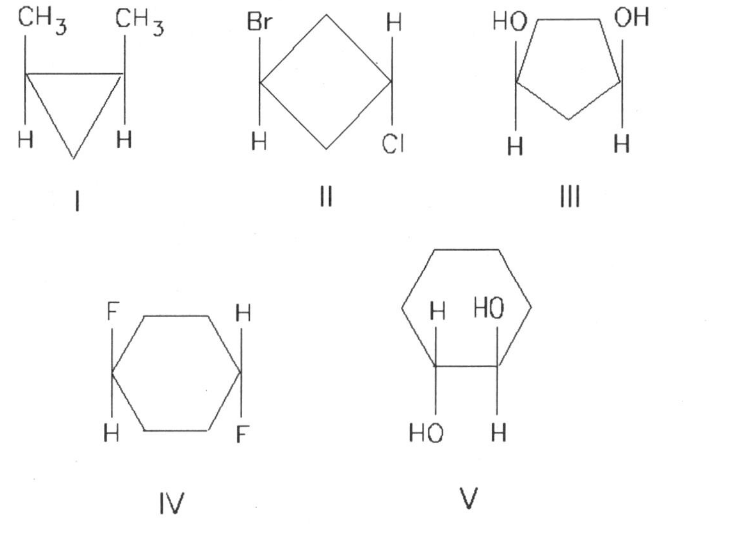

A) I B) II C) III D) IV * E) V

24. Which of the following is NOT true of enantiomers? They have the same:

A) boiling point.
B) melting point.
* C) specific rotation.
D) density.
E) chemical reactivity toward achiral reagents.

25. What is the percent composition of a mixture of (S)-(+)-2-butanol, $[!]_D^{25}$ = +13.52x, and (R)-(-)-2-butanol, $[!]_D^{25}$ = -13.52x, with a specific rotation $[!]_D^{25}$ = +6.76x?

A) 75%(R) 25%(S)
* B) 25%(R) 75%(S)
C) 50%(R) 50%(S)
D) 67%(R) 33%(S)
E) 33%(R) 67%(S)

26. Which molecule is a meso compound?

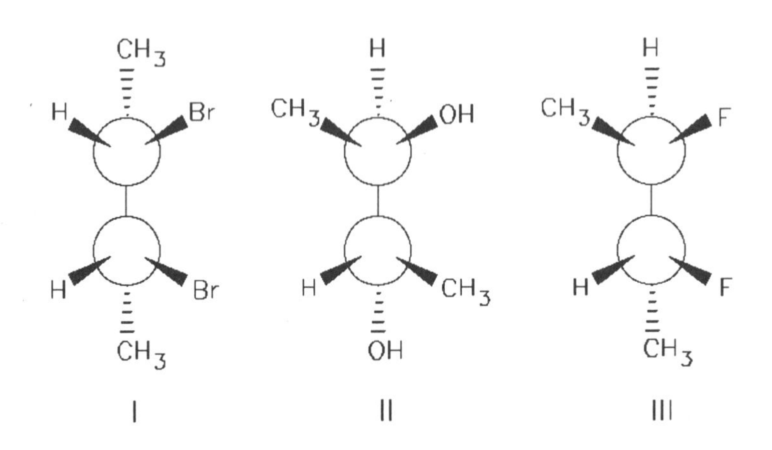

A) I
B) II
C) III
* D) More than one of the above
E) None of the above

27. The molecules shown are:

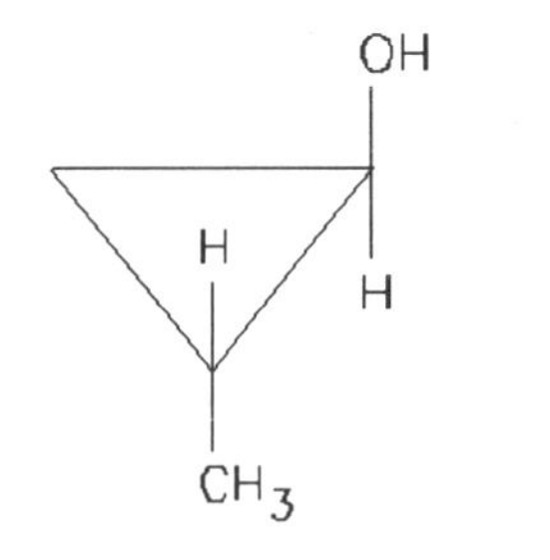

A) constitutional isomers.
* B) enantiomers.
C) diastereomers.
D) identical.
E) None of these

28. How many chiral stereoisomers can be drawn for $CH_3CHClCHBrCH_3$?
A) 1
B) 2
C) 3
* D) 4
E) 8

29. Which of the following is(are) meso compound(s)?

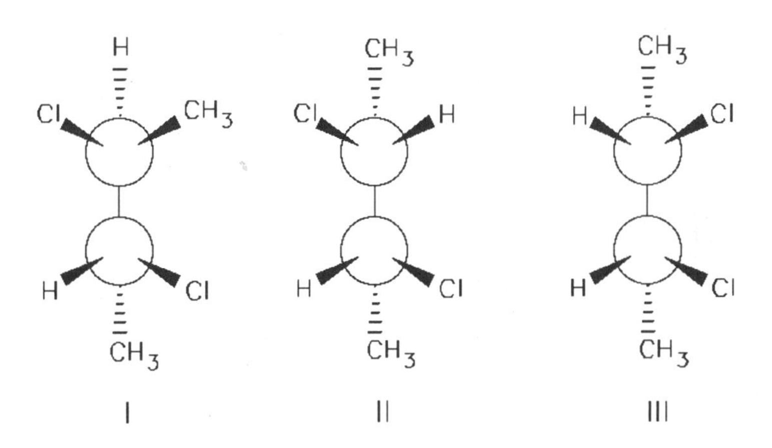

A) I
B) II
C) III
D) Both II and III
* E) Both I and III

30. The compounds whose molecules are shown below would have:

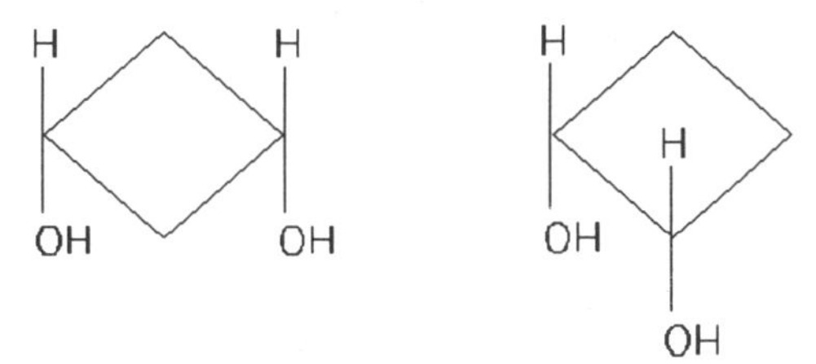

A) the same melting point.
* B) different melting points.
C) equal but opposite optical rotations.
D) More than one of these
E) None of these

31. Which molecule is a meso compound?

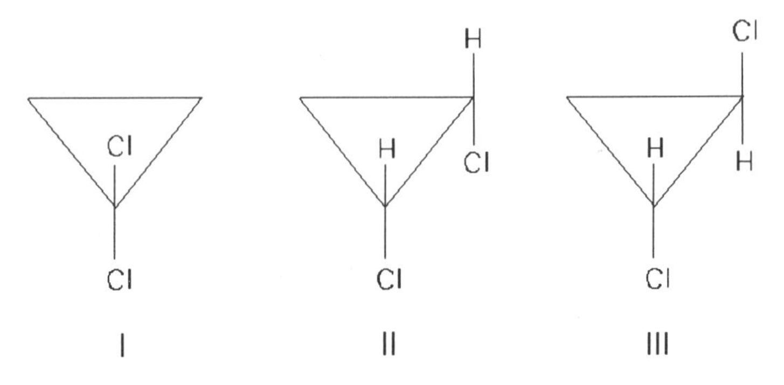

A) I
* B) II
C) III
D) More than one of the above
E) None of the above

32. The molecules shown are:

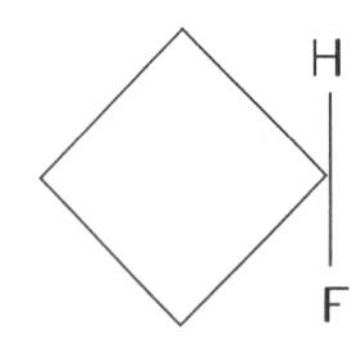

H
F

A) constitutional isomers.
B) enantiomers.
C) diastereomers.
D) identical.
* E) None of these

33. The molecules below are:

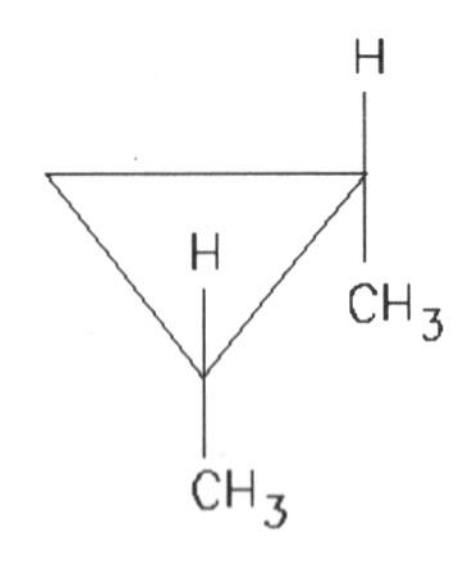

CH3 H
H CH3

A) structural isomers.
B) enantiomers.
* C) diastereomers.
D) identical.
E) None of these

34. <u>Cis</u>-<u>trans</u> isomers are:

A) diastereomers.
B) enantiomers.
C) stereoisomers.
D) constitutional isomers.
* E) More than one of these

35. The molecules below are:

A) constitutional isomers.
B) enantiomers.
C) diastereomers.
* D) identical.
E) None of these

36. The molecules below are:

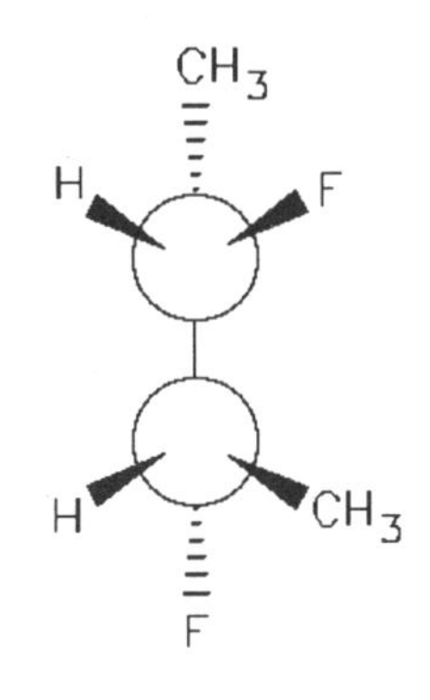

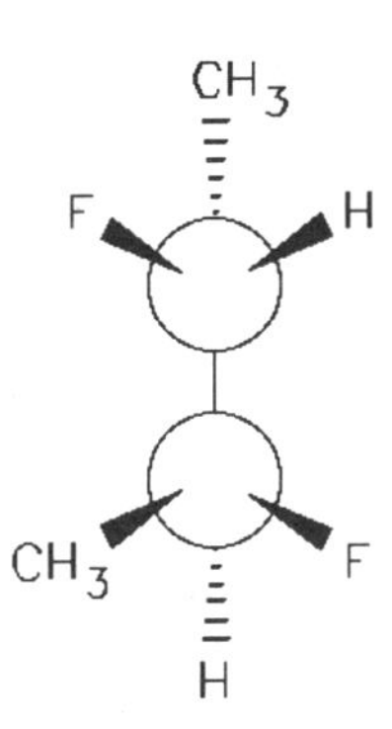

A) constitutional isomers.
B) enantiomers.
* C) diastereomers.
D) identical.
E) None of these

37. Which structure represents (S)-1-chloro-1-fluoroethane?

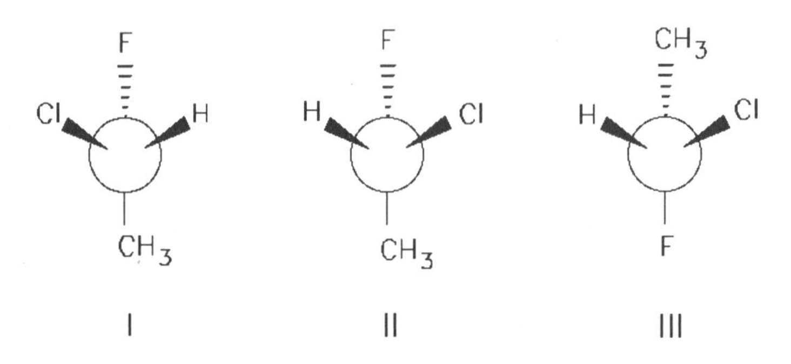

A) I
B) II
C) III
* D) More than one of the above
E) None of the above

38. The molecules below are:

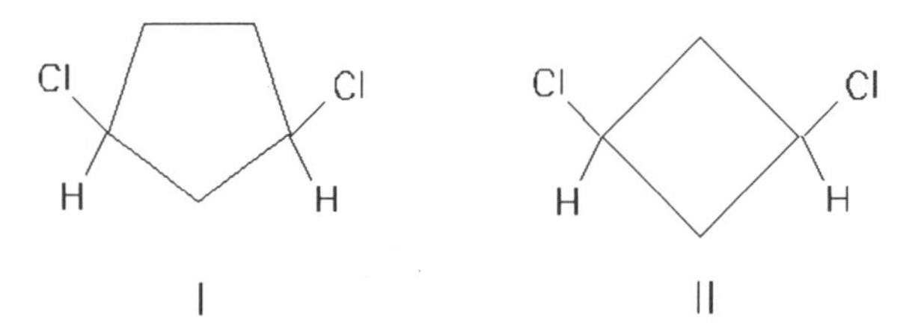

A) constitutional isomers.
B) enantiomers.
C) diastereomers.
D) identical.
* E) None of these

39. Which structure represents (R)-1-chloro-1-fluoroethane?

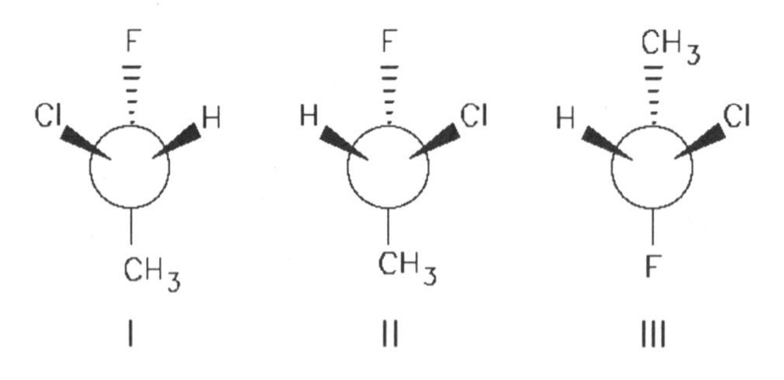

A) I
* B) II
C) III
D) More than one of the above
E) None of the above

40. The compounds whose structures are shown below would have:

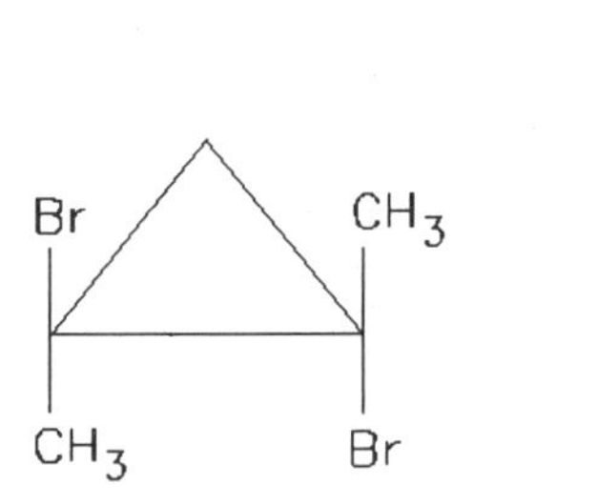

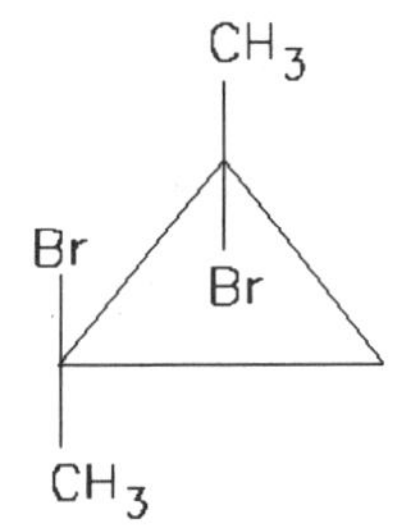

A) the same melting point.
B) different melting points.
C) equal but opposite optical rotations.
* D) More than one of these
E) None of these

41. Which molecule is a meso compound?

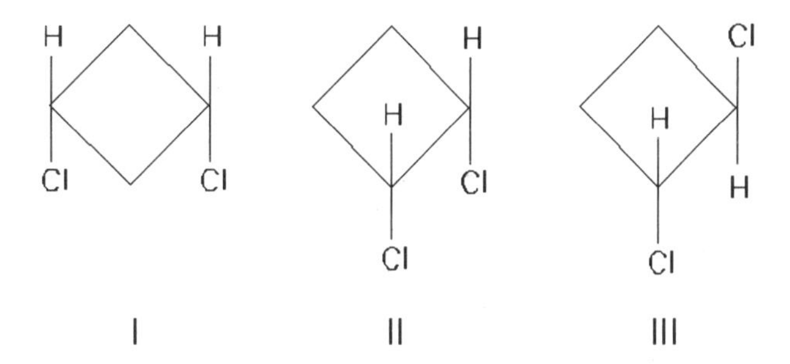

A) I
B) II
C) III
* D) More than one of these
E) None of these

42. The molecules below are:

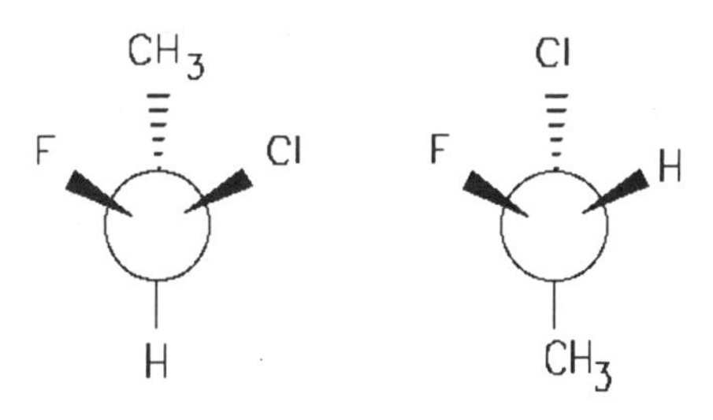

A) constitutional isomers.
B) enantiomers.
C) diastereomers.
* D) identical.
E) None of these

43. The molecules below are:

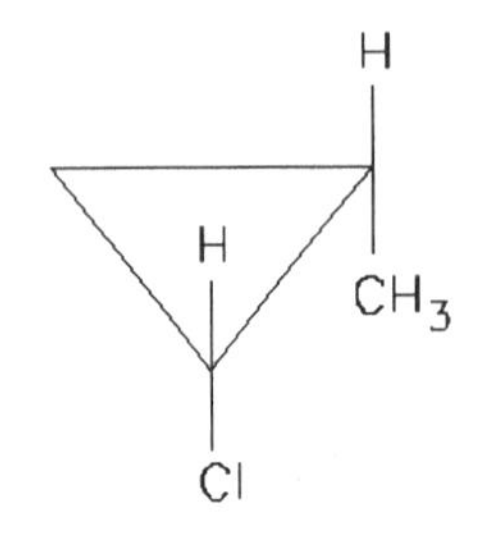

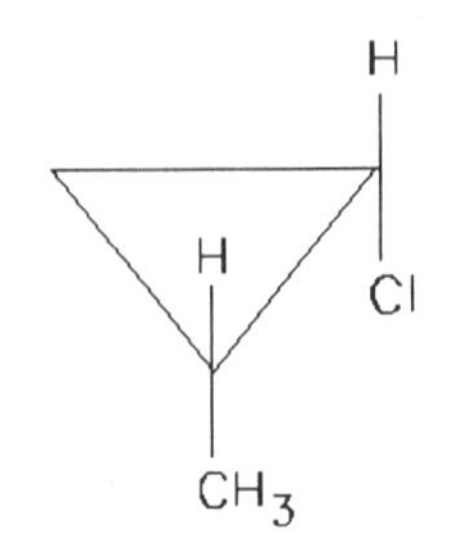

A) constitutional isomers.
* B) enantiomers.
C) diastereomers.
D) identical.
E) None of these

44. Which molecule is achiral?

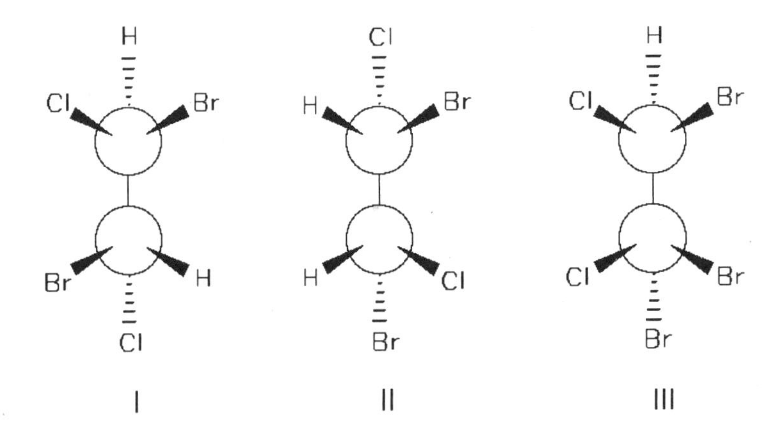

* A) I
B) II
C) III
D) More than one of these
E) None of these

45. Which of the following molecules is achiral?
A) (2R,3R)-2,3-Dichloropentane
B) (2R,3S)-2,3-Dichloropentane
C) (2S,3S)-2,3-Dichlorobutane
* D) (2R,3S)-2,3-Dichlorobutane
E) None of these

46. The molecules below are:

* A) constitutional isomers.
  B) enantiomers.
  C) diastereomers.
  D) identical.
  E) None of these

47. Which structure represents (S)-2-bromobutane?

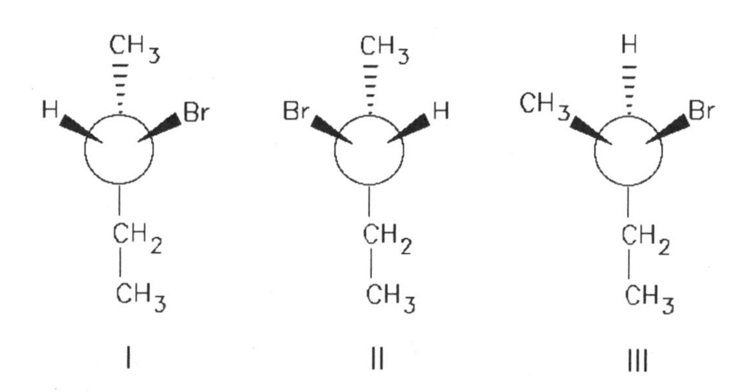

* A) I
  B) II
  C) III
  D) More than one of the above
  E) None of the above

48. Which molecule is achiral?

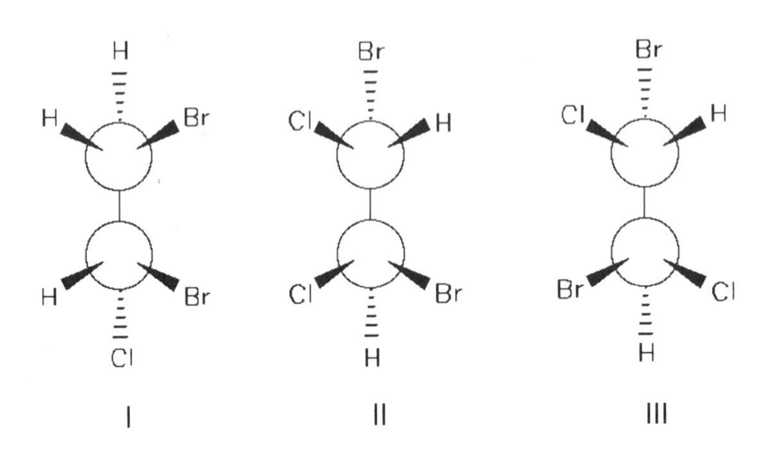

A) I
B) II
* C) III
D) More than one of the above
E) None of the above

49. Which reaction must take place with retention of configuration at the stereocenter?

A)
```
        OH                                 Cl
        3                                  3
     CH3CCH2Br  +  HCl   DDDDDD4   CH3CCH2Br
        3                                  3
        CH2F                               CH2F
```

B)
```
     CH3CHCH2OH  +  PCl3   DDDDDD4   CH3CHCH2Cl
        3                                 3
        CH2F                              CH2F
```

C)
```
                         O                            O            O
                         :                            :            :
     CH3CHCH2OH  +  (CH3C)2O   DDDDDD4   CH3CHCH2OCCH3  +  CH3COH
        3                                   3
        CH2F                                CH2F
```

* D) More than one of the above
E) None of the above

50. Which statement is <u>not</u> true for a meso compound?
A) The specific rotation is 0x.
B) There are one or more planes of symmetry.
C) A single molecule is identical to its mirror image.
D) More than one stereocenter must be present.
* E) The stereochemical labels, (R) and (S), must be identical for each stereocenter.

51. The structures

and

represent:

* A) a single compound.
  B) enantiomers.
  C) meso forms.
  D) diastereomers.
  E) conformational isomers.

52. Which one of the following can exist in optically active forms?

  A) cis-1,3-Dichlorocyclohexane
* B) trans-1,3-Dichlorocyclohexane
  C) cis-1,4-Dichlorocyclohexane
  D) trans-1,4-Dichlorocyclohexane
  E) cis-1,2-Dichlorocyclohexane

53. A solution of which of these allenes will rotate plane-polarized light?

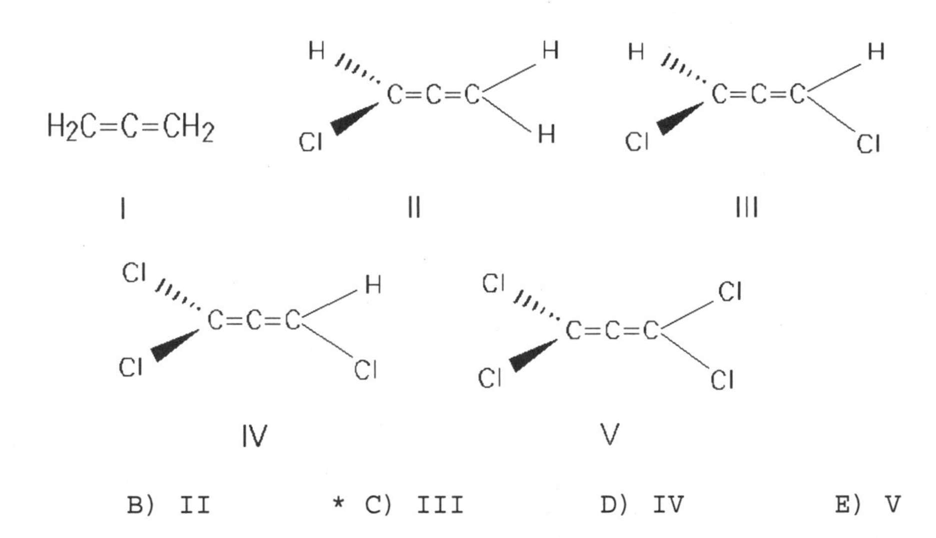

  A) I
  B) II
* C) III
  D) IV
  E) V

54. $CH_3CHBrCHBrCHBrCH_3$ is the generalized representation of what number of stereoisomers?

* A) 4
  B) 5
  C) 6
  D) 7
  E) 8

55. For the generalized structure $ClCH_2CHClCH_2CHClCH_2Cl$ there exists what number of stereoisomers?

  A) 2
* B) 3
  C) 4
  D) 6
  E) 8

56. Which of these is a comparatively insignificant factor affecting the magnitude of specific optical rotation?
  A) Concentration of the substance of interest
  B) Purity of the sample
* C) Temperature of the measurement
  D) Length of the sample tube
  E) All of the above are equally significant.

57. How many discrete dimethylcyclopropanes are there?
  A) 2   B) 3   * C) 4   D) 5   E) 6

58.

Br Br and Br Br

are:
  A) identical.
* B) enantiomers.
  C) diastereomers.
  D) conformational isomers.
  E) meso forms.

59.

```
    CH3
     3
    CH2
     3
  HDCDCl
     3
 ClDCDH
     3
 ClDCDH
     3
    CH3
```

is properly named:

  A) (3R,4S,5R)-3,4,5-Trichlorohexane*
  B) (3S,4S,5S)-3,4,5-Trichlorohexane
  C) (3S,4R,5R)-3,4,5-Trichlorohexane
  D) (2S,3S,4S)-2,3,4-Trichlorohexane
  E) (2S,3S,4R)-2,3,4-Trichlorohexane

60. The Cahn-Ingold-Prelog stereochemical designations used for

are:
A) 2R,4S
B) 2S,4R
* C) 2R,4R
D) 2S,4S
E) The R,S terminology doesn't apply in this case.

61. Which of the following is a meso compound?

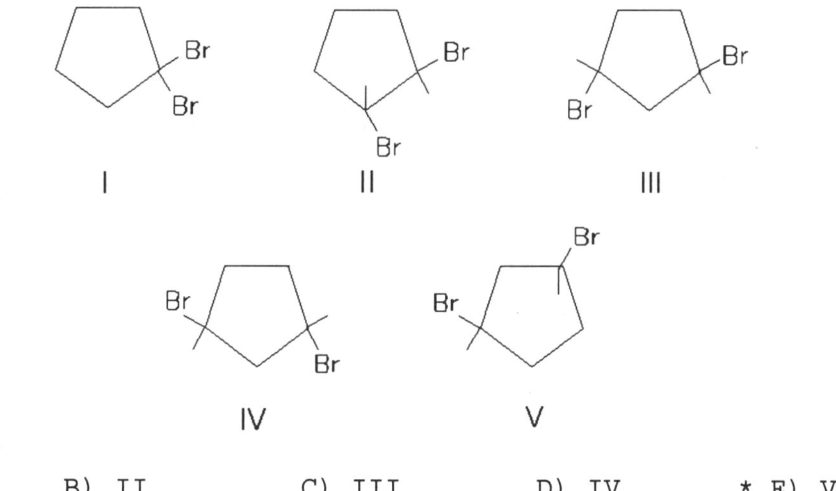

A) I B) II C) III D) IV * E) V

62. What can be said with certainty if a compound has $[!]_D^{25}$ = -9.25x ?
A) The compound has the (S) configuration.
B) The compound has the (R) configuration.
* C) The compound is not a meso form.
D) The compound possesses only one stereocenter.
E) The compound has an optical purity of less than 100%.

63. O
:
The reaction of $CH_3CH_2CCH_2CH_3$ with H_2/Ni forms:
A) pentane.
B) one particular chiral pentanol.
C) an equimolecular mixture of two chiral pentanols.
* D) one achiral pentanol.
E) no product.

64. In the absence of specific data, it can only be said that (R)-2-bromopentane is:
A) dextrorotatory (+).
B) levorotatory (-).
C) optically inactive.
D) achiral.
* E) analogous in absolute configuration to (R)-2-chloropentane.

65. Which of these is <u>not</u> a correct representation of the (S) form of
$Cl_2CHDCHDCH_2Br$?
3
OH

A) $CHCl_2$
3
HDCDOH
3
CH_2Br

B) CH_2Br
3
HODCDH
3
$CHCl_2$

C) H
3
$Cl_2CHDCDCH_2Br$
3
OH

* D) OH
3
$HDCDCHCl_2$
3
CH_2Br

E) OH
3
$BrCH_2DCDCHCl_2$
3
H

66. An alkane which can exhibit optical activity is:
A) Neopentane.
B) Isopentane.
C) 3-Methylpentane.
* D) 3-Methylhexane.
E) 2,3-Dimethylbutane.

67. (2R,3S)-2,3-Dichlorobutane and (2S,3R)-2,3-dichlorobutane are:
A) enantiomers.
B) diastereomers.
* C) identical.
D) conformational isomers.
E) constitutional isomers.

68. If a solution of a compound (30.0 g/100 mL of solution) has a measured rotation of +15x in a 2 dm tube, the specific rotation is:
A) +50x * B) +25x C) +15x D) +7.5x E) +4.0x

69. Which compound would show optical activity?

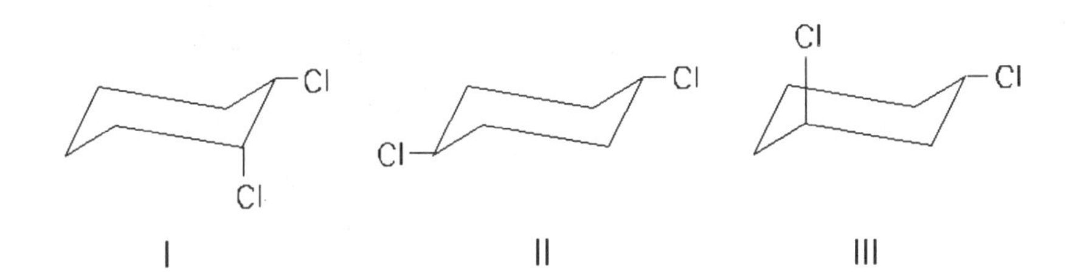

A) I
B) II
* C) III
D) More than one of these
E) None of these

70. Which statement is true of 1,3-dimethylcyclobutane?
A) Only one form of the compound is possible.
* B) Two diastereomeric forms are possible.
C) Two sets of enantiomers are possible.
D) Two enantiomeric forms and one meso compound are possible.
E) None of the previous statements is true.

71. Which is a meso compound?
A) (2R,3R)-2,3-Dibromobutane
B) (2R,3S)-2,3-Dibromopentane
C) (2R,4R)-2,4-Dibromopentane
* D) (2R,4S)-2,4-Dibromopentane
E) None of these

72. What is the molecular formula for the alkane of smallest molecular weight which possesses a stereocenter?
A) C_4H_{10} B) C_5H_{12} C) C_6H_{14} * D) C_7H_{16} E) C_8H_{18}

73. Which of the following is true of any (S)-enantiomer?
A) It rotates plane-polarized light to the right.
B) It rotates plane-polarized light to the left.
C) It is a racemic form.
* D) It is the mirror image of the corresponding (R)-enantiomer.
E) It has the highest priority group on the left.

74. Which pair of structures represents a single compound?

I	II	III	IV	V
CH_3	CH_3	CH_3	CH_3	CH_3
3	3	3	3	3
H-C-OH	H-C-OH	H-C-OH	HO-C-H	H-C-OH
3	3	3	3	3
H-C-OH	HO-C-H	HO-C-H	HO-C-H	H-C-OH
3	3	3	3	3
H-C-OH	H-C-OH	HO-C-H	H-C-OH	HO-C-H
3	3	3	3	3
CH_3	CH_3	CH_3	CH_3	CH_3

A) I and II
B) II and III
C) III and IV
* D) III and V
E) IV and V

75. What is the relationship between the compounds shown below?

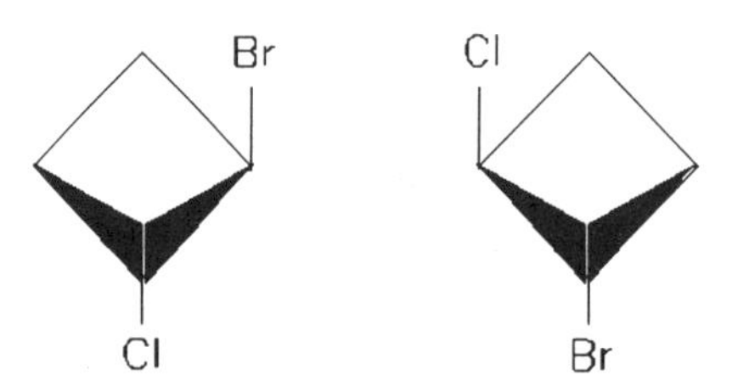

A) They are identical.
* B) They are enantiomers.
C) They are diastereomers.
D) They are constitutional isomers.
E) They are conformational isomers.

76. The two compounds shown below are:

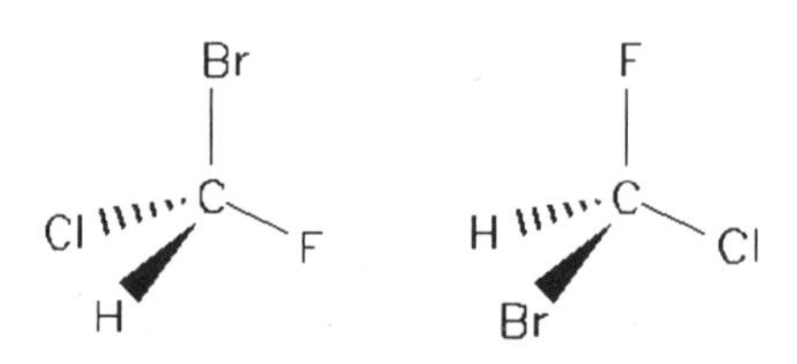

* A) enantiomers.
B) diastereomers.
C) constitutional isomers.
D) identical.
E) different but not isomeric.

Consider the following compounds:

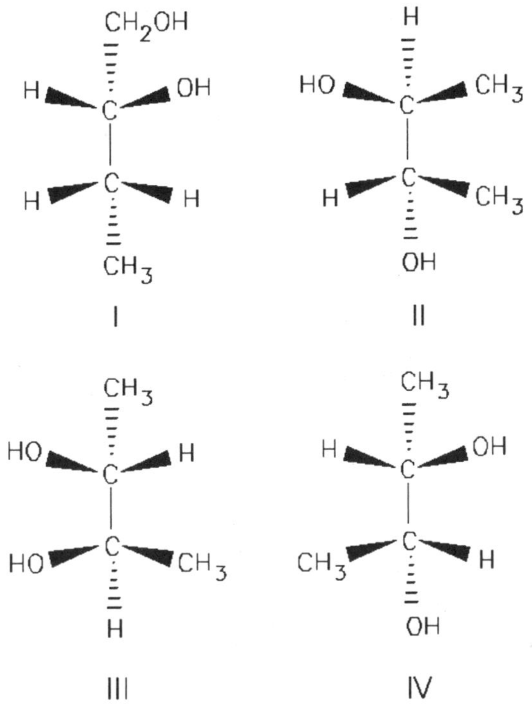

[Structures I, II, III, IV: wedge-dash drawings of substituted two-carbon compounds]

I: CH_2OH (top), H and OH on upper C; H and H on lower C, CH_3 (bottom)

II: H (top), HO and CH_3 on upper C; H and CH_3 on lower C, OH (bottom)

III: CH_3 (top), HO and H on upper C; HO and CH_3 on lower C, H (bottom)

IV: CH_3 (top), H and OH on upper C; CH_3 and H on lower C, OH (bottom)

77. Which of the compounds above (I-IV) represent enantiomers?
  A) I and II
* B) II and III
  C) III and IV
  D) II and IV
  E) III and IV

78. Which compound above (I-IV) is a meso compound?
  A) I
  B) II
  C) III
* D) IV
  E) None of these

79. Which compound above (I-IV) is (2R,3R)-2,3-butanediol?
  A) I
  B) II
* C) III
  D) IV
  E) None of these

80. Which compounds above (I-IV) form a set of stereoisomers?
  A) I, II and III
* B) II, III and IV
  C) II and III
  D) I, III and IV
  E) I, II, III and IV

81. How many optically active compounds are represented by the following generalized formula?

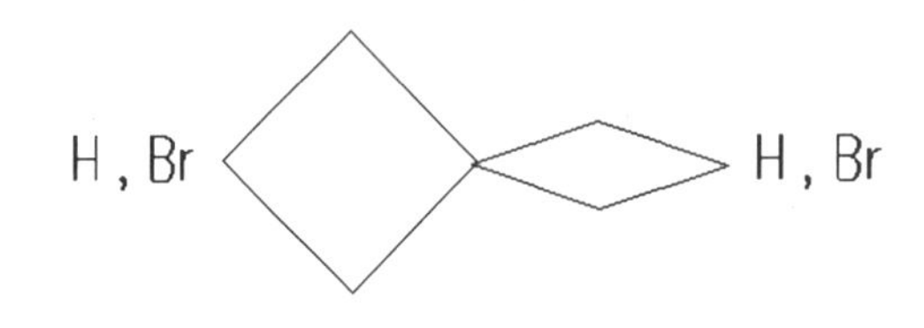

A) 0 B) 1 * C) 2 D) 3 E) 4

82. How many different compounds are there which correspond to the general name "1-<u>sec</u>-butyl-3-chlorocyclobutane"?

A) 1 B) 2 * C) 4 D) 6 E) 8

83. How many stereocenters are there in the anabolic steriod methenolone?

A) 4 B) 5 C) 6 * D) 7 E) 8

84. What is the total number of compounds, stereoisomers included, designated by the general name "dichlorocyclopentane"?

A) 4 B) 5 * C) 7 D) 8 E) 9

85. Of the compounds which correspond to the general name "dichlorocyclobutane", how many are optically active?

A) 0 B) 1 * C) 2 D) 3 E) 4

1. Consider the S_N2 reaction of butyl bromide with OH^- ion.

   $CH_3CH_2CH_2CH_2Br$ + OH^- DDDDDD4 $CH_3CH_2CH_2CH_2OH$ + Br^-

   Assuming no other changes, what effect on the rate would result from simultaneously doubling the concentrations of both butyl bromide and OH^- ion?

   A) No effect.
   B) It would double the rate.
   C) It would triple the rate.
   * D) It would increase the rate four times.
   E) It would increase the rate six times.

2. Which alkyl halide would you expect to undergo S_N1 hydrolysis most rapidly?

   * A) $(CH_3)_3CI$
   B) $(CH_3)_3CBr$
   C) $(CH_3)_3CCl$
   D) $(CH_3)_3CF$
   E) They would all react at the same rate.

3. Which of the following pentyl halides would you expect to give the highest yield of substitution product under conditions for a bimolecular reaction with ethoxide ion?

   * A) $CH_3CH_2CH_2CH_2CH_2Br$

   B) $CH_3CH_2CHCH_2Br$
   3
   CH_3

   C) CH_3
   3
   CH_3CCH_2Br
   3
   CH_3

   D) $CH_3CH_2CH_2CHCH_3$
   3
   Br

   E) Br
   3
   $CH_3CH_2CCH_3$
   3
   CH_3

4. Which of the following alkyl halides would you expect to be most reactive in a unimolecular reaction?

A) $C_6H_{11}CH_2CH_2CH_2Br$

B) $C_6H_{11}CHCH_2Br$
   3
   CH_3

C) CH_3
   3
   $C_6H_{11}CCH_2Br$
   3
   CH_3

D) $C_6H_{11}CH_2CHCH_3$
   3
   Br

* E) Br
   3
   $C_6H_{11}CCH_3$
   3
   CH_3

5. Which of the following statements is (are) true of S_N1 reactions of alkyl halides in general?

A) The rate of an S_N1 reaction depends on the concentration of the alkyl halide.
B) The rate of an S_N1 reaction depends on the concentration of the nucleophile.
C) S_N1 reactions of alkyl halides are favored by polar solvents.
* D) Answers A) and C) only are true.
E) Answers A), B) and C) are true.

6. Which S_N2 reaction will occur most rapidly in aqueous acetone solution? Assume concentrations and temperature are the same in each instance.

* A) HO^- + CH_3-Cl DDDDDD4 CH_3OH + Cl^-

B) HO^- + CH_3CH_2-Cl DDDDDD4 CH_3CH_2OH + Cl^-

C) HO^- + $(CH_3)_2CH$-Cl DDDDDD4 $(CH_3)_2CHOH$ + Cl^-

D) HO^- + $(CH_3)_3C$-Cl DDDDDD4 $(CH_3)_3COH$ + Cl^-

E) HO^- + $(CH_3)_3CCH_2$-Cl DDDDDD4 $(CH_3)_3CCH_2OH$ + Cl^-

7. What would you expect to be the chief organic product(s) when tert-butyl bromide reacts with sodium acetylide, i.e.,

```
        CH3
        3          +  -
    CH3DCDBr   +   Na :CpCH   DDDDDD4   ?
        3
        CH3
```

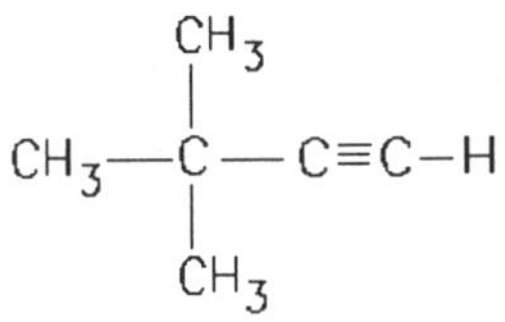

$$CH_3-\underset{CH_3}{\overset{CH_3}{C}}-C\equiv C-H$$

I

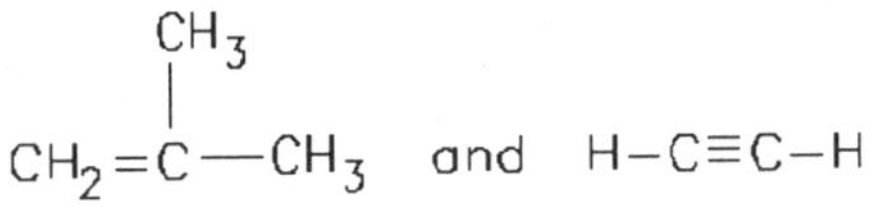

$$CH_2=\overset{CH_3}{C}-CH_3 \text{ and } H-C\equiv C-H$$

II

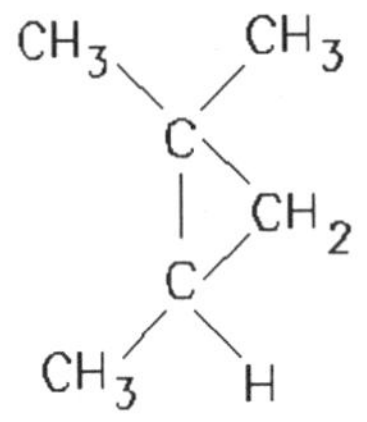

$$(CH_3)_2C\!<\!CH_2\!>\!C(CH_3)H \text{ (three-membered ring: } C(CH_3)_2\text{, } CH_2\text{, } C(CH_3)H)$$

III

$$H-C\equiv C-CH_2-\overset{CH_3}{CH}-CH_3$$

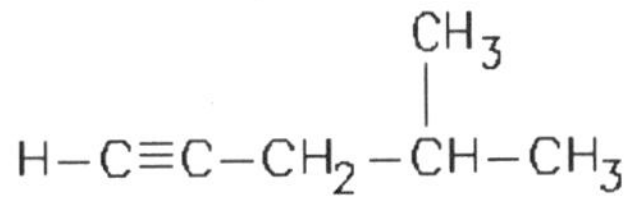

IV

A) I
* B) II
C) III
D) IV
E) None of these

8. S_N2 reactions of the type, $Nu^- + RL$ DDDDDD4 $Nu\text{-}R + L^-$, are favored:
A) when tertiary substrates are used.
* B) by using a high concentration of the nucleophile.
C) by using a solvent of high polarity.
D) by the use of weak nucleophiles.
E) by none of the above.

9. What product(s) would you expect to obtain from the following S_N2 reaction?

H H CH$_3$ Br + CH_3O^- $\xrightarrow{CH_3OH}$?

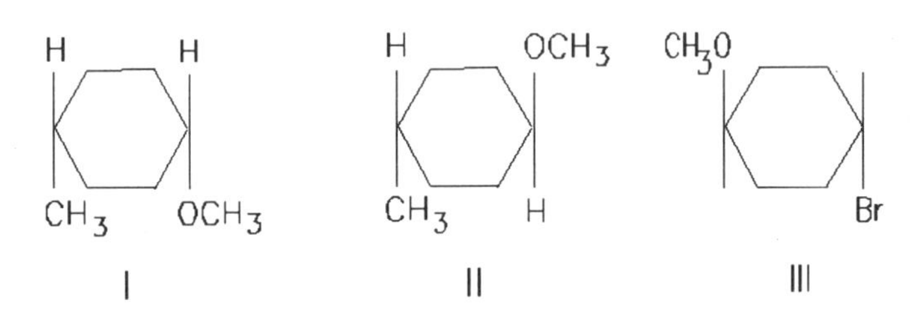

A) I
* B) II
C) An equimolar mixture of I and II.
D) III
E) None of these

10. Identify the nucleophile in the following reaction.

$2\ H_2O$ + RX DDDDDD4 ROH + H_3O^+ + X^-

A) X^- B) H_3O^+ C) ROH * D) H_2O E) RX

11. An increase in the kinetic energy of reacting molecules results in:
A) a decrease in reaction rate.
B) an increase in the probability factor.
C) a decrease in the probability factor.
* D) an increase in the reaction rate.
E) no changes.

12. Identify the leaving group in the following reaction.

$C_6H_5S^-Na^+$ + CH_3CH_2I DDDDDD4 $C_6H_5SCH_2CH_3$ + Na^+ + I^-

A) $C_6H_5S^-$
B) Na^+
C) CH_3CH_2I
D) $C_6H_5SCH_2CH_3$
* E) I^-

13. Select the rate law for a second order reaction.
A) Rate = k [RX]
* B) Rate = k [RX] $[OH^-]$
C) Rate = k $[RX]^2$ $[OH^-]$
D) Rate = k [RX] $[OH^-]^2$
E) Rate = k $[RX]^2$ $[OH^-]^2$

14. Increasing the temperature of a chemical reaction usually increases greatly the rate of the reaction. The most important reason for this is that increasing the temperature increases:
A) the collision frequency.
B) the probability factor.
* C) the fraction of collisions with energy greater than E_{act}.
D) the energy of activation.
E) the amount of heat released in the reaction.

15. Which is the weakest nucleophile in polar aprotic solvents?
* A) I^- B) Br^- C) Cl^- D) F^-

16. Which of the following is the poorest leaving group?
* A) H^- B) Cl^- C) H_2O D) O : ⁻ODSDR : O E) O : ⁻ODSDODR : O

17. Treating $(CH_3)_3C\text{-}Cl$ with a mixture of H_2O and CH_3OH at room temperature would yield:
A) $CH_2MC(CH_3)_2$
B) $(CH_3)_3COH$
C) $(CH_3)_3COCH_3$
* D) All of these
E) None of these

18. S_N1 reactions of the type, Nu^- + RL DDDDDD4 NuDR + L^-, are favored:
* A) when tertiary substrates are used.
B) by using a high concentration of the nucleophile.
C) when L^- is a strong base.
D) by use of a non-polar solvent.
E) by none of the above.

19. Which of the following is a substitution reaction?
A) CH_3CH_2Cl + I^- DDDDDD4 CH_3CH_2I + Cl^-

B) CH_3CH_3 + Cl_2 DDDDDD4 CH_3CH_2Cl + HCl

C) CH_3CH_2Cl + OH^- DDDDDD4 CH_2MCH_2 + H_2O + Cl^-

* D) More than one of the above
E) None of the above

20. Which of the following is a substitution reaction?

A) CH_3CH_3 + Br_2 DDDDDD4 CH_3CH_2Br + HBr

B) CH_3CH_2Br + OH^- DDDDDD4 CH_2MCH_2 + H_2O + Br^-

C) CH_3CH_2Br + OH^- DDDDDD4 CH_3CH_2OH + Br^-

* D) More than one of the above
E) None of the above

21. Select the potential energy diagram that represents an exothermic reaction.

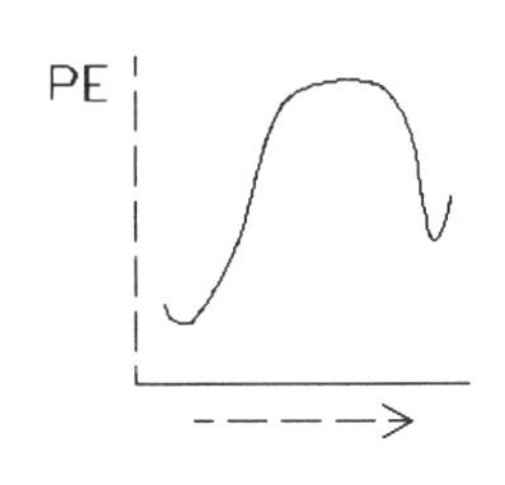

I

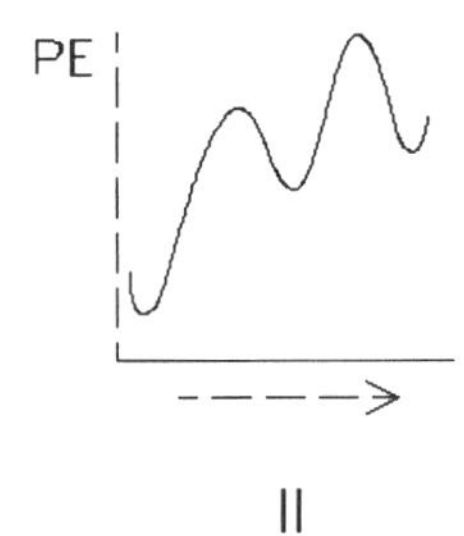

II

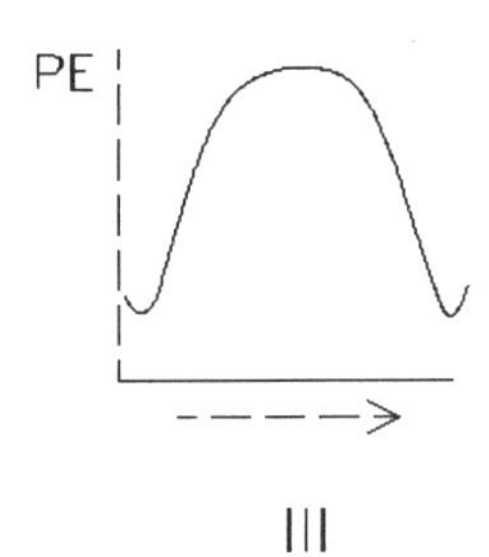

III

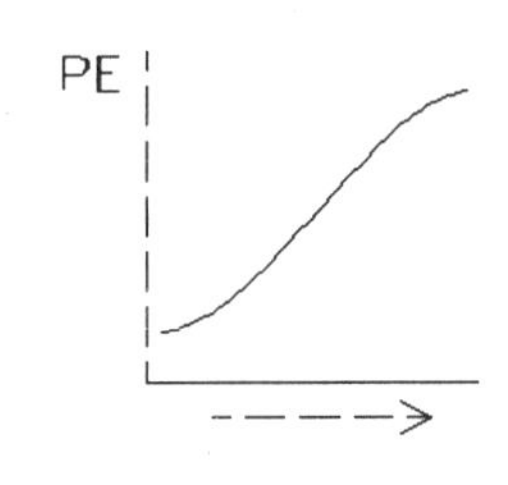

IV

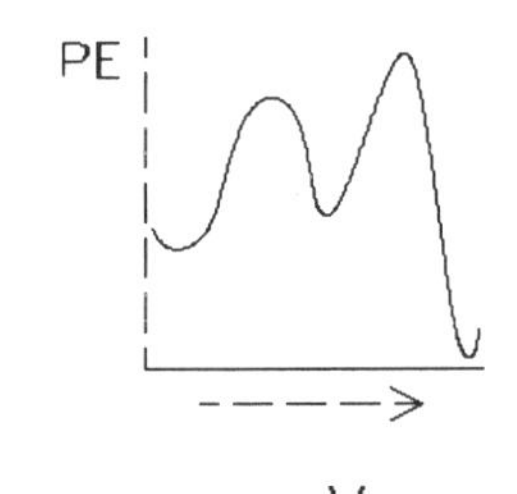

V

A) I   B) II   C) III   D) IV   * E) V

22. Which S_N2 reaction will occur most rapidly in aqueous acetone solution?

A) HO^- + CH_3-F DDDDDD4 CH_3OH + F^-
B) HO^- + CH_3-Cl DDDDDD4 CH_3OH + Cl^-
C) HO^- + CH_3-Br DDDDDD4 CH_3OH + Br^-
* D) HO^- + CH_3-I DDDDDD4 CH_3OH + I^-
E) They will all occur at the same rate.

23. Which alkyl halide would give the highest yield of the elimination product when treated with sodium ethoxide in ethanol?

A) CH_3CH_2Br

B) $CH_3CH_2CH_2Br$

C) $CH_3CH_2CH_2CH_2Br$

D) CH_3CH_2CHBr
3
CH_3

* E) CH_3
3
CH_3CBr
3
CH_3

24. Which alkyl halide would you expect to undergo an S_N2 reaction most slowly?

A) $CH_3CH_2CH_2CH_2CH_2Br$

B) $CH_3CHCH_2CH_2Br$
3
CH_3

C) $CH_3CH_2CHCH_2Br$
3
CH_3

D) $CH_3CH_2CH_2CHBr$
3
CH_3

* E) CH_3
3
CH_3CCH_2Br
3
CH_3

25. Which alkyl halide would you expect to react most rapidly in an S_N1 reaction?

A) $CH_3CH_2CH_2CH_2CH_2Cl$

B) $CH_3CHCH_2CH_2Cl$
3
CH_3

C) $CH_3CH_2CHCH_2Cl$
3
CH_3

D) CH_3
3
CH_3CCH_2Cl
3
CH_3

* E) CH_3
3
CH_3CH_2CCl
3
CH_3

26. Which nucleophilic substitution reaction would be unlikely to occur?

A) Br^- + $CH_3DOH_2^+$ DDDDDD4 CH_3DBr + H_2O

* B) NH_3 + $CH_3DOH_2^+$ DDDDDD4 $CH_3DNH_3^+$ + H_2O

C) I^- + CH_3DCl DDDDDD4 CH_3DI + Cl^-

D) More than one of the above
E) None of the above

27. Which of the following reactions proceeds with inversion of configuration at the carbon bearing the leaving group?

* A) S_N2
B) S_N1
C) E2
D) E1
E) All of these

28. Which nucleophilic substitution reaction would be unlikely to occur?

A) HO^- + CH_3CH_2DI DDDDDD4 CH_3CH_2DOH + I^-

* B) I^- + CH_3CH_2DH DDDDDD4 CH_3CH_2DI + H^-

C) CH_3S^- + CH_3DBr DDDDDD4 CH_3SDCH_3 + Br^-

D) All of the above
E) None of the above

29. Which ion is the strongest nucleophile in aqueous solution?

A) F^-
B) Cl^-
C) Br^-
* D) I^-
E) All of these are equally strong.

30. Which of the following is <u>not</u> a good leaving group?

* A) OH^-
B) Cl^-
C) Br^-
D)

```
   O
   :
 -ODSDR
   :
   O
```

E)

```
    O
    :
 -O-S-O-R
    :
    O
```

31. Identify the leaving group in the following reaction:

```
                               +
CH3OH  +  CH3OH2+  DDDDDD4  CH3OCH3  +  H2O
                               3
                               H
```

A) CH_3OH
B) $CH_3OH_2^+$
C) CH_3OCH_3
* D) H_2O
E) None of these

32. Which alkyl halide would be the most reactive in an S_N2 reaction?

* A) $CH_3CH_2CH_2CH_2Cl$

B)
```
CH3CHCH2Cl
    3
    CH3
```

C)
```
    CH3
    3
CH3CCH2Cl
    3
    CH3
```

D)
```
CH3CH2CHCH3
      3
      Cl
```

E)
```
  Cl
  3
CH3CCH3
  3
  CH3
```

33. Which of the following is not a nucleophile?

A) H_2O
B) CH_3O^-
C) NH_3
* D) NH_4^+
E) All are nucleophiles.

34. Which of the following statements is (are) true of an S_N2 reaction of (R)-2-bromobutane with hydroxide ion?

A) Doubling the hydroxide ion concentration would double the rate of the reaction. (Assume that all other experimental conditions are unchanged.)
B) The reaction would occur with inversion of configuration.
C) Doubling the concentration of (R)-2-bromobutane would double the rate of the reaction. (Assume that all other experimental conditions are unchanged.)
* D) All of the above
E) None of the above

35. Select the strongest nucleophile for an S_N2 reaction.

A) H_2O B) ROH C) RCO_2^- D) OH^- * E) RO^-

36. Which is the strongest nucleophile?

A) OH^- * B) $CH_3CH_2O^-$ C)
```
   O
   :
CH3CO-
```
D) CH_3CH_2OH E) H_2O

37. Which of the following is not a good leaving group?

A) Cl^- B) Br^- C) I^- * D) CH_3O^- E)
```
     O
     :
CF3DSDO-
     :
     O
```

38. O
:
Which is the most reactive nucleophile in DMF, i.e., in $HCN(CH_3)_2$?

* A) F^-
B) Cl^-
C) Br^-
D) I^-
E) They are all equally reactive.

39. Which alkyl halide would undergo solvolysis most rapidly in a mixture of methanol and water?

A)
CH_3
3
CH_3CH_2CDF
3
CH_3

B)
CH_3
3
CH_3CH_2CDCl
3
CH_3

C)
CH_3
3
CH_3CH_2CDBr
3
CH_3

* D)
CH_3
3
CH_3CH_2CDI
3
CH_3

E) They would all react at the same rate.

40. The major product of the following reaction would be:

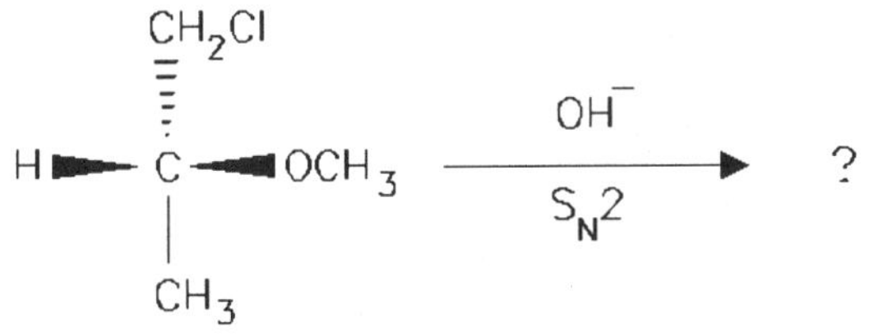

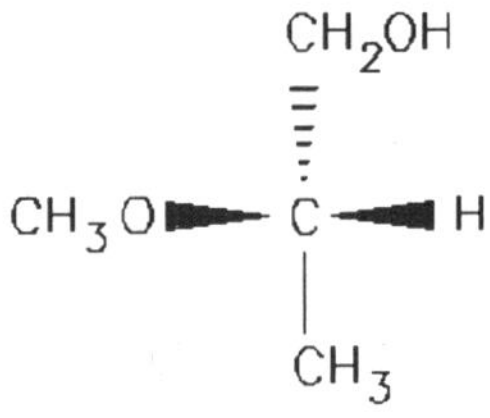

$$\text{H} \blacktriangleright \overset{\text{CH}_2\text{Cl}}{\underset{\text{CH}_3}{\text{C}}} \blacktriangleleft \text{OCH}_3 \xrightarrow[\text{S}_N2]{\text{OH}^-} ?$$

I: $\text{CH}_3\text{O} \blacktriangleright \overset{\text{CH}_2\text{OH}}{\underset{\text{CH}_3}{\text{C}}} \blacktriangleleft \text{H}$

II: $\text{H} \blacktriangleright \overset{\text{CH}_2\text{OH}}{\underset{\text{CH}_3}{\text{C}}} \blacktriangleleft \text{OCH}_3$

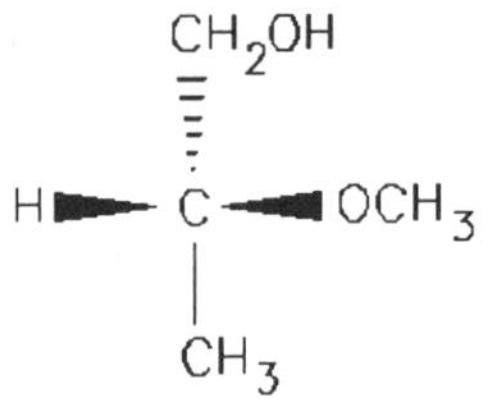

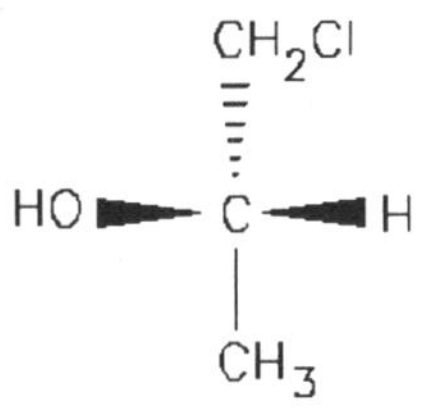

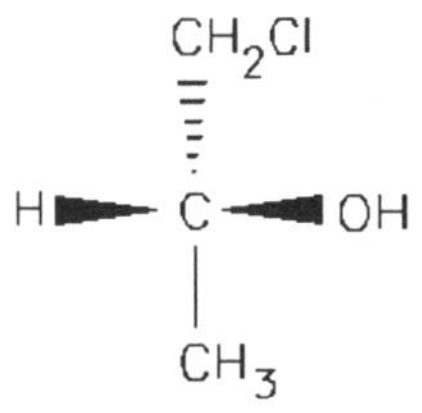

III: $\text{HO} \blacktriangleright \overset{\text{CH}_2\text{Cl}}{\underset{\text{CH}_3}{\text{C}}} \blacktriangleleft \text{H}$

IV: $\text{H} \blacktriangleright \overset{\text{CH}_2\text{Cl}}{\underset{\text{CH}_3}{\text{C}}} \blacktriangleleft \text{OH}$

A) I
* B) II
C) III
D) IV
E) An equimolar mixture of I and II.

41. Which S_N2 reaction would take place most rapidly?

A)

$$HO^- \quad + \quad CH_3DCl \quad \overset{H_2O}{\underset{25xC}{DDDDDD4}} \quad CH_3OH \quad + \quad Cl^-$$

B)

$$\overset{\overset{O}{:}}{CH_3CO^-} \quad + \quad CH_3DCl \quad \overset{CH_3CO_2H}{\underset{25xC}{DDDDDDD4}} \quad \overset{\overset{O}{:}}{CH_3OCCH_3} \quad + \quad Cl^-$$

C)

$$H_2O \quad + \quad CH_3DCl \quad \overset{H_2O}{\underset{25xC}{DDDDDD4}} \quad CH_3OH_2{}^+ \quad + \quad Cl^-$$

D)

$$CH_3OH \quad + \quad CH_3DCl \quad \overset{CH_3OH}{\underset{25xC}{DDDDDD4}} \quad \underset{\underset{H}{3}}{CH_3O^+CH_3} \quad + \quad Cl^-$$

* E)

$$HS^- \quad + \quad CH_3DCl \quad \overset{H_2O}{\underset{25xC}{DDDDDD4}} \quad CH_3SH \quad + \quad Cl^-$$

42. Increasing the concentration of either of the reactants of an S_N2 reaction increases the rate of the reaction. The reason for this is that increasing the concentration increases:

* A) the collision frequency.
B) the orientation factor.
C) the fraction of collisions with energy greater than E_{act}.
D) the energy of activation.
E) the amount of heat released in the reaction.

43. Heating <u>tert</u>-butyl chloride with 1.0 <u>M</u> NaOH in a mixture of water and methanol would yield mainly:

A) $(CH_3)_3COH$ through an S_N1 reaction.
B) $(CH_3)_3COCH_3$ through an S_N1 reaction.
C) $(CH_3)_3COH$ through an S_N2 reaction.
D) $(CH_3)_3COCH_3$ through an S_N2 reaction.
* E) $CH_2{=}C(CH_3)_2$ through an E2 reaction.

44. For the typical S_N2 reaction

$$Y^- \quad + \quad RX \quad DDDDD4 \quad RY \quad + \quad X^D$$

it can be predicted that $(S^E$ will be:

A) positive.
B) zero.
* C) negative.
D) either positive or negative.
E) unpredictable as to algebraic sign.

45. Which alkyl halide would you expect to react most slowly when heated in aqueous solution?

* A) $(CH_3)_3C-F$
B) $(CH_3)_3C-Cl$
C) $(CH_3)_3C-Br$
D) $(CH_3)_3C-I$
E) They would all react at the same rate.

46. S_N1 reactions of the following type,

$$Nu:^- + R-X \text{ DDDDDD4 } R-Nu + :X^-$$

are favored:

A) by the use of tertiary substrates (as opposed to primary or secondary substrates).
B) by increasing the concentration of the nucleophile.
C) by increasing the polarity of the solvent.
D) by use of a weak nucleophile.
* E) by more than one of the above.

47. Which would be formed in the following reaction?

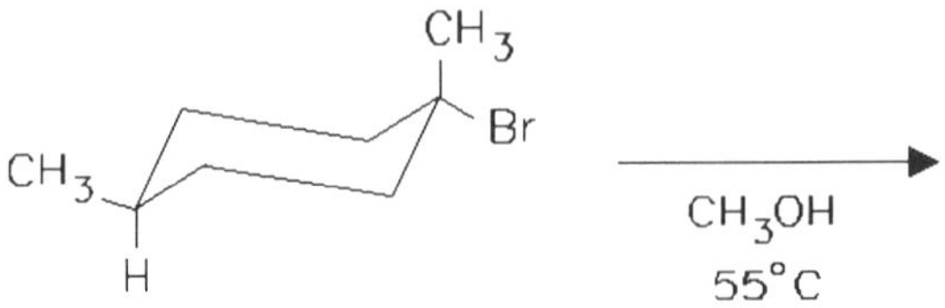

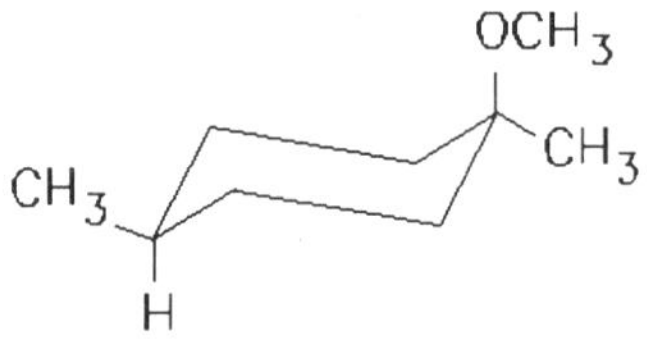

CH_3OH
55°C

I

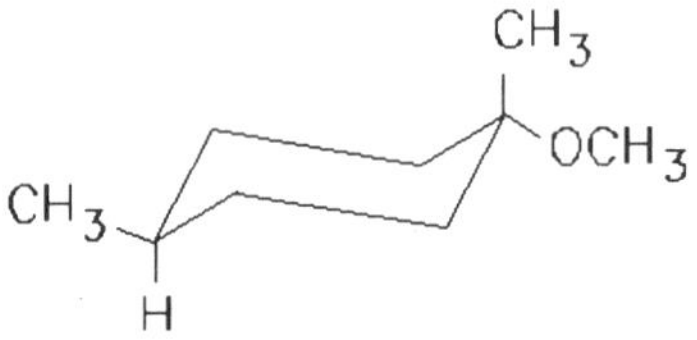

II

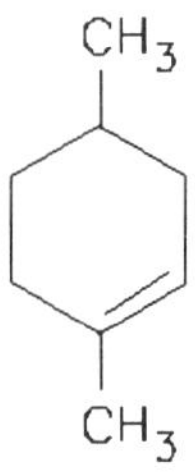

III

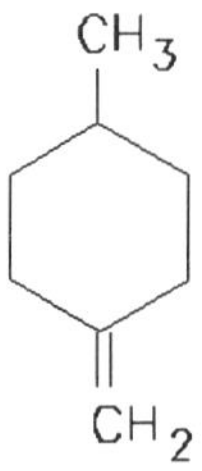

IV

A) I
B) II
C) III
D) IV
* E) All of the above

48. Which S_N2 reaction will occur most rapidly in a mixture of water and ethanol?

* A) I^- + CH_3CH_2DBr DDDDDD4 CH_3CH_2DI + Br^-

B) I^- + CH_3CH_2DCl DDDDDD4 CH_3CH_2DI + Cl^-

C) I^- + CH_3CH_2DF DDDDDD4 CH_3CH_2DI + F^-

D) Br^- + CH_3CH_2DCl DDDDDD4 CH_3CH_2DBr + Cl^-

E) Br^- + CH_3CH_2DF DDDDDD4 CH_3CH_2DBr + F^-

49. Which would be the major product of the following reaction?

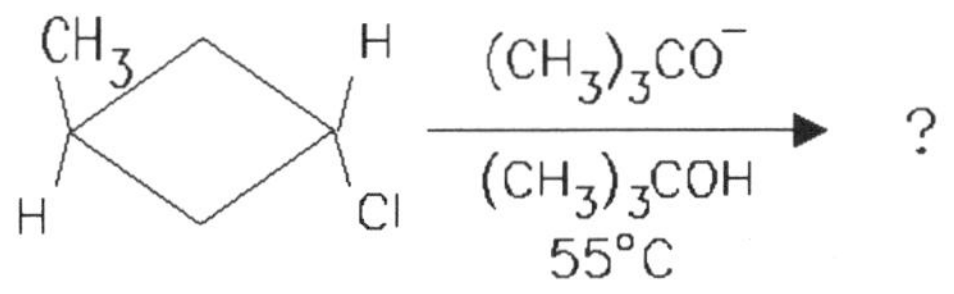

CH3
H
H
Cl
(CH3)3CO⁻
(CH3)3COH
55°C
?

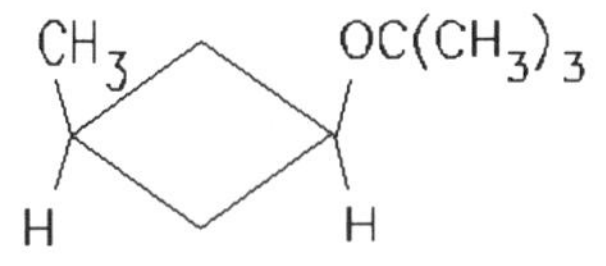

I

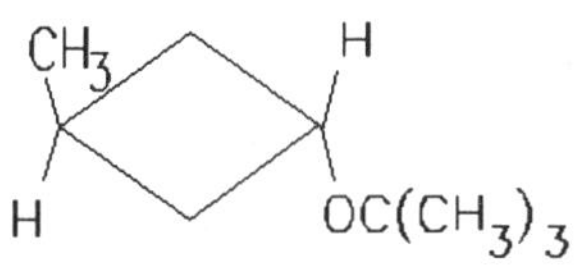

II

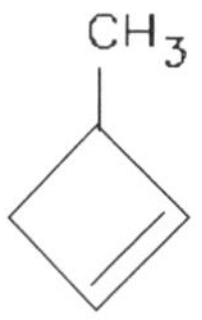

III

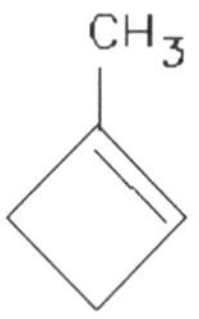

IV

A) I
B) II
* C) III
D) IV
E) None of the above

50. Which S_N2 reaction would you expect to take place most rapidly? Assume the concentrations of the reactants and the temperature are the same in each instance:

* A) CH_3S^- + CH_3DI DDDDDD4 CH_3SCH_3 + I^-

B) CH_3SH + CH_3DI DDDDDD4 $CH_3S^+CH_3$ + I^-
3
H

C) CH_3O^- + CH_3DI DDDDDD4 CH_3OCH_3 + I^-

D) CH_3OH + CH_3DI DDDDDD4 $CH_3O^+CH_3$ + I^-
3
H

E) CH_3S^- + CH_3DCl DDDDDD4 CH_3SCH_3 + Cl^-

51. Which of these species, acting in a protic solvent, exhibits greater nucleophilic activity than expected on the basis of its basicity?

A) OH^- B) CH_3O^- * C) SH^- D) Cl^- E) H_2O

52. Which nucleophilic substitution reaction is not likely to occur?

A) I^- + CH_3CH_2DCl DDDDDD4 CH_3CH_2DI + Cl^-

B) I^- + CH_3CH_2DBr DDDDDD4 CH_3CH_2DI + Br^-

* C) I^- + CH_3CH_2DOH DDDDDD4 CH_3CH_2DI + OH^-

D) CH_3O^- + CH_3CH_2DBr DDDDDD4 $CH_3CH_2DOCH_3$ + Br^-

E) OH^- + CH_3CH_2DCl DDDDDD4 CH_3CH_2DOH + Cl^-

53. Which of the following would be most reactive in an S_N2 reaction?

* A) CH_3Br B) CH_2MCHBr C) CH_3CH_2Br D) C_6H_5Br E) CH_3CHBr
3
CH_3

54. The rate equation for an S_N1 reaction of an alkyl bromide (R-Br) with I^- ion would be:

* A) Rate = k [RBr]
B) Rate = k $[I^-]$
C) Rate = k [RBr]$[I^-]$
D) Rate = k $[RBr]^2[I^-]$
E) Rate = k [RBr]$[I^-]^2$

55. Consider the S_N1 reaction of <u>tert</u>-butyl chloride with iodide ion.

$(CH_3)_3CDCl$ + I^- DDDDDD4 $(CH_3)_3CDI$ + Cl^-

Assuming no other changes, how would it affect the rate if one simultaneously doubled the concentration of <u>tert</u>-butyl chloride and iodide ion?

A) No effect
* B) It would double the rate.
C) It would triple the rate.
D) It would quadruple the rate.
E) It would increase the rate five times.

56. An increase in the kinetic energy of reacting molecules increases:

A) the collision frequency.
B) the fraction of molecules with proper orientation.
C) the fraction of molecules with energy greater than E_{act}.
* D) More than one of the above
E) None of the above

57. The major product(s) of the following reaction is(are):

H, Cl, CH_3, H + I^- $\xrightarrow[25^\circ C]{CH_3CH_2OH}$?

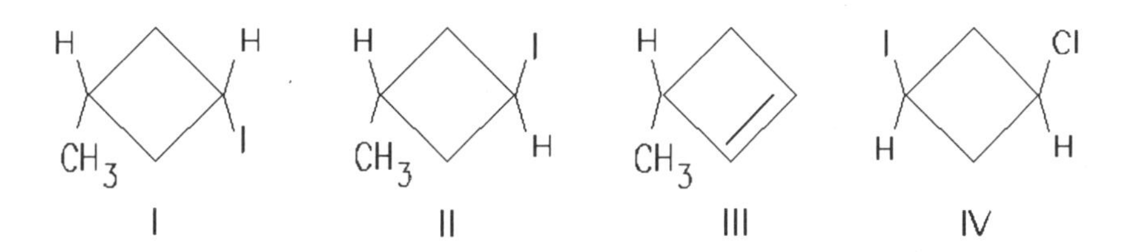

* A) I
  B) II
  C) III
  D) IV
  E) Equal amounts of I and II

58. The hybridization state of the charged carbon in a carbocation is
  A) sp^4 B) sp^3 * C) sp^2 D) sp E) s

59. The p-orbital of a methyl cation, CH_3^+, contains how many electrons?
  A) 1 B) 2 C) 3 D) 4 * E) 0

60. The reaction,

CH_3Cl + OH^- DDDDDD4 CH_3OH + Cl^-
H_2O

has the following thermodynamic values at 27xC: (H = -18 kcal mol^{-1}. (S = 13 cal K^{-1} mol^{-1}. What is the value of (G for this reaction?
  A) -3918 kcal mol^{-1}
  B) -14.1 kcal mol^{-1}
  C) -369 kcal mol^{-1}
  D) +21.9 kcal mol^{-1}
  * E) -21.9 kcal mol^{-1}

61. Which alkyl chloride, though primary, is essentially unreactive in S_N2 reactions?

A) $CH_3CH_2CH_2CH_2CH_2CH_2Cl$

B) $CH_3CH_2CH_2CHCH_2Cl$
   3
   CH_3

C) $CH_3CH_2CHCH_2CH_2Cl$
   3
   CH_3

D) CH_3
   3
   $CH_3CCH_2CH_2Cl$
   3
   CH_3

* E) CH_3
   3
   $CH_3CH_2CCH_2Cl$
   3
   CH_3

62. Which is a true statement concerning the transition state of the rate-determining step of an S_N1 reaction?

* A) Structurally, it closely resembles the carbocation intermediate.
B) Both covalent bond-breaking and bond-making are occurring.
C) Formation of the transition state is an exothermic reaction.
D) Necessarily, the transition state has zero charge overall.
E) More than one of the above

63. Which ion is the strongest nucleophile in an aprotic solvent such as dimethylsulfoxide?

A) I^-
B) Br^-
C) Cl^-
* D) F^-
E) These are all equal.

64. Which alkyl halide would be most reactive in an S_N1 reaction?

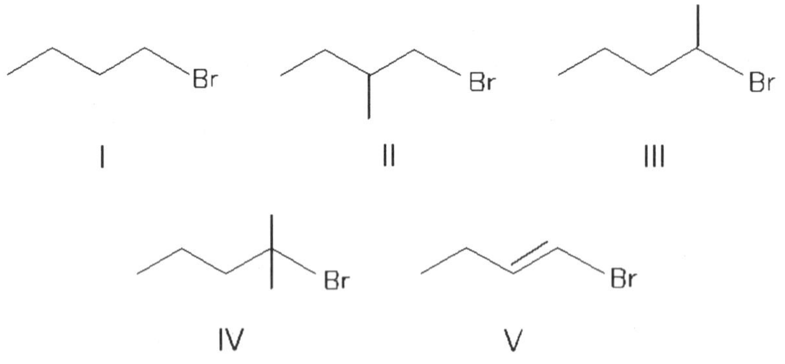

A) I   B) II   C) III   * D) IV   E) V

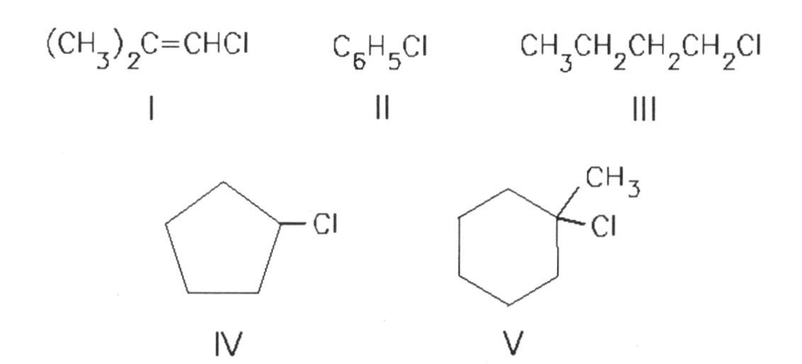

65. Which alkyl halide would be most reactive in an S_N1 reaction?
A) I B) II C) III D) IV * E) V

66. Which alkyl halide would be most reactive in an S_N2 reaction?
A) I B) II * C) III D) IV E) V

67. Which S_N2 reaction would be expected to occur most rapidly?
A) $CH_3CH_2F + CN^-$
B) $CH_3CH_2Cl + CN^-$
C) $CH_3CH_2Br + CN^-$
* D) $CH_3CH_2I + CN^-$
E) $CH_3CH_2OH + CN^-$

68. Which reaction would be expected to occur most slowly?
A) $CH_3CH_2F + CN^-$
B) $CH_3CH_2Cl + CN^-$
C) $CH_3CH_2Br + CN^-$
D) $CH_3CH_2I + CN^-$
* E) $CH_3CH_2OH + CN^-$

69. Consider the substitution reaction that takes place when (R)-3-bromo-3-methylhexane is treated with methanol. Which of the following would be true?
A) The reaction would take place <u>only</u> with inversion of configuration at the stereocenter.
B) The reaction would take place <u>only</u> with retention of configuration at the stereocenter.
* C) The reaction would take place with racemization.
D) No reaction would take place.
E) The alkyl halide does not possess a stereocenter.

70. S_N2 reactions of the type, $Nu^- + RDL$ DDDDDD4 $NuDR + L^-$ are favored:
A) when tertiary substrates are used.
* B) when primary substrates are used.
C) when the leaving group, L^-, is a strong base.
D) by more than one of the above.
E) by none of the above.

71. What would be the major product of the following reaction?

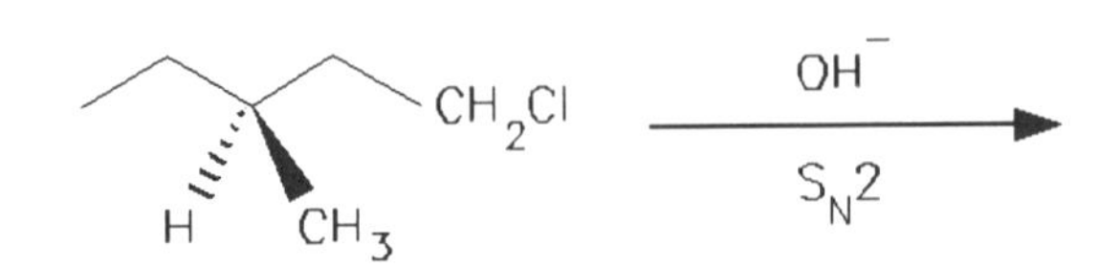

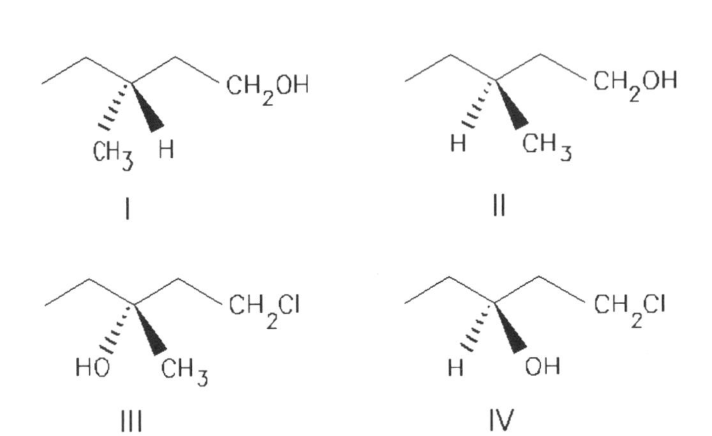

A) I
* B) II
C) III
D) IV
E) An equimolar mixture of I and II

72. Your task is to convert 2-chloropentane into 1-pentene. Which reagents would you choose?

A) $NaOH/H_2O$
B) KOH/CH_3OH
C) CH_3ONa/CH_3OH
D) CH_3CH_2ONa/CH_3CH_2OH
* E) $(CH_3)_3COK/(CH_3)_3COH$

73. When <u>tert</u>-pentyl chloride undergoes solvolysis in aqueous ethanol at room temperature, there is/are formed:

A) OH
3
$CH_3CH_2C(CH_3)_2$
B) OC_2H_5
3
$CH_3CH_2C(CH_3)_2$
C) $CH_3CH{=}C(CH_3)_2$
D) CH_3
3
$CH_3CH_2C{=}CH_2$
* E) All of these

74. What would be the major product of the following reaction?

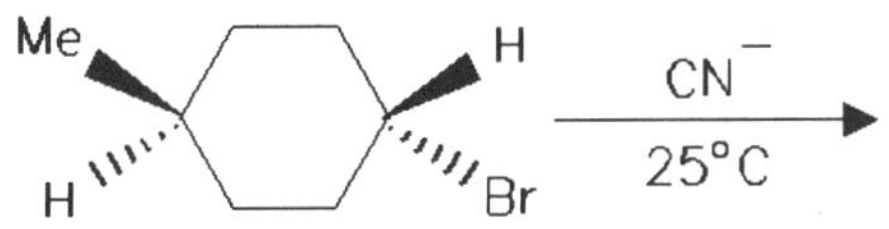

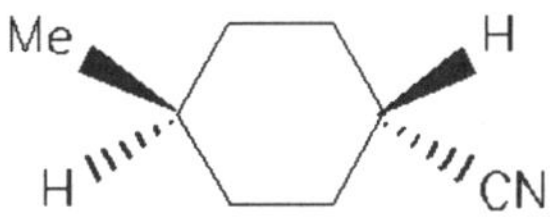

$\xrightarrow[25°C]{CN^-}$

I

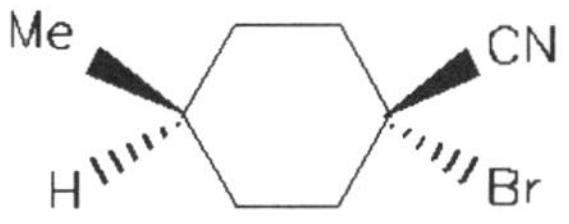

II

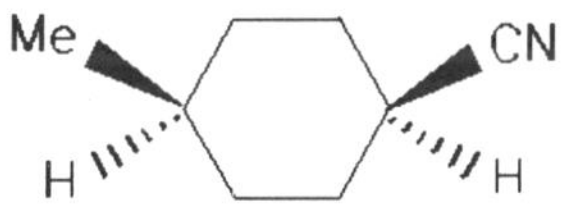

III

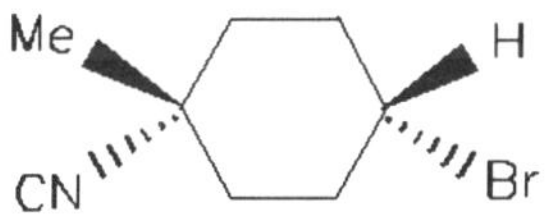

IV

A) I
B) II
* C) III
D) IV
E) Equal amounts of I and III

75. You want to synthesize 2-methyl-1-butene from 2-chloro-2-methylbutane. Which reagent would you use?

A) $NaOH/H_2O$
B) KOH/H_2O
C) CH_3ONa/CH_3OH
D) CH_3CH_2ONa/CH_3CH_2OH
* E) $(CH_3)_3COK/(CH_3)_3COH$

76. What would be the <u>major</u> product of the following reaction?

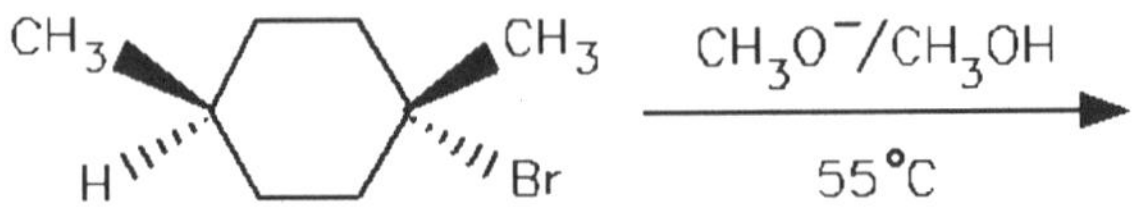

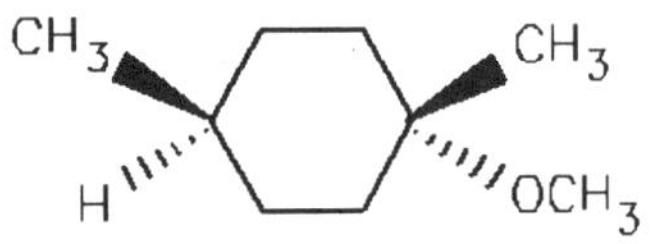

CH_3O^-/CH_3OH, 55°C

I 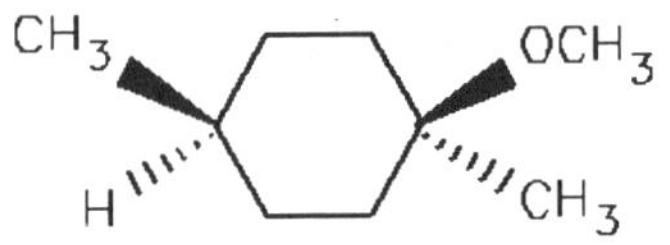 II 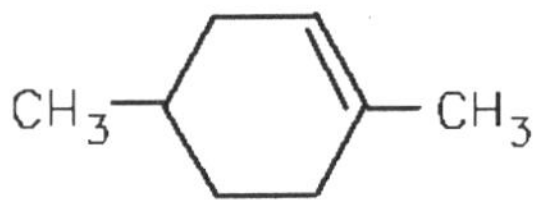 III 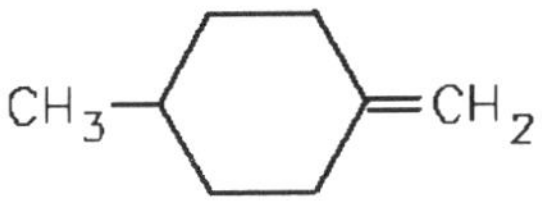 IV

A) I
B) II
* C) III
D) IV
E) An equimolar mixture of I and II.

77. Reaction of sodium ethoxide with 1-bromopentane at 50xC yields primarily:

A) $CH_3CH_2CH_2CHMCH_2$
B) $CH_3CH_2CHMCHCH_3$
C) $CH_3CH_2CH_2CH_2CH_3$
D) $CH_3CH_2CH_2CH_2CH_2OH$
* E) $CH_3CH_2CH_2CH_2CH_2OCH_2CH_3$

78. If 0.10 mol of $HSCH_2CH_2OH$ reacts at 25xC, sequentially, with 0.20 mol of NaH, 0.10 mol of CH_3CH_2Br and H_2O, which is the major product?

A) $HSCH_2CH_2OCH_2CH_3$
* B) $CH_3CH_2SCH_2CH_2OH$
C) $CH_3CH_2SCH_2CH_2OCH_2CH_3$
D) $CH_2=CH_2$
E) CH_3CH_3

79. The difference in the bond energies of reactants and the transition state of a reaction is designated by the notation:

A) (Hx * B) (H^E C) (Gx D) (G^E E) (S^E

80. A true statement about the transition state(s) of an S_N2 reaction is:

A) the two transition states are of unequal energy.
B) the transition states precede and follow an unstable reaction intermediate.
* C) the single transition state represents the point of maximum free energy of the reaction.
D) existence of this transition state implies an exothermic reaction.
E) the transition state will always have a net charge of -1.

81. Elimination reactions are favored over nucleophilic substitution reactions:

A) at high temperatures.
B) when tert-butoxide ion is used.
C) when 3x alkyl halides are used as substrates.
D) when nucleophiles are used which are strong bases and the substrate is a 2x alkyl halide.
* E) in all of these cases.

82. For the nucleophilic substitution reaction

$$Br^- \ + \ CH_3CH_2CH_2CH_2CH_2OH \quad DDDD4$$

to be successful, it is necessary that:

A) the reaction be carried out at high temperature.
B) a large excess of Br^- be used.
* C) the reaction be carried out at low pH.
D) a polar aprotic solvent be used.
E) Under none of these conditions can this synthesis be accomplished.

83.

```
               CH3
               3
When ICH2CCH2CH2I (0.10 mol) is treated with 0.10 mol of NaCN
               3
               CH3
```

in dimethyl sulfoxide at 30xC, the product formed is:

A)
```
        CH3
        3
NCCH2CCH2CH2I
        3
        CH3
```

D)
```
        CH3
        3
NCCH2CCH2CH2CN
        3
        CH3
```

* B)
```
       CH3
       3
ICH2CCH2CH2CN
       3
       CH3
```

E)
```
       CH3
       3
ICH2CCHMCH2
       3
       CH3
```

C) both A) and B).

84. When 0.10 mol of $ICH_2CH_2CH_2CH_2Cl$ reacts with 0.10 mol of $NaOCH_3$ in CH_3OH at 40xC, the major product is:

* A) $CH_3OCH_2CH_2CH_2CH_2Cl$
  B) $CH_3OCH_2CH_2CH_2CH_2I$
  C) $CH_3OCH_2CH_2CH_2CH_2OCH_3$
  D) $CH_2MCHCH_2CH_2Cl$
  E) $CH_2MCHCH_2CH_2I$

85. Which is not a polar aprotic solvent?

  A) O : CH_3CCH_3
  B) CH_3CN
  C) O : CH_3SCH_3
  D) O : $HCN(CH_3)_2$
* E) CH_3OH

86. The relative nucleophilicities of species do not necessarily parallel the relative basicities of the same species because:

  A) not all nucleophiles are bases, and vice versa.
  B) experimental measurements of sufficient accuracy are not available to make the comparisons.
  C) nucleophilicity is a thermodynamic matter; basicity is a matter of kinetics.
* D) basicity is a thermodynamic matter; nucleophilicity is a matter of kinetics.
  E) Actually, the relative values do parallel one another.

87. Ambident nucleophiles are ones which can react with a substrate at either of two nucleophilic sites. Which of the following is not an ambident nucleophile?

I: $:CN:^-$ II: $:\ddot{O}H^-$ III: $^-:\ddot{O}CH_2CH_2\ddot{S}:^-$

IV: $[:\ddot{O}:N::\ddot{O}]^-$ V: $H_2\ddot{N}:\ddot{O}:H$

A) I * B) II C) III D) IV E) V

88. Considering the relative solvation of reactants and the transition states of S_N reactions of these reactants, predict which general type of reaction would be most favored by the use of a polar solvent.

* A) $Y: + RX$ DDDDD4 $RY^+ + X:^-$
  B) $Y:^- + RX$ DDDDD4 $RY + X:^-$
  C) $Y: + RX^+$ DDDDD4 $RY^+ + X:$
  D) $Y:^- + RX^+$ DDDDD4 $RY + X:$
  E) RX^+ DDDDD4 $R^+ + X:$

89. The Hammond-Leffler postulate when applied to nucleophilic substitutions and elimination reactions states that:
  A) a negatively-charged nucleophile is stronger than its conjugate acid.
  B) polar aprotic solvents strongly accelerate the rate of S_N2 processes.
  C) bimolecular nucleophilic substitutions are 2nd. order kinetically.
* D) the transition state for an endergonic reaction step (one accompanied by an increase in free energy) resembles the product of that step.
  E) elimination reactions will always compete with nucleophilic substitution reactions.

90. Which will be true for any actual or potential nucleophilic substitution reaction?
  A) (Hx is positive.
  B) (Hx is negative.
* C) $(G^E$ is positive.
  D) (Gx is positive.
  E) (Gx is negative.

91. Which of these compounds would give the largest $E2/S_N2$ product ratio on reaction with sodium ethoxide in ethanol at 55xC?
  A) $CH_3CH_2CH_2CH_2CH_2Cl$
  B) $CH_3CH_2CHCH_2Cl$
         3
         CH_3
  C) $CH_3CHCH_2CH_2Cl$
       3
       CH_3
* D) $CH_3CH_2CHCH_2CH_3$
         3
         Cl
  E) $(CH_3)_3CCH_2Cl$

92. In a 60:40 mixture of ethanol and water at room temperature are dissolved 0.10 mol of $CH_3CH_2CH_2OSO_2CH_3$ and 0.10 mol each of NaCl, NaI, NaF and NaBr. What is the principal product of the subsequent reaction?
  A) $CH_3CH_2CH_2Cl$
* B) $CH_3CH_2CH_2I$
  C) $CH_3CH_2CH_2F$
  D) $CH_3CH_2CH_2Br$
  E) $CH_3CH_2CH_2OCH_2CH_3$

Chapter 7: Alkenes and Alkynes I

1. Which of the following correctly lists the compounds in order of decreasing acidity?

* A) H_2O > HCpCH > NH_3 > CH_3CH_3
B) HCpCH > H_2O > NH_3 > CH_3CH_3
C) CH_3CH_3 > HCpCH > NH_3 > H_2O
D) CH_3CH_3 > HCpCH > H_2O > NH_3
E) H_2O > NH_3 > HCpCH > CH_3CH_3

2. A correct IUPAC name for the following compound is:

$CH_3CHCH_2CCH_2CH_3$
3 :
CH_3 CH_2

A) 4-Ethyl-2-methyl-4-pentene
B) 2-*sec*-Butyl-1-butene
C) 2-Isobutyl-1-butene
* D) 2-Ethyl-4-methyl-1-pentene
E) 2-Methyl-4-methylenehexane

3. Which one of the following alcohols would dehydrate most rapidly when treated with sulfuric acid?

A)
CH_3
3
CH_3CHCCH_3
3 3
HO CH_3

* B)
CH_3
3
$CH_3CDDCHCH_3$
3 3
OH CH_3

C)
CH_3
3
$CH_3CCH_2CH_2OH$
3
CH_3

D) $CH_3CHDDCHCH_3$
3 3
CH_3 OH

E) $CH_3CH_2CH_2CH_2OH$

4. Which of the following carbocations would NOT be likely to undergo rearrangement?

A) $CH_3CHCHCH_3$
+ 3
CH_3

B) CH_3
3
CH_3CHCCH_3
+ 3
CH_3

* C) CH_3
3
$CH_3CCH_2CH_3$
+

D) CH_3
3
$CH_3CHCH_2{}^+$

E) CH_3
3
$CH_3CCHCH_2CH_3$
3+
CH_3

5. Which of the following reactions would yield CH_3 / 3 / $CH_3CCHMCH_2$ / 3 / CH_3 in a reasonable percentage yield (i.e., greater than 50%)?

A) CH_3
3
$CH_3CDDDCHCH_3$ $\quad H_2SO_4$ DDDDDD4 heat
3 3
CH_3 OH

B) CH_3
3
$CH_3CDDDCHCH_3$ $\quad (CH_3)_3CO^-K^+$ DDDDDDDDDDD4 $(CH_3)_3COH$
3 3
CH_3 Br

C) CH_3
3
$CH_3CDDDCHCH_2Br$ $\quad$ Zn DDDDDDDDDDD4 acetic acid
3 3
CH_3 Br

D) All of these

* E) Answers B) and C) only

6. The correct IUPAC name for the following compound is:

Br
3
$CH_3DCHDCH_2DCDCH_2DCH_3$
:
CH_2

A) 2-Bromo-4-methylenehexane
B) 2-(2-Bromopropyl)-1-butene
* C) 4-Bromo-2-ethyl-1-pentene
D) 2-Bromo-4-ethyl-1-pentene
E) 2-Bromo-4-ethyl-4-pentene

7. Which reaction would you expect to liberate the least heat?

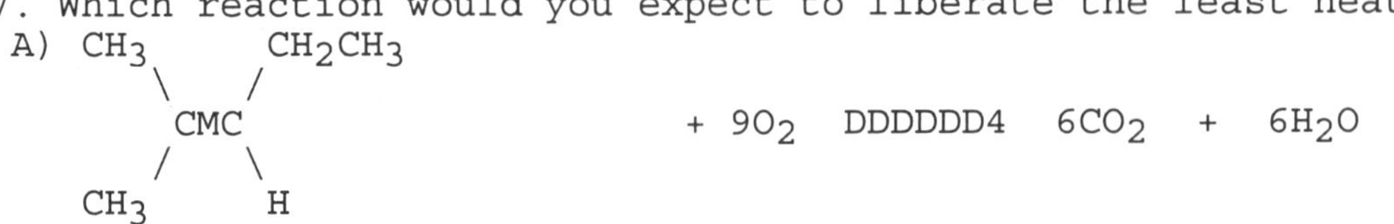

```
A) CH3      CH2CH3
      \    /
       CMC                       + 9O2  DDDDDD4  6CO2  +  6H2O
      /    \
   CH3      H
```

```
B)          CH3
            3
   CH3      CHCH3
      \    /
       CMC                       + 9O2  DDDDDD4  6CO2  +  6H2O
      /    \
     H      H
```

```
* C)   CH3      CH3
          \    /
           CMC                   + 9O2  DDDDDD4  6CO2  +  6H2O
          /    \
       CH3      CH3
```

```
D) CH2MCHCH2CH2CH2CH3  +  9O2  DDDDDD4  6CO2  +  6H2O
```

```
E) CH2MCCH2CH2CH3  +  9O2  DDDDDD4  6CO2  +  6H2O
       3
       CH3
```

8. Which compound listed below would you expect to be the major product of this reaction?

```
             CH3
             3           ethanol
      CH3CH2CBr  +  KOH  DDDDDDDD4  ?
             3           reflux
             CH3
```

```
A)       CH3                  * D) CH3CHMCCH3
         3                                3
   CH3CH2COH                              CH3
         3
         CH3                    E) CH2MCHCHCH3
                                          3
B)       CH3                              CH3
         3
   CH3CH2CDODCH2CH3
         3
         CH3
C) CH3CH2CMCH2
         3
         CH3
```

9. Which product(s) would be produced by acid-catalyzed dehydration of the following alcohol?

```
        CH3
        3             H+, heat
     CH3CCH2CH2CH3    DDDDDDDD4
        3              (-H2O)
        OH
```

A)
```
     CH2
     :
  CH3CCH2CH2CH3
```

* B)
```
      CH3                 CH3
      3                   3
  CH2MCCH2CH2CH3 and CH3CMCHCH2CH3
```

C)
```
     CH3                CH3
     3                  3
  CH3CHCHMCHCH3 and CH3CHCH2CHMCH2
```

D)
```
     CH3
     3
  CH3CHCH2CHMCH2
```

E)
```
           CH3 CH3
           3   3
  CH3CH2CH2C-O-CCH2CH2CH3
           3   3
           CH3 CH3
```

10. The correct IUPAC name for the following compound is:

```
  CH3CH2CCH2CH2CH2Cl
        :
        CH2
```

A) 1-Chloro-4-methylenehexane
B) 1-Chloro-4-ethyl-4-pentene
C) 5-Chloro-2-ethyl-1-hexene
* D) 5-Chloro-2-ethyl-1-pentene
E) 2-(3-Chloropropyl)-1-butene

11. Which alcohol would initially produce the most stable carbocation when treated with concentrated H_2SO_4?

```
* A)     CH3
         3
     CH3CCH2CH3
         3
         OH

  B)     CH3
         3
     CH3CHCHCH3
            3
            OH

  C)     CH3
         3
     CH3CHCH2CH2OH

  D)        CH3
            3
     HOCH2CHCH2CH3

  E)     CH2OH
         3
     CH3CHCH2CH3
```

12. Which alkene would yield 3-methylpentane when subjected to catalytic hydrogenation?

```
  A) H       H
      \     /
       CMC
      /     \
     H       CH2CH2CH2CH3

* B) H       H
      \     /
       CMC
      /     \
     H       CHCH2CH3
             3
             CH3

  C) CH3       H
        \     /
         CMC
        /     \
       H       CH2CH2CH3

  D) CH3       CH2CH3
        \     /
         CMC
        /     \
     CH3       H

  E) H       H
      \     /
       CMC
      /     \
     H       CH2CHCH3
                3
                CH3
```

13. Give the IUPAC name for

```
CH3CHCpCCH3
   3
   CH2CH3
```

A) 3-Methyl-4-hexyne
* B) 4-Methyl-2-hexyne
C) 2-Ethyl-3-pentyne
D) 4-Ethyl-2-pentyne
E) 3-Methyl-2-hexyne

14. The correct IUPAC name for the following compound is:

```
          CH3    CH2DCH2DCH3
          3      3
      CH3DCHDCHDCHDCHMCH2
             3
             CH3
```

  A) 4,5-Dimethyl-3-propyl-2-hexene
* B) 4,5-Dimethyl-3-propyl-1-hexene
  C) 3-(2,3-Dimethylpropyl)-1-hexene
  D) 2,3-Dimethyl-4-isopropyl-5-hexene
  E) 2,3-Dimethyl-4-propyl-5-hexene

15. A correct IUPAC name for the following compound is:

```
        CH3
        3
   CH3DCHDCDCH2DCH2DCH3
           :
           CH2
```

  A) 2-Methyl-3-methylenehexane
* B) 2-Isopropyl-1-pentene
  C) 3-Methyl-2-propyl-1-butene
  D) 2-Methyl-3-propyl-3-butene
  E) 2-Methyl-3-propyl-1-butene

16. What is the simplest alkene, i.e., the one with the smallest molecular weight, which can exhibit optical activity?

* A) 3-methyl-1-pentene
  B) 3-methyl-2-pentene
  C) 4-methyl-1-pentene
  D) 3-methyl-1-butene
  E) 4-methyl-2-hexene

17. One mol of each of the following alkenes is subjected to complete combustion. Which would you expect to liberate the LEAST heat?

  A) 1-Pentene
  B) 2-Pentene
  C) 2-Methyl-1-butene
* D) 2-Methyl-2-butene
  E) 3-Methyl-1-butene

18. What is the major product of the following reaction?

```
                    CH3
                    3                    C2H5OH
          CH3DCH2DCDBr   +   C2H5ONa   DDDDDDD4   ?
                    3
                    CH3
```

```
  A) CH2MCHDCHDCH3
              3
              CH3
```

```
  B)           CH3
               3
     CH3DCH2DCDOH
               3
               CH3
```

```
* C)          CH3
              3
     CH3DCHMCDCH3
```

```
  D)              CH3
                  3
     CH3DCH2DCDODCH2DCH3
                  3
                  CH3
```

```
  E)                CH3
                   /
     CH3DCH2DCH
                   \
                    CH3
```

19. Select the structure of 4-ethyl-2,3-dimethyl-2-heptene.

```
  A)        CH3
            3
     H2CMCDCHDCHDCH2DCH2DCH3
            3     3
            CH3   CH2CH3
```

```
* B)        CH3
            3
     CH3DCMCDCHDCH2DCH2DCH3
          3   3
          CH3 CH2CH3
```

```
  C)          CH3
              3
     CH3DCHDCHDCMCHDCH2DCH3
          3      3
          CH3    CH2CH3
```

```
  D)            CH3
                3
     CH3DCHDCHDCHDCHMCHDCH3
          3      3
          CH3    CH2CH3
```

```
  E)            CH3
                3
     CH3DCHDCHDCHDCH2DCHMCH2
          3      3
          CH3    CH2CH3
```

20. Which of the following carbocations would be likely to undergo rearrangement?

A) CH3DCHDCHDCH3
        +   3
            CH3

B)
            CH3
            3
   CH3DCHDCDCH3
        +   3
            CH3

C)
        CH3
        3
   CH3DCDCH2DCH3
        +

* D) More than one of the above

E) All of the above

21. Compute the index of hydrogen deficiency for the molecule $C_{10}H_8$.

A) 3 B) 4 C) 5 D) 6 * E) 7

22. Upon catalytic hydrogenation, a compound C_6H_6 absorbs three moles of hydrogen. Select a structure for C_6H_6.

$CH_2=CH-C\equiv C-CH=CH_2$

I

II

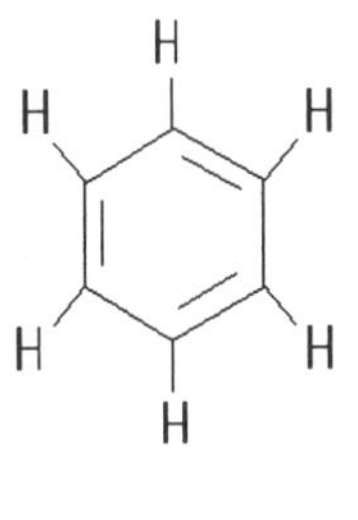

III

$H-C\equiv C-C\equiv C-CH_2-CH_3$

IV

$H-C\equiv C-CH_2-CH_2-C\equiv CH$

V

A) I B) II * C) III D) IV E) V

23. Which molecule would have the lowest heat of hydrogenation?

I II III IV V

* A) I B) II C) III D) IV E) V

24. Which mechanistic step in the acid-catalyzed dehydration of 3,3-dimethyl-2-butanol is the rate determining step?

A) Step 1.

```
      CH3                                  CH3
      3                                    3
  CH3DCDDDCHDCH3   +  H3O+  QMO  CH3DCDDDCHDCH3   +   H2O
      3    3                               3    3
      CH3  OH                              CH3  OH2
                                                +
```

* B) Step 2.

```
      CH3                      CH3
      3                        3
  CH3DCDDDCHDCH3   QMO   CH3DCDDDCHDCH3   +   H2O
      3    3                   3     +
      CH3  OH2                 CH3
           +
```

C) Step 3.

```
      CH3                   CH3
      3                     3
  CH3DCDCHDCH3   QMO   CH3DCDCHDCH3
      3 +                 + 3
      CH3                   CH3
```

D) Step 4.

```
      CH3
      3
  CH3DCDCHDCH3   +   H2O   QMO   (CH3)2CMC(CH3)2   +   H3O+
      + 3
        CH3
```

25. Select a reagent that could be used in the separation of cyclohexane from cyclohexene.

A) CrO_3/H^+
B) H_2/Zn
C) Zn/acetone
D) $KMnO_4$
* E) Br_2/CCl_4

26. Which of the following methods could be used to synthesize 4,4-dimethyl-2-hexyne?

A)

CH_3
3
$C_2H_5DCDCpC{:}^-Na^+$ + CH_3DI DDDDDD4
3
CH_3

B)

CH_3
3
$CH_3DCpC{:}^-Na^+$ + C_2H_5DCDBr DDDDDD4
3
CH_3

C)

CH_3
3 $NaNH_2$ (excess)
$CH_3DCH_2DCBr_2DCDC_2H_5$ DDDDDDDDDDDDDD4
3 liq. NH_3
CH_3

* D) More than one of these
E) None of these

27. Your task is to convert 2-bromobutane to 1-butene in highest yield. Which reagents would you use?

A) KOH/H_2O
B) KOH/CH_3OH
C) CH_3ONa/CH_3OH
D) CH_3CH_2ONa/CH_3CH_2OH
* E) $(CH_3)_3COK/(CH_3)_3COH$

28. Which of these alkenes has the smallest heat of hydrogenation?

A) CH_2MCH_2

B) $CH_3DCHMCH_2$

C)

H
3
$CH_3DCMCDCH_3$
3
H

D)

CH_3
3
$CH_3DCMCHDCH_3$

* E)

CH_3
3
$CH_3DCMCDCH_3$
3
CH_3

29. Which compound has an index of hydrogen deficiency equal to three?

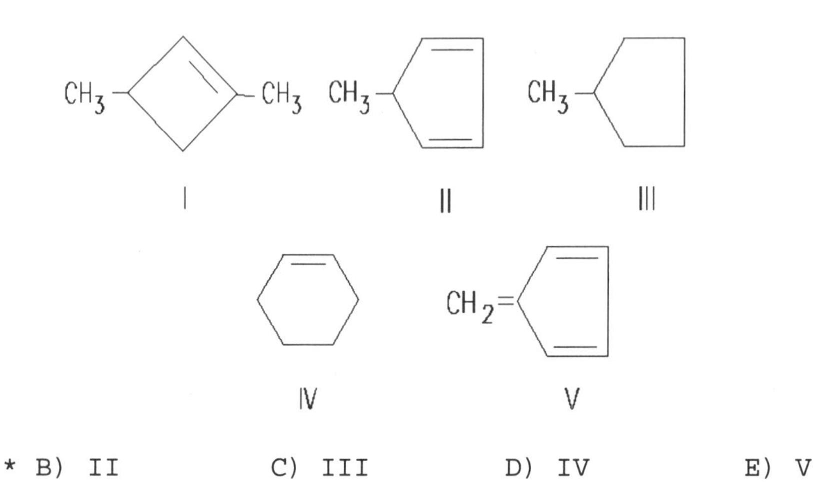

A) I   * B) II   C) III   D) IV   E) V

30. Rearrangements are likely to occur in which of the following reaction types?

A) S_N1 reactions
B) S_N2 reactions
C) E1 reactions
D) E2 reactions
* E) Both S_N1 and E1 reactions

31. What is the total number of stereoisomers which corresponds to this general structure:

Cl
3
$CH_3CHCHCHMCHCH_3$?
3
CH_3

A) 4   B) 6   * C) 8   D) 10   E) 12

32. Which alcohol would be most easily dehydrated?

* A)
CH_3
3
$CH_3CH_2CCH_2CH_3$
3
OH

B)
CH_3
3
$CH_3CH_2CHCHCH_3$
3
OH

C)
CH_3
3
$CH_3CH_2CHCH_2CH_2OH$

D)
CH_3
3
$HOCH_2CHCH_2CH_2CH_3$

E)
CH_2OH
3
$CH_3CH_2CHCH_2CH_3$

33. Which can exist as cis-trans isomers?

A) 1-Pentene
* B) 3-Hexene
C) Cyclopentene
D) 2-Methyl-2-butene
E) 3-Ethyl-2-pentene

34. Which compound would have the smallest heat of combustion?

I II III IV V

* A) I B) II C) III D) IV E) V

35. On hydrogenation, a compound C_9H_{12} absorbs 2 mol of hydrogen. Which of the following is a possible structure for the compound?

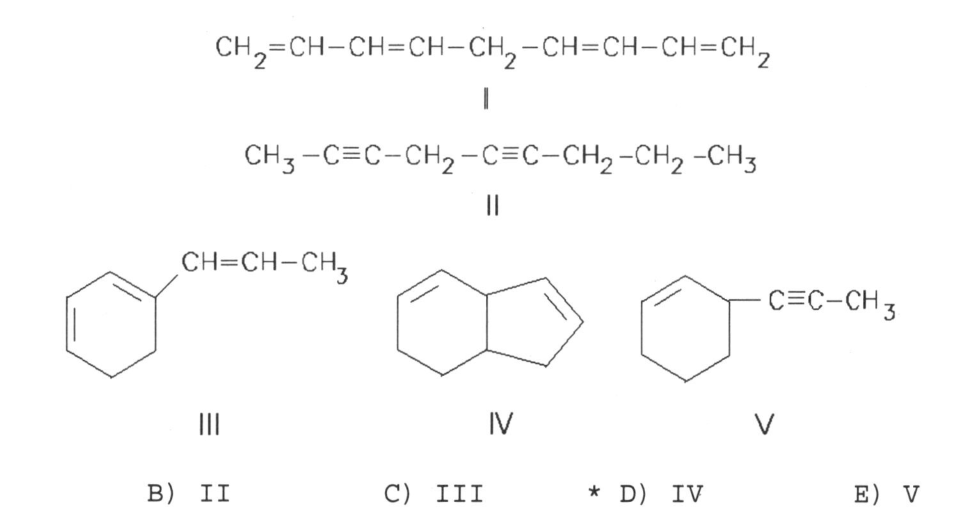

A) I B) II C) III * D) IV E) V

36. Which compound(s) would be produced by the following reaction?

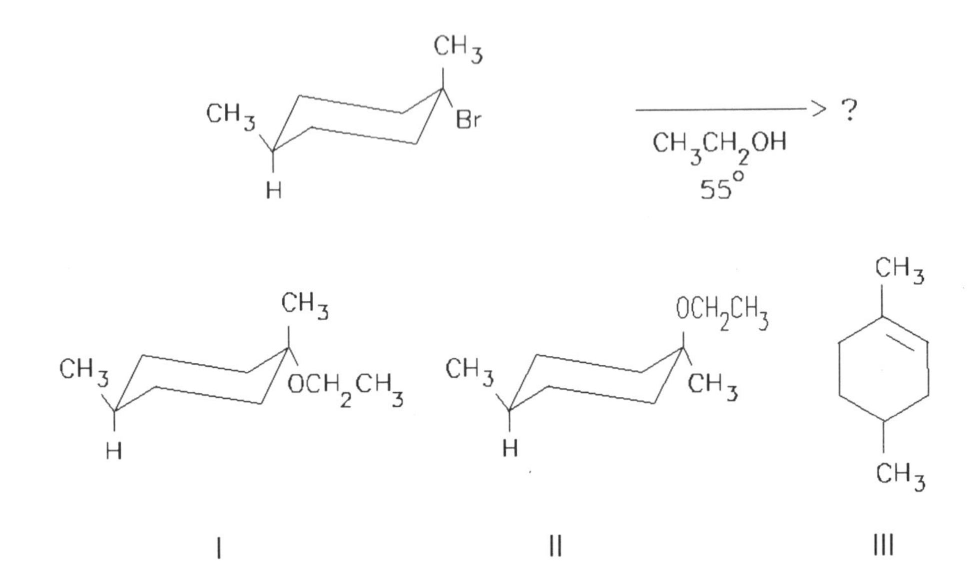

A) I
B) II
C) III
D) More than one of the above
* E) All of the above

37. A compound with the formula $C_{10}H_{14}$ reacts with excess hydrogen and a catalyst to yield a compound with the formula $C_{10}H_{18}$. The compound could have:

A) one ring and three double bonds.
B) two rings and two double bonds.
C) two rings and a triple bond.
D) no rings and two double bonds.
* E) More than one of the above

38. Which is the rate-limiting step in dehydration of 2-butanol?

A) $CH_3CH_2CHCH_3$ + H_3O^+ QMO $CH_3CH_2CHCH_3$ + H_2O
3 3
OH $^+OH_2$

* B)
+
$CH_3CH_2CHCH_3$ QMO $CH_3CH_2CHCH_3$ + H_2O
3
$^+OH_2$

C)
+
$CH_3CH_2CHCH_3$ + H_2O QMO CH_3 \ / CH_3 CMC / \ H H + H_3O^+

D)
+
$CH_3CH_2CHCH_3$ + H_2O QMO CH_3 \ / H CMC / \ H CH_3 + H_3O^+

E)
+
$CH_3CH_2CHCH_3$ + H_2O QMO $CH_3CH_2CH{=}CH_2$ + H_3O^+

39. For which of the following is <u>cis</u>-<u>trans</u> isomerism impossible?

A) 2-Hexene
B) 3-Methyl-2-pentene
C) 3-Hexene
* D) 2-Methyl-2-butene
E) 2-Pentene

40. Which product (or products) would be formed in appreciable amount(s) when <u>trans</u>-1-bromo-2-methylcyclohexane undergoes dehydrohalogenation upon treatment with sodium ethoxide in ethanol?

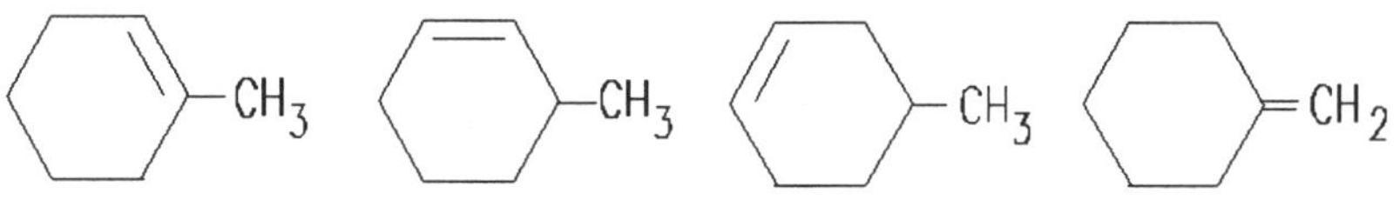

CH_3 CH_3 CH_3 CH_2

I II III IV

A) I
* B) II
C) III
D) IV
E) More than one of these

41. Which would be the major product of the following reaction?

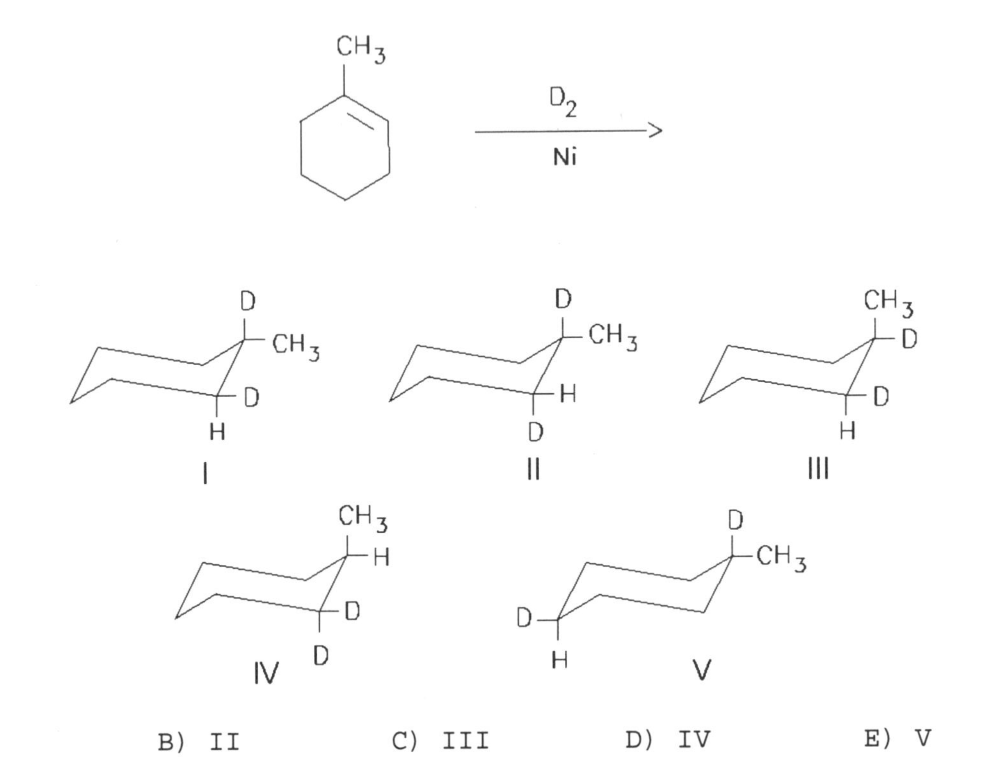

* A) I B) II C) III D) IV E) V

42. One mole of each of the following alkenes is subjected to complete combustion. Which would you expect to liberate the most heat?

* A) 1-Pentene
B) <u>cis</u>-2-Pentene
C) 2-Methyl-1-butene
D) 2-Methyl-2-butene
E) <u>trans</u>-2-Pentene

43. What is the index of hydrogen deficiency of bicyclo[2.2.2]octane?

A) 1 * B) 2 C) 3 D) 4 E) 5

44. Which reaction would you expect to liberate the most heat?

A)
```
CH3      CH2DCH3
   \    /
    CMC                          + 9O2   DDDDDD4   6CO2   +   6H2O
   /    \
CH3      H
```

B)
```
         CH3
         3
CH3      CHDCH3
   \    /
    CMC                          + 9O2   DDDDDD4   6CO2   +   6H2O
   /    \
  H      H
```

C)
```
CH3      CH3
   \    /
    CMC                          + 9O2   DDDDDD4   6CO2   +   6H2O
   /    \
CH3      CH3
```

* D) $CH_2MCHCH_2CH_2CH_2CH_3$ + $9O_2$ DDDDDD4 $6CO_2$ + $6H_2O$

E)
```
CH2MCCH2CH2CH3   +   9O2   DDDDDD4   6CO2   +   6H2O
    3
    CH3
```

45. Which of the following compounds can exhibit cis-trans isomerism?

A) 1-Pentene
* B) 2-Pentene
C) 2-Methyl-2-pentene
D) 3-Methyl-1-pentene
E) 1-Hexene

46. What is the major product of the reaction,

```
     CH3 Br
     3   3          (CH3)3CO-
CH3DCDDDCHDCH3      DDDDDDDDD4     ?
     3              (CH3)3COH
     CH3
```

A) $(CH_3)_2CMC(CH_3)_2$
* B) $(CH_3)_3CDCHMCH_2$
C) $(CH_3)_2CMCHCH_3$
D) $(CH_3)_2CMCHCH_2CH_3$
E) None of these

47. Select the strongest base.

A) OH^- B) $RCpC^-$ C) NH_2^- D) CH_2MCH^- * E) $CH_3CH_2^-$

48. Select the structure for cis-3-methyl-6-vinylcyclohexene.

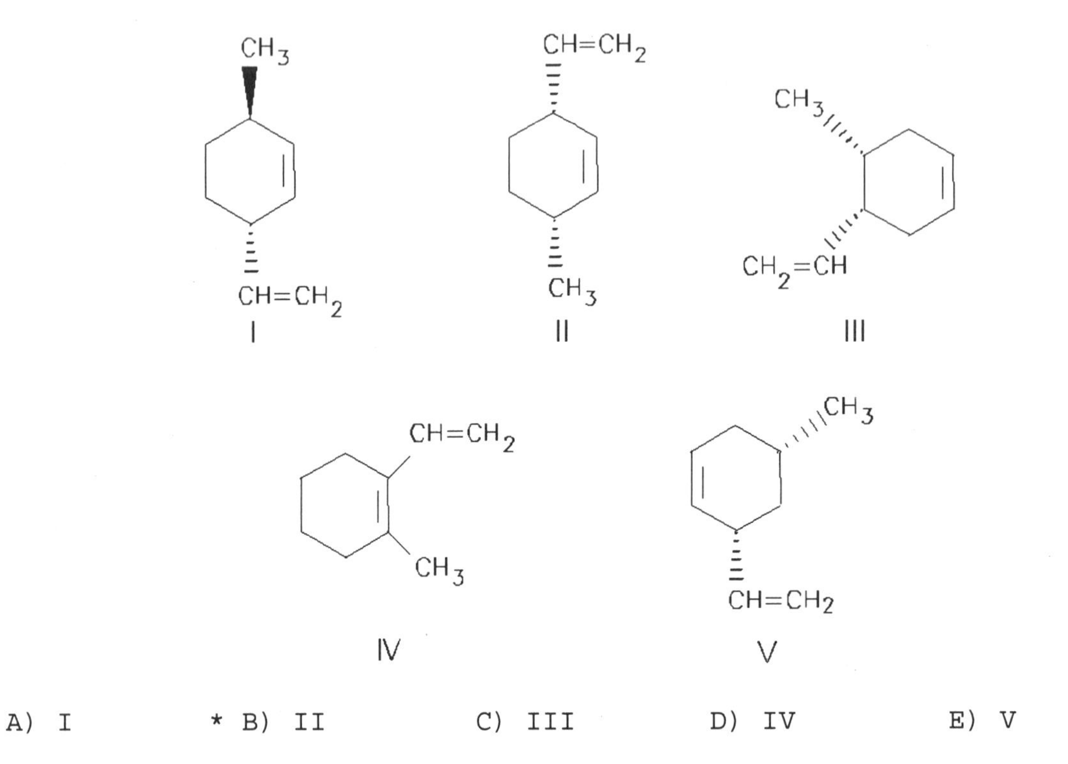

A) I * B) II C) III D) IV E) V

49. Which of the following compounds can exhibit cis-trans isomerism?

I. 1-Pentene
II. 2-Pentene
III. 2-Methyl-2-pentene
IV. 1,2-Dibromoethene
V. 1,1-Dichloroethene

A) II, III, IV
B) I, III, V
* C) II, IV
D) II, III
E) None of these

50. How many sigma bonds are there in $CH_2MCHDCpCH$?
A) 6 * B) 7 C) 8 D) 9 E) 10

51. What characteristic(s) of alkynes would make it difficult to prepare cyclohexyne?
* A) The requirement for linearity at the triple bond center
B) The large electron density between carbons of a triple bond
C) The short carbon-carbon triple bond length
D) The need that the carbon-carbon triple bond be internal in the chain
E) All of these

52. What would be the major product of the following reaction?

```
                    CH3
                    3              CH3OH
      CH3DCH2DCDCl   +   CH3ONa   DDDDD4    ?
                    3               55x
                    CH3
```

```
A) CH2MCHDCHDCH3                        D) CH3DCH2DCMCH2
           3                                       3
           CH3                                     CH3

B)          CH3                         E)               CH3
            3                                            /
   CH3DCH2DCDODCH3                         CH3DCH2DCH
            3                                            \
            CH3                                          CH3

* C)       CH3
           3
   CH3DCHMCDCH3
```

53. Which ion is the weakest base?

```
A) CH3CH2:-    B) CH2MCH:-    C) HCpC:-    D) :ṄH2-    * E) :ȮH-
                                                             .
```

54. Which compound would be the major product?

```
                    CH3
                    3      (CH3)3COK
              CH3CH2CBr   DDDDDDDDD4        ?
                    3      (CH3)3COH
                    CH3
```

```
A)         CH3                          D)         CH3
           3                                       3
   CH3CH2COC(CH3)3                         CH3CH2COH
           3                                       3
           CH3                                     CH3

B) CH3CHMCCH3                         * E) CH3CH2CMCH2
           3                                       3
           CH3                                     CH3

C) CH2MCHCHCH3
           3
           CH3
```

55. The correct IUPAC name for the following compound is:

```
         CH3     Cl
           \    /
            CMC
           /    \
         Br      CH2CHCH3
                     3
                    CH3
```

* A) (E)-2-Bromo-3-chloro-5-methyl-2-hexene
B) (E)-2-Bromo-3-chloro-5-methyl-3-hexene
C) (Z)-2-Bromo-3-chloro-5-methyl-3-hexene
D) (Z)-2-Bromo-3-chloro-5-methyl-2-hexene
E) (E)-2-Methyl-5-bromo-4-chloro-4-hexene

56. Which statement(s) is (are) true of acid-catalyzed alcohol dehydrations?

A) Protonation of the alcohol is a fast step.
B) Formation of a carbocation from the protonated alcohol is a slow step.
C) Rearrangements of less stable carbocations to more stable carbocations are common.
D) Loss of a proton by the carbocation is a fast step.
* E) All of the above

57. Predict the major product.

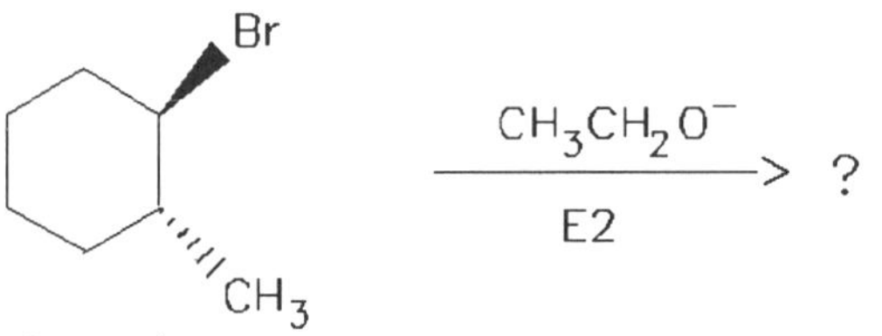

Br, CH_3 (trans) $\xrightarrow[E2]{CH_3CH_2O^-}$?

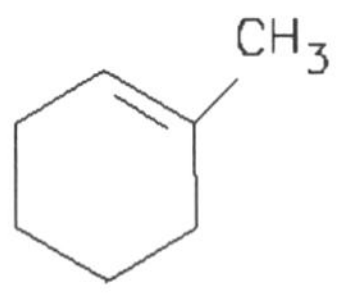

CH_3 I CH_3 II CH_3 III

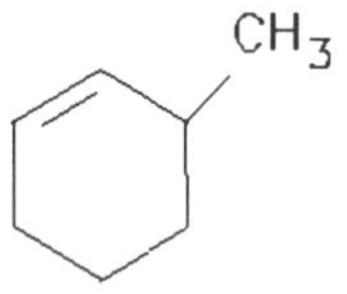

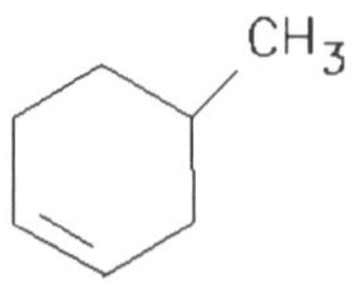

A) I
* B) II
C) III
D) Equal amounts of I and II
E) No E2 reaction will occur.

58. Which alkene would liberate the most heat per mole when subjected to complete combustion?

```
* A) H       H                           D) CH3CH2       CH2CH3
      \     /                                     \     /
       CMC                                         CMC
      /     \                                     /     \
     H       CH2CH2CH2CH3                        H       H

  B) CH3      H                          E) CH3CH2       H
        \    /                                    \     /
         CMC                                       CMC
        /    \                                    /     \
       H      CH2CH2CH3                          H       CH2CH3

  C) CH3      CH2CH2CH3
        \    /
         CMC
        /    \
       H      H
```

59. Given:

```
                    H2,Pt
        X(C8H14)   DDDDDD4   Y(C8H16)
                    25xC
```

One can conclude that X has:

A) no rings and no double bonds.
B) no rings and one double bond.
* C) one ring and one double bond.
D) two rings and no double bonds.
E) one triple bond.

60. The major product of the following reaction would be:

```
           Br
           3                   KCN
        CH3CCH2CH2CH3   DDDDDDDDDDDDDD4     ?
           3              CH3CH2OH/H2O
           CH3                70xC
```

```
  A) CH2MCCH2CH2CH3                      D) CH3CHCHMCHCH3
         3                                     3
         CH3                                   CH3

* B) CH3CMCHCH2CH3                       E)    OCH2CH3
        3                                      3
        CH3                                 CH3CCH2CH2CH3
                                               3
  C)    CN                                     CH3
        3
     CH3CCH2CH2CH3
        3
        CH3
```

61. Which alkene would liberate the most heat per mole when subjected to catalytic hydrogenation?

* A)
```
H      H
 \    /
  CMC
 /    \
H      CH2CH2CH2CH3
```

B)
```
CH3      H
   \    /
    CMC
   /    \
  H      CH2CH2CH3
```

C)
```
CH3      CH2CH2CH3
   \    /
    CMC
   /    \
  H      H
```

D)
```
CH3CH2      CH2CH3
      \    /
       CMC
      /    \
     H      H
```

E)
```
CH3CH2      H
      \    /
       CMC
      /    \
     H      CH2CH3
```

62. What will be the major product of the following reaction?

```
      CH3 OH
      3    3         85% H3PO4
CH3DCDDDCHDCH3       DDDDDDDDD4
      3                 heat
      CH3
```

* A)
```
        CH3
        3
CH3DCMCDCH3
           3
           CH3
```

B)
```
           CH3
           3
CH2MCDCHDCH3
        3
        CH3
```

C)
```
        CH3
        3
CH3DCDCHMCH2
        3
        CH3
```

D)
```
      CH3 CH3          CH3
      3   3            3
CH3DCDDDCHDODCHDDDCDCH3
      3          3     3
      CH3        CH3   CH3
```

E)
```
      CH3
      3
CH3DCHDCHMCHDCH3
```

63. Which is a possible structure for a compound with an index of hydrogen deficiency equal to 3 and which absorbs one molar equivalent of hydrogen when treated with hydrogen and a platinum catalyst?

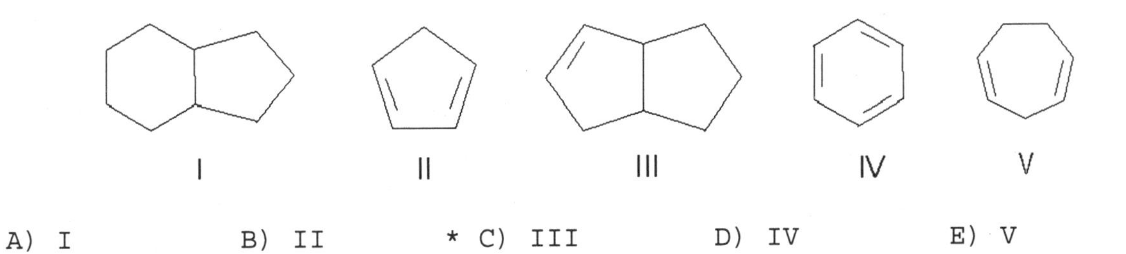

A) I   B) II   * C) III   D) IV   E) V

64. Which alcohol would be most easily dehydrated?

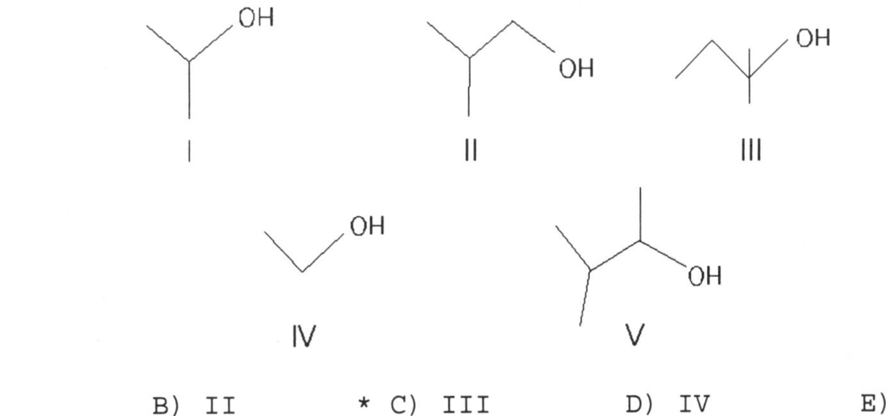

A) I   B) II   * C) III   D) IV   E) V

65. Which alkene would you expect to be the major product of the following dehydration?

OH

$\xrightarrow[(-H_2O)]{H_2SO_4/heat}$?

I II III

IV V

A) I B) II * C) III D) IV E) V

66. One mole of each of the following alkenes is subjected to complete combustion. Which would liberate the least heat?

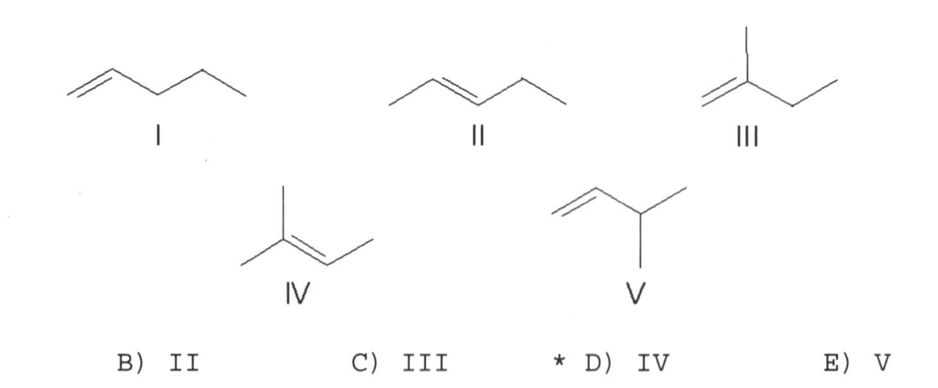

A) I B) II C) III * D) IV E) V

67. Which is a correct name for the following compound?

* A) (E)-1-Bromo-1-chloro-2-methyl-1-hexene
B) (Z)-1-Bromo-1-chloro-2-methyl-1-hexene
C) (E)-2-Bromochloromethylenehexane
D) (Z)-2-Bromochloromethylenehexane
E) 2-(E,Z)-Bromochloromethyl-1-hexene

68. Which alkene is most stable?

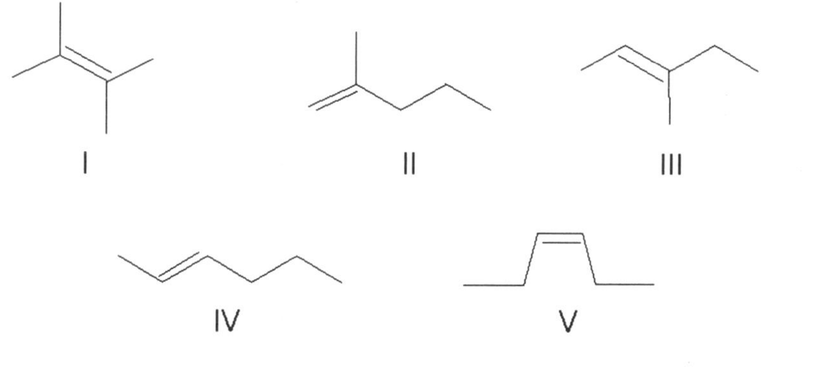

* A) I B) II C) III D) IV E) V

69. Which of the following reactions would produce the following alkene in a reasonable percentage yield, i.e., greater than 50%?

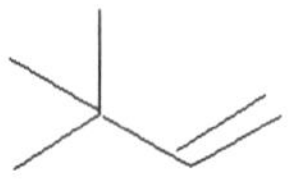

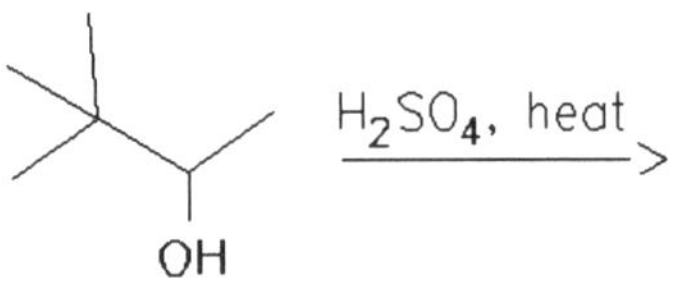

OH $\xrightarrow{H_2SO_4,\ heat}$

I

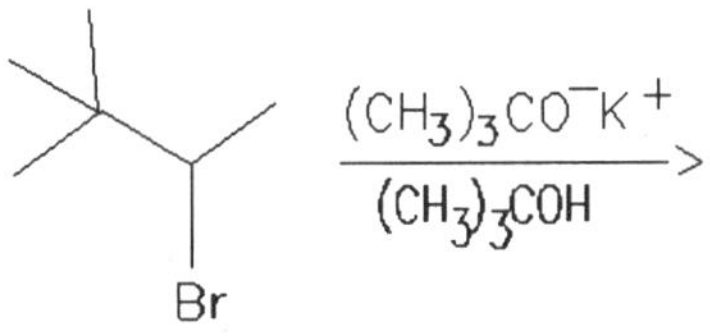

Br $\xrightarrow[(CH_3)_3COH]{(CH_3)_3CO^-K^+}$

II

CH_2Br, Br $\xrightarrow[acetone]{NaI}$

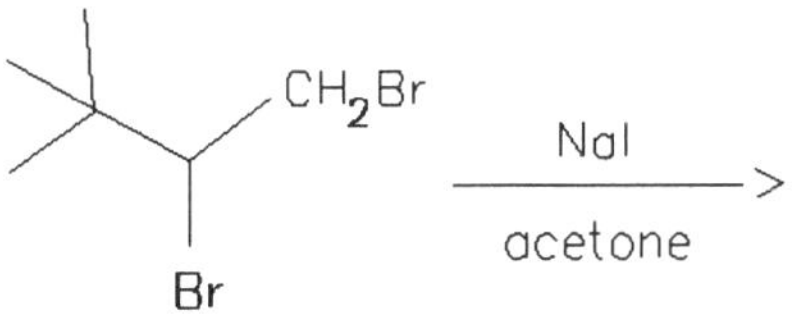

III

A) I
B) II
C) III
* D) More than one of these reactions
E) None of these reactions

70. Neopentyl alcohol, $(CH_3)_3CCH_2OH$, cannot be dehydrated to an alkene without rearrangement. What is the chief product of dehydration?

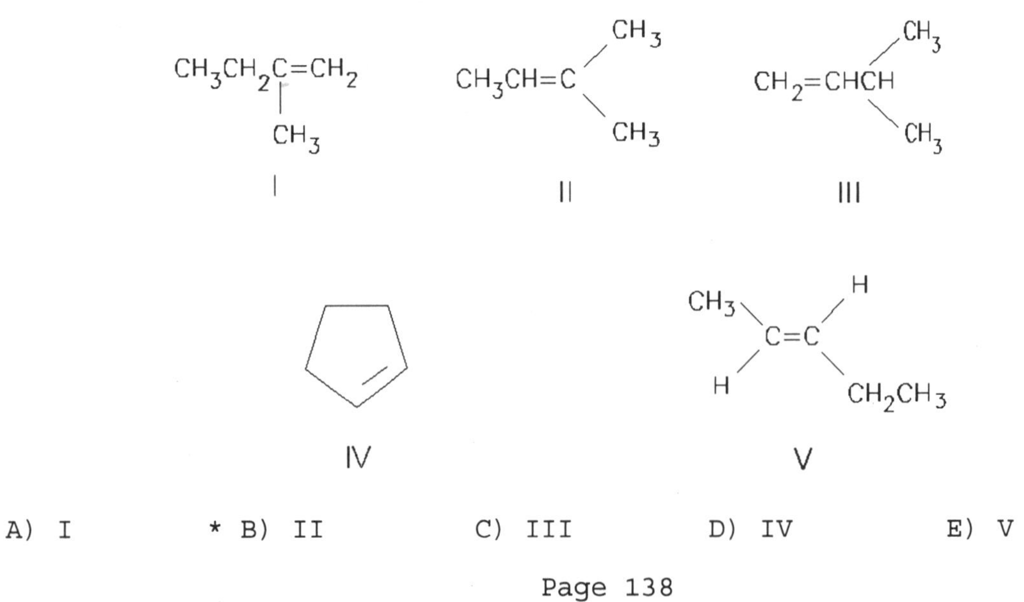

A) I
* B) II
C) III
D) IV
E) V

71. What is the total number of pentene isomers, including stereoisomers?

A) 2 B) 3 C) 4 D) 5 * E) 6

72. A compound, C_6H_{10}, which reacts with 2 mol of hydrogen over a platinum catalyst and which gives a precipitate with $Ag(NH_3)_2^+$ could be:

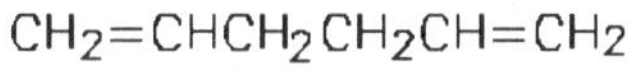

I: $CH_2{=}CHCH_2CH_2CH{=}CH_2$

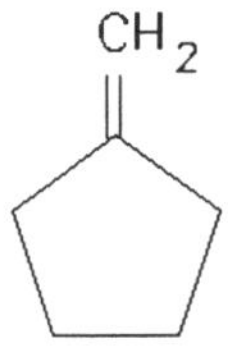

II: methylenecyclopentane (cyclopentane ring with $=CH_2$)

III: $CH_3C(CH_3)_2C{\equiv}CH$ (central C bearing two CH_3 groups, drawn above and below)

IV: vinylcyclobutane (cyclobutane ring with $CH{=}CH_2$)

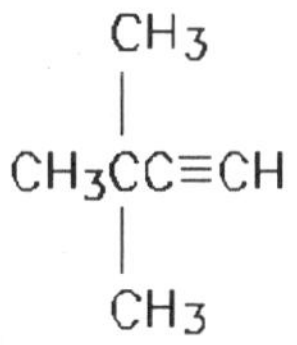

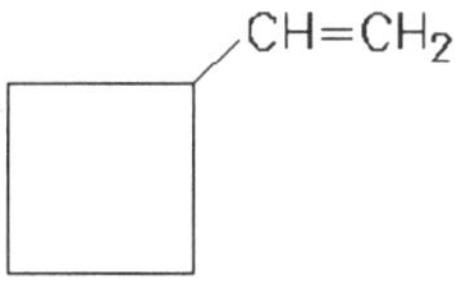

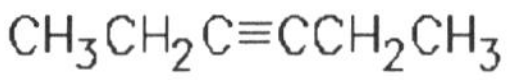

V: $CH_3CH_2C{\equiv}CCH_2CH_3$

A) I B) II * C) III D) IV E) V

73. For which alkene does cis not equal (Z) or trans not equal (E)?

A) CH_3 and CH_2CH_3 (top), H and H (bottom), on C=C

B) CH_3 and Cl (top), Cl and CH_3 (bottom), on C=C

* C) Br and Br (top), I and H (bottom), on C=C

D) Cl and Br (top), F and F (bottom), on C=C

E) CH_3CH_2 and COOH (top), $ClCH_2$ and CH_2CH_3 (bottom), on C=C

74. The most specific term used to designate the relationship of *cis*-3-hexene to *trans*-3-hexene is:

A) stereoisomers.
B) enantiomers.
* C) diastereomers.
D) constitutional isomers.
E) conformational isomers.

75. Which of these chloro derivatives of ethene has a dipole moment of zero?

A)

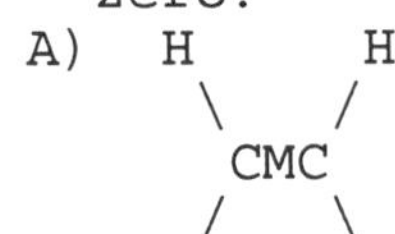

B)
```
H       Cl
  \    /
   CMC
  /    \
H       Cl
```

C)
```
Cl      Cl
  \    /
   CMC
  /    \
H       H
```

* D)
```
Cl      H
  \    /
   CMC
  /    \
H       Cl
```

E)
```
Cl      Cl
  \    /
   CMC
  /    \
H       Cl
```

76. What is the structure of a compound with formula C_6H_{10} which forms a precipitate when mixed with ammoniacal silver nitrate and which can be catalytically reduced with hydrogen to 2-methylpentane?

A)
```
CH3CHCpCCH3
    3
    CH3
```

B)
```
CH2MCCHMCHCH3
     3
     CH3
```

C)
```
CH3CH2CHCpCH
        3
        CH3
```

* D)
```
HCpCCH2CHCH3
           3
           CH3
```

E)
```
     CH2
      :
CH3CCMCH2
     3
     CH3
```

77. Dehydrohalogenation of *tert*-pentyl bromide will produce 2-methyl-1-butene as the chief product when:

A) CH_3COONa is employed as the base.
B) KOH/C_2H_5OH is employed as the base.
C) CH_3CH_2ONa/CH_3CH_2OH is employed as the base.
* D) $(CH_3)_3COK/(CH_3)_3COH$ is employed as the base.
E) any base is used, as long as the temperature is sufficiently high.

78. Which is not a satisfactory procedure for the synthesis of 3-methyl-1-butene?

* A) $(CH_3)_2CHCHCH_3$ + concd. H_2SO_4
  3
  OH

B) $(CH_3)_2CHCpCH$ + $Li/liq.NH_3$

C) $(CH_3)_2CHCHBrCH_2Br$ + NaI/acetone

D) $(CH_3)_2CHCHBrCH_3$ + $(CH_3)_3COK/(CH_3)_3COH$

E) $(CH_3)_2CHCpCH$ + H_2/Ni_2B (P-2)

79. Which of these is the most satisfactory method for the preparation of cis-2-pentene?

A) $CH_3CHBrCH_2CH_2CH_3$ + $(CH_3)_3COK/(CH_3)_3COH$
B) $CH_3CpCCH_2CH_3$ + H_2, Pt
* C) $CH_3CpCCH_2CH_3$ + H_2, Ni_2B (P-2)
D) $CH_3CpCCH_2CH_3$ + Li/liq. NH_3
E) $CH_3CH_2CHBrCH_2CH_3$ + CH_3CH_2ONa/CH_3CH_2OH

80. The ambiguous name "methylcyclohexene" does not differentiate among this number of compounds (ignoring stereoisomers):

A) 2 * B) 3 C) 4 D) 5 E) 6

81. Concerning the relative stabilities of alkenes, which is an untrue statement?

A) Unless hydrogenation of the alkenes gives the same alkane, heats of hydrogenation cannot be used to measure their relative stabilities.
B) In general, the greater the number of alkyl groups attached to the carbon atoms of the double bond, the greater the stability of the alkene.
C) The greater the quantity of heat liberated on combustion or hydrogenation of an alkene, the greater its energy content.
* D) trans-Cycloalkenes are always more stable than the cis-isomers.
E) Heats of combustion can be used to measure the relative stabilities of isomeric alkenes, even though their hydrogenation products are not identical.

82. Zaitsev's rule states that:

A) In electrophilic addition of an unsymmetrical reagent to an unsymmetrical alkene, the more positive portion of the reagent will become attached to the carbon of the double bond bearing the greater number of hydrogen atoms.

B) An equatorial substituent in cyclohexane results in a more stable conformation than if that substituent were axial.

C) E2 reactions occur only if the a-hydrogen and leaving group can assume an anti-periplanar arrangement.

* D) When a reaction forms an alkene, and several possibilities exist, the more (or most) stable isomer is the one which predominates.

E) The order of reactivity of alcohols in dehydration reactions is 3x > 2x > 1x.

83. Regarding the use of potassium <u>tert</u>-butoxide as a base in E2 reactions, it is incorrect to state that:

A) this base is more effective than ethoxide ion, because it (KO-<u>t</u>-Bu) is the more basic of the two.

B) it tends to give the anti-Zaitsev, i.e., Hofmann, product.

C) it is more reactive in dimethyl sulfoxide than it is in <u>tert</u>-butyl alcohol.

D) it favors E2 reactions over competing S_N2 reactions.

* E) it will form, predominantly, the more stable alkene.

84. A compound <u>X</u> with the formula C_7H_{10} undergoes catalytic hydrogenation to produce a compound <u>Y</u> with the formula C_7H_{14}. What might be true of <u>X</u>?

A) X might have one triple bond and one ring.

B) X might have two double bonds and one ring.

C) X might have one double bond and two rings.

D) X might have one double bond and one triple bond.

* E) More than one of the above

85. The IUPAC name for diisobutylacetylene is

A) 2,7-Dimethyl-4-octene

* B) 2,7-Dimethyl-4-octyne

C) 3,6-Dimethyl-4-octyne

D) 2,5-Diethyl-3-hexyne

E) 2,2,5,5-Tetramethyl-3-hexyne

86. 2-Pentyne is a constitutional isomer of which of these?

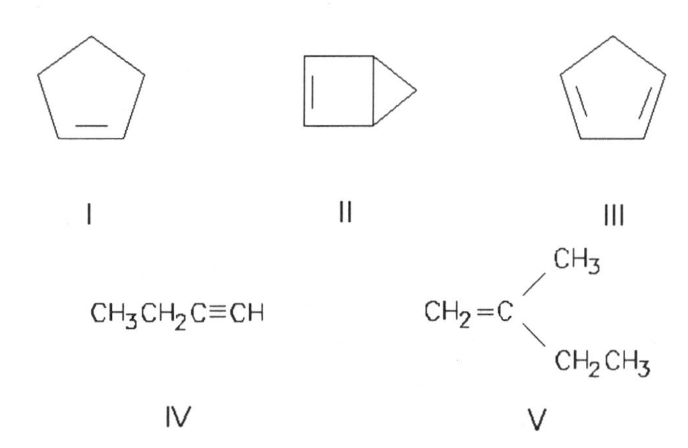

* A) I B) II C) III D) IV E) V

87. Which reaction would not be a method for preparing propyne?

A) (1) 2 $NaNH_2$, liq NH_3
$CH_3CH{=}CHBr$ DDDDDDDDDDDDDDDDDDDDD4
(2) H^+

B) HCpCNa + CH_3I DDDDDDDDDD4

* C) CH_3Na + HCpCH DDDDDDDDDD4

D) (1) 3 $NaNH_2$, liq NH_3
$CH_3CHBrCH_2Br$ DDDDDDDDDDDDDDDDDDDDD4
(2) H^+

E) CH_3CO_2H
$CH_3CBr_2CHBr_2$ + Zn DDDDDDDDDDDDDD4

88. Which reaction could be used as a method of synthesis of 1-butyne?

A) $CH_3CH_2CH{=}CHBr$ + 2 $NaNH_2$/liq. NH_3 DDDDDD4
(2) H^+

B) Br
3
$CH_3CH_2CHCH_2Br$ + 3 $NaNH_2$/liq. NH_3 DDDDDD4
(2) H^+

C) $CH_3CH_2CH_2CHBr_2$ + 3 $NaNH_2$/liq. NH_3 DDDDDD4
(2) H^+

* D) All of these

E) None of these

89. In the following hydrocarbon, which hydrogen would have the smallest value for pK_a?

$CH_3CH_2CH=CHCH_2CH_2CpCH$

A) $CH_3CH_2CH=CHCH_2CH_2CpCH$
(t)

B) $CH_3CH_2CH=CHCH_2CH_2CpCH$
(t)

C) $CH_3CH_2CH=CHCH_2CH_2CpCH$
(t)

D) $CH_3CH_2CH=CHCH_2CH_2CpCH$
(t)

* E) $CH_3CH_2CH=CHCH_2CH_2CpCH$
(t)

90. Which reaction would yield 2-butyne?

* A) $CH_3CpC{:}^-Na^+$ + CH_3Br DDDDD4

B) CH_3CH_2Br + $HCpC{:}^-Na^+$ DDDDD4

C) $CH_3{:}^-Na^+$ + $HCpCCH_3$ DDDDD4

D) More than one of these

E) None of these

91. Which of the following statements is true when ethane, ethene and acetylene are compared with one another?

A) Acetylene is the weakest acid and has the longest C-H bond length.

* B) Acetylene is the strongest acid and has the shortest C-H bond length.

C) Ethane is the strongest acid and has the longest C-H bond length.

D) Ethene is the strongest acid and has the shortest C-H bond length.

E) Ethene is the weakest acid and has the longest C-H bond length.

92. The structure of the product, C, of the following sequence of reactions would be:

C_6H_5CpCH DDDDDDDD4 ($NaNH_2$, liq. NH_3) A DDDDDDDD4 (CH_3CH_2Br) B DDDDDDDDD4 (H_2, Ni_2B [P-2]) C

* A) cis-$CH_3CH_2CH=CHC_6H_5$

B) cis-$CH_3CH=CHC_6H_5$

C) trans-$CH_3CH_2CH=CHC_6H_5$

D) $C_6H_5CpCCH_2CH_2Br$

E) $C_6H_5CpCCH_2CH_3$

93. Which of the following reductions of an alkyne is NOT correct?

A)
```
                      2H2
       CH3CH2CpCCH3   DDD4   CH3CH2CH2CH2CH3
                      Pt
```
B)
```
                                CH3CH2       CH3
                       H2             \     /
       CH3CH2CpCCH3   DDDD4            C=C
                      Ni2B            /     \
                                     H       H
```
* C)
```
                                  CH3CH2       CH3
                         Li             \     /
       CH3CH2CpCCH3   DDDDDDDDD4         C=C
                      liq. NH3          /     \
                                       H       H
```
D) All of the above are correct.
E) None of the above is correct.

94. Which of the following reactions would yield 2-pentyne?

A)
```
                NaNH2     CH3CH2CH2I
       HCpCH    DDDDD4    DDDDDDDDDD4
                1 mol
```
* B)
```
                  NaNH2    CH3CH2I
       CH3CpCH    DDDDD4   DDDDDDD4
```
C)
```
             Br
             3                   Zn
       CH3CHCHCH2CH3         DDDDDDD4
             3               CH3CO2H
             Br
```
D)
```
                                 H+
       CH3CH2CHCH2CH3           DDDD4
                3               heat
                OH
```
E)
```
          Br
          3                      NaOC2H5
       CH3CHCH2CH2CH3            DDDDDDD4
                                 C2H5OH
```

95. The structure of the product obtained from 2-butyne and $Li/C_2H_5NH_2$ is:

* A)
```
       CH3       H
          \     /
           CMC
          /     \
         H       CH3
```
B)
```
       CH3       CH3
          \     /
           CMC
          /     \
         H       H
```
C)
```
       CH3       NHC2H5
          \     /
           CMC
          /     \
         H       CH3
```
D)
```
          CH3       CH3
             \     /
              CMC
             /     \
       C2H5NH       H
```
E) $CH_2MCHCH_2CH_3$

96. Carbocations are frequent intermediates in acidic reactions of alkenes, alcohols, etc. What do carbocations usually do? They may:

A) rearrange to a more stable carbocation.
B) lose a proton to form an alkene.
C) combine with a nucleophile.
D) react with an alkene to form a larger carbocation.
* E) do all of the above.

Chapter 8: Alkenes and Alkynes II

1. An alkene adds hydrogen in the presence of a catalyst to give 3,4-dimethylhexane. Ozonolysis of the alkene followed by treatment with zinc and water gives a single organic product. The structure of the alkene is:

A) CH_3 CH_3
3 3
$CH_3CHMCDDDCHCH_2H_3$ (cis or trans)

* B) CH_3 CH_2CH_3
3 3
$CH_3CH_2CMMMCCH_3$ (cis or trans)

C) CH_3
3
$CH_2MCHCHCHCH_2CH_3$
3
CH_3

D) CH_2
:
$CH_3CH_2CCHCH_2CH_3$
3
CH_3

E) $CH_3CH_2CHDDDCHCHMCH_2$
3 3
CH_3 CH_3

2. What would be the major product of the following reaction?

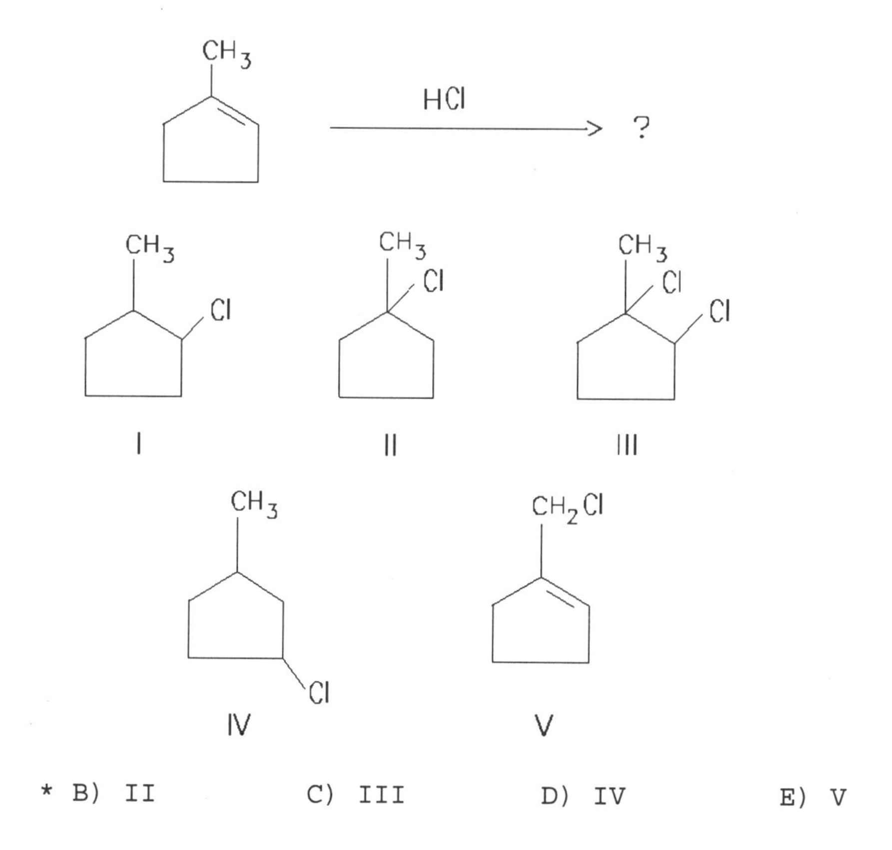

A) I * B) II C) III D) IV E) V

3. Treating 1-methylcyclohexene with HCl would yield primarily which of these?

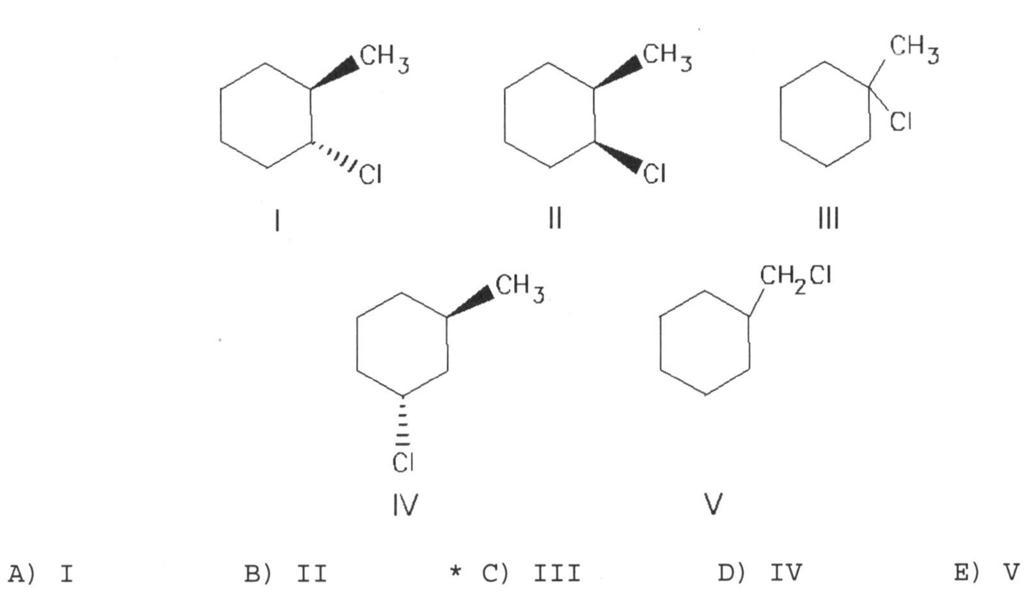

A) I B) II * C) III D) IV E) V

4. Which product would you expect from the acid-catalyzed addition of water to

?

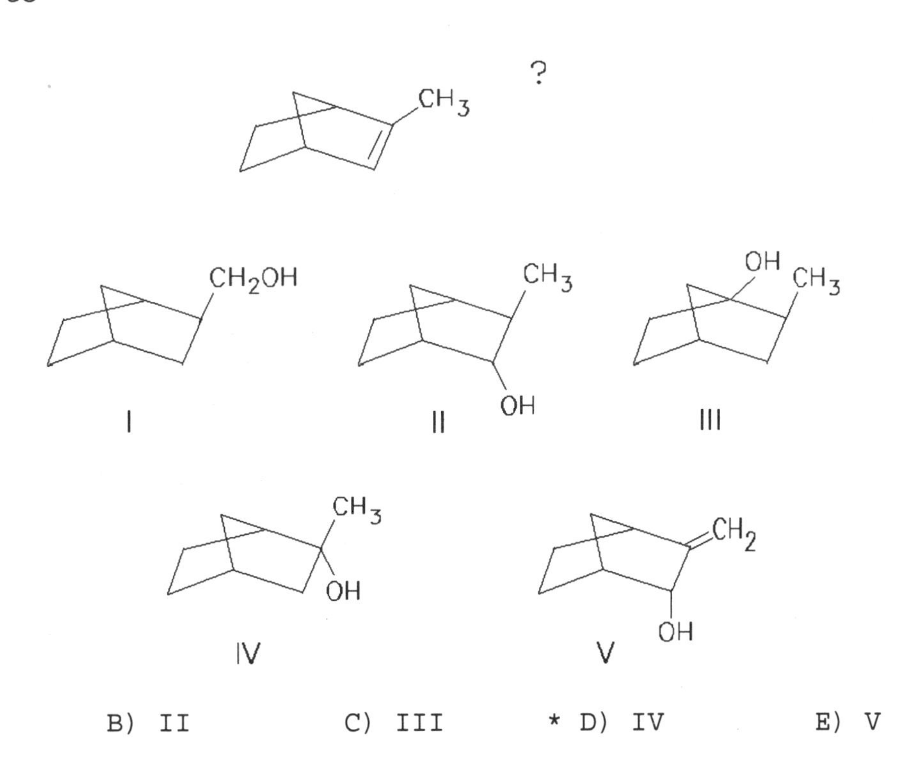

A) I B) II C) III * D) IV E) V

5. Cyclohexene is treated with cold dilute alkaline $KMnO_4$. Assuming syn addition, the spatial arrangement of the two hydroxyl groups in the product would be:

* A) equatorial-axial
  B) axial-axial
  C) equatorial-equatorial
  D) coplanar
  E) trans

6. The addition of bromine to cyclohexene would produce the compound(s) represented by structure(s):

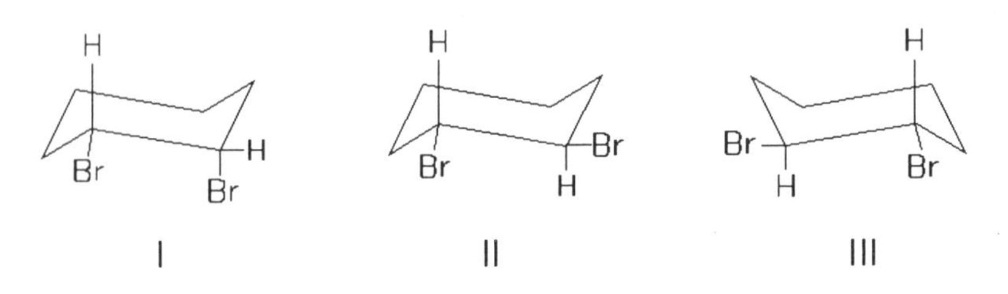

  A) I alone
  B) II alone
* C) II and III
  D) III alone
  E) I and II

7. The interaction of the c bond of an alkene with an electrophile can initially result in the formation of a species termed a c complex. Which of these cannot combine with an alkene to form a c complex?

  A) H^+ * B) NH_3 C) Ag^+ D) Hg^{2+} E) BF_3

8. A method for converting <u>trans</u>-2-butene into pure <u>cis</u>-2-butene would be:

A) <u>trans</u>-2-Butene + H_2 DDDDDD4 (Ni_2B (P-2))

B) <u>trans</u>-2-Butene + Br_2 DDDDDD4 (CCl_4) product DDDDDDD4 ($2NaNH_2$, liq. NH_3) product DDDDDDD4 (Na, liq. NH_3)

C) <u>trans</u>-2-Butene + H^+ DDDDDD4 (heat)

* D) <u>trans</u>-2-Butene + Br_2 DDDDDD4 (CCl_4) product DDDDDDD4 ($2NaNH_2$, liq. NH_3) product DDDDDDDD4 (H_2, Ni_2B(P-2))

E) None of these

9. What is the structure of the compound Z that yields 2 mol of formaldehyde ($\overset{O}{\overset{:}{HCH}}$) and 1 mol of $CH_3\overset{O}{\overset{:}{C}}CH_2\overset{O}{\overset{:}{C}}H$ upon ozonolysis followed by treatment with zinc in water?

$$\underline{Z} \xrightarrow[\text{2) Zn, } H_2O]{\text{1) } O_3} 2H\overset{O}{\overset{:}{C}}H + CH_3\overset{O}{\overset{:}{C}}CH_2\overset{O}{\overset{:}{C}}H$$

(arrow printed as: DDDDDDDDDDDD4)

CH_2

I

$H_2C=C(CH_3)-CH_2-CH=CH_2$

II

$CH_3-\overset{O}{\overset{\|}{C}}-CH_2-C(=CH_2)-CH=CH_2$

III

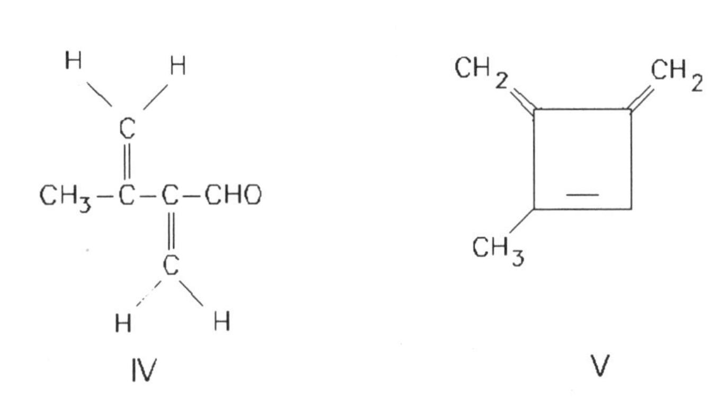

A) I   * B) II   C) III   D) IV   E) V

10. What product would you expect from addition of deuterium chloride to 2-cyclohexyl-4-methyl-2-pentene?

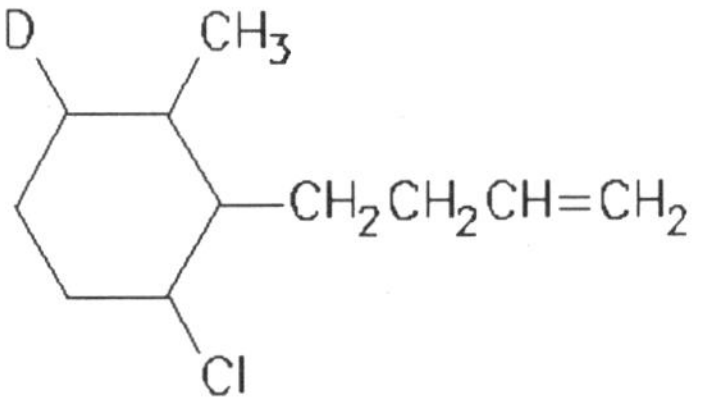

D CH$_3$
—CH$_2$CH$_2$CH=CH$_2$
Cl
I

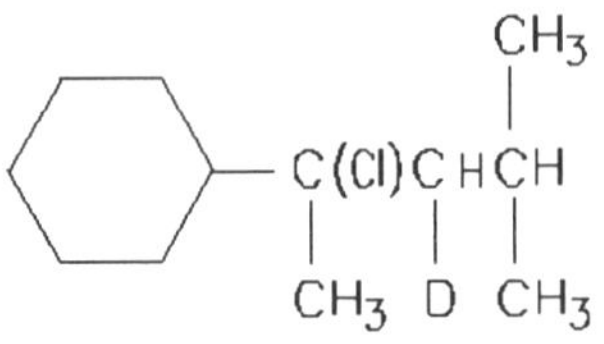

CH$_3$
—C(Cl)CHCH
CH$_3$ D CH$_3$
II

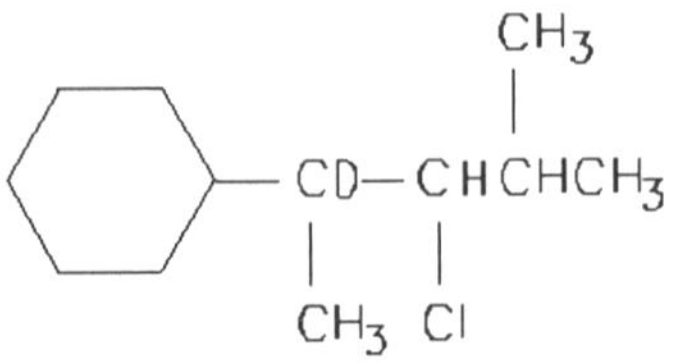

CH$_3$
—CD—CHCHCH$_3$
CH$_3$ Cl
III

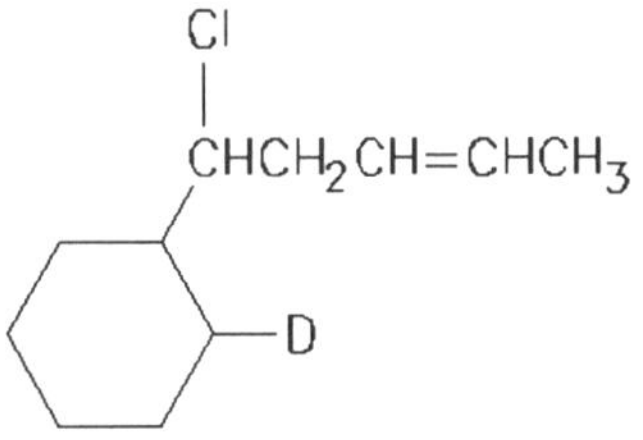

Cl
CHCH$_2$CH=CHCH$_3$
—D
IV

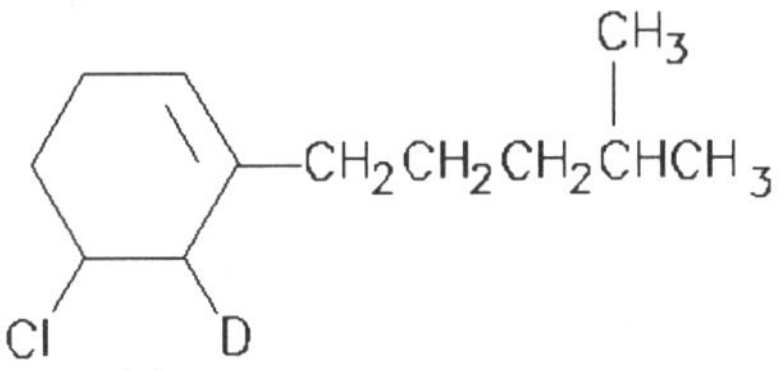

CH$_3$
—CH$_2$CH$_2$CH$_2$CHCH$_3$
Cl D
V

A) I * B) II C) III D) IV E) V

11. What is the chief product of the reaction of IBr with 1-butene?

* A) $CH_3CH_2CHCH_2I$
3
Br

B) $CH_3CH_2CHCH_2Br$
3
I

C) $CH_3CH_2CHCH_2Br$
3
Br

D) I
3
CH_3CCH_2Br
3
CH_3

E) I
3
$CH_3CHCHCH_3$
3
Br

12. Which of the following reactions would yield a meso product?

A) Cyclobutene + H^+ DDDDDD4
H_2O

B) Cyclobutene + Br_2 DDDDDD4
CCl_4

C)
O
:
RCOOH H+
Cyclobutene DDDDDD4, then DDDDDD4
H_2O

D) Cyclobutene + Cl_2 DDDDDD4
CCl_4

* E)
Pt
Cyclobutene + D_2 DDDDDD4

13. What is the chief product of the acid-catalyzed hydration of 2-methyl-2-butene?

A) CH_3
3
$CH_3CHCH_2CH_2OH$

B) CH_3
3
$CH_3CHCHCH_3$
3
OH

* C) OH
3
$(CH_3)_2CCH_2CH_3$

D) CH_3
3
$HOCH_2CHCH_2CH_3$

E) CH_3
3
CH_3C-$CHCH_3$
3 3
HO OH

14. What product would result from the following reaction?

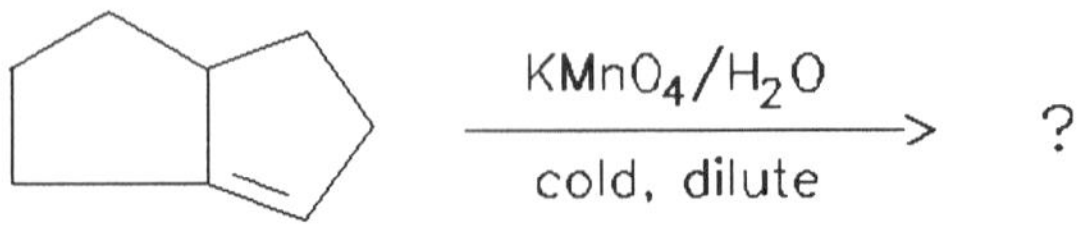

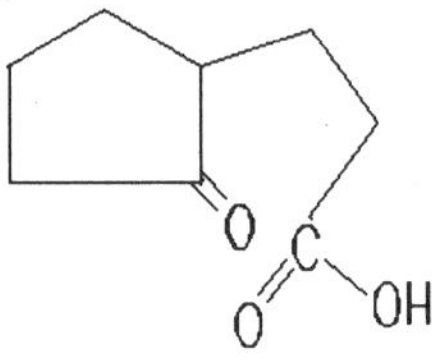

$$\xrightarrow[\text{cold, dilute}]{KMnO_4/H_2O} \ ?$$

O C OH O

I

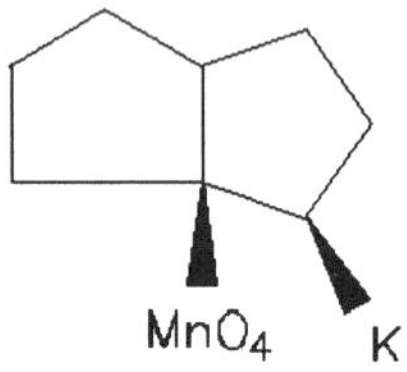

MnO_4 K

II

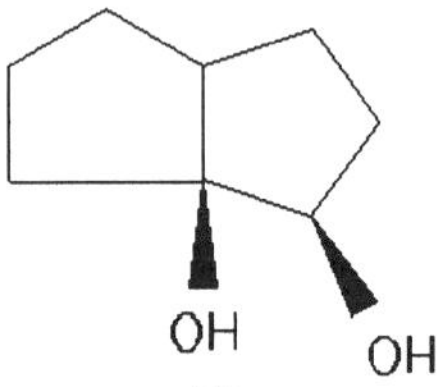

OH OH

III

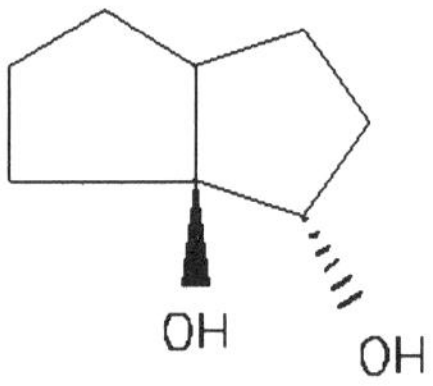

OH OH

IV

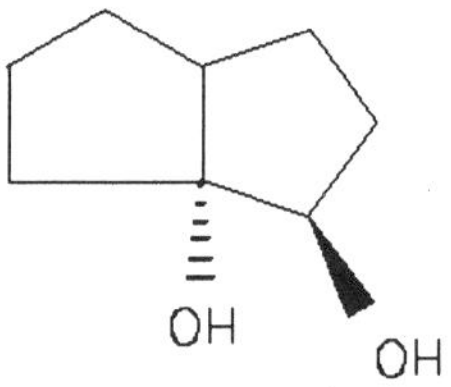

OH OH

V

A) I   B) II   * C) III   D) IV   E) V

15. Which alkene would undergo the following reaction?

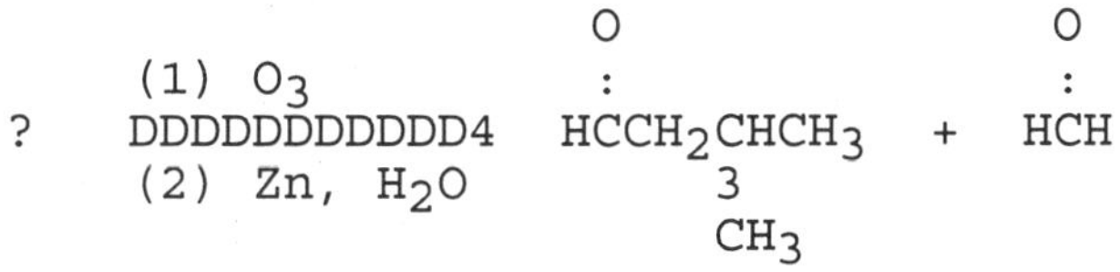

```
                                O                 O
          (1) O3                :                 :
     ?    DDDDDDDDDDD4   HCCH2CHCH3    +   HCH
          (2) Zn, H2O           3
                                CH3
```

```
A) H       H                        D) CH3       CH2CH2CH3
    \     /                               \     /
     CMC                                   CMC
    /     \                               /     \
   H       CH2CH2CH2CH3                  H       H

B) H       H                      * E) H       H
    \     /                               \     /
     CMC                                   CMC
    /     \                               /     \
   H       CHCH2CH3                      H       CH2CHCH3
           3                                        3
           CH3                                      CH3

C) CH3       H
      \     /
       CMC
      /     \
     H       CH2CH2CH3
```

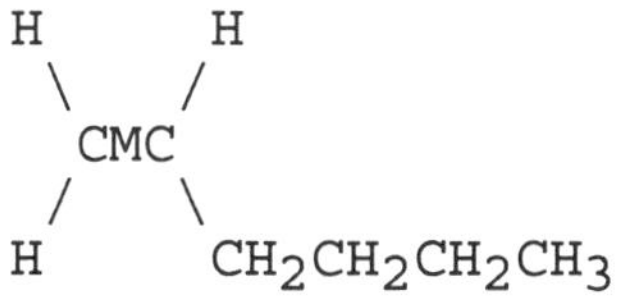

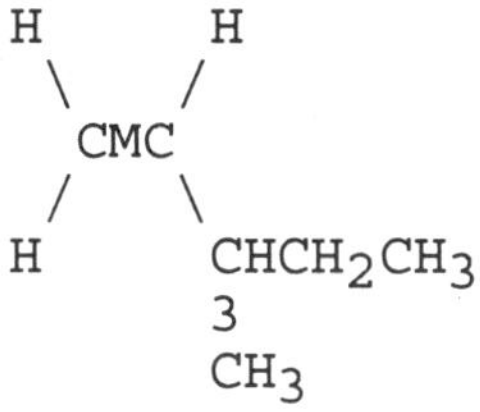

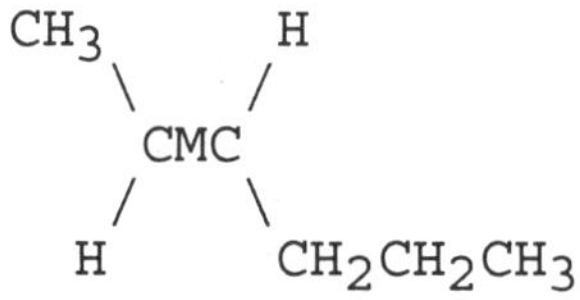

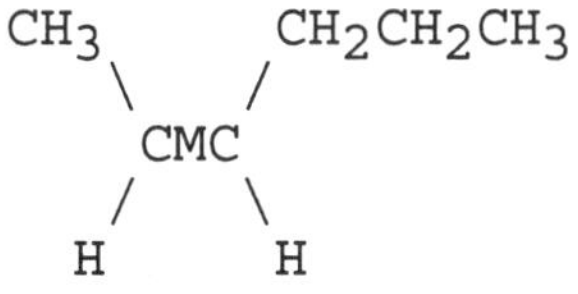

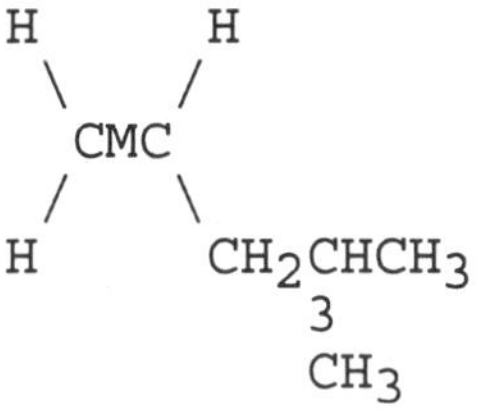

16. Which alkene would yield <u>only</u> CH_3CH_2COOH on oxidation with hot alkaline potassium permanganate?

```
A) CH3       H                    * D) CH3CH2       H
      \     /                                \     /
       CMC                                    CMC
      /     \                                /     \
     H       CH2CH2CH3                      H       CH2CH3

B) CH3       CH2CH2CH3               E) CH3       H
      \     /                               \     /
       CMC                                   CMC
      /     \                               /     \
     H       H                             H       CHCH3
                                                   3
C) CH3       CH2CH3                                CH3
      \     /
       CMC
      /     \
   CH3       H
```

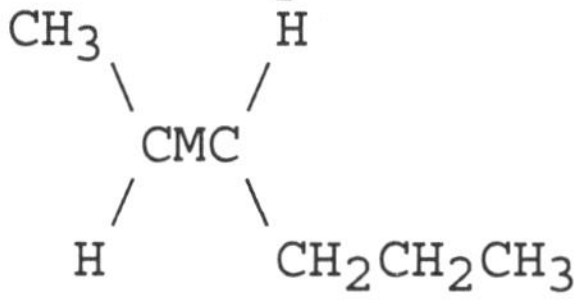

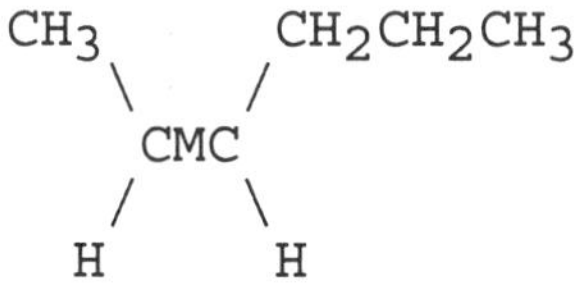

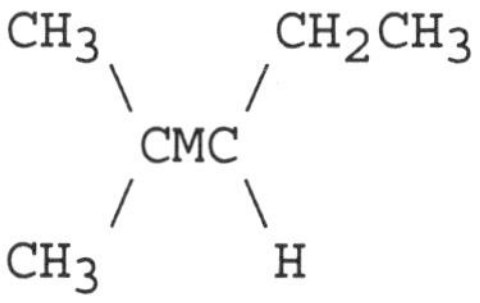

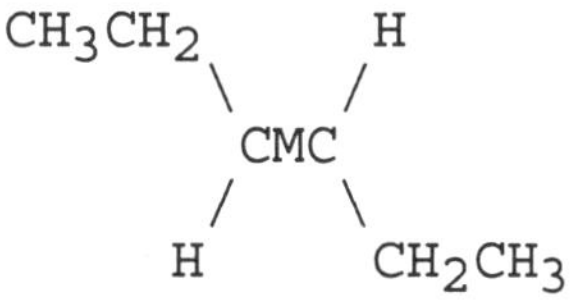

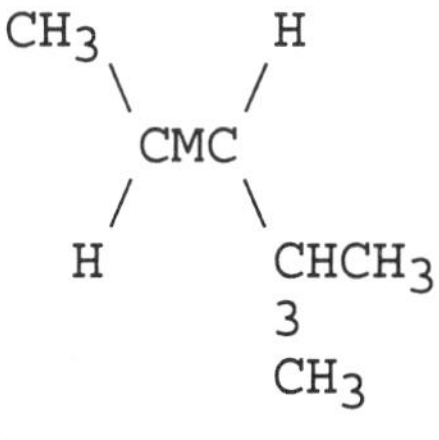

17. Acid-catalyzed hydration of 2-methyl-1-butene would yield which of the following?

A)
   CH_3
   3
$CH_2CHCH_2CH_3$
3
OH

* B)
   CH_3
   3
$CH_3CCH_2CH_3$
   3
   OH

C)
   CH_3
   3
$CH_3CHCHCH_3$
      3
      OH

D)
   CH_3
   3
$CH_3CHCH_2CH_2OH$

E) $CH_3CH_2CHCH_2OH$
        3
        CH_3

18. Which of these is not formed when cyclopentene reacts with an aqueous solution of bromine?

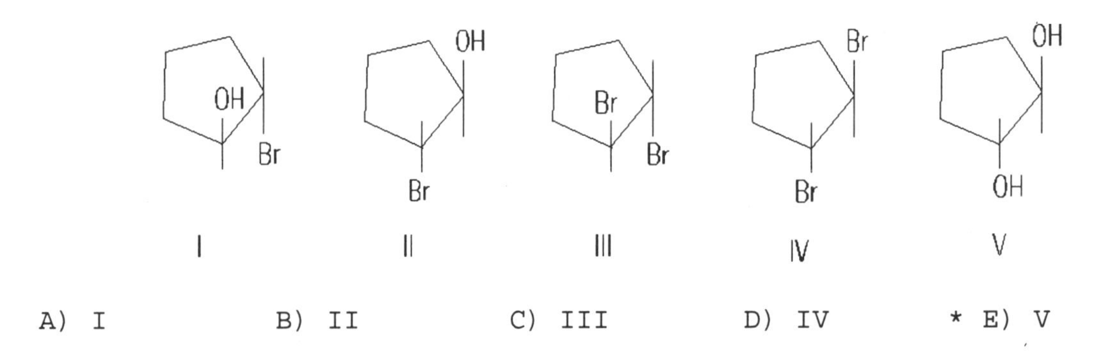

A) I   B) II   C) III   D) IV   * E) V

19. Which reagent(s) given below could be used to synthesize cis-1,2-cyclopentanediol from cyclopentene?

* A) $KMnO_4$
B) O
   :
HCOOH
C) H_2SO_4
D) All of these
E) None of these

20. Compound, C, has the molecular formula C_7H_{12}. On catalytic hydrogenation, 1 mol of C absorbs 1 mol of hydrogen and yields a compound with the molecular formula C_7H_{14}. On ozonolysis and subsequent treatment with zinc and water, C yields only:

$$CH_3\overset{O}{\overset{\|}{C}}CH_2CH_2CH_2CH_2\overset{O}{\overset{\|}{C}}H$$

The structure of C is:

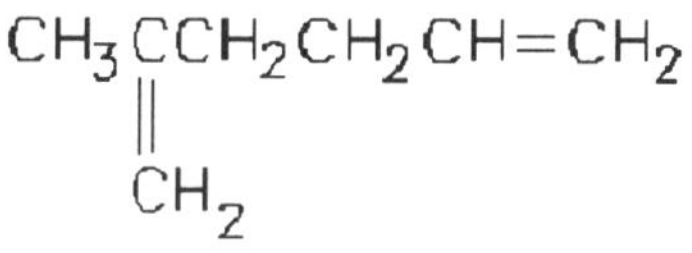

$CH_3C(=CH_2)CH_2CH_2CH=CH_2$

I

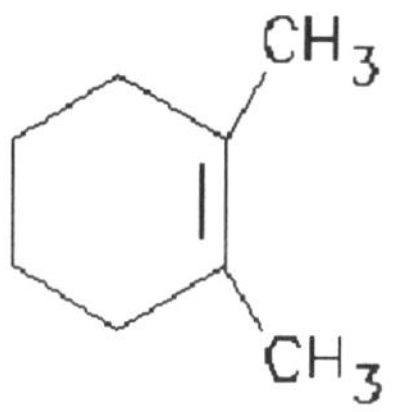

CH_3 CH_3

II

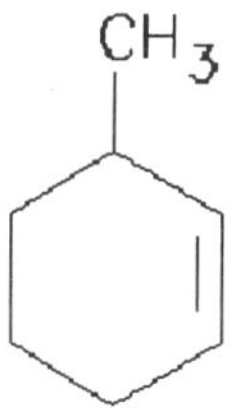

CH_3

III

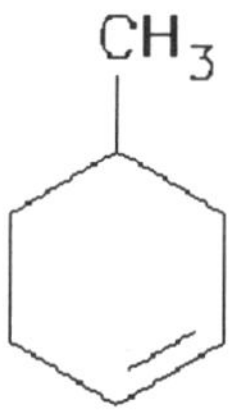

CH_3

IV

CH_3

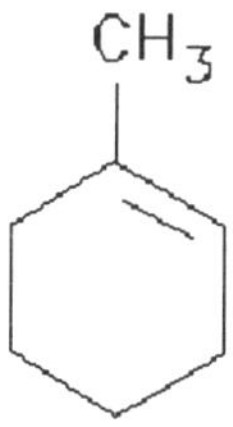

V

A) I B) II C) III D) IV * E) V

21. Which alkene would you expect to be most reactive toward acid-catalyzed hydration?

A) $CH_2MCHCH_2CH_3$

B)
```
CH3      H
   \    /
    CMC
   /    \
  H      CH3
```

C)
```
CH3      CH3
   \    /
    CMC
   /    \
  H      H
```

* D)
```
CH3      H
   \    /
    CMC
   /    \
CH3      H
```

E) All of these would be equally reactive.

22. How many compounds are possible from the addition of bromine to $CH_2MCHCH_2CH_3$ (including stereoisomers)?

A) One * B) Two C) Three D) Four E) Five

23. Which alkene would react with cold dilute alkaline permanganate solution to form an optically inactive and irresolvable product?

A)

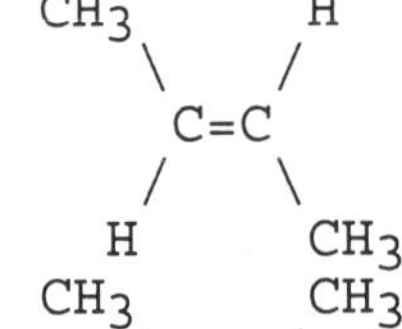

B)
```
CH3      CH3
   \    /
    C=C
   /    \
CH3      H
```

* C)
```
CH3CH2       CH2CH3
      \     /
       C=C
      /    \
     H      H
```

D)
```
CH3CH2      CH3
      \    /
       C=C
      /    \
     H      H
```

E)
```
CH3CH2      H
      \    /
       C=C
      /    \
     H      CH3
```

24. Markovnikov addition of HCl to propene involves:

A) initial attack by a chlorine ion.
B) initial attack by a chlorine atom.
C) isomerization of 1-chloropropane.
D) formation of a propyl cation.
* E) formation of an isopropyl cation.

25. An unknown compound, B, has the molecular formula C_7H_{12}. On catalytic hydrogenation 1 mol of B absorbs 2 mol of hydrogen and yields 2-methylhexane. When B is treated with a solution of $Ag(NH_3)_2OH$, a precipitate forms; analysis of the precipitate shows that it contains silver, and treatment of the precipitate with dilute HNO_3 regenerates B. Which compound best represents B?

A) $CH_3CH_2CHCH_2CpCH$
      3
      CH_3

B) $CH_3CpCCH_2CHCH_3$
      3
      CH_3

C) $CH_2MCHCHMCHCHCH_3$
      3
      CH_3

* D) $CH_3CHCH_2CH_2CpCH$
      3
      CH_3

E) $CH_2MCCH_2CH_2CHMCH_2$
      3
      CH_3

26. Which of these compounds will react with <u>each</u> of these reagents?

cold concd. H_2SO_4 $AgNO_3/NH_3/H_2O$
Br_2 in CCl_4

A) $CH_3CH_2CH{=}CHCH_3$
B) $CH_3CH_2CH_2CH{=}CH_2$
C) $CH_3CH_2CpCCH_3$
* D) $(CH_3)_2CHCpCH$
E) $(CH_3)_2CHCpCCH_3$

27. Which reaction is NOT stereospecific?

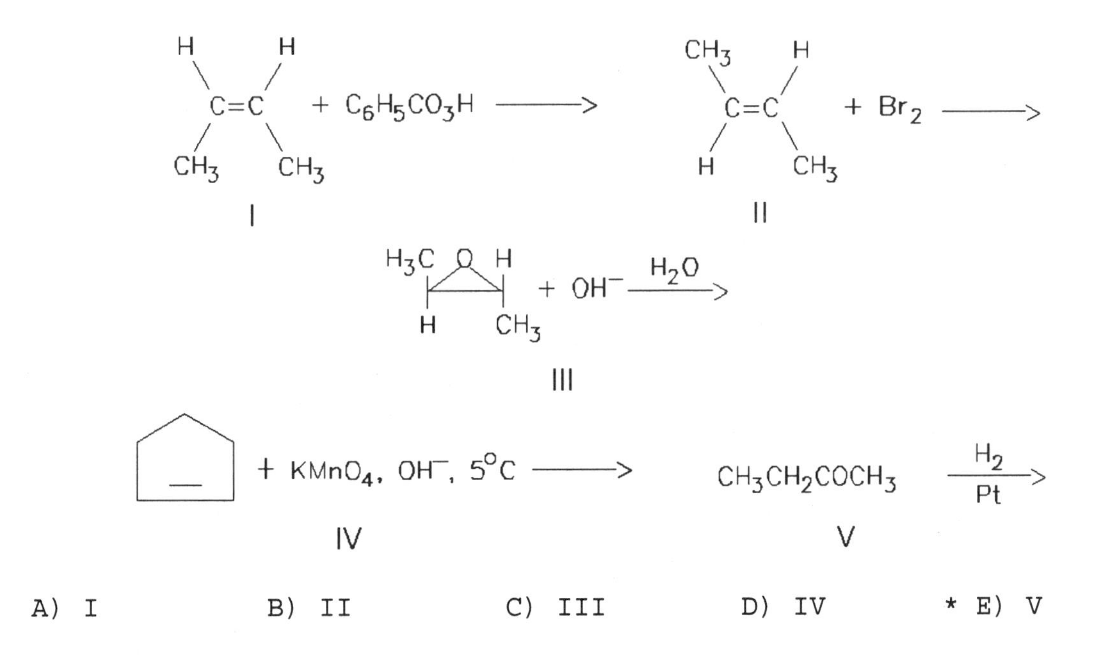

A) I   B) II   C) III   D) IV   * E) V

28. Which reaction is regioselective?

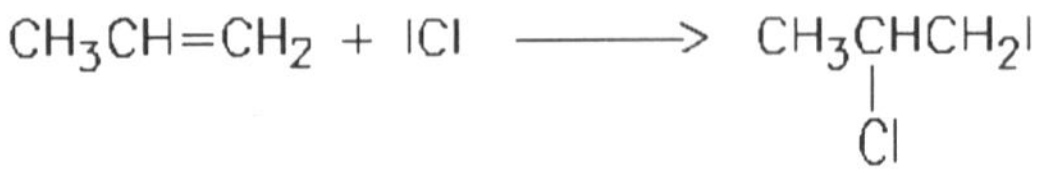

$CH_3CH{=}CH_2 + ICl \longrightarrow CH_3CH(Cl)CH_2I$

I

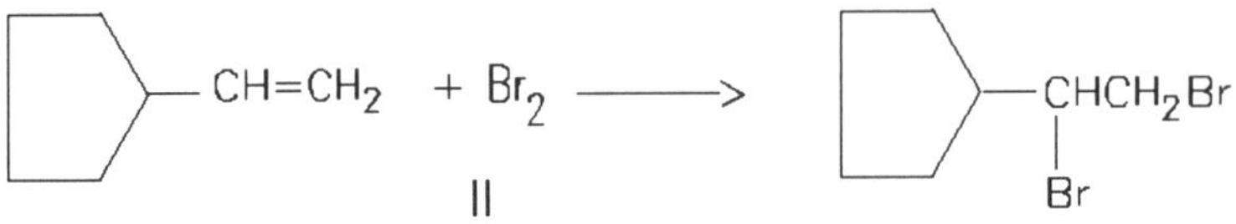

(cyclopentyl)$-CH{=}CH_2 + Br_2 \longrightarrow$ (cyclopentyl)$-CH(Br)CH_2Br$

II

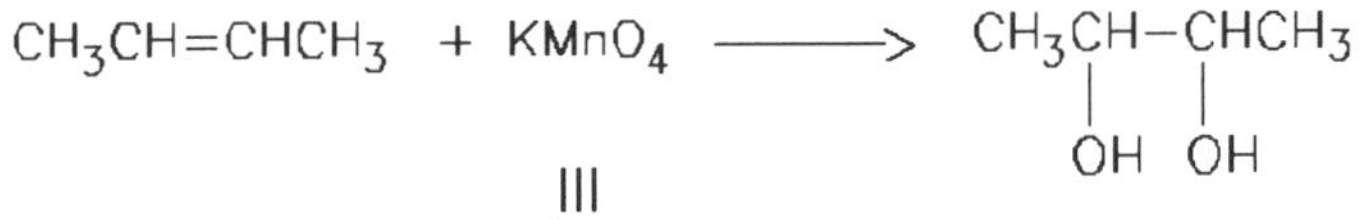

$CH_3CH{=}CHCH_3 + KMnO_4 \longrightarrow CH_3CH(OH){-}CH(OH)CH_3$

III

cyclohexene $+ D_2 \xrightarrow{Ni}$ (cyclohexane with H, D; D, H)

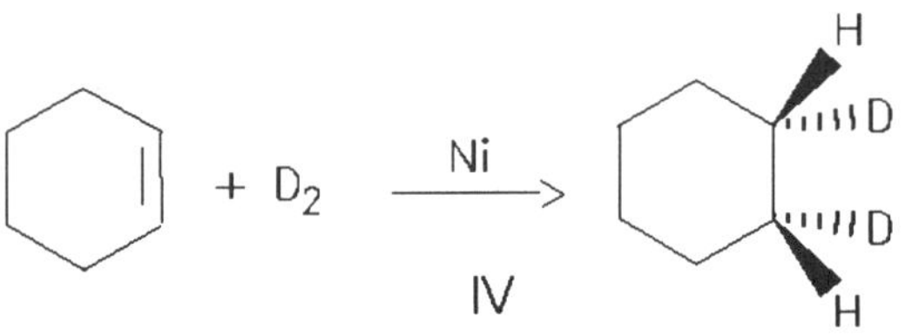

IV

* A) I
B) II
C) III
D) IV
E) None of these

29. Which reaction of an alkene proceeds with anti addition?
A) Hydroboration/oxidation
* B) Bromination
C) Permanganate oxidation
D) Hydrogenation
E) Epoxidation

30. A pair of enantiomers results from which of these reactions?
A) 5xC
cyclopentene + dil. $KMnO_4$, OH^- DDDDD4
B) trans-2-butene + Br_2 DDDD4
* C) 1-pentene + HCl DDDD4
D) cis-2-butene + D_2/Pt DDDD4
E) cyclobutene + OsO_4 DDDDDDDDDDDD4
(2) $NaHSO_3$

31. Which compound is reasonably anticipated as a by-product in the hydroxylation of cyclopentene with cold alkaline permanganate?

OH OH

I

OH

II

O

III

$O=C(H)(CH_2)_3C(H)=O$

IV

$HO_2C(CH_2)_3CO_2H$

V

A) I B) II C) III D) IV * E) V

32. Select the structure of the major product formed in the following reaction.

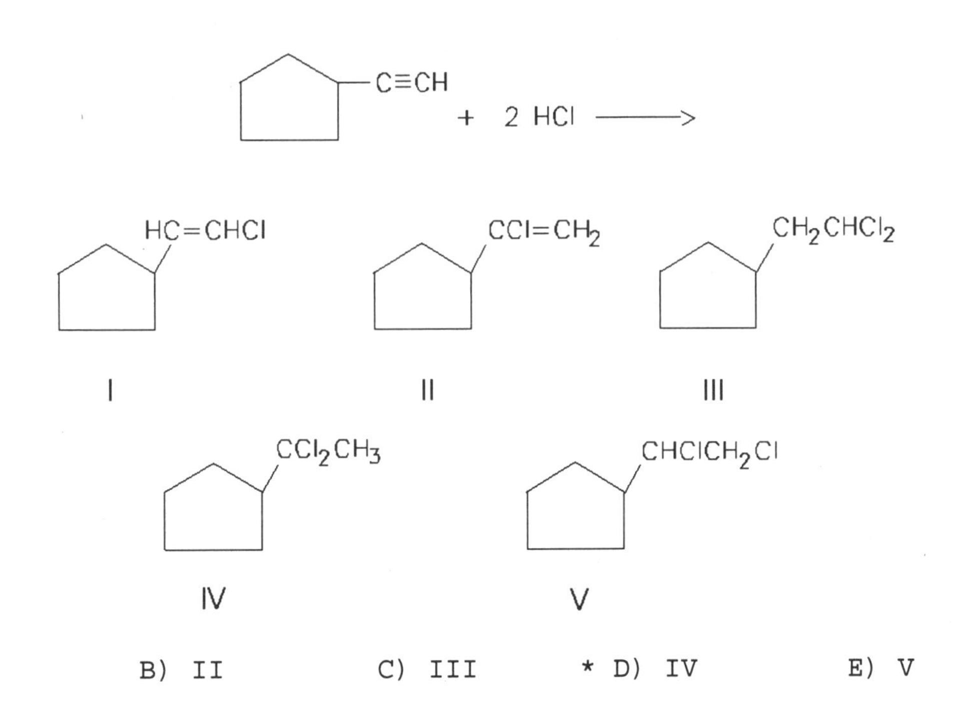

A) I B) II C) III * D) IV E) V

33. Which of these compounds is <u>not</u> formed when gaseous ethene is bubbled into a aqueous solution of bromine, sodium chloride and sodium nitrate?

A) $BrCH_2CH_2Br$
B) $BrCH_2CH_2Cl$
C) $BrCH_2CH_2OH$
* D) $ClCH_2CH_2OH$
E) $BrCH_2CH_2ONO_2$

34. Which of the following would be soluble in cold concentrated sulfuric acid, decolorize bromine in carbon tetrachloride, and give a precipitate with silver nitrate in aqueous ammonia?

A) $CH_3CHMCHCH_3$
B) CH_3CHMCH_2
* C) CH_3CH_2CpCH
D) $CH_3CH_2CH_2CH_3$
E) All of these would give positive tests with all reagents.

35. (Z)-3-Chloro-3-heptene is:

A)
```
                Cl
                3
  CH3CH2        CHCH2CH3
        \      /
          CMC
        /     \
      H         H
```

B)
```
        CH3CH2        H
              \      /
                CMC
              /     \
  CH3CH2CH2          Cl
```

C)
```
  CH3CH2        CH2CH2CH3
        \      /
          CMC
        /     \
      Cl        H
```

* D)
```
  CH3CH2        H
        \      /
          CMC
        /     \
      Cl        CH2CH2CH3
```

E)
```
  CH3CH2        Cl
        \      /
          CMC
        /     \
      H         CH2CH2CH3
```

36. Addition of hydrogen chloride to the following molecule would produce:

$$\mathrm{Cl \blacktriangleright C(CH{=}CH_2)(CH_3) \blacktriangleleft H} \xrightarrow{\mathrm{HCl}} ?$$

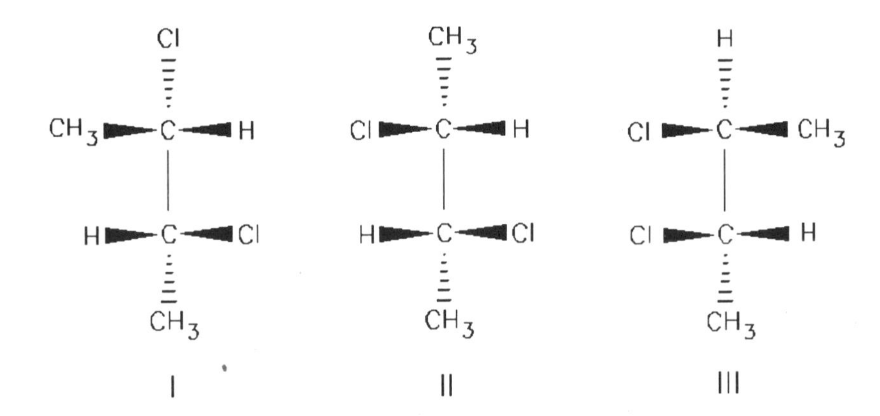

A) I and II
* B) I and III
C) II and III
D) I, II and III
E) III

37. The reaction

$$CH_2{=}CH_2 \quad + \quad HCl \quad \underset{\text{phase}}{\overset{\text{gas}}{\text{DDDDDDD4}}} \quad CH_3CH_2Cl$$

has the following thermodynamic parameters at 298 K: (Hx = -15.5 kcal mol^{-1}, (Sx = -31.3 cal K^{-1} mol^{-1}, (Gx = -6.17 kcal mol^{-1}. Which of the following statements is true of the reaction?

A) Both (Hx and (Sx favor product formation.
B) Neither (Hx nor (Sx favors product formation.
* C) The entropy term is unfavorable but the formation of ethyl chloride is favored.
D) The entropy term is favorable but the formation of ethyl chloride is not favored.
E) The sign of (Gx indicates that the reaction cannot occur as written.

38. Which of the following reagents might serve as the basis for a simple chemical test that would distinguish between pure 1-hexene and pure hexane?

A) Bromine in carbon tetrachloride
B) Dilute aqueous potassium permanganate
C) Concentrated sulfuric acid
* D) All of the above
E) Answers A) and B) only

39. What would be the major product of the following reaction?

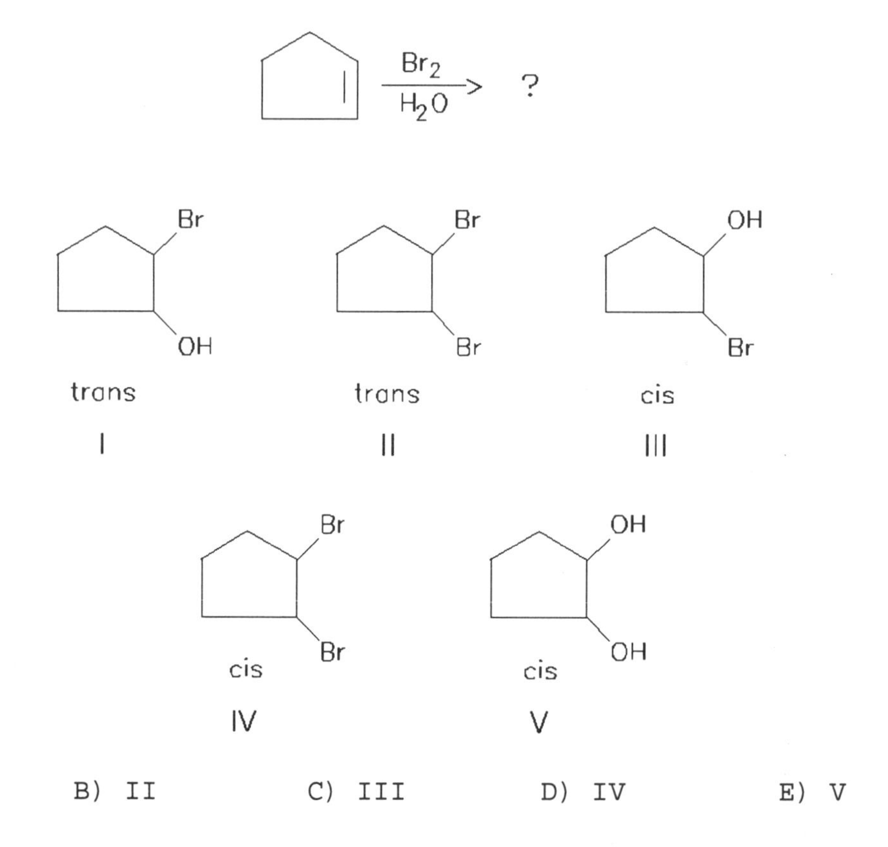

* A) I B) II C) III D) IV E) V

40. Compound X has the molecular formula C_6H_{10}. X decolorizes bromine in carbon tetrachloride. X also reacts with Ag^+ in ammonia to form an insoluble salt. When treated with excess hydrogen and a nickel catalyst, X yields 2-methylpentane. The most likely structure for X is:

A) $CH_3CMCHCHMCH_2$
   3
   CH_3

B) $CH_2MCCHMCHCH_3$
   3
   CH_3

C) $CH_3CHCpCCH_3$
   3
   CH_3

* D) CH_3CHCH_2CpCH
   3
   CH_3

E) $CH_3CH_2CHCpCH$
   3
   CH_3

41. Addition of hydrogen chloride to the following molecule would produce:

CH_2
||
CH

Cl H

CH_3

HCl ⟶ ?

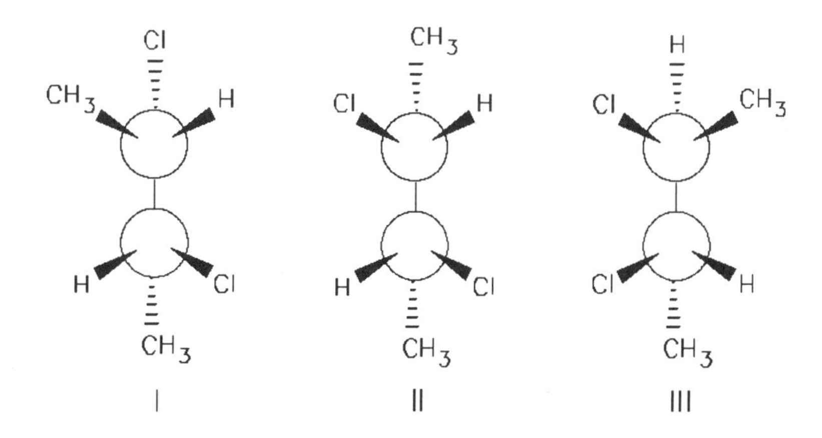

A) I and II
* B) I and III
C) II and III
D) I, II and III
E) III

42. Which reagent could be used to distinguish between 2-pentyne and 1-pentyne?
A) Br_2/CCl_4
* B) $Ag(NH_3)_2OH$
C) Concd. H_2SO_4
D) $KMnO_4, OH^-$
E) None of these

43. Which reagent could be used to distinguish between 1-bromohexane and hexane?
A) Br_2/CCl_4
* B) $AgNO_3/C_2H_5OH$
C) dilute $KMnO_4$
D) cold concd. H_2SO_4
E) None of these

44. Which reaction would yield a racemic modification?
A) Cyclopentene + D_2 DDDDDD4 (Pt)

B) Cyclopentene + OsO_4, then Na_2SO_3 DDDDDD4

* C) Cyclopentene + RCOOH (O, :), then H_3O^+ DDDDDD4

D) Cyclopentene + cold, dilute $KMnO_4$ DDDDDD4
E) Cyclopentene + dilute H_2SO_4 DDDDDD4

45. Which of the following could be used to distinguish between 1-octyne and 2-octyne?
A) Treatment with 2 mol of HX
B) Addition of water
C) Reaction with $KMnO_4$
D) Decolorization of bromine in CCl_4
* E) Treatment with $Ag(NH_3)_2OH$

46. An unknown compound, A, has the molecular formula C_7H_{12}. On oxidation with hot aqueous potassium permanganate, A yields $CH_3CHCOOH$ and CH_3CH_2COOH. Which of the following structures best represents A?
3
CH_3

A) $CH_3C{=}CHCH{=}CHCH_3$
3
CH_3

B) $CH_3CHCHMCHCH_2CH_3$
3
CH_3

* C) $CH_3CHCpCCH_2CH_3$
3
CH_3

D) $CH_3CHCH_2CpCCH_3$
3
CH_3

E) $CH_3CHCH_2CH_2CpCH$
3
CH_3

47. An alkene, X, with the formula C_7H_{10} adds one mole of hydrogen on catalytic hydrogenation. On treatment with hot basic $KMnO_4$ followed by acidification, X yields

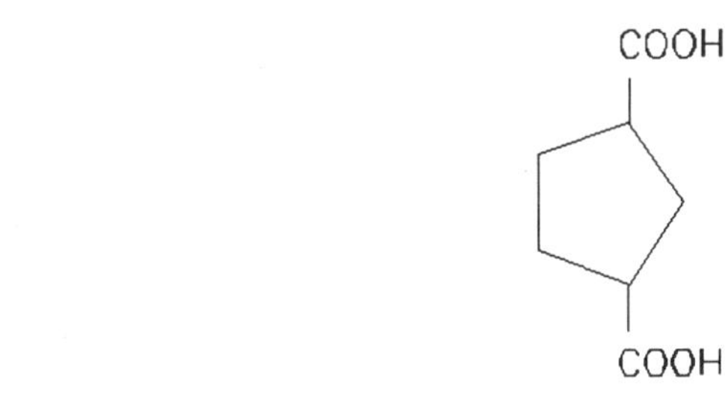

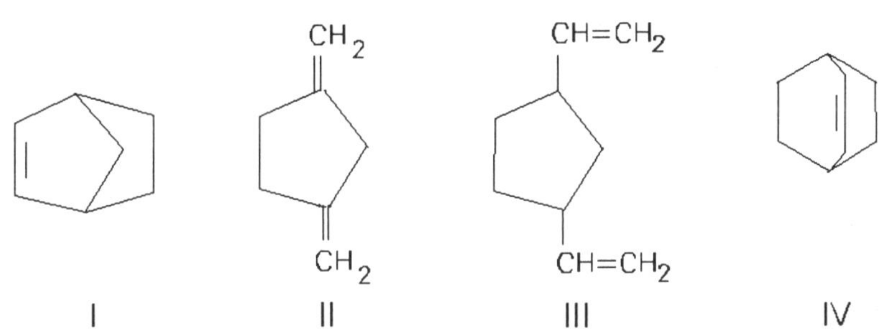

Which is a possible structure for X?

* A) I
B) II
C) III
D) IV
E) None of these

48. An optically active compound, Y, with the molecular formula C_7H_{12} gives a positive test with cold dilute $KMnO_4$ and gives a precipitate when treated with $Ag(NH_3)_2OH$. On catalytic hydrogenation, Y yields $Z(C_7H_{16})$ and Z is also optically active. Which is a possible structure for Y?

A) $CH_3CH_2CH_2CH_2CH_2CpCH$

B)
```
    H
    3
CH3CCH2CH2CpCH
    3
    CH3
```

* C)
```
        H
        3
CH3CH2CCH2CpCH
        3
        CH3
```

D)
```
        H
        3
CH3CH2CCpCCH3
        3
        CH3
```

E)
```
        H
        3
CH2MCHCCH2CHMCH2
        3
        CH3
```

49. Select the structure of the major product formed in the following reaction.

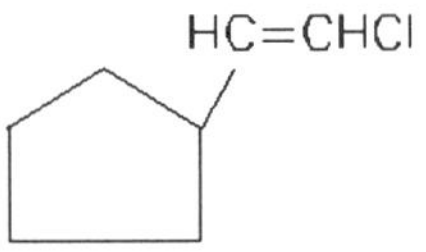

Cyclopentyl–C≡CH + 2 Cl_2 ——>

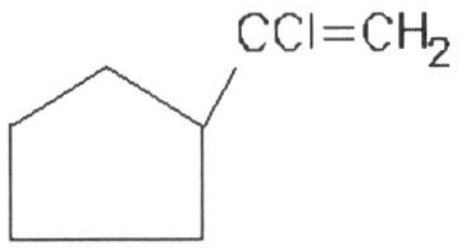

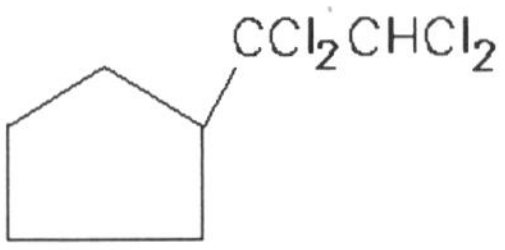

I: cyclopentyl–HC=CHCl

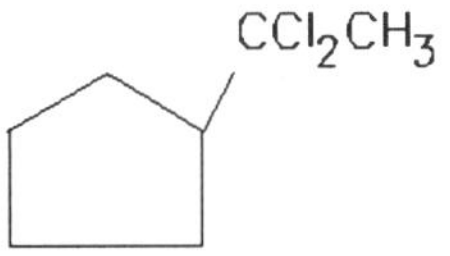

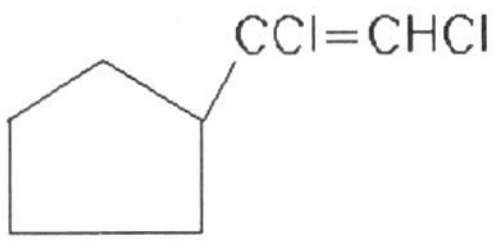

II: cyclopentyl–CCl=CH_2

III: cyclopentyl–CCl_2CHCl_2

IV: cyclopentyl–CCl_2CH_3

V: cyclopentyl–CCl=CHCl

A) I B) II * C) III D) IV E) V

50. Which reaction sequence would convert cis-2-butene to trans-2-butene?
  A) Br_2/CCl_4; then $2NaNH_2$; then H_2/Ni_2B(P-2)
* B) Br_2/CCl_4; then $2NaNH_2$; then Li/liq. NH_3
  C) H_3O^+/H_2O, heat; then cold dilute $KMnO_4$
  D) HBr; then $NaNH_2$; then H_2, Pt
  E) None of these

51. A reagent that could be used as the basis for a simple chemical test that will distinguish between 1-pentene and 1-pentyne would be:
  A) Bromine in carbon tetrachloride
  B) Dilute aqueous potassium permanganate
  C) CrO_3 in H_2SO_4
  D) H_2SO_4
* E) $Ag(NH_3)_2OH$

52. What compound would yield $CH_3CH_2CH_2CH(=O)$ + $CH_3CH(=O)$ upon treatment with O_3, followed by Zn/H_2O?
  A) 1-Hexene
  B) cis-2-Hexene
  C) trans-2-Hexene
* D) More than one of these
  E) None of these

53. What is the product of the following reaction?

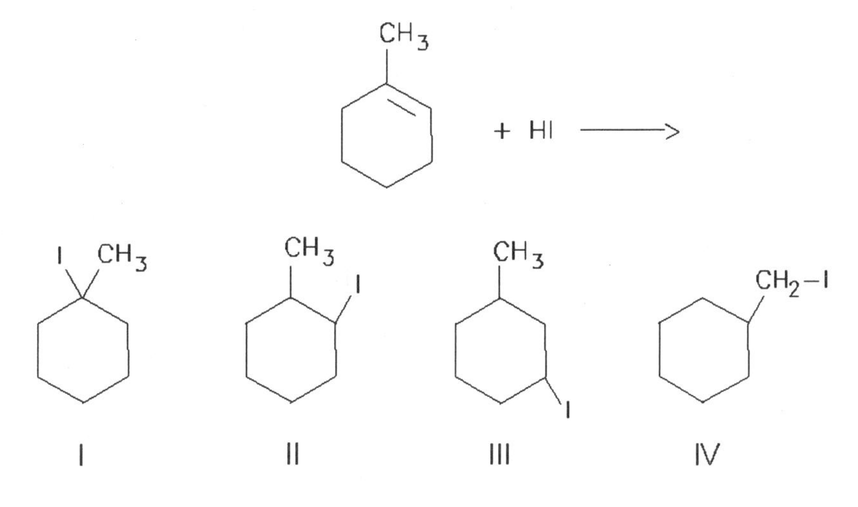

* A) I
B) II
C) III
D) IV
E) None of these

54. What would be the major product of the following reaction?

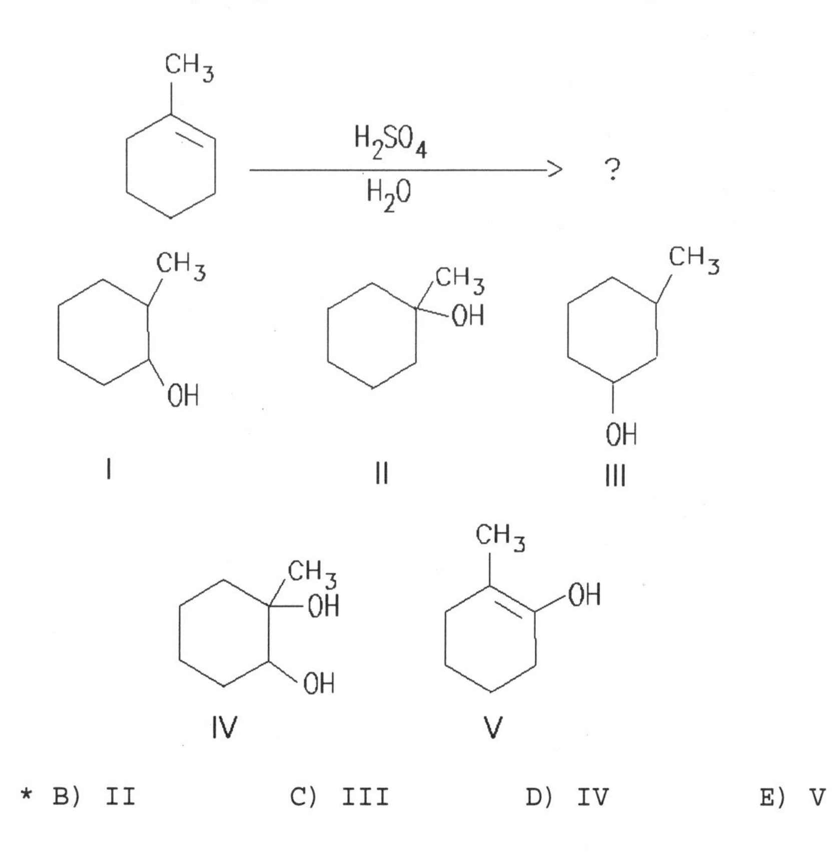

A) I
* B) II
C) III
D) IV
E) V

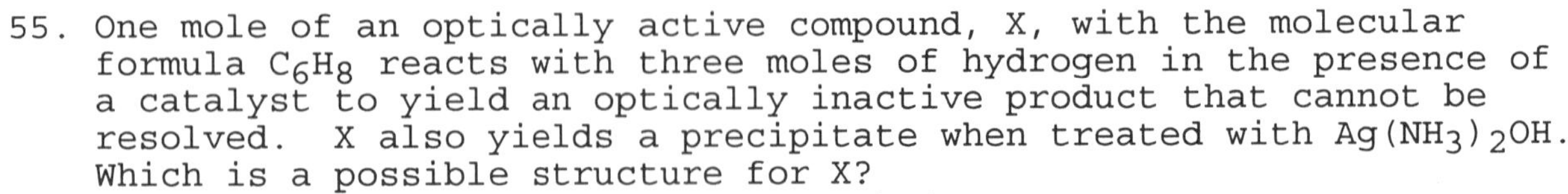

55. One mole of an optically active compound, X, with the molecular formula C_6H_8 reacts with three moles of hydrogen in the presence of a catalyst to yield an optically inactive product that cannot be resolved. X also yields a precipitate when treated with $Ag(NH_3)_2OH$. Which is a possible structure for X?

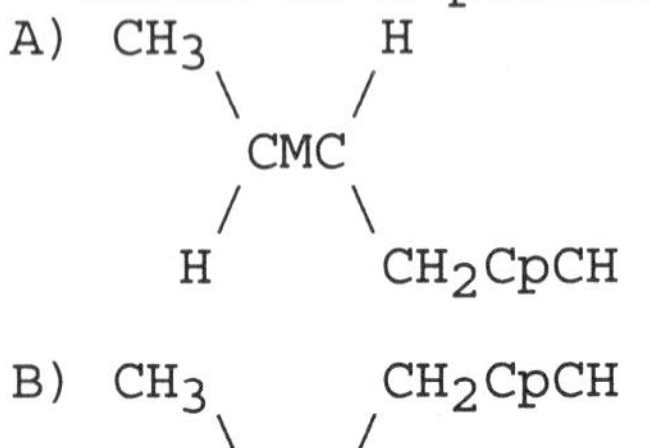

```
A) CH3      H
      \   /
       CMC
      /   \
    H      CH2CpCH

B) CH3      CH2CpCH
      \   /
       CMC
      /   \
    H      H

C) CH3   H
     \ /
      C
       [
      CDCpCCH3
      3
      H
```

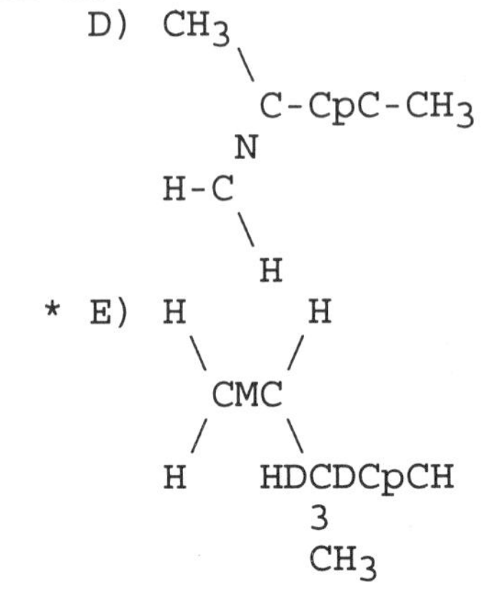

```
D) CH3
      \
       C-CpC-CH3
      N
   H-C
      \
       H

* E) H     H
       \   /
        CMC
       /   \
     H    HDCDCpCH
             3
             CH3
```

56. Which of the following reagents could be used as the basis for a simple chemical test that would distinguish betwen 1-pentyne and pentane?
A) $Ag(NH_3)_2OH$
B) Br_2/CCl_4
C) $KMnO_4/H_2O$
D) Two of these
* E) All of these

57. Cyclohexene is subjected to epoxidation followed by acid-catalyzed hydrolysis to yield 1,2-cyclohexanediol. In the most stable conformation of the product, the hydroxyl groups would be:
A) both axial
* B) both equatorial
C) axial-equatorial
D) coplanar
E) None of these

58. Which reaction would yield a meso compound?

A) <u>cis</u>-2-Butene DDDDDDDD4 (Br_2/CCl_4)

B) <u>cis</u>-2-Butene DDDDDDDDDDDD4 ((1) RCOOH with O above; (2) H_3O^+/H_2O)

* C) <u>cis</u>-2-Butene DDDDDDDDDD4 ((1) OsO_4; (2) $NaHSO_3$)

D) <u>trans</u>-2-Butene DDDDDDDDD4 (dil $KMnO_4$; 5x)

E) None of these

59. An alkene with the molecular formula C_8H_{16} undergoes ozonolysis to yield $CH_3\overset{O}{\overset{:}{C}}CH_3$ and $(CH_3)_3\overset{O}{\overset{:}{C}}H$. The alkene is:

A) 2,2-Dimethyl-2-hexene
B) 2,3-Dimethyl-2-hexene
C) 2,4-Dimethyl-2-hexene
* D) 2,4,4-Trimethyl-2-pentene
E) More than one of the above is a possible answer.

60. Addition of 2 mol of HCl to 1-butyne would yield:

A) $CH_3CH_2CH_2CHCl_2$
* B) $CH_3CH_2CCl_2CH_3$
C) $CH_3CH_2CHClCH_2Cl$
D) $CH_3CH_2CHMCHCl$
E) $CH_3CHClCHClCH_3$

61. Cyclohexene reacts with bromine to yield 1,2-dibromocyclohexane. Molecules of the product would:

* A) be a racemic form and, in their most stable conformation, they would have both bromine atoms equatorial.
B) be a racemic form and, in their most stable conformation, they would have one bromine atom equatorial and one axial.
C) be a meso compound and, in its most stable conformation, it would have both bromine atoms equatorial.
D) be a meso compound and, in its most stable conformation, it would have one bromine atom equatorial and one axial.
E) be a pair of diastereomers and, in their most stable conformation, one would have the bromines equatorial and axial, and the other would have the bromines equatorial and equatorial.

62. Consider the addition of HCl to 3-methyl-1-butene. The major product of the reaction would be:

A) 1-Chloro-2-methylbutane
B) 1-Chloro-3-methylbutane
* C) 2-Chloro-2-methylbutane
D) 2-Chloro-3-methylbutane
E) 1-Chloropentane

63. Which of the following reactions would yield the final product as a racemic form?

* A) Cyclohexene + a peroxy acid, then H_3O^+/H_2O
B) Cyclohexene + cold, dilute $KMnO_4$ and OH^-
C) Cyclohexene + HCl
D) Cyclohexene + OsO_4, then $NaHSO_3$
E) Cyclohexene + D_2/Pt

64. (R)-3-Chloro-1-butene reacts with HCl by Markovnikov addition, and the products are separated by gas-liquid chromatography. How many total fractions would be obtained and how many would be optically active?

A) One optically active fraction only
* B) One optically active fraction and one optically inactive
C) Two optically active fractions
D) One optically active fraction and two optically inactive
E) Two optically active fractions and one optically inactive

65. What is the final product of the following synthesis?

$$CH_3CpCCH_3 \xrightarrow{H_2/Ni_2B\ (P\text{-}2)} C_4H_8 \xrightarrow[(2)\ NaHSO_3]{(1)\ OsO_4} \text{Final Product}$$

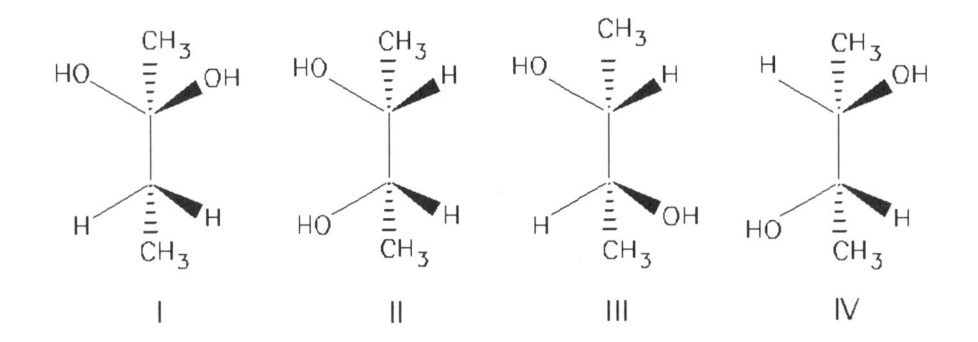

A) I
* B) II
C) III
D) IV
E) An equimolar mixture of III and IV

66. Which reagent(s) could be used as the basis of a simple chemical test to distinguish between cyclopentane and cyclopentene?

* A) Br_2/CCl_4
B) $AgNO_3$/EtOH
C) $Ag(NH_3)_2{}^+OH^-$ in NH_3/H_2O
D) More than one of the above
E) None of the above

67. Which reaction would give a meso compound as the product?

A) Cyclopentene + Br_2/CCl_4
* B) Cyclopentene + OsO_4, then $NaHSO_3$
C) Cyclopentene + RCO_3H, then H_3O^+/H_2O
D) Cyclopentene + Cl_2, H_2O
E) More than one of these

68. What is the major product of the following reaction?

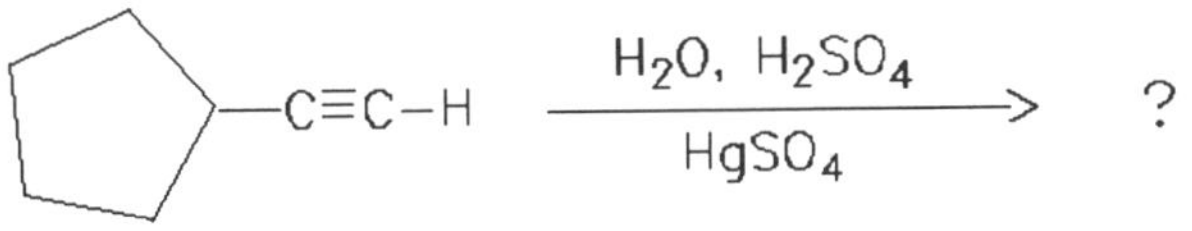

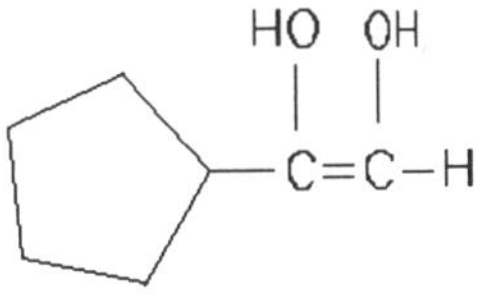

$$\text{C}_5\text{H}_9\text{–C}\equiv\text{C–H} \xrightarrow[\text{HgSO}_4]{\text{H}_2\text{O, H}_2\text{SO}_4} \ ?$$

HO OH
C=C–H

I

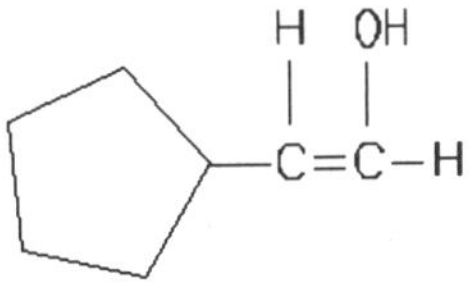

H OH
C=C–H

II

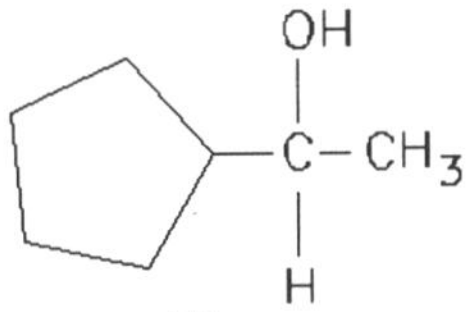

OH
C–CH_3
H

III

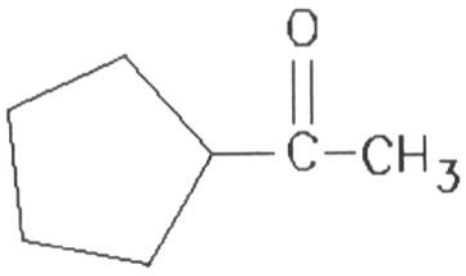

O
C–CH_3

IV

H O
C–C–H
H

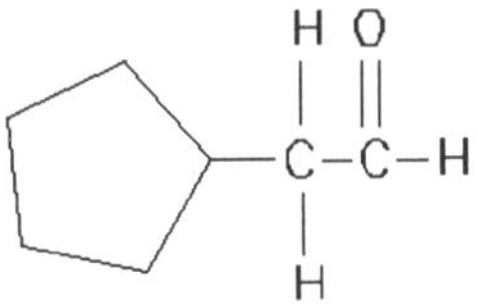

V

A) I B) II C) III * D) IV E) V

69. An alkene with the molecular formula C_8H_{14} is treated with ozone and then with zinc and water. The product isolated from these reactions is

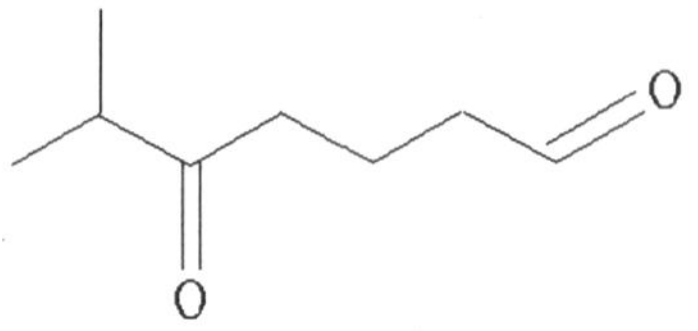

What is the structure of the alkene?

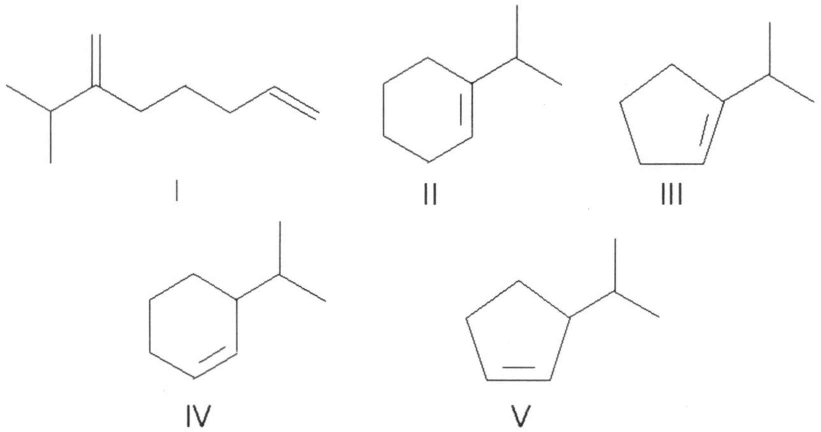

I II III

IV V

A) I B) II * C) III D) IV E) V

70. An optically active compound, A, with the molecular formula C_7H_{12} reacts with cold dilute $KMnO_4$ and gives a precipitate when treated with $Ag(NH_3)_2OH$. On catalytic hydrogenation, A is converted to B (C_7H_{16}) and B is also optically active. Which is a possible structure for A?

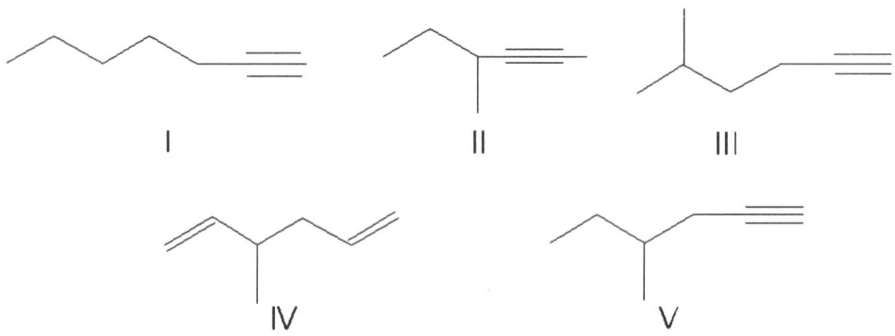

I II III

IV V

A) I B) II C) III D) IV * E) V

71. The conversion of ethylene to vinyl bromide can be accomplished by use of these reagents in the order indicated.
A) (1) HBr; (2) $NaOC_2H_5$
* B) (1) Br_2; (2) $NaOC_2H_5$
C) (1) Br_2; (2) H_2O
D) (1) $NaNH_2$; (2) HBr
E) (1) HBr; (2) H_2SO_4

72. Acid-catalyzed dimerization of propene is predicted to yield chiefly:
A) 4-Methyl-1-pentene
* B) 4-Methyl-2-pentene
C) Hexane
D) 2-Methylpentane
E) 2,3-Dimethyl-2-butene

73. The most resistant compound to the action of hot alkaline $KMnO_4$ is:
* A) Pentane
B) 1-Pentene
C) 2-Pentene
D) 2-Pentyne
E) Cyclopentene

74. 2-Pentyne will <u>not</u> react with:
A) H_2, Pt B) Br_2 * C) $Ag(NH_3)_2^+$ D) H_2SO_4 E) $KMnO_4/H_2O$

75. The ozonolysis (followed by hydrolysis) of an unsymmetrical and unbranched alkene forms:
A) a single aldehyde.
B) an aldehyde and a ketone.
C) two different ketones.
* D) two different aldehydes.
E) a single ketone.

76. Hydroxylation of <u>cis</u>-2-butene with cold alkaline $KMnO_4$ yields

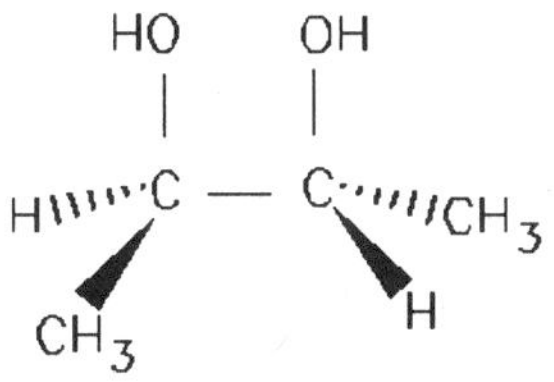

I 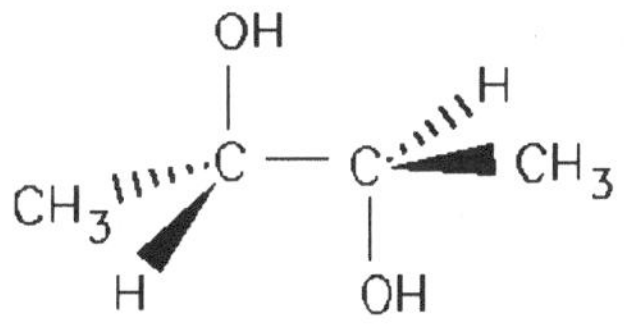 II 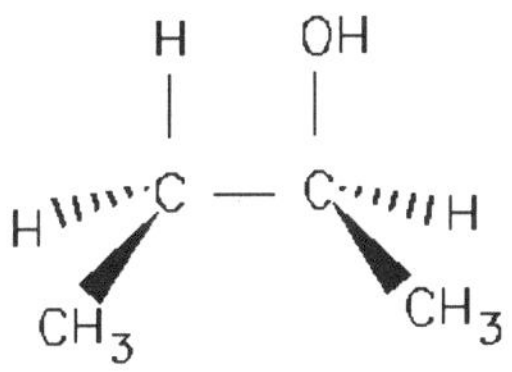 III

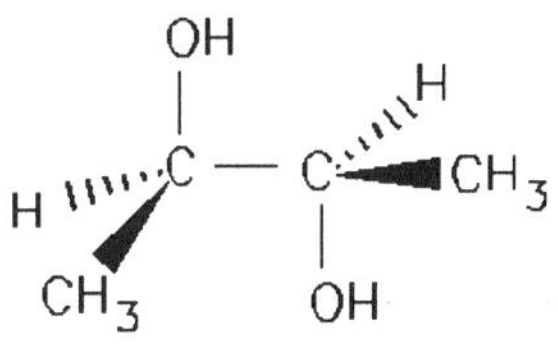

IV 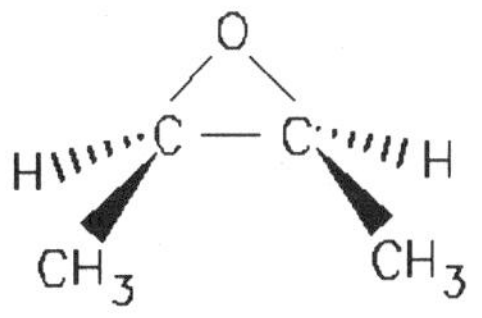 V

A) I * B) II C) III D) IV E) V

77. The reaction of BrCl (bromine monochloride) with 1-methylcyclopentene will produce as the predominant product:

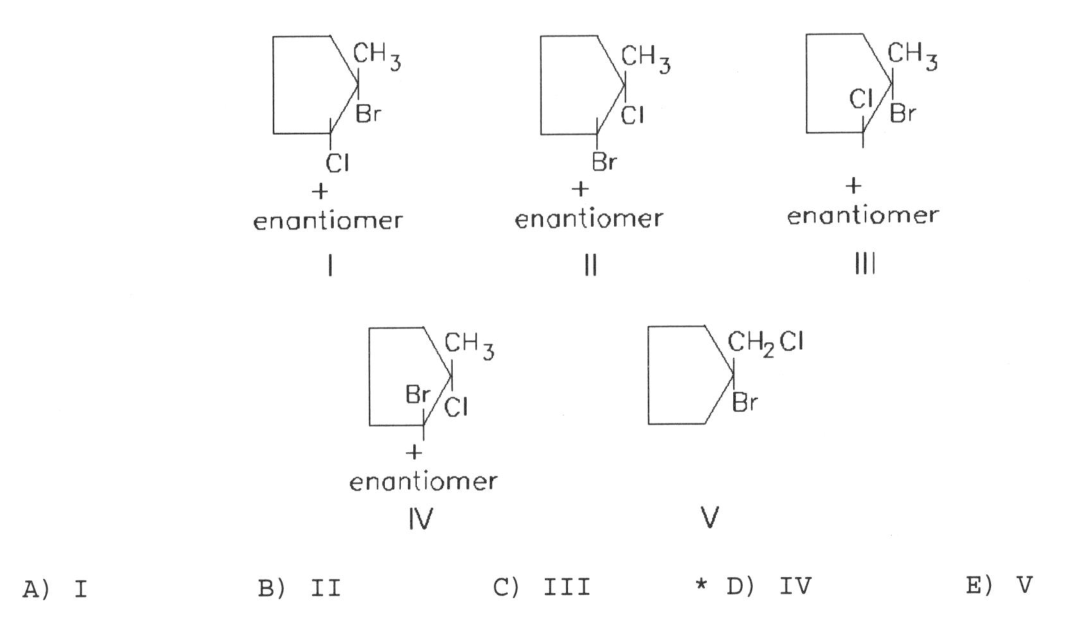

A) I B) II C) III * D) IV E) V

78. Reaction of trans-2-pentene with a solution of Br_2 in CCl_4 produces:

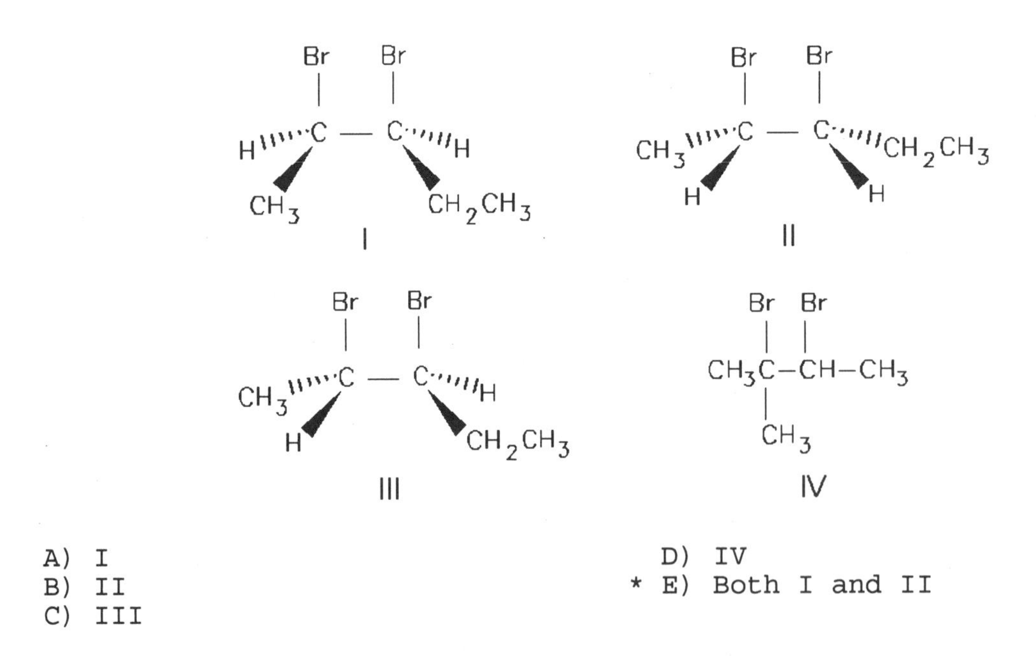

A) I
B) II
C) III
D) IV
* E) Both I and II

79. Which of these compounds belongs to the class of substances commonly known as "halohydrins"?

A) $BrCH_2CH_2Cl$

B)
```
      O
      :
ClCH2COH
```

* C) ICH_2CH_2OH

D) $FCH_2CH_2NH_2$

E)
```
HOCH2CMO
       3
       Cl
```

80. When either <u>cis</u>- or <u>trans</u>-2-butene is treated with hydrogen chloride in ethanol, the product mixture that results includes:

A) $CH_3CH_2CH_2CH_2Cl$

B) $CH_3CH_2CH_2CH_2OCH_2CH_3$

* C)
```
CH3CH2CHOCH2CH3
      3
      CH3
```

D) $(CH_3)_3CCl$

E) $(CH_3)_2CHCH_2OCH_2CH_3$

81. <u>In general</u>, when the addition of an unsymmetrical electrophilic reagent to an unsymmetrical alkene forms the product predicted by Markovnikov's rule, that occurs because:

A) the product is statistically favored.
B) steric hindrance favors its formation.
* C) it is formed via the more/most stable carbocation.
D) it is the more/most stable product.
E) All of the above are reasons.

Chapter 9: Free Radical Reactions

1. A chain reaction is one that:
 A) involves a series of steps.
 B) involves two steps of equal activation energy.
 C) is one that can be initiated by light.
* D) involves a series of steps, each of which generates a reactive intermediate that brings about the next step.
 E) involves free radicals that have an unusual stability and thereby cause a large quantum yield.

2. For the following reaction,

$(CH_3)_3CDH$ + ClDCl DDDDDD4 $(CH_3)_3CDCl$ + HDCl

(DHxM91 kcal mol^{-1}) (DHxM58 kcal mol^{-1}) (DHxM78.5 kcal mol^{-1}) (DHxM103 kcal mol^{-1})

the overall (Hx is:
 A) +58 kcal mol^{-1}
* B) -32.5 kcal mol^{-1}
 C) +32.5 kcal mol^{-1}
 D) -57.5 kcal mol^{-1}
 E) +181.5 kcal mol^{-1}

3. For the following reaction,

$CH_3CH_2CH_2DH$ + BrDBr DDDDDD4 $CH_3CH_2CH_2DBr$ + HDBr

(DHxM98 kcal mol^{-1}) (DHxM46 kcal mol^{-1}) (DHxM69 kcal mol^{-1}) (DHxM87.5 kcal mol^{-1})

the overall (Hx is:
 A) +144 kcal mol^{-1}
 B) -23 kcal mol^{-1}
 C) -41.5 kcal mol^{-1}
 D) +12.5 kcal mol^{-1}
* E) -12.5 kcal mol^{-1}

4. The p-orbital of a methyl radical carbon, CH_3y, contains how many electrons?
* A) 1 B) 2 C) 3 D) 4 E) 0

5. For the following reaction,

CH_3CH_3 + Cl_2 DDDDDD4 CH_3CH_2Cl + HCl

(DHx) (98 kcal mol^{-1}) (58 kcal mol^{-1}) (81.5 kcal mol^{-1}) (103 kcal mol^{-1})

the overall (Hx is:

* A) -28.5 kcal mol^{-1}
  B) +28.5 kcal mol^{-1}
  C) +58 kcal mol^{-1}
  D) +156 kcal mol^{-1}
  E) -184.5 kcal mol^{-1}

6. For the following reaction,

$(CH_3)_2CHDH$ + FDF DDDDDD4 $(CH_3)_2CHDF$ + HDF

(DHxM94.5 kcal mol^{-1}) (DHxM38 kcal mol^{-1}) (DHxM105 kcal mol^{-1}) (DHxM136 kcal mol^{-1})

the overall (Hx is:

  A) +108.5 kcal mol^{-1}
* B) -108.5 kcal mol^{-1}
  C) -241 kcal mol^{-1}
  D) +132.5 kcal mol^{-1}
  E) -373.5 kcal mol^{-1}

7. Carbocations are NOT intermediates in which one of the following reactions?

A) CH_3CHMCH_2 + HBr DDDDDD4 CH_3CHCH_3
3
Br

B) CH_3CHCH_3 + H_2SO_4 DDDDDD4 CH_3CHMCH_2 + H_2O
3
OH

C) CH_3CHMCH_2 + $Hg(OAc)_2$ $\xrightarrow{H_2O}$ DDDDDD4 CH_3CHCH_2HgOAc + CH_3COOH
3
OH

D) CH_3CHMCH_2 + H_2SO_4 DDDDDD4 CH_3CHCH_3
3
OSO_3H

* E) CH_3CHMCH_2 + HBr $\xrightarrow{RDODODR}$ DDDDDDD4 $CH_3CH_2CH_2Br$

8. The hybridization state of the carbon of a methyl radical is:

A) sp * B) sp^2 C) sp^3 D) sp^4 E) p^3

9. Which of the following statements is true when used to compare the reaction of chlorine with isobutane and the reaction of bromine with isobutane?
* A) Bromine is the less reactive and the more selective.
  B) Chlorine is the less reactive and the more selective.
  C) Chlorine is the more reactive and the more selective.
  D) Bromine is the more reactive and the more selective.
  E) None of the above

10. Which reaction would you expect to have the smallest energy of activation?

(Hx(kcal mol^{-1})

* A) CH_3y + CH_3y DDDDDD4 CH_3CH_3 -88
  B) CH_4 + Fy DDDDDD4 CH_3y + HF -32
  C) CH_4 + Iy DDDDDD4 CH_3y + HI +33
  D) CH_4 + Bry DDDDDD4 CH_3y + HBr +16.5
  E) CH_4 + Cly DDDDDD4 CH_3y + HCl +1

11. Which of the following reactions would have an activation energy equal to zero?
  A) CH_3DCH_3 DDDDDD4 $2CH_3$y
  B) Hy + CH_3CH_3 DDDDDD4 CH_3CH_3 + Hy
* C) $2CH_3CH_2$y DDDDDD4 $CH_3CH_2CH_2CH_3$
  D) CH_3y + CH_3CH_3 DDDDDD4 CH_3CH_3 + CH_3y
  E) None of the above

12. Which of the following reactions would have the smallest energy of activation?
  A) CH_4 + Bry DDDDDD4 CH_3y + HBr
  B) CH_3CH_3 + Bry DDDDDD4 CH_3CH_2y + HBr
  C) CH_3CHCH_3 (3 CH_3) + Bry DDDDDD4 CH_3CHCH_2y (3 CH_3) + HBr
* D) CH_3CHCH_3 (3 CH_3) + Bry DDDDDD4 CH_3CCH_3 (y above; 3 CH_3 below) + HBr
  E) CH_3DCDCH_3 (CH_3 3 above; 3 CH_3 below) + Bry DDDDDD4 CH_3DCDCH_2y (CH_3 3 above; 3 CH_3 below) + HBr

13. An example of a reaction having an E_{act} = 0 would be:

A) Bry + BrDBr DDDDDD4 BrDBr + Bry

B) Fy + CH_4 DDDDDD4 HDF + CH_3y

C) CH_3y + CH_3CH_3 DDDDDD4 CH_4 + CH_3CH_2y

D) Bry + HDBr DDDDDD4 HDBr + Bry

* E) CH_3y + CH_3y DDDDDD4 CH_3DCH_3

14. Which of the following reactions should have the smallest energy of activation?

A) CH_4 + Cly DDDDDD4 CH_3y + HCl

B) CH_3CH_3 + Cly DDDDDD4 CH_3CH_2y + HCl

C) CH_3CHCH_3 + Cly DDDDDD4 CH_3CHCH_2y + HCl
3 3
CH_3 CH_3

* D)
y
CH_3CHCH_3 + Cly DDDDDD4 CH_3CCH_3 + HCl
3 3
CH_3 CH_3

E) CH_3 CH_3
3 3
CH_3DCDCH_3 + Cly DDDDDD4 CH_3DCDCH_2y + HCl
3 3
CH_3 CH_3

15. Which reaction would you expect to have the largest energy of activation?

	(Hx (kcal mol^{-1})
A) CH_3y + CH_3y DDDDDD4 CH_3CH_3	-88
B) CH_3y + Bry DDDDDD4 CH_3Br	-70
* C) CH_4 + Iy DDDDDD4 CH_3y + HI	+33
D) CH_4 + Bry DDDDDD4 CH_3y + HBr	+16.5
E) CH_4 + Cly DDDDDD4 CH_3y + HCl	+1

16. An example of a reaction having an E_{act} = 0 would be:
  A) Bry + BrDBr DDDDDD4 BrDBr + Bry

  B) Fy + CH_4 DDDDDD4 HDF + CH_3y

* C) CH_3y + Cly DDDDDD4 CH_3Cl

  D) More than one of these

  E) None of these

17. Which of the following reactions would have an activation energy equal to zero?
  A) HDH DDDDDD4 2Hy

  B) Hy + CH_3DH DDDDDD4 CH_3DH + Hy

* C) CH_3y + CH_3y DDDDDD4 CH_3DCH_3

  D) CH_3y + CH_3DH DDDDDD4 CH_3DH + CH_3y

  E) All of the above

18. Which of the reactions listed below would have a value of (Hx equal to zero?
  A) HDH DDDDDD4 2Hy

  B) Hy + CH_3DH DDDDDD4 CH_3DH + Hy

  C) CH_3y + CH_3y DDDDDD4 CH_3DCH_3

  D) CH_3y + CH_3DH DDDDDD4 CH_3DH + CH_3y

* E) Reactions (B) and (D)

19. Which of the reactions listed below would be exothermic?
  A) CH_3DCH_3 DDDDDD4 2CH_3y

  B) CH_3y + CH_4 DDDDDD4 CH_4 + CH_3y

* C) 2$(CH_3)_2$CHy DDDDDD4 $(CH_3)_2$CHDCH$(CH_3)_2$

  D) Hy + $(CH_3)_3$CH DDDDDD4 $(CH_3)_3$CH + Hy

  E) None of the above

20. Which of the reactions listed below would be exothermic?

A) HDH DDDDDD4 2Hy

B) Hy + CH_3DH DDDDDD4 CH_3DH + Hy

* C) CH_3y + CH_3y DDDDDD4 CH_3DCH_3

D) CH_3y + CH_3DH DDDDDD4 CH_3DH + CH_3y

E) All of the above

21. What is the product of the reaction

peroxides
$CH_3CH_2CH{=}CH_2$ + CBr_4 DDDDDDDDDDD4 ?

A) $CH_3CH_2CH{=}CHCBr_3$

* B) $CH_3CH_2CHCH_2CBr_3$
3
Br

C) $CH_3CH_2CHCH_2Br$
3
CBr_3

D) $CH_3CH_2CH_2CH_2CBr_3$

E) No reaction occurs.

22. Which of the following free radicals is the most stable?

A) CH_2y
3
$CH_3CHCH_2CH_3$

B) CH_3
3
$CH_3CHCH_2CH_2$y

C) CH_3
3
y$CH_2CHCH_2CH_3$

D) CH_3
3
$CH_3CHCHCH_3$
y

* E) CH_3
3
$CH_3CCH_2CH_3$
y

23. Select the structure of the major product formed in the following reaction.

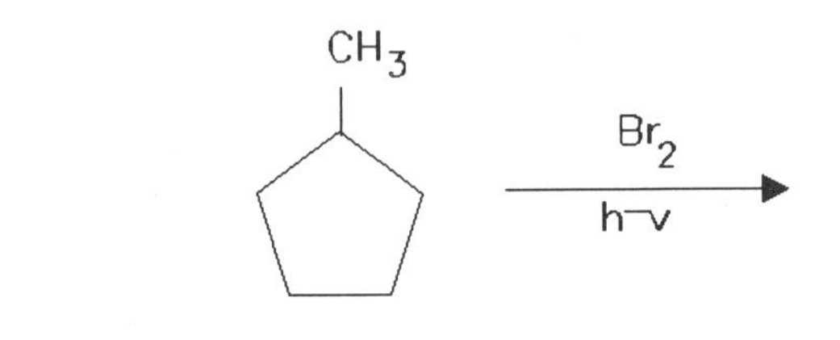

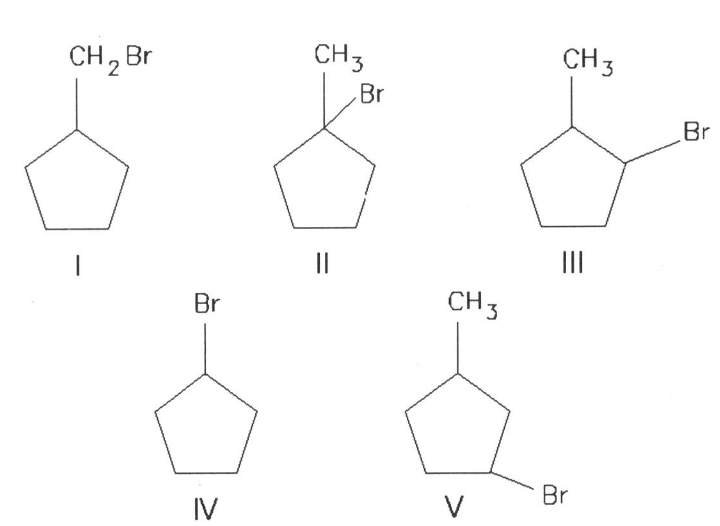

A) I   * B) II   C) III   D) IV   E) V

24. Which free radical is most stable relative to the hydrocarbon from which it is formed?

A) $CH_3CH_2CH_2y$
B) $(CH_3)_2CHy$
C) $(CH_3)_3Cy$
D) CH_2MCHy
* E) $\overset{y}{CH_2MCHCHCH_3}$

25. An alternate mechanism for the chlorination of methane is:

Cl_2 DDDDDD4 2Cly	ClDCl, DHx = 58 kcal mol^{-1}
Cly + CH_4 DDDDDD4 CH_3Cl + Hy	CH_3DH, DHx = 104 kcal mol^{-1}
Hy + Cl_2 DDDDDD4 HCl + Cly	CH_3DCl, DHx = 83.5 kcal mol^{-1}
	HDCl, DHx = 103 kcal mol^{-1}

This mechanism is unlikely because:

A) The overall (Hx is highly endothermic.
B) The probability factor is low.
C) One of the chain propagating steps is non-productive.
* D) One of the chain propagating steps has a very high E_{act}.
E) One of the chain propagating steps is highly exothermic.

26. Given the following bond dissociation energies:

	DHx (kcal mol^{-1})
CH_3CH_2DH	98
HDF	136
HDCl	103
HDBr	87.5
HDI	71

predict which of the following reactions would have the highest energy of activation.

A) CH_3CH_3 + Fy DDDDDD4 CH_3CH_2y + HF

B) CH_3CH_3 + Cly DDDDDD4 CH_3CH_2y + HCl

C) CH_3CH_3 + Bry DDDDDD4 CH_3CH_2y + HBr

* D) CH_3CH_3 + Iy DDDDDD4 CH_3CH_2y + HI

27. What would be the major product of the following reaction?

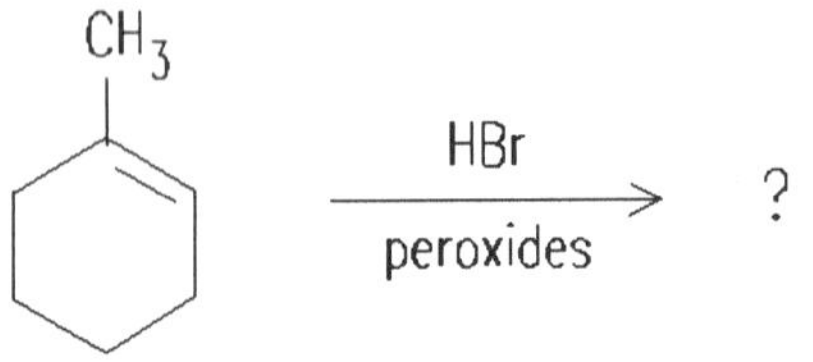

CH_3

HBr

peroxides

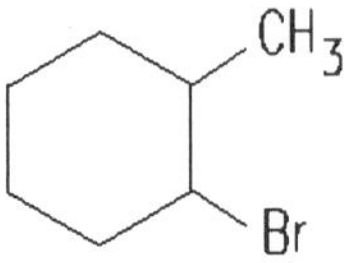

?

CH_3 Br CH_3 Br CH_3 OR

I 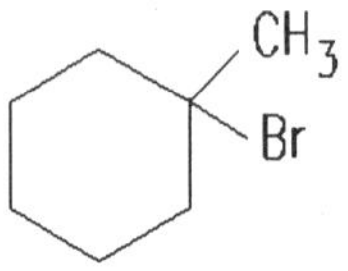 II 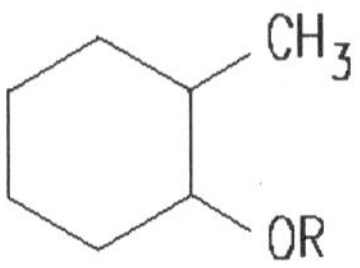 III

CH_3 Br Br CH_3

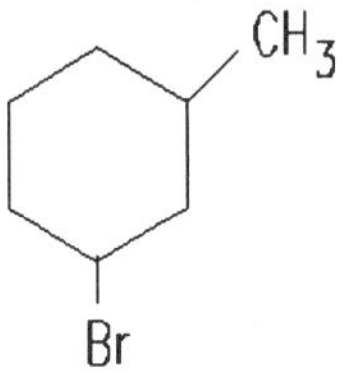

IV 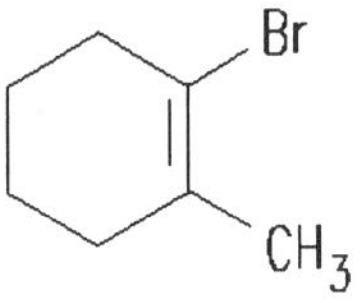 V

* A) I B) II C) III D) IV E) V

28. Given the following bond dissociation energies:

	DHx(kcal mol^{-1})
CH_3DH	104
HDF	136
HDCl	103
HDBr	87.5
HDI	71

predict which of the following reactions would have the highest energy of activation.

A) CH_4 + Fy DDDDDD4 CH_3y + HF

B) CH_4 + Cly DDDDDD4 CH_3y + HCl

C) CH_4 + Bry DDDDDD4 CH_3y + HBr

* D) CH_4 + Iy DDDDDD4 CH_3y + HI

29. In the presence of light at 25xC, isobutane (1 mol) and bromine (1 mol) yield which monobromo product(s)?

A) 2-Methyl-1-bromopropane (almost exclusively)
* B) 2-Methyl-2-bromopropane (almost exclusively)
C) A mixture of 50% (A) and 50% (B)
D) A mixture of 90% (A) and 10% (B)
E) Butyl bromide

30. Which statement(s) is(are) true about the reaction of bromine with isobutane?

A) Bromine selectively abstracts the tertiary hydrogen.
B) The transition state of the rate determining step is product-like.
C) The major product formed from this reaction is 1-bromo-2-methylpropane.
* D) A) and B)
E) A), B) and C)

31. For which reaction would the transition state be most product-like?

* A) CH_4 + Bry DDDDDD4 CH_3y + HBr

B) CH_3CH_3 + Bry DDDDDD4 CH_3CH_2y + HBr

C) CH_3CHCH_3 + Bry DDDDDD4 CH_3CHCH_2y + HBr
3 3
CH_3 CH_3

D)
y
CH_3CHCH_3 + Bry DDDDDD4 CH_3CCH_3 + HBr
3 3
CH_3 CH_3

E) CH_3 CH_3
3 3
CH_3CCH_3 + Bry DDDDDD4 CH_3CCH_2y + HBr
3 3
CH_3 CH_3

32. For which of the following reactions would the transition state most resemble the products? The following bond dissociation energies may be useful.

$(CH_3)_2CHDH$ (94.5 kcal mol^{-1}), $CH_3CH_2CH_2DH$ (98 kcal mol^{-1}),

HDF (136 kcal mol^{-1}), HDCl (103 kcal mol^{-1}), HDBr (87.5 kcal mol^{-1})

A)
y
$CH_3CH_2CH_3$ + Fy DDDDDD4 CH_3CHCH_3 + HF

B) $CH_3CH_2CH_3$ + Fy DDDDDD4 $CH_3CH_2CH_2$y + HF

C)
y
$CH_3CH_2CH_3$ + Cly DDDDDD4 CH_3CHCH_3 + HCl

D) $CH_3CH_2CH_3$ + Cly DDDDDD4 $CH_3CH_2CH_2$y + HCl

* E) $CH_3CH_2CH_3$ + Bry DDDDDD4 $CH_3CH_2CH_2$y + HBr

33. In the presence of light, ethane (1 mol) reacts with chlorine (1 mol) to form which product(s)?

A) $CH_2ClCHCl_2$
B) CH_3CHCl_2
C) CH_3CH_2Cl
D) $ClCH_2CH_2Cl$
* E) All of these

34. How many different monochlorobutanes (including stereoisomers) are formed in the free radical chlorination of butane?

A) 1   B) 2   * C) 3   D) 4   E) 5

35. The reaction of 2-methylbutane with chlorine would yield how many monochloro derivatives (include stereoisomers)?

A) 2   B) 3   C) 4   D) 5   * E) 6

36. The reaction of 2,2-dimethylbutane with chlorine would yield how many monochloro derivatives (include stereoisomers)?

A) 1   B) 2   C) 3   * D) 4   E) 5

37. The free radical chlorination of pentane produces this number of monochloro compounds, including stereoisomers.

A) 2   B) 3   * C) 4   D) 5   E) 6

38. Which of the following gas-phase reactions is a possible chain-terminating step in the light-initiated chlorination of methane?

A) ClDCl DDDDDD4 2Cly

B) Cly + CH_4 DDDDDD4 CH_3y + HDCl

* C) CH_3y + CH_3y DDDDDD4 CH_3DCH_3

D) CH_3y + ClDCl DDDDDD4 CH_3Cl

E) More than one of the above

39. The reaction of 1-butene with HBr in the presence of peroxides yields 1-bromobutane. The mechanism for the reaction involves:

A) attack on the alkene by a Br^+ ion.
B) attack on the alkene by a H^+ ion.
* C) attack on the alkene by a bromine atom, Bry.
D) attack on the alkene by a hydrogen atom, Hy.
E) isomerization of the 2-bromobutane produced initially.

40. For which of the following gas-phase reactions would the E_{act} be equal to (Hx?

* A) ClDCl DDDDDD4 2Cly

B) 2 Cly DDDDDD4 ClDCl

C) Cly + CH_4 DDDDDD4 CH_3y + HDCl

D) CH_3y + CH_3y DDDDDD4 CH_3DCH_3

E) CH_3y + ClDCl DDDDDD4 CH_3DCl + Cly

41. Free radicals can be produced by:

A) use of high temperatures.
B) irradiation with light.
C) reaction of a molecule with another free radical.
D) both A) and B).
* E) all of A), B) and C).

42. Free radical chlorination will produce but one monochloro derivative in the case of:

A) Propane.
B) Butane.
C) Isobutane.
D) Isopentane.
* E) Neopentane.

43. Free radical chlorination of hexane produces this number of monochloro derivatives (including stereoisomers):

A) 3 B) 4 * C) 5 D) 7 E) 8

44. More than one monochloro compound can be obtained from the free radical chlorination of:

A) Cyclopentane
B) Neopentane
* C) Isobutane
D) Ethane
E) Methane

45. The DHx value is expected to be least for which indicated C-H bond of isopentane?

A) $HDCH_2CHCH_2CH_3$
3
CH_3

B) $CH_3CHCH_2CH_3$
3
CH_2DH

* C) H
3
$CH_3CCH_2CH_3$
3
CH_3

D) H
3
$CH_3CHCHCH_3$
3
CH_3

E) $CH_3CHCH_2CH_2DH$
3
CH_3

46. If chlorocyclopentane were chlorinated to form all possible dichloro compounds and the product mixture subjected to precise fractional distillation, how many fractions would be obtained (ideally)?

A) 3 B) 4 * C) 5 D) 7 E) 9

47. The free radical chlorination of (R)-2-chloropentane forms a mixture of dichloropentanes which includes:

A) three optically active compounds.
* B) two achiral compounds.
C) two meso compounds.
D) one pair of diastereomers.
E) one racemic mixture.

48. Which is true for a chain-terminating step?

A) A new free radical is formed.
B) The process is endothermic.
* C) $E_{act} = 0$.
D) (Hx is positive.
E) A product is formed which is immune to further reaction.

49. The reaction of Cl_2 with a methyl radical has a positive (Hx. Which of these drawings is the best representation of the transition state of this reaction?

A)
```
   H   H
ky  \ /  ky
Cl···C···Cl
     3
     H
```

B)
```
        H
       /
ClDClDCDH
       \
        H
```

C)
```
                  H
ky             ky/
Cl··Cl·······CDH
               \
                H
```

D)
```
                H
               /
Cly   +   ClDCDH
               \
                H
```

* E)
```
                        H
ky                   ky/
Cl···········Cl···CDH
                    \
                     H
```

50. Which of these molecules is not expected to arise as a product of the high temperature chlorination of methane?

A) CCl_4 B) $HCCl_3$ C) CH_2Cl_2 D) CH_3CH_3 * E) CH_2MCH_2

51. In a competition reaction, equimolar amounts of five alkanes compete for a deficiency of chlorine at 300xC. The greatest amount of reaction would occur in the case of which of these alkanes?

A) Ethane B) Propane C) Butane * D) Isobutane E) Pentane

52. At some temperature, the relative reactivities of 3x, 2x and 1x alkane hydrogens in free radical chlorination are in the ratio of 5:3:1. Thus, monochlorination of isopentane should produce these percentages of 2-chloro-2-methylbutane (A), combined 1-chloro-2-methylbutane and 1-chloro-3-methylbutane (B), and 2-chloro-3-methylbutane (C):

A) 8% A, 75% B, 17% C
* B) 25% A, 45% B, 30% C
C) 29% A, 53% B, 18% C
D) 30% A, 35% B, 35% C
E) 36% A, 43% B, 21% C

53. The free radical chlorination of 3-chloropentane forms a mixture of dichloropentanes which, on precise fractional distillation, affords these fractions:

* A) 4 fractions, none optically active
B) 4 fractions, 2 optically active
C) 7 fractions, 4 optically active
D) 7 fractions, 6 optically active
E) 7 fractions, all optically active

54. Consider the light-initiated chlorination of (S)-2-chlorobutane followed by careful fractional distillation (or separation by GLC) of all of the products with the formula $C_4H_8Cl_2$. How many fractions (in total) would be obtained and how many of these fractions would be optically active?

A) Three fractions total; all optically active
B) Four fractions total; three fractions optically active
C) Five fractions total; all optically active
D) Five fractions total; four fractions optically active
* E) Five fractions total; three fractions optically active

55. Which of the following combinations of reactants can provide a demonstrable example of anti-Markovnikov addition?

A) $CH_2{=}CHCH_3$ + HCl + ROOR
B) $CH_3CH{=}CH_2$ + H_2O + Cl_2
C) $CH_3CH{=}CHCH_3$ + HBr + ROOR
* D) $CH_3CH_2CH{=}CH_2$ + HBr + ROOR
E) $CH_3CH_2CH{=}CH_2$ + Br_2 + ROOR

56. 2-Methyl-2-butene reacts with HBr in the presence of peroxide to give (chiefly):

A) $(CH_3)_2CHCH_2CH_2Br$

* B) $(CH_3)_2CHCHCH_3$
3
Br

C) $(CH_3)_2CCH_2CH_3$
3
Br

D) CH_3
3
$BrCH_2CHCH_2CH_3$

E) Br
3
$(CH_3)_2CCHCH_3$
3
Br

57. What product would result from the following reaction?

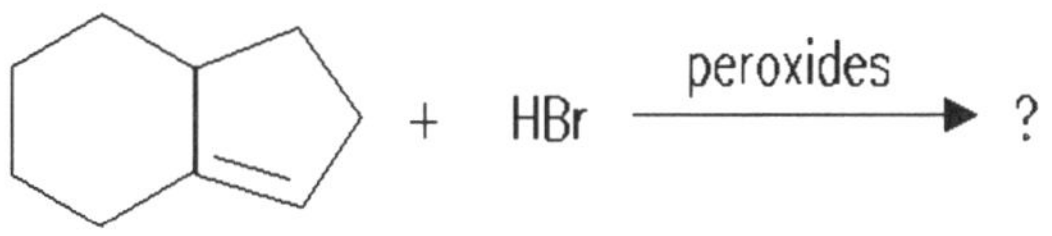

+ HBr —peroxides→ ?

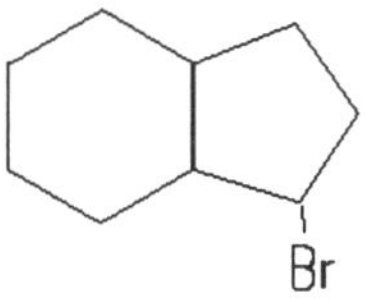

I 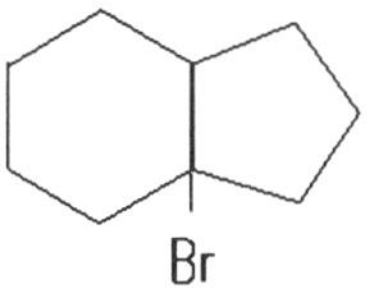 II 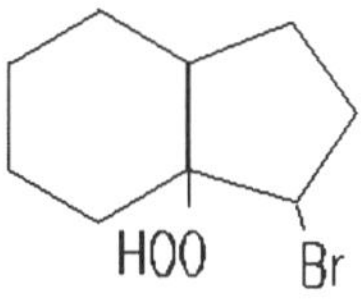 III 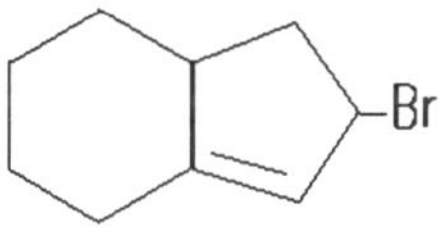 IV 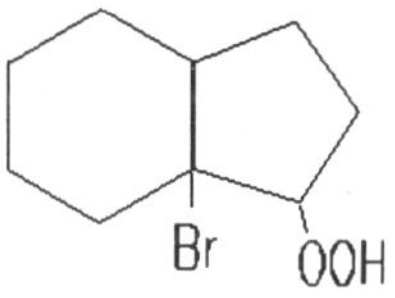 V

* A) I B) II C) III D) IV E) V

58. How many monochloro derivatives, including stereoisomers, can be formed in the chlorination of 1-bromobutane?

A) 4 B) 5 C) 6 * D) 7 E) 8

59. What is the total number of trichloropropanes which can be produced by free radical chlorination of propane. Include all stereoisomers.

A) 4 B) 5 * C) 6 D) 7 E) 8

60. According to the present explanation of the role of atmospheic chlorofluorocarbons in ozone depletion, it is this species which destroys, i.e., reacts irreversibly with, ozone.

A) Fy * B) Cly C) Oy D) ClOy E) FOy

61. If propene polymerization is initiated by the use of diacyl peroxide, this is an intermediate species formed early in the process.

* A)
```
     CH3  CH3
      3    3
RCH2CHCH2CHy
```
B)
```
CH3  CH3
 3    3
RCHCH2CHCH2y
```
C)
```
CH3      CH3
 3        3
RCHCH2CH2CHy
```
D)
```
     CH3
      3
RCH2CHCHCH2y
        3
        CH3
```
E)
```
 CH3 CH3
  3   3
RCCH2CHy
 3
 CH3
```

62. As the term "peroxide" is used in Chapter 9, it can refer to which structure(s)?

A) ROOR
B) ROOH
C)
```
 O  O
 :  :
RCOOCR
```
D) Answers A) and B) only
* E) Answers A), B) and C)

63. When an alkane in which all hydrogen atoms are not equivalent is monosubstituted, use of this halogen produces a ratio of isomers which is essentially statistical, i.e., dependent only on the number of each type of hydrogen.

* A) F_2　B) Cl_2　C) Br_2　D) I_2

64. How could the following synthesis be accomplished?

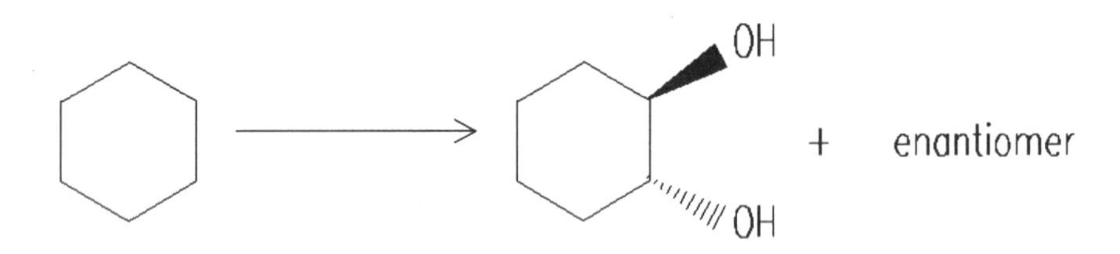

* A) (1) Cl_2/h0; (2) t-BuOK; (3) peroxy acid; (4) H_3O^+
B) (1) t-BuOK; (2) Cl_2/h0; (3) peroxy acid; (4) H_3O^+
C) (1) H_3O^+; (2) t-BuOK; (3) peroxy acid; (4) H_2O
D) (1) Cl_2/h0; (2) peroxy acid; (3) t-BuOK; (4) H_3O^+
E) (1) Cl_2/h0; (2) H_3O^+; (3) t-BuOK; (4) peroxy acid

65. Which of the following would serve as the best synthesis of 2-bromohexane?

A) $CH_2=CHCH_2CH_2CH_2CH_3$ + HBr $\xrightarrow{\text{peroxides, heat}}$

* B) $CH_2=CHCH_2CH_2CH_2CH_3$ + HBr $\xrightarrow{\text{heat}}$

C) $CH_3CH=CHCH_2CH_2CH_3$ + HBr $\xrightarrow{\text{heat}}$

D) $CH_3CH=CHCH_2CH_2CH_3$ + HBr $\xrightarrow[\text{heat}]{\text{peroxides}}$

E) All of the above would be equally suitable.

66. Which would be the best way to carry out the following synthesis?

$$(CH_3)_3COH \xrightarrow{?} (CH_3)_2CHCH_2Br$$

A) (1) H^+, heat; (2) HBr
B) (1) HBr and peroxides; (2) Br_2/CCl_4
* C) (1) H^+, heat; (2) HBr and peroxides
D) (1) Br_2/CCl_4; (2) H^+, heat
E) (1) H^+, heat; (2) Br_2/CCl_4

67. What sequence of reactions could be used to prepare the compound below from cyclopentane?

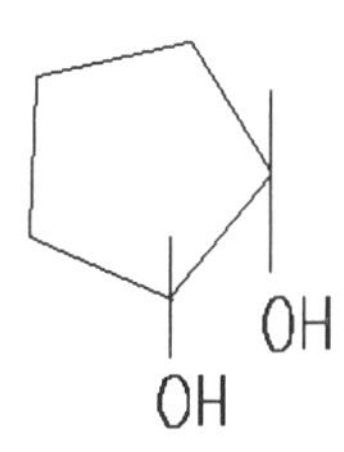

* A) (1) Cl_2, hν; (2) t-BuOK/t-BuOH; (3) OsO_4; (4) $NaHSO_3/H_2O$
B) (1) t-BuOK/t-BuOH; (2) Cl_2, hν; (3) $NaOH/H_2O$
C) (1) Cl_2, hν; (2) t-BuOK/t-BuOH; (3) H_2O_2
D) (1) $NaOH/H_2O$; (2) Br_2; (3) $NaNH_2$(2eq.)/liq.NH_3; (4) $KMnO_4$, $NaOH/H_2O$, 5°C
E) (1) Cl_2, hν; (2) t-BuOK/t-BuOH; (3) $RCOO_2H$; (4) H_3O^+

Chapter 10: Alcohols and Ethers

1. A correct IUPAC name for

OH
3
$CH_3CHCH_2CHCH_3$
3
CH_2
3
CH_3

is:

A) 2-Ethyl-4-pentanol
B) 4-Ethyl-2-pentanol
C) 3-Methyl-5-pentanol
D) 5-Methyl-2-hexanol
* E) 4-Methyl-2-hexanol

2. A correct IUPAC name for isobutyl alcohol is:
* A) 2-Methyl-1-propanol
B) 2-Methyl-1-butanol
C) 1-Methyl-1-propanol
D) 1,1-Dimethyl-1-ethanol
E) 3-Methyl-1-propanol

3. The correct structure for benzyl alcohol is:

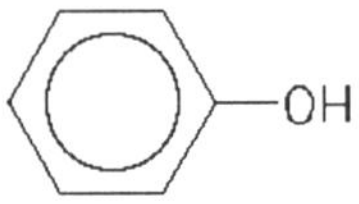

I: C_6H_5–OH
II: C_6H_5–CH_2OH
III: C_6H_5–OCH_3
IV: CH_3–C_6H_4–OH
V: CH_3–C_6H_4–CH_2OH

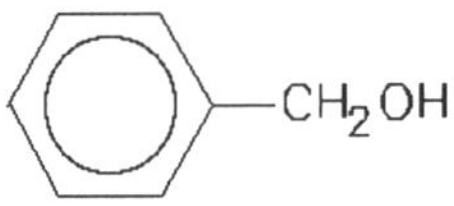

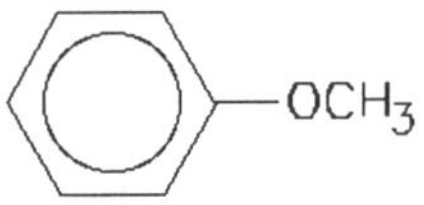

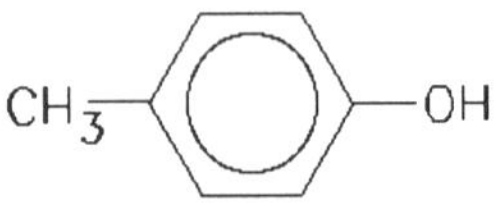

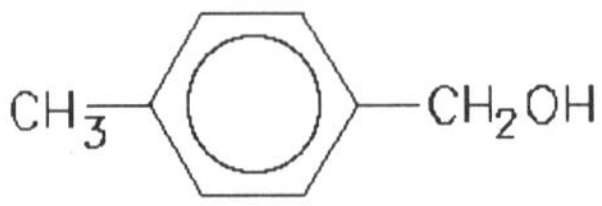

A) I
* B) II
C) III
D) IV
E) V

4. The correct IUPAC substitutive name for

CH_3
3
$CH_3CCH_2CHMCH_2$
3
OH

is:

A) 4-Penten-2-methyl-2-ol
B) 4-Methyl-1-penten-2-ol
* C) 2-Methyl-4-penten-2-ol
D) 4-Methyl-1-penten-4-ol
E) 4-Hydroxy-4-methyl-1-pentene

5. How many ethers with the formula $C_4H_{10}O$ are possible?
A) 1 B) 2 * C) 3 D) 4 E) 5

6. Which compound would have the highest boiling point?
A) $CH_3CH_2CH_2CH_3$
B) $CH_3CH_2OCH_3$
C) $CH_3CH_2CH_2OH$
D) CH_3CHCH_3 (3) OH
* E) $HOCH_2CH_2OH$

7. Which is a correct IUPAC name for $CH_3OCH_2CH_2OCH_3$?
A) 1,4-Dioxane
B) Ethylene glycol dimethyl ether
C) 1,4-Dioxapentalene
* D) 1,2-Dimethoxyethane
E) 1,2-Diethoxymethane

8. What is the electrophilic species involved in the initial step of the reaction below?

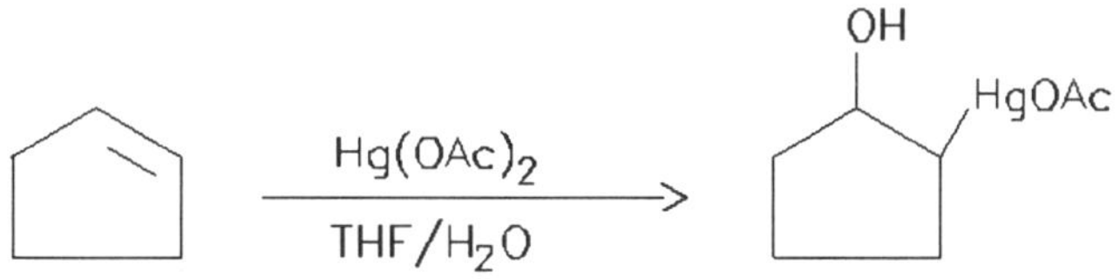

Cyclohexene $\xrightarrow[THF/H_2O]{Hg(OAc)_2}$ 2-(HgOAc)cyclohexanol (OH, HgOAc)

A) ^+OH
* B) ^+HgOAc
C) H_3O^+
D) THF
E) the THF/H_2O complex

9. Which of the following reactions would serve as a synthesis of butyl bromide?
A) $CH_3CH_2CH_2CH_2OH$ + HBr DDDDDD4 (reflux)

B) $CH_3CH_2CH_2CH_2OH$ + PBr_3 DDDDDD4

C) $CH_3CH_2CH_2CH_2OH$ + NaBr DDDDDD4 (reflux)

D) $CH_3CH_2CH_2CH_2OH$ + Br_2 DDDDDD4

* E) Answers A) and B) only

10. Which of the reagents listed below would serve as the basis for a simple chemical test to distinguish between

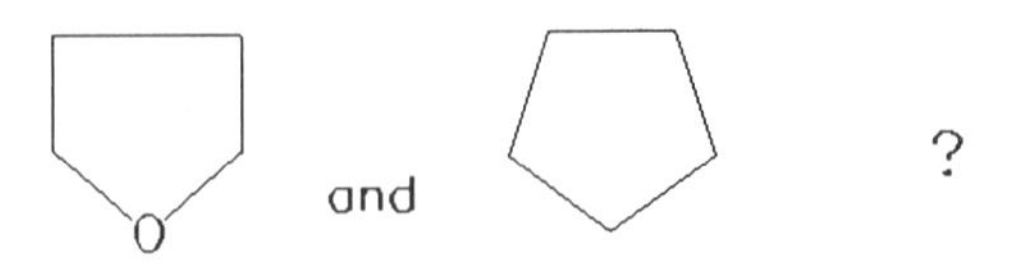

A) $AgNO_3$ in alcohol
B) NaOH in H_2O
C) Br_2 in CCl_4
* D) Cold concd. H_2SO_4
E) $KMnO_4$ in H_2O

11. Which of the following statements is NOT true of ethers?
A) Ethers are generally unreactive molecules toward reagents other than strong acids.
B) Ethers generally have lower boiling points than alcohols of a corresponding molecular weight.
* C) Ethers generally have much lower water solubilities than alcohols with a corresponding molecular weight.
D) Ethers can generally be cleaved by heating them with strong acids.
E) Ethers form peroxides when allowed to stand in the presence of oxygen.

12. The product(s) of the following reaction

CH_2—CH_2
CH_2 CH_2
O

$\xrightarrow[\text{heat}]{\text{excess HBr}}$ is/are:

$CH_3CH_2OCH_2CH_3$

I

$CH_3CH_2CH_2CH_2OH$ and $CH_3CH_2CH_2CH_2Br$

II

$BrCH_2CH_2CH_2CH_2OH$ and $BrCH_2CH_2CH_2CH_2Br$

III

CH_2—$CHBr$
CH_2 CH_2
O

IV

A) I
B) II
* C) III
D) IV
E) None of these

13. Which product(s) would you expect to obtain from the following sequence of reactions?

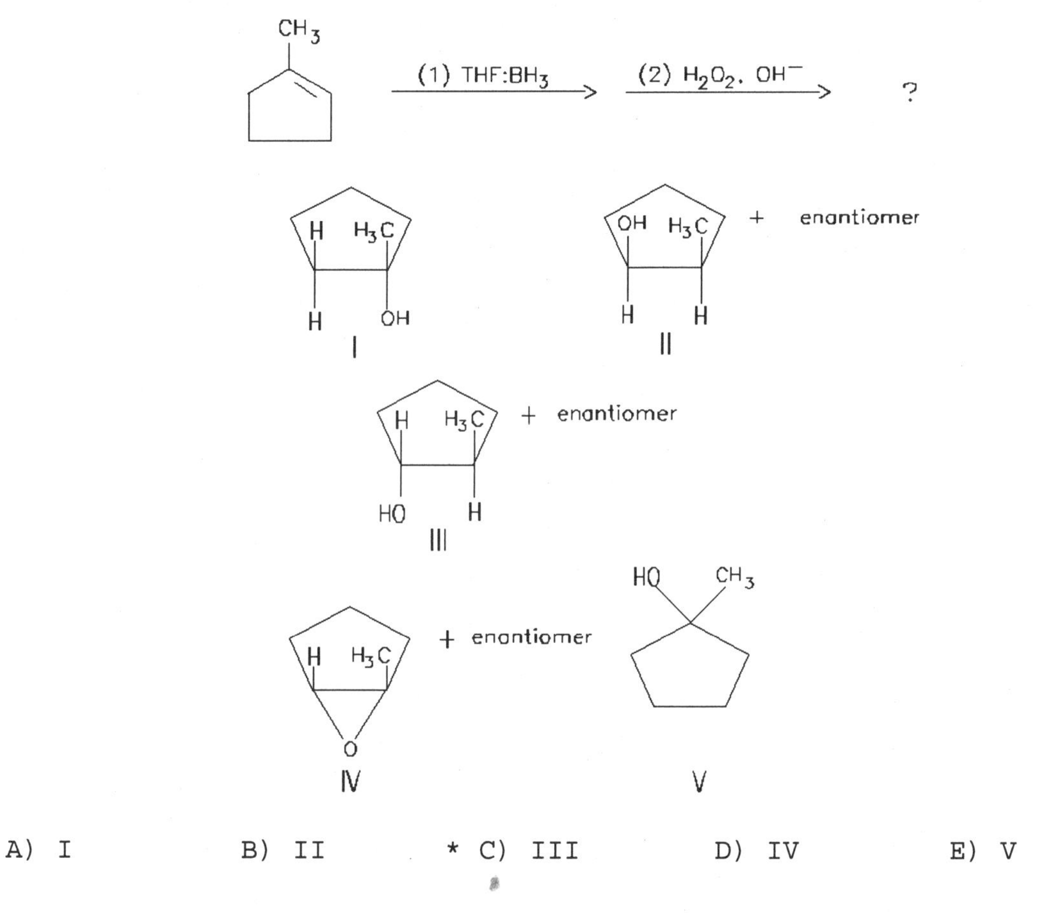

A) I B) II * C) III D) IV E) V

14. Which would be the best way to carry out the following synthesis?

$$CH_3CH_2CHCH_3 \quad \overset{?}{\text{DDDDDD4}} \quad CH_3CH_2CH_2CH_2OH$$

3
Br

A) (1) H^+, heat; (2) H_3O^+, H_2O, heat
* B) (1) $(CH_3)_3COK$ / $(CH_3)_3COH$; (2) THF:BH_3, then H_2O_2, OH^-
C) (1) $(CH_3)_3COK$ / $(CH_3)_3COH$; (2) H_3O^+, then H_2O, heat
D) (1) KOH, C_2H_5OH; (2) THF:BH_3, then H_2O_2, OH^-
E) (1) KOH, C_2H_5OH; (2) H^+, heat; (3) H_3O^+, H_2O, heat

15. Select the structure of the major product formed from the following reaction.

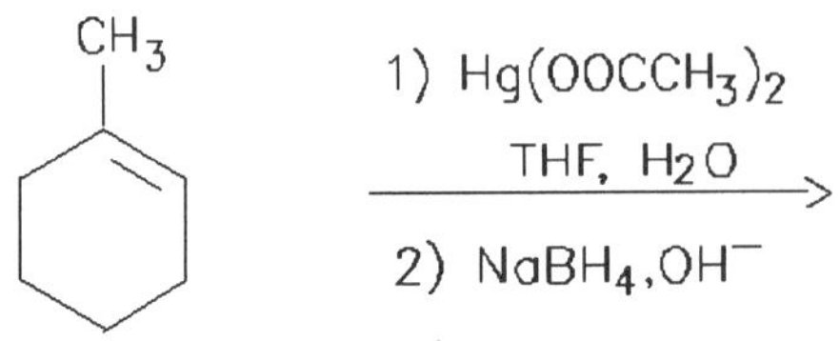
CH3
1) Hg(OOCCH3)2
THF, H2O
2) NaBH4,OH−

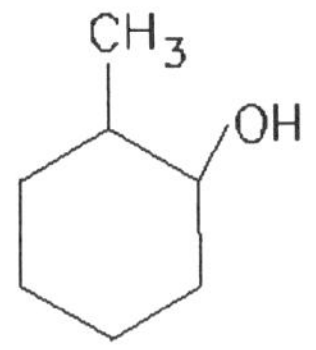
CH3
OH

I

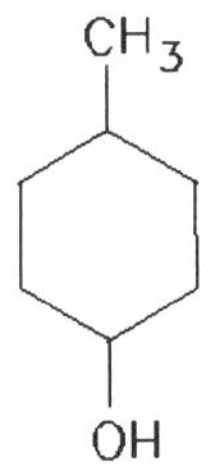
CH3
OH

II

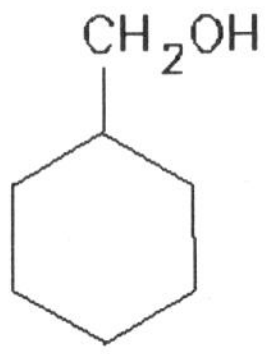
CH2OH

III

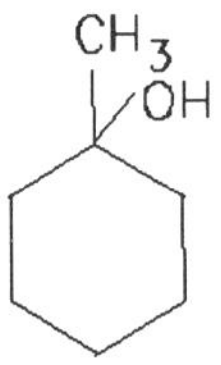
CH3
OH

IV

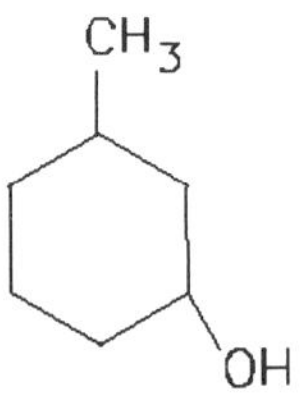
CH3
OH

V

A) I B) II C) III * D) IV E) V

16. What product would you expect from the following reaction?

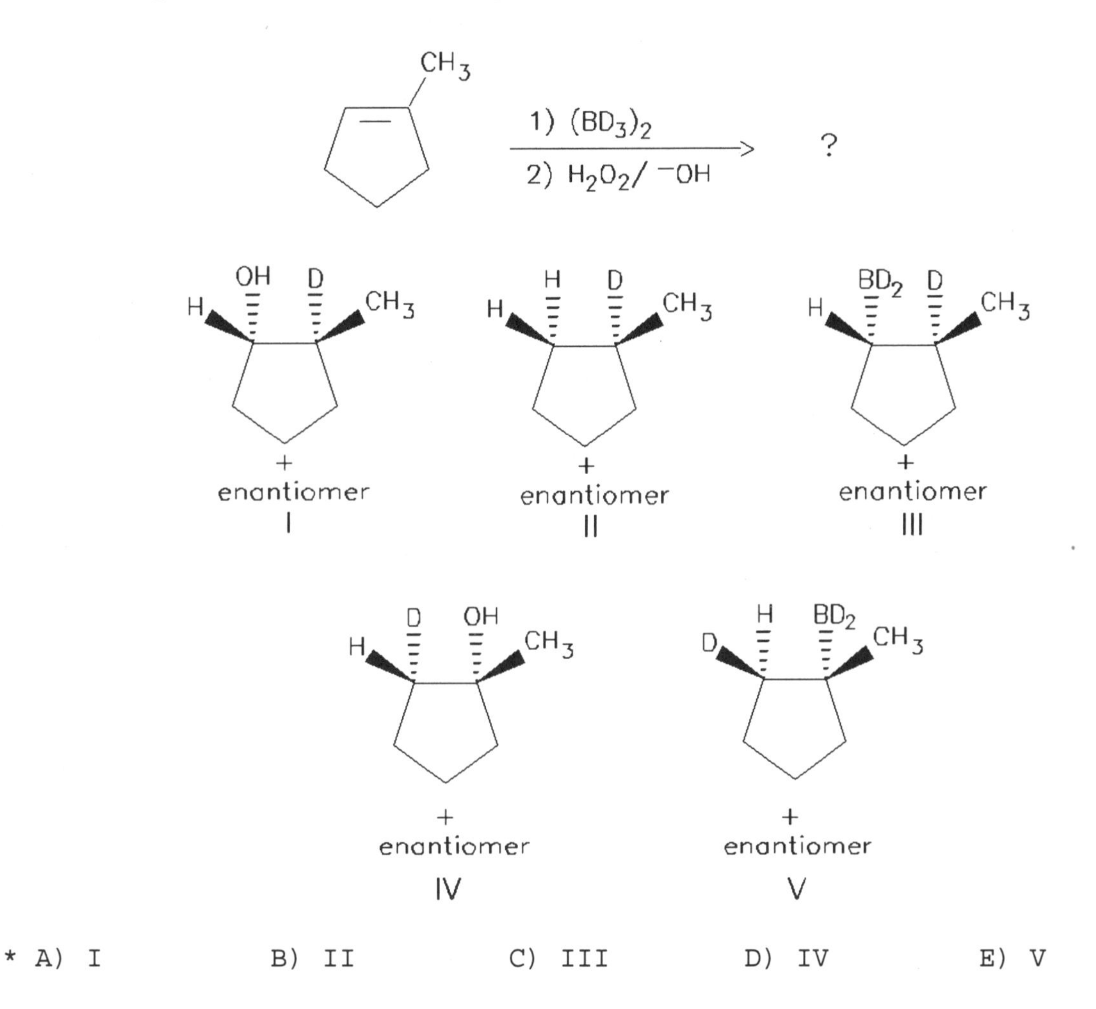

* A) I B) II C) III D) IV E) V

17. The following reaction,

$$2CH_3CH_2CH_2CH_2OH \xrightarrow[\text{heat}]{H_2SO_4} (CH_3CH_2CH_2CH_2)_2O + H_2O$$

(the arrow is printed as "DDDDDD4")

is probably:

A) an S_N1-type reaction involving the protonated alcohol as the substrate.

* B) an S_N2-type reaction involving the protonated alcohol as the substrate.

C) an E1-type reaction involving the protonated alcohol as the substrate.

D) an E2-type reaction involving the protonated alcohol as the substrate.

E) an epoxidation reaction.

18. Which of the following reagents might serve as the basis for a simple chemical test that would distinguish between pure 1-butene and $CH_3CH_2CH_2CH_2OH$?

A) Bromine in carbon tetrachloride
B) Dilute aqueous potassium permanganate
C) Chromic oxide in aqueous sulfuric acid
* D) All of these
E) None of these

19. Which of the reagents listed below would serve as the basis for a simple chemical test to distinguish between

(tetrahydrofuran structure) and (bromocyclopentane structure) ?

* A) $AgNO_3$ in C_2H_5OH
B) Dilute HCl
C) Br_2 in CCl_4
D) NaOH in H_2O
E) $KMnO_4$ in H_2O

20. Which compound is a tosylate?

I: $CH_3-C_6H_4-S(=O)_2-O-CH_3$

II: $Br-C_6H_4-S(=O)_2-O-CH_2-CH_3$

III: $CH_3-C_6H_4-S(=O)-O-CH_3$

IV: $CH_3-C_6H_4-S(=O)-CH_3$

V: $CH_3-C_6H_4-O-S(=O)_2-CH_3$

* A) I
B) II
C) III
D) IV
E) V

21. Which of the compounds listed below would you expect to have the highest boiling point? (They all have approximately the same molecular weight.)

A) $CH_3CH_2CH_2CH_2CH_3$
* B) $CH_3CH_2CH_2CH_2OH$
C) $CH_3CH_2CH_2OCH_3$
D) $CH_3CH_2CH_2Cl$
E) $CH_3CH_2OCH_2CH_3$

22. Select the structure of the major product formed in the following reaction.

```
                 H+
CH3CHDCH2     DDDDDD4
    \ /        H2 18O
     O
```

A) $CH_3CH_2CH_2{}^{18}OH$

B) CH_3CHCH_3
   3
   ^{18}OH

* C) CH_3CHCH_2OH
   3
   ^{18}OH

D) $CH_3CHDDCH_2$
   3 3
   HO ^{18}OH

E) $CH_3CHCH_2{}^{18}OH$
   3
   ^{18}OH

23. What would be the major product of the following reaction?

$$C_6H_5CH_2OCH_3 \quad \xrightarrow[\text{heat}]{\text{Concd. HBr}_{(xs)}} \quad ?$$

(arrow printed as DDDDDDDDDDDDDD4)

A) C_6H_5Br + CH_3OH
* B) $C_6H_5CH_2Br$ + CH_3Br
C) $C_6H_5CH_2OH$ + CH_3Br
D) $C_6H_5CH_2Br$ + CH_3OH
E) $C_6H_5CH_2CH_2Br$

24. Which is the best way to prepare isopropyl methyl ether via the Williamson method?

A) CH_3OH + $(CH_3)_2CHOH$ + H_2SO_4, 140xC
B) CH_3OH + $(CH_3)_2CHCH_2OH$ + H_2SO_4, 140xC
C) CH_3ONa + $(CH_3)_2CHBr$
* D) CH_3I + $(CH_3)_2CHONa$
E) CH_3I + $(CH_3)_2CHCH_2ONa$

25. Which is the best method to prepare <u>tert</u>-butyl methyl ether?

A) CH_3ONa + $(CH_3)_3CBr$
B) CH_3ONa + $(CH_3)_2CHCH_2Br$
* C) CH_3OH + H_2SO_4; then $(CH_3)_2CMCH_2$
D) CH_3OH + $(CH_3)_3CBr$
E) CH_3OH + $(CH_3)_3COH$ + H_2SO_4, 140xC

26. Which of the following could be used to synthesize $CH_3CH_2CMCH_2$?
3
Br

A) $CH_3CH_2CHMCH_2$ + Br_2 DDDDDD4 H_2O

B) $CH_3CH_2CHCH_3$ + PBr_3 DDDDDD4
3
OH

* C) CH_3CH_2CpCH + HBr DDDDDD4

D) CH_3CH_2CpCH + Br_2 DDDDDD4

E) CH_3CH_2CpCH + HBr, ROOR DDDDDD4

27. Which of the following could be used to synthesize 2-bromopropene?

A) CH_3CHMCH_2 + Br_2 DDDDDD4 H_2

B) CH_3CHCH_3 + PBr_3 DDDDDD4
3
OH

* C) CH_3CpCH + HBr DDDDDD4

D) CH_3CpCH + Br_2 DDDDDD4

E) CH_3CHMCH_2 + HBr DDDDDD4

28. Which of the alcohols listed below would you expect to react most rapidly with HBr?

A) $C_6H_5CH_2CH_2CH_2OH$

* B) $C_6H_5CHCH_2CH_3$
3
OH

C) $C_6H_5CH_2CHCH_3$
3
OH

D) $CH_3CH_2CH_2CH_2CH_2OH$

E) CH_3
3
CH_3CCH_2OH
3
CH_3

29. The following reaction,

$$CH_3CH_2CH_2CH_2OH \xrightarrow[\text{heat}]{\text{HBr}} CH_3CH_2CH_2CH_2Br + H_2O$$

is probably:

A) an S_N1-type reaction involving the protonated alcohol as the substrate.
* B) an S_N2-type reaction involving the protonated alcohol as the substrate.
C) an E1-type reaction involving the protonated alcohol as the substrate.
D) an E2-type reaction involving the protonated alcohol as the substrate.
E) an epoxidation reaction.

30. Which of the following could be used to synthesize 1-bromobutane?

A) $CH_3CH_2CH{=}CH_2$ + HBr $\longrightarrow$

B) $CH_3CH_2CH{=}CH_2$ + HBr $\xrightarrow{\text{peroxides}}$

C) $CH_3CH_2CH_2CH_2OH$ + HBr $\longrightarrow$

D) $CH_3CH_2CH_2CH_2OH$ + Br_2 $\longrightarrow$

* E) More than one of these

31. The IUPAC name of compound $CH_3CH_2C(OH)(CH_2CH_3)_2$ is:

A) 1,1,1-Triethylmethanol
B) 1,1-Diethyl-1-propanol
C) 2-Ethyl-3-pentanol
* D) 3-Ethyl-3-pentanol
E) tert-Heptanol

32. What is the relationship between alcohols I and II?

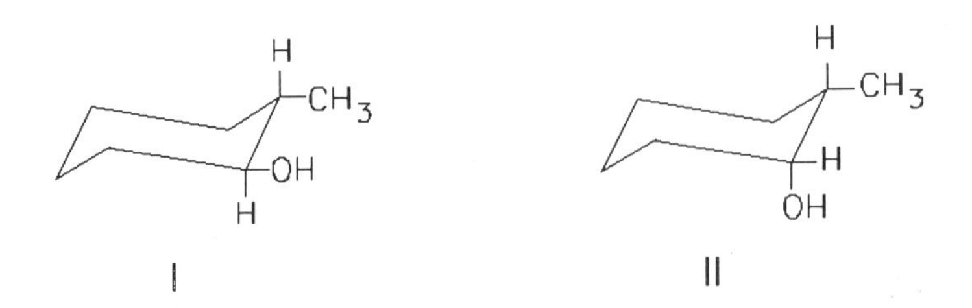

They are:

A) different conformations of the same compound.
B) constitutional isomers.
C) enantiomers.
* D) diastereomers.
E) identical.

33. Which of the following would be a reasonable synthesis of 2-butanol?

A) 1-Butene $\xrightarrow{H_3O^+,\ H_2O,\ heat}$

B) 1-Butene $\xrightarrow[(2)\ H_2O_2,\ OH^-]{(1)\ THF:BH_3}$

C) 1-Butene $\xrightarrow[(2)\ NaBH_4,\ OH^-]{(1)\ Hg(OAc)_2/THF\text{-}H_2O}$

* D) More than one of these
E) None of these

34. What would be the final product?

$(CH_3)_2C{=}CH_2 \xrightarrow{RCO_3H}$ product $\xrightarrow{CH_3OH,\ H^+}$ final product

(printed as: CH_3CMCH_2 with CH_3 above; RCOOH with O above)

A) $(CH_3)_2CHCH_2OCH_3$

B) $(CH_3)_2C(OCH_3)CH_3$

* C) $(CH_3)_2C(OCH_3)CH_2OH$

D) $(CH_3)_2C(OH)CH_2OCH_3$

E) $(CH_3)_2C(OCH_3)CH_2OCH_3$

35. Which of the following would be a reasonable synthesis of $CH_3CH_2CH_2CH_2OH$?

A) 1-Butene DDDDDDDDDDDDDDD4 (H_3O^+, H_2O, heat)

* B) 1-Butene DDDDDDDDDDDDDD4 ((1) $THF:BH_3$; (2) H_2O_2, OH^-)

C) 1-Butene DDDDDDDDDDDDDDDDDDDDD4 ((1) $Hg(OAc)_2/THF\text{-}H_2O$; (2) $NaBH_4$, OH^-)

D) More than one of these

E) None of these

36. Which would be the best method for converting A into B?

```
       CH3                       CH3
       3                         3
    CH3CCHMCH2   DDDDDD4   CH3CDDDCHCH3
       3                         3   3
       CH3                       CH3 OH

        A                          B
```

A) H_3O^+/H_2O, heat

B) $THF:BH_3$; then H_2O_2, OH^-

C) concd. H_2SO_4; then H_2O, heat

* D) $Hg(OAc)_2/THF\text{-}H_2O$; then $NaBH_4$, OH^-

E) HBr; then $NaOH/H_2O$

37. trans-3-Methylcyclopentanol is treated with CH_3SO_2Cl in the presence of base. The product of this reaction is then heated with KI in methanol. What is the final product?

A) trans-1-Iodo-3-methylcyclopentane

* B) cis-1-Iodo-3-methylcyclopentane

C) 1-Methylcyclopentene

D) 2-Methylcyclopentene

E) 3-Methylcyclopentene

38. Which method would provide the best synthesis of ethyl isopropyl ether?

* A) $(CH_3)_2CHONa$ + CH_3CH_2Br DDDDDD4

B) CH_3CH_2ONa + $(CH_3)_2CHBr$ DDDDDD4

C) CH_3CH_2OH + $(CH_3)_2CHOH$ DDDDDDDDDDDDD4 (H_2SO_4, 140xC)

D) CH_3CH_2OH + $(CH_3)_2CHOH$ DDDDDDDDDDDD4 (H_2SO_4, 180xC)

E) CH_3CH_2ONa + $(CH_3)_2CHOH$ DDDDDD4

39. What is the relationship between alcohols I and II?

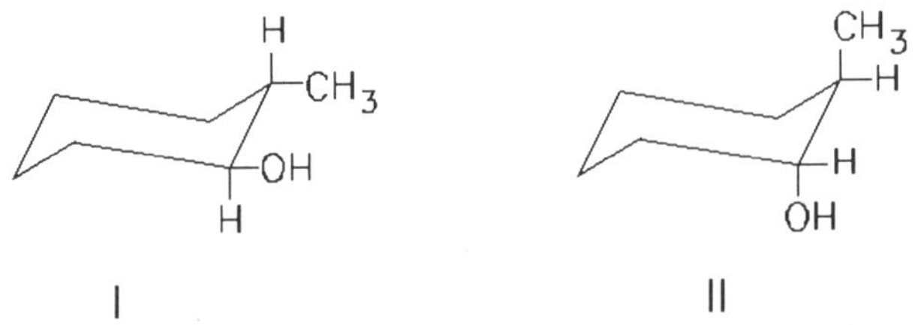

I: [cyclohexane chair; H and CH_3 on one carbon, OH and H on adjacent carbon]

II: [cyclohexane chair; CH_3 and H on one carbon, H and OH on adjacent carbon]

They are:
A) different conformations of the same compound.
B) constitutional isomers.
* C) enantiomers.
D) diastereomers.
E) identical.

40. Select the structure of benzyl methyl ether.

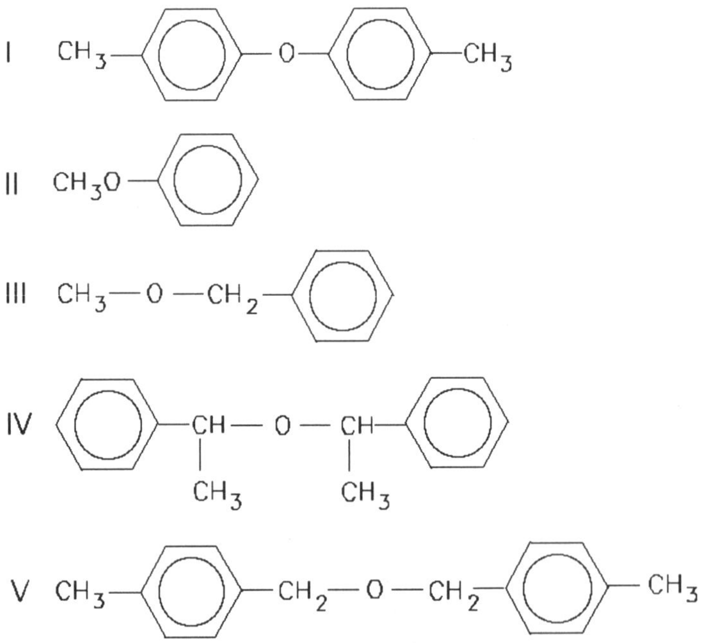

I $CH_3-C_6H_4-O-C_6H_4-CH_3$

II $CH_3O-C_6H_5$

III $CH_3-O-CH_2-C_6H_5$

IV $C_6H_5-CH(CH_3)-O-CH(CH_3)-C_6H_5$

V $CH_3-C_6H_4-CH_2-O-CH_2-C_6H_4-CH_3$

A) I B) II * C) III D) IV E) V

41. Which of the following could be used to synthesize 1-bromopentane?

A) $CH_3CH_2CH_2CHMCH_2$ + HBr DDDDDD4

* B) $CH_3CH_2CH_2CHMCH_2$ + HBr DDDDDDDDD4 peroxides

C) $CH_3CH_2CH_2CH_2CH_2OH$ + NaBr DDDDDD4

D) $CH_3CH_2CH_2CH_2CH_2OH$ + Br_2 DDDDDD4

E) $CH_3CH_2CH_2CHMCH_2$ + Br_2 DDDDDD4

42. Which compound (or compounds) would be produced when trans-2-butene is treated first with a peroxy acid to form an epoxide, and then the epoxide is subjected to acid-catalyzed hydrolysis?

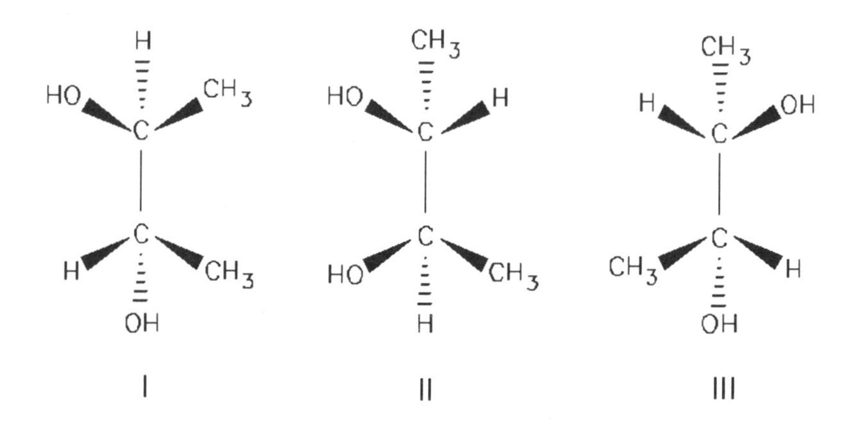

A) An equimolar mixture of I and II
B) An equimolar mixture of II and III
C) I alone
D) II alone
* E) III alone

43. Heating $CH_3CH_2OCH_2CH_2CH_3$ with excess concentrated HBr would produce:

A) $CH_3CH_2OCH_2CH_2CH_2Br$
B) $BrCH_2CH_2OCH_2CH_2CH_3$
C) CH_3CH_2OH and $CH_3CH_2CH_2Br$
D) CH_3CH_2Br and $CH_3CH_2CH_2OH$
* E) CH_3CH_2Br and $CH_3CH_2CH_2Br$

44. Which compound would have the lowest solubility in water?

A) Diethyl ether
B) Methyl propyl ether
C) 1-Butanol
D) 2-Butanol
* E) Pentane

45. Which is the best method for the synthesis of <u>tert</u>-butyl methyl ether?

A) CH_3ONa + $(CH_3)_3CBr$ DDDDDD4

* B) $(CH_3)_3CONa$ + CH_3I DDDDDD4

C) CH_3OH + $(CH_3)_3COH$ + H_2SO_4 at 140x C DDDDDD4

D) $(CH_3)_3CONa$ + CH_3OCH_3 DDDDDD4

E) CH_3ONa + $(CH_3)_3COH$ DDDDDD4

46. The correct IUPAC name for <u>tert</u>-butyl alcohol is:

A) 1-Butanol
B) 2-Methyl-1-propanol
* C) 2-Methyl-2-propanol
D) 2-Butanol
E) 1,1-Dimethyl-1-ethanol

47. Which alcohol would undergo acid-catalyzed dehydration most rapidly?

A)
```
          CH3
          3
   CH3DCDCH2DCH2OH
          3
          CH3
```

B)
```
          CH3
          3
   CH3DCDCH2DCH3
          3
          CH2OH
```

C)
```
          CH3
          3
   CH3DCDDDCHDCH3
          3   3
          CH3 OH
```

* D)
```
          CH3
          3
   CH3DCDCH2DCH2DCH3
          3
          OH
```

E)
```
          CH2OH
          3
   CH3DCDCH2DCH3
          3
          CH3
```

48. Which of these ethers is most resistant to peroxide formation on exposure to atmospheric oxygen?

A) $CH_3OCH_2CH_3$
B) $CH_3CH_2OCH_2CH_3$
C) $(CH_3)_2CHOCH(CH_3)_2$
D) $(CH_3)_2CHOCH_2CH_3$
* E) $CH_3OC(CH_3)_3$

49. What would be the major product of the following reaction?

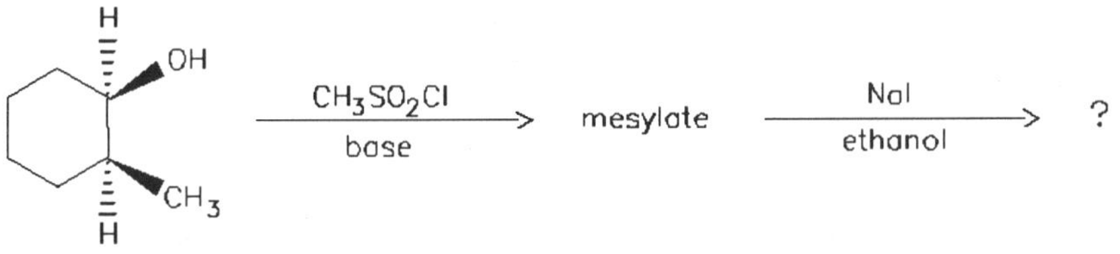

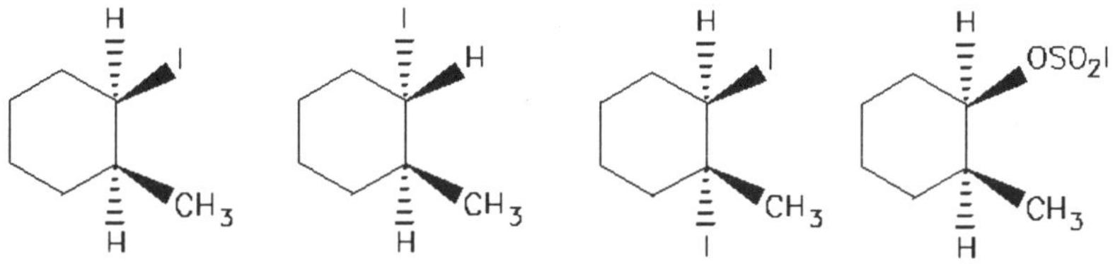

$\xrightarrow[\text{base}]{CH_3SO_2Cl}$ mesylate $\xrightarrow[\text{ethanol}]{NaI}$?

I II III IV

A) I
* B) II
C) III
D) IV
E) An equimolar mixture of I and II

50. Which compound would have the lowest boiling point?

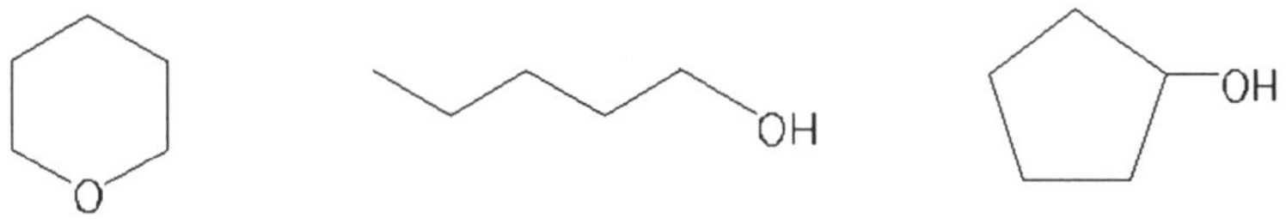

I II III

IV V

* A) I B) II C) III D) IV E) V

51. cis-3-Methylcyclopentanol is treated with CH_3SO_2Cl in the presence of a base. The product of the reaction then is allowed to react with KI in methanol. What is the final product?

* A) trans-1-Iodo-3-methylcyclopentane
B) cis-1-Iodo-3-methylcyclopentane
C) 1-Methylcyclopentene
D) 2-Methylcyclopentene
E) 3-Methylcyclopentene

52. The major product of the following reaction would be:

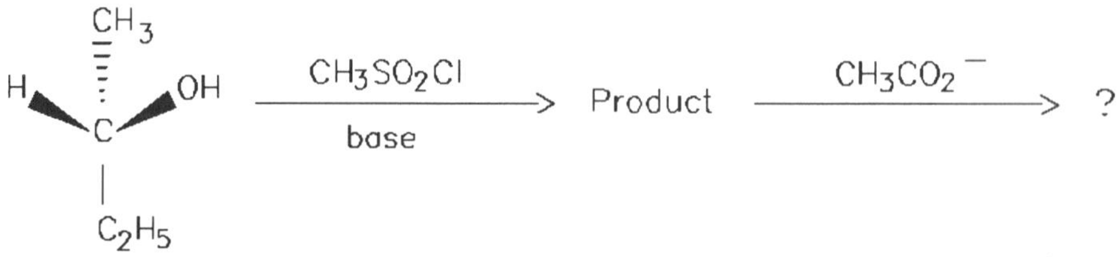

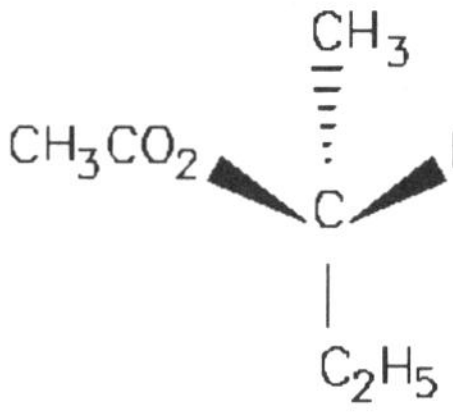

CH_3, H, OH, C_2H_5 stereocenter $\xrightarrow[\text{base}]{CH_3SO_2Cl}$ Product $\xrightarrow{CH_3CO_2^-}$?

I: CH_3 (dashed), CH_3CO_2 (wedge), H (wedge), C_2H_5

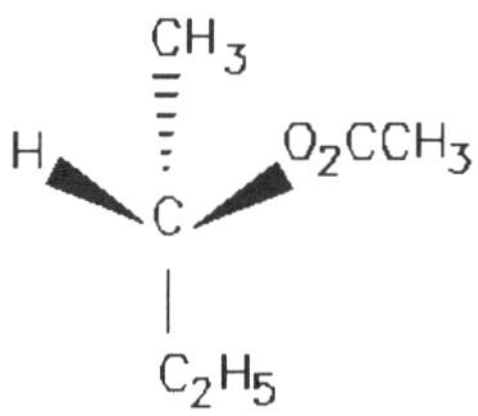

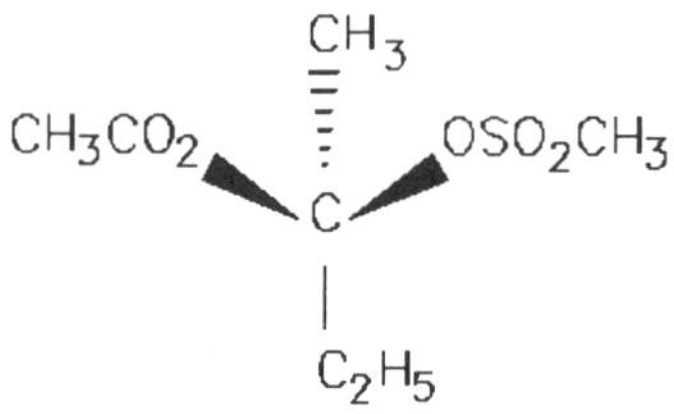

II: CH_3 (dashed), H (wedge), O_2CCH_3 (wedge), C_2H_5

III: CH_3 (dashed), CH_3CO_2 (wedge), OSO_2CH_3 (wedge), C_2H_5

* A) I
B) II
C) III
D) Equal amounts of I and II
E) None of these

53. Which would be the major product of the reaction shown?

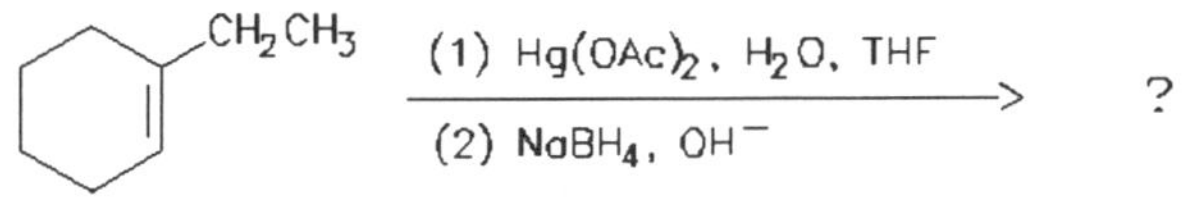

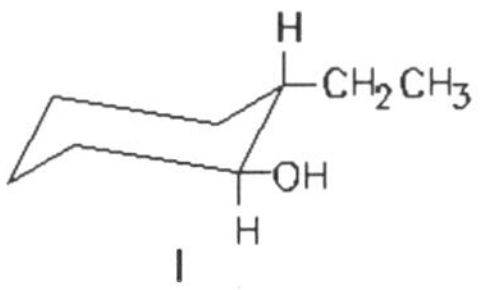

$$\text{1-ethylcyclohexene} \xrightarrow[(2)\ NaBH_4,\ OH^-]{(1)\ Hg(OAc)_2,\ H_2O,\ THF} ?$$

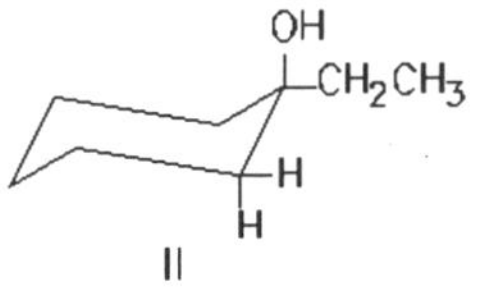

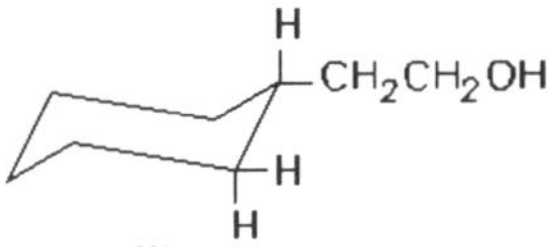

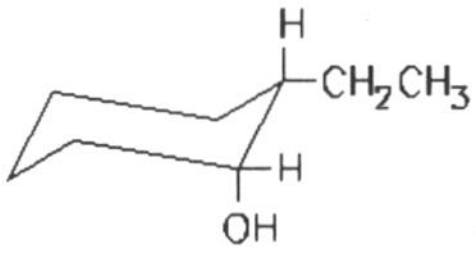

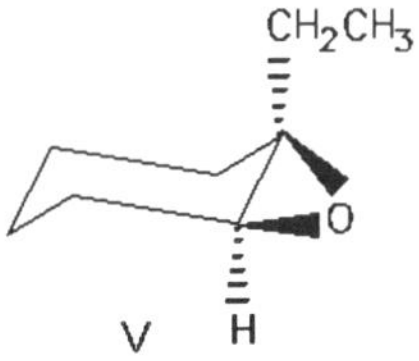

I: H, CH_2CH_3 / OH, H

II: OH, CH_2CH_3 / H, H

III: H, CH_2CH_2OH / H, H

IV: H, CH_2CH_3 / H, OH

V: CH_2CH_3, O, H

A) I * B) II C) III D) IV E) V

54. Which reagent(s) would transform propyl alcohol into propyl bromide?
A) Concd. HBr and heat
B) PBr_3
C) $NaBr/H_2O$ and heat
* D) More than one of these
E) All of these

55. The major industrial process in use today for the production of methanol is the:
A) hydration of ethyne.
B) distillation of wood.
C) hydrogenation of carbon dioxide.
D) reduction of methanal.
* E) catalytic reduction of carbon monoxide.

56. Today, most industrial ethanol is made in the U.S. by the:
A) fermentation of grain.
B) hydrolysis of ethyl bromide.
* C) hydration of ethylene.
D) reduction of acetaldehyde.
E) hydration of acetylene.

57. The number of optically active pentyl alcohols, i.e., the total number of individual enantiomers, is:
A) 0 B) 2 C) 3 D) 4 * E) 6

58. What is the major product of the reaction of C_6H_5CpCH with Sia_2BH (a hindered dialkylborane), followed by reaction with alkaline hydrogen peroxide?

A) $C_6H_5CH_2CH_3$

B) OH
3
$C_6H_5CHCH_3$

* C) H
3
$C_6H_5CH_2CMO$

D) $C_6H_5CHMCH_2$

E) O
:
$C_6H_5CCH_3$

59. Epoxidation followed by reaction with aqueous base converts cyclopentene into which of these?

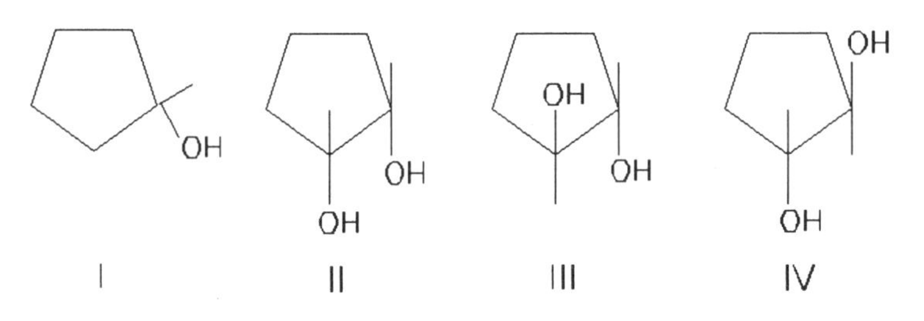

A) I
B) II
C) III
D) IV
* E) Equal amounts of III and IV

60. Which of these ethers is least likely to undergo significant cleavage by hot aqueous H_2SO_4 ?

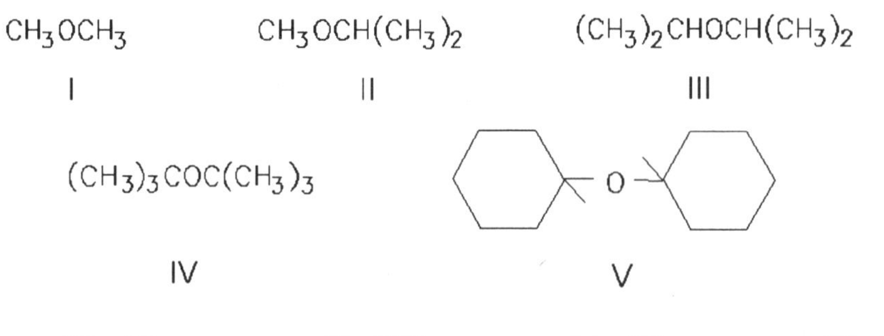

* A) I
B) II
C) III
D) IV
E) V

61.

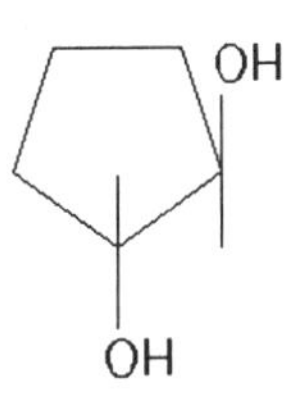

is properly named:

A) <u>cis</u>-1,2-Cyclopentanediol
B) <u>meso</u>-1,2-Cyclopentanediol
* C) (1R,2R)-1,2-Cyclopentanediol
D) (1R,2S)-1,2-Cyclopentanediol
E) (1S,2S)-1,2-Cyclopentanediol

62. 2,2-Dimethyl-1-propanol has the common name:

A) Isoamyl alcohol
B) Isopentyl alcohol
C) <u>tert</u>-Pentyl alcohol
* D) Neopentyl alcohol
E) 2-Methylisobutyl alcohol

63. Which of these is the most accurate name for the molecule

H
3
$CH_3CH_2COC_6H_5$?
3
CH_3

A) <u>sec</u>-Butyl phenyl ether
B) Isobutyl phenyl ether
C) <u>tert</u>-Butyl phenyl ether
D) (R)-2-Phenoxybutane
* E) (S)-2-Phenoxybutane

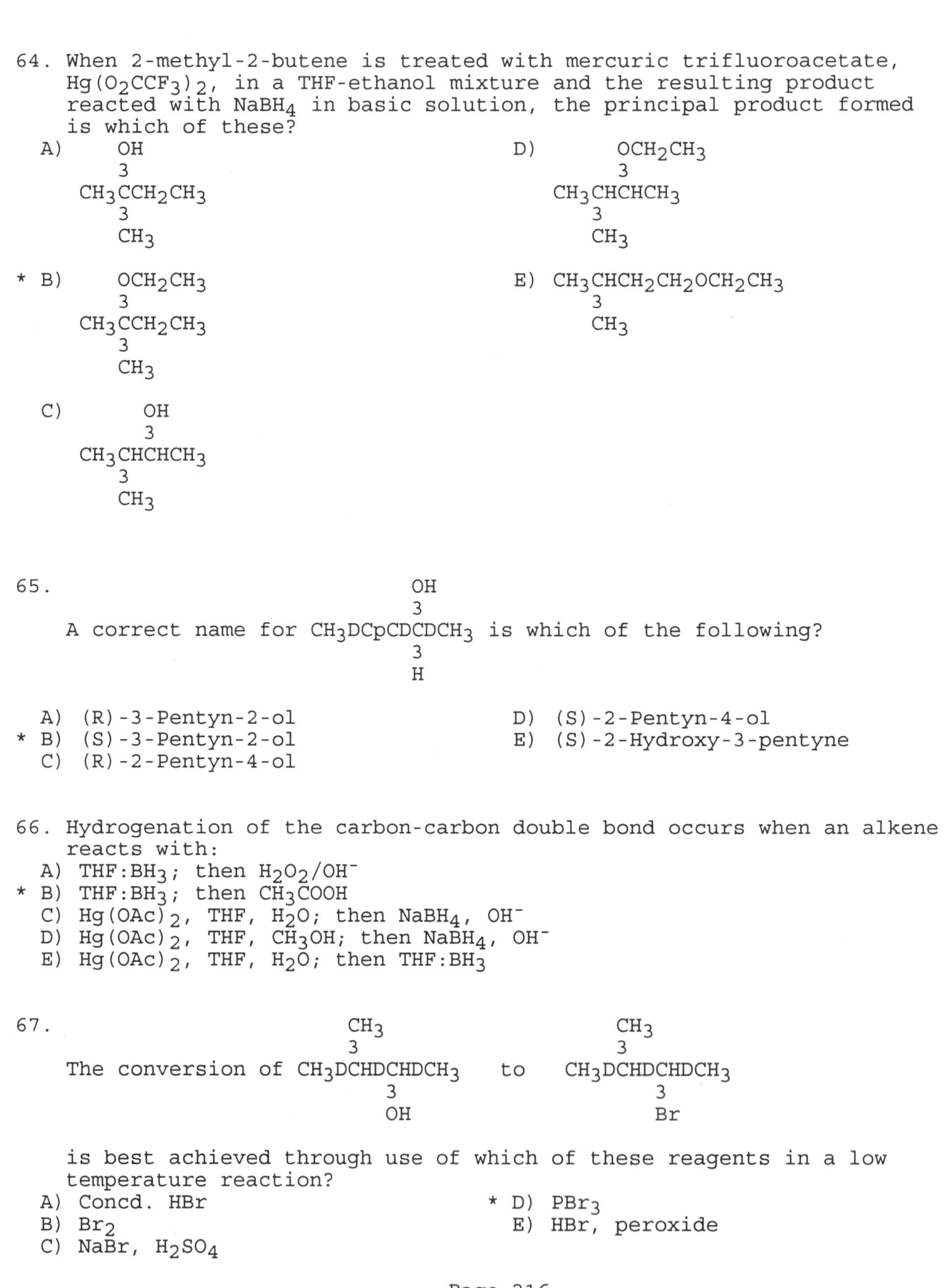

64. When 2-methyl-2-butene is treated with mercuric trifluoroacetate, $Hg(O_2CCF_3)_2$, in a THF-ethanol mixture and the resulting product reacted with $NaBH_4$ in basic solution, the principal product formed is which of these?

A)
```
    OH
    3
CH3CCH2CH3
    3
    CH3
```

* B)
```
    OCH2CH3
    3
CH3CCH2CH3
    3
    CH3
```

C)
```
      OH
      3
CH3CHCHCH3
    3
    CH3
```

D)
```
      OCH2CH3
      3
CH3CHCHCH3
    3
    CH3
```

E)
```
CH3CHCH2CH2OCH2CH3
    3
    CH3
```

65.
```
                                OH
                                3
A correct name for CH3DCpCDCDCH3 is which of the following?
                                3
                                H
```

A) (R)-3-Pentyn-2-ol
* B) (S)-3-Pentyn-2-ol
C) (R)-2-Pentyn-4-ol
D) (S)-2-Pentyn-4-ol
E) (S)-2-Hydroxy-3-pentyne

66. Hydrogenation of the carbon-carbon double bond occurs when an alkene reacts with:
A) $THF:BH_3$; then H_2O_2/OH^-
* B) $THF:BH_3$; then CH_3COOH
C) $Hg(OAc)_2$, THF, H_2O; then $NaBH_4$, OH^-
D) $Hg(OAc)_2$, THF, CH_3OH; then $NaBH_4$, OH^-
E) $Hg(OAc)_2$, THF, H_2O; then $THF:BH_3$

67.
```
                        CH3                    CH3
                        3                      3
The conversion of CH3DCHDCHDCH3    to    CH3DCHDCHDCH3
                           3                      3
                           OH                     Br
```

is best achieved through use of which of these reagents in a low temperature reaction?
A) Concd. HBr
B) Br_2
C) NaBr, H_2SO_4
* D) PBr_3
E) HBr, peroxide

68. Which of these alkyl halide syntheses is predicted to occur at the greatest rate?

A) $CH_3CH_2CH_2CH_2OH$ + HI DDDDD4

B) $(CH_3)_2CHCH_2OH$ + HBr DDDDD4

C) $CH_3CHCH_2CH_3$ + HCl DDDDD4
3
OH

D) $CH_3CHCH_2CH_3$ + HBr DDDDD4
3
OH

* E) $(CH_3)_3COH$ + HI DDDDD4

69. Which of these, though commonly used, is an incorrect name for $CH_3CHOHCH_3$?

A) Isopropyl alcohol

B) sec-Propyl alcohol

C) 2-Propanol

* D) Isopropanol

E) More than one of these.

70. Which of these, though commonly used, is an incorrect name for $(CH_3)_3COH$?

A) tert-Butyl alcohol

* B) tert-Butanol

C) 2-Methyl-2-propanol

D) More than one is incorrect.

E) Each is a correct name.

71. What is the product of the reaction of propyl alcohol with $(CH_3)_3SiCl$ in the presence of a tertiary amine.

A) $CH_3CH_2CH_2Si(CH_3)_3$

B) $(CH_3)_2CHSi(CH_3)_3$

* C) $CH_3CH_2CH_2OSi(CH_3)_3$

D) $(CH_3)_2CHOSi(CH_3)_3$

E) $(CH_3CH_2CH_2)_3SiOH$

72. What is the major product of the reaction:

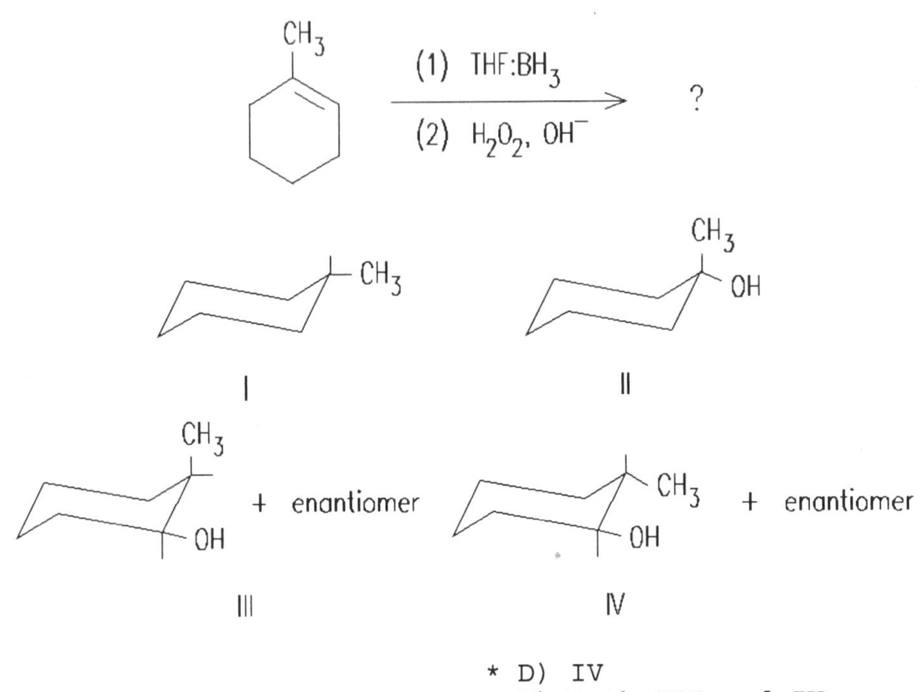

A) I
B) II
C) III
* D) IV
E) Both III and IV

73. "Amyl" is an archaic designation of a five-carbon, i.e., pentyl, group. What structure corresponds to the once-used name "primary active amyl alcohol"?

A) $CH_3CH_2CH_2CH_2CH_2OH$
B) $(CH_3)_2CHCH_2CH_2OH$
* C)
```
            H
            3
  CH3CH2CCH2OH
            3
            CH3
```
(or enantiomer)

D)
```
            H
            3
  (CH3)2CHCCH3
            3
            OH
```
(or enantiomer)

E)
```
              H
              3
  CH3CH2CH2COH
              3
              CH3
```
(or enantiomer)

74. What is the total number of pentyl alcohols, including stereoisomers?

A) 7 B) 8 C) 9 D) 10 * E) 11

75. Oxymercuration-demercuration of 3-methylcyclopentene produces this/these product(s):

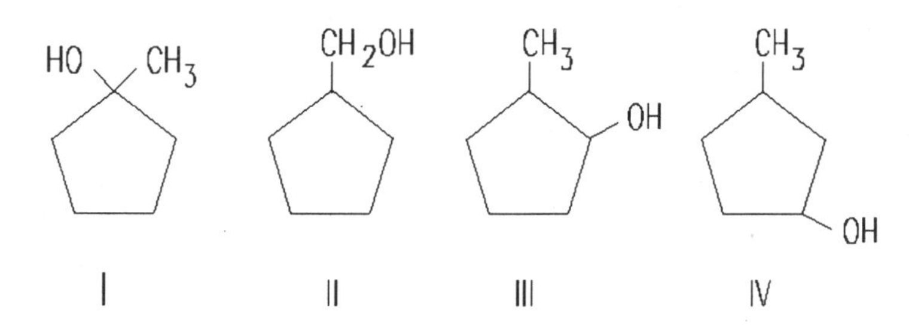

A) I
B) II
C) III
D) IV
* E) Both III and IV

76. Which statement is true concerning the formation of alcohols by the hydroboration-oxidation sequence?
* A) Overall, the process results in syn addition and anti-Markovnikov orientation.
B) Overall, the process results in anti addition and anti-Markovnikov orientation.
C) Overall, the process results in syn addition and Markovnikov orientation.
D) Overall, the process results in anti addition and Markovnikov orientation.
E) The stereochemistry and orientation are unpredictable.

77.

```
                                    O
                                    :
Methanesulfonic acid, CH3DSDOH, is treated, in turn, with PCl5 and
                                    :
                                    O
```

(R)-2-butanol. What is the final product?

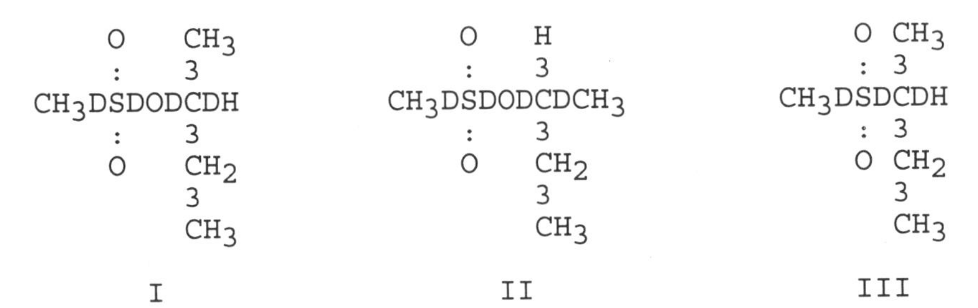

```
        O H                       O CH3
        : 3                       : 3
    CH3DSDCDCH3              CH3DSDCDCl
        : 3                       : 3
        O CH2                     O CH2
          3                         3
          CH3                       CH3

        IV                          V
```

* A) I B) II C) III D) IV E) V

78. cis-3-Hexene is treated with magnesium monoperoxyphthalate and the product is then subjected to acid-catalyzed hydrolysis. What is the final product?

```
       OH                  HO H               H OH               HO OH
       3                    3 3               3 3                 3 3
CH3CH2CCH2CH2CH3    CH3CH2CDCCH2CH3    CH3CH2CDCCH2CH3    CH3CH2CDCCH2CH3
       3                    3 3               3 3                 3 3
       H                    H OH              HO H                H H

  (+ enantiomer)

       I                    II                III                 IV
```

A) I
B) II
C) III
* D) equal amounts, II and III
E) IV

Chapter 11: Alcohols from Carbonyl Compounds

1. Which reagent(s) will distinguish between cyclopentanol and cyclopentane?

A) Br_2/CCl_4
B) $KMnO_4$ (cold)
* C) CrO_3/aqueous H_2SO_4
D) NaOH (aq)
E) A) and B)

2. Which of the following reagents might serve as the basis for a simple chemical test that would distinguish between $CH_3CH{=}CHCH_3$ and $CH_3CHCH_2CH_3$?

3
OH

A) CrO_3 in H_2SO_4
B) Dilute aqueous $KMnO_4$
C) Br_2 in CCl_4
D) Two of these
* E) All of these

3. Which of the reagents listed below would serve as the basis for a simple chemical test to distinguish between

$CH_3CHMCHCH_2OH$ and $CH_3CH_2CH_2CH_2OH$?

A) CrO_3 in H_2SO_4
B) Cold concd. H_2SO_4
* C) Br_2 in CCl_4
D) $NaOH/H_2O$
E) $Ag(NH_3)_2OH$

4. The final product, D, in the following reaction sequence,

CH_2CH_2
\ /
O

CH_3CHOH $\xrightarrow{PBr_3}$ A $\xrightarrow[\text{ether}]{Mg}$ B $\xrightarrow{CH_2CH_2O}$ C $\xrightarrow{H_3O^+}$ D, would be?

3
CH_3

(Arrows printed as "DDDDDD4".)

A) $CH_3CHOCH_2CH_2OH$
3
CH_3

B) $CH_3CHCH_2CH_2Br$
3
CH_3

* C) $CH_3CHCH_2CH_2OH$
3
CH_3

D) $CH_3CHOCH_2CH_3$
3
CH_3

E) $CH_3CHCH_2CH_3$
3
CH_3

5. Your task is to synthesize $CH_3CH_2CH_2C(C_6H_5)(OH)CH_3$ through a Grignard synthesis. Which pair(s) of compounds listed below would you choose as starting materials?

```
                C6H5
                3
CH3CH2CH2CCH3
                3
                OH
```

* A) $CH_3CH_2CH_2Br$ and $CH_3C(=O)C_6H_5$

B) $CH_3CH_2CH_2CH(=O)$ and C_6H_5Br

C) $C_6H_5CH(=O)$ and $CH_3CH_2CH(Br)CH_3$

D) More than one of these

E) None of these

6. Which of the following synthetic procedures would be employed most effectively to transform ethanol into ethyl propyl ether?

A) Ethanol + HBr, then Mg/ether, then H_3O^+, then Na, then CH_3CH_2Br

* B) Ethanol + HBr, then Mg/ether, then HCHO, then H_3O^+, then Na, then CH_3CH_2Br

C) Ethanol + $CH_3CH_2CH_2OH$ + H_2SO_4/140xC

D) Ethanol + Na, then HCHO, then H_3O^+, then HBr, then Mg/ether, then $CH_3CH_2CH_2Br$

E) Ethanol + H_2SO_4/180xC, then $CH_3CH_2CH_2Br$

7. What would be the major product of the following reaction?

```
                          H+
CH3CH2OH  +  CH2CH2   DDDDDD4
              \ /
               O
```

* A) $CH_3CH_2OCH_2CH_2OH$

B) $CH_3CH_2OCH_2CH_2OCH_2CH_3$

C) $HOCH_2CH_2OH$

D) $HOCH_2CH_2OCH_2CH_2OH$

E) $CH_3CH_2OCH(OH)CH_3$

8. Which reagent(s) is/are incapable of reducing $CH_3CH_2CH_2C(=O)OCH_2CH_3$?

A) $LiAlH_4$/ether

* B) $NaBH_4/H_2O$

C) Na, C_2H_5OH

D) H_2, catalyst, high pressure

E) All can be used successfully.

9. Which of the reagents listed below would serve as the basis for a simple chemical test to distinguish between

$$(CH_3)_3COH \text{ and } (CH_3)_2CHCH_2OH \quad ?$$

A) $Ag(NH_3)_2OH$
B) $NaOH/H_2O$
C) Br_2 in CCl_4
D) Cold concd. H_2SO_4
* E) CrO_3 in H_2SO_4

10. Your task is to synthesize 2-phenyl-2-hexanol through a Grignard synthesis. Which pair(s) of compounds listed below would you choose as starting materials?

A) $CH_3CH_2CH_2CH_2Br$ and $CH_3C(=O)C_6H_5$

B) $CH_3CH(CH_3)CH_2Br$ and $CH_3C(=O)C_6H_5$

C) $CH_3CH_2CH_2CH_2C(=O)CH_3$ and C_6H_5Br

D) Answers A) or B)
* E) Answers A) or C)

11. What is the product, A, that would be obtained from the following reaction sequence?

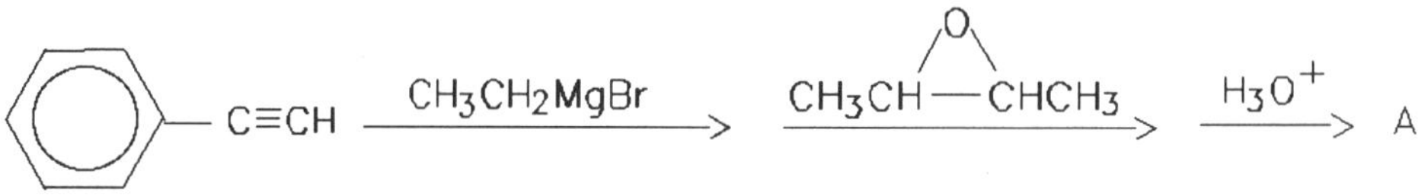

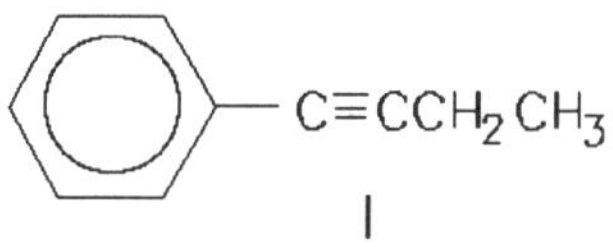

$$C_6H_5-C\equiv CH \xrightarrow{CH_3CH_2MgBr} \xrightarrow{CH_3CH(O)CHCH_3 \text{ (epoxide)}} \xrightarrow{H_3O^+} A$$

I: $C_6H_5-C\equiv CCH_2CH_3$

II: $C_6H_5-C\equiv CCH(CH_3)-CH(CH_3)OCH_2CH_3$

III: $CH_3CH_2-C_6H_4-C\equiv CCH(CH_3)-CH(CH_3)OH$

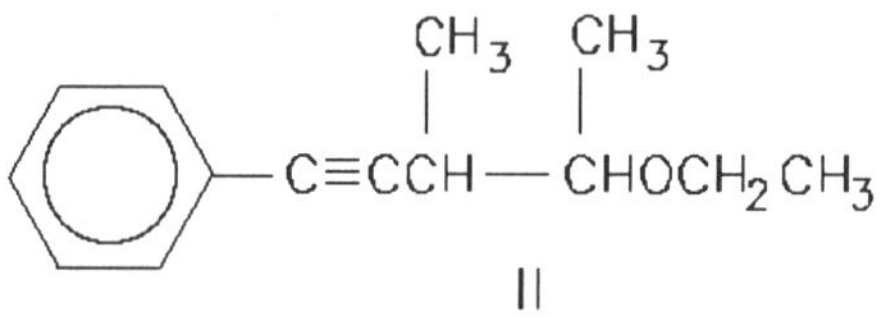

IV: $C_6H_5-C\equiv CCH(CH_3)-CH(CH_3)OH$

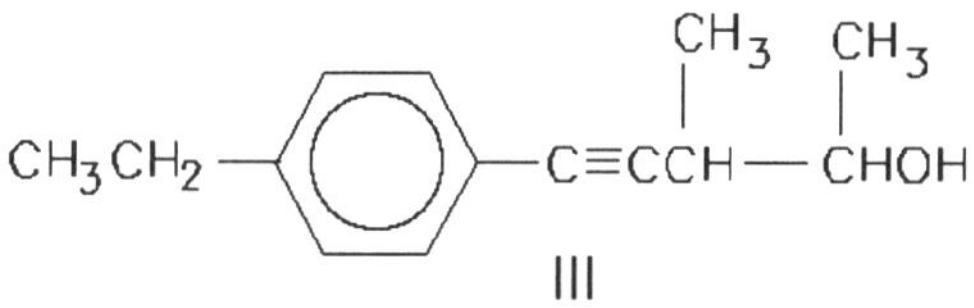

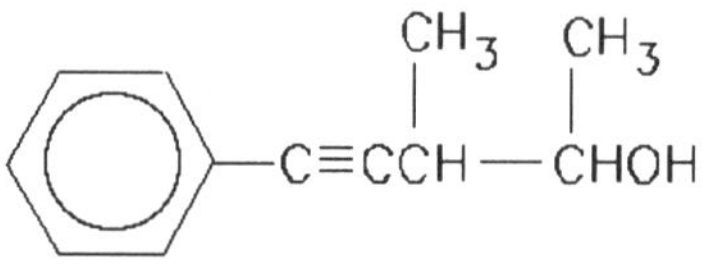

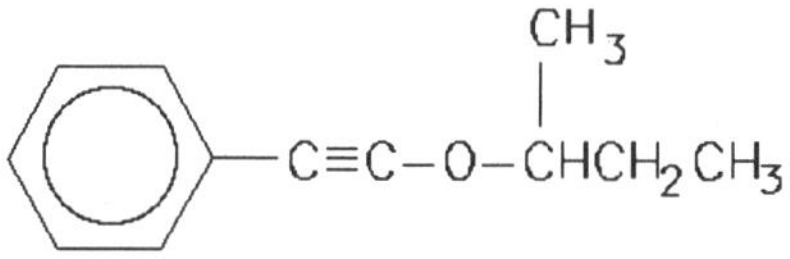

V: $C_6H_5-C\equiv C-O-CH(CH_3)CH_2CH_3$

A) I B) II C) III * D) IV E) V

12. The principal product(s) formed when 1 mol of methylmagnesium iodide reacts with 1 mol of p-hydroxyacetophenone, i.e., p-$HOC_6H_4CCH_3$, is/are:

```
p-HOC6H4CCH3
         :
         O
```

CH_4 + $COCH_3$ / OMgI
I

OMgI / $H_3C-C-CH_3$ / OH
II

$COCH_3$ / OCH_3
III

$COCH_3$ / MgI / OCH_3
IV

$COCH_3$ / CH_3 / OMgI
V

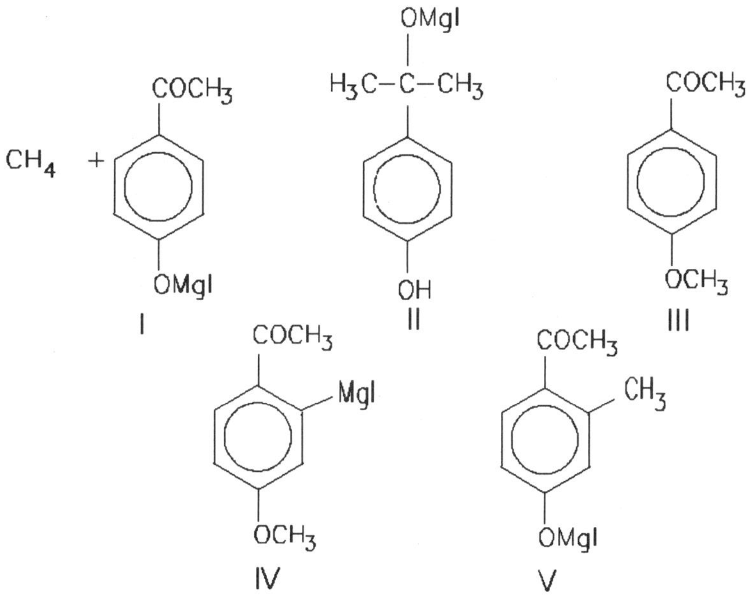

* A) I B) II C) III D) IV E) V

13. The product, B, of the following reaction,

```
                              H2O
CH3CCH3  +  NaBD4  DDDDDD4  A  DDDDDD4  B
   :
   O
```

would be:

```
A) CH3CHCH3  * B) CH3CDCH3   C) CH3CDCH3   D) CH3CHCH3   E) CH3CCH2D
      3              3             3             3             :
      OD             OH            OD            OH            O
```

14. What is the product, A, that would be obtained from the following reaction sequence?

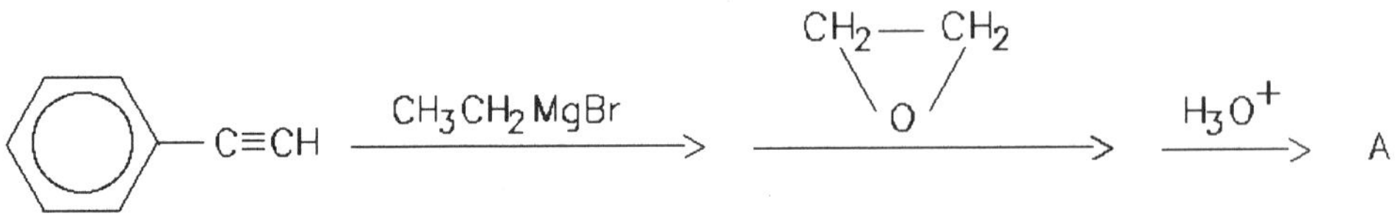

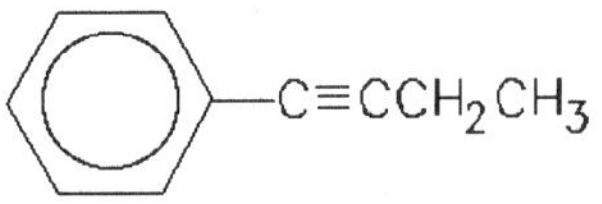

$$C_6H_5-C\equiv CH \xrightarrow{CH_3CH_2MgBr} \xrightarrow{CH_2-CH_2 \text{ (epoxide, O)}} \xrightarrow{H_3O^+} A$$

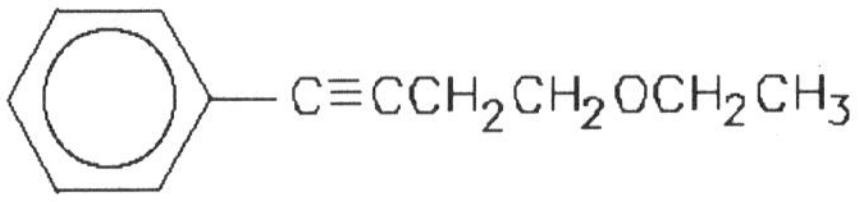

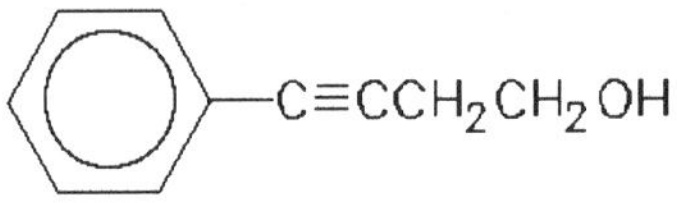

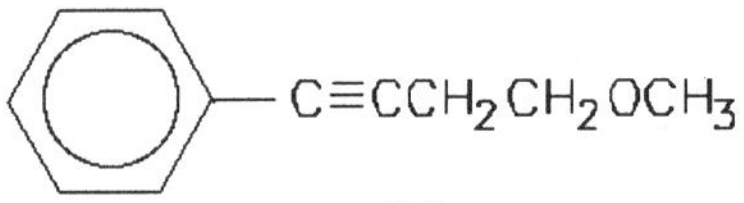

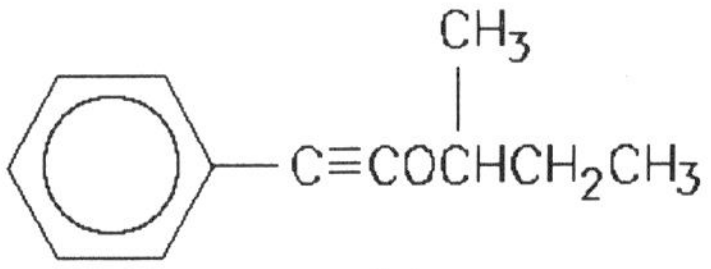

I: $C_6H_5-C\equiv CCH_2CH_3$

II: $C_6H_5-C\equiv CCH_2CH_2OCH_2CH_3$

III: $C_6H_5-C\equiv CCH_2CH_2OH$

IV: $C_6H_5-C\equiv CCH_2CH_2OCH_3$

V: $C_6H_5-C\equiv COCH(CH_3)CH_2CH_3$

A) I B) II * C) III D) IV E) V

15. What would be the product, C, of the following reaction sequence?

```
    CH3                                       CH3
    3                                         3
              Li          CuI         CH3CHCH2CH2Br
CH3CCH2Br  DDDDDD4  A  DDDDDD4  B  DDDDDDDDDDDDDD4  C
    3
    CH3
```

A) 2,6-Dimethylheptane
B) 2,2-Dimethylpropane
C) 2-Methylpentane
D) 2,2,5-Trimethylhexane
* E) 2,2,6-Trimethylheptane

16. Which of the following is the strongest acid?
A) RMgX B) Mg(OH)X C) RH * D) H_2O

17. Which of the following is the strongest base?
* A) RMgX B) Mg(OH)X C) RH D) H_2O

18. In which of the following series are the compounds arranged in order of decreasing basicity?

A) CH_3CH_2MgBr > $NaNH_2$ > HCpCNa > NaOH > CH_3CH_2ONa
* B) CH_3CH_2MgBr > $NaNH_2$ > HCpCNa > CH_3CH_2ONa > NaOH
C) HCpCNa > CH_3CH_2MgBr > $NaNH_2$ > CH_3CH_2ONa > NaOH
D) $NaNH_2$ > CH_3CH_2MgBr > HCpCNa > CH_3CH_2ONa > NaOH
E) None of these

19. Select the correct reagent(s) for the following reaction:

O   OH
:   3
$CH_3CCH_2CH_2CO_2CH_3$ DDDDDD4 $CH_3CHCH_2CH_2CH_2OH$

* A) $LiAlH_4$; then H^+
B) $NaBH_4$; then H^+
C) H_2 with Pt/C
D) A) and B)
E) A), B) and C)

20. Select the correct reagent(s) for the following reaction:

O   OH
:   3
$CH_3CCH_2CH_2CO_2CH_3$ DDDDDD4 $CH_3CHCH_2CH_2CO_2CH_3$

A) $LiAlH_4$; then H^+
* B) $NaBH_4$; then H^+
C) H_2 with Pt/C
D) A) and B)
E) A), B) and C)

21. Which of the following would serve as a synthesis of racemic:

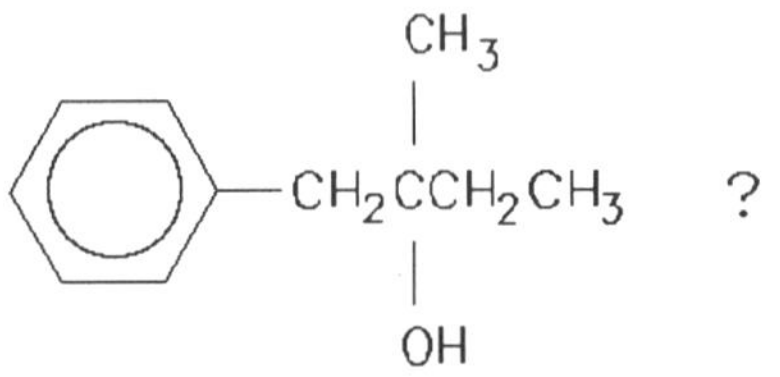

$C_6H_5-CH_2C(CH_3)(OH)CH_2CH_3$?

I $CH_3CH_2C(=O)CH_3$ + $C_6H_5-CH_2MgCl$ $\xrightarrow[(2)\ H_3O^+]{(1)\ Et_2O}$

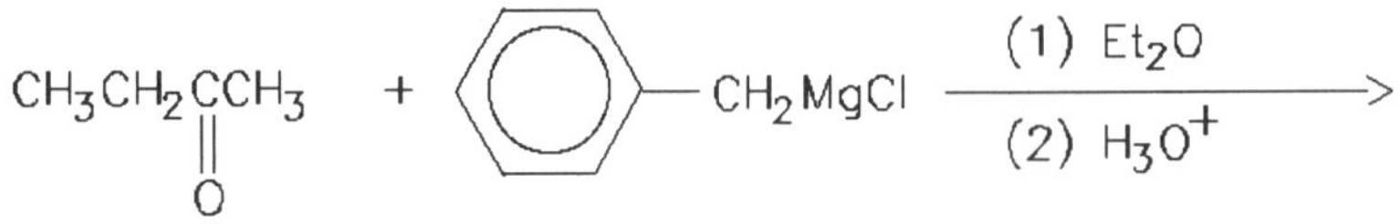

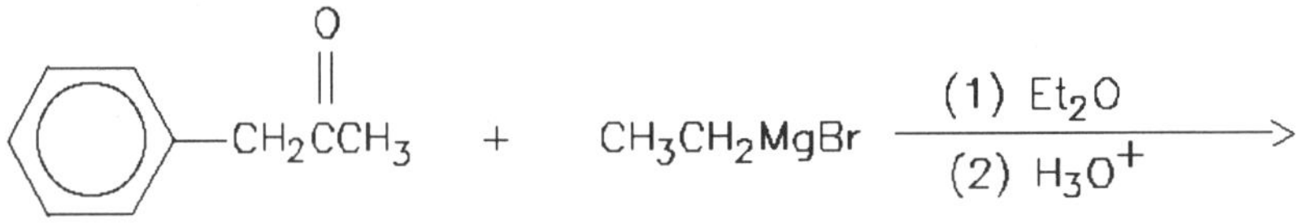

II $C_6H_5-CH_2C(=O)CH_3$ + CH_3CH_2MgBr $\xrightarrow[(2)\ H_3O^+]{(1)\ Et_2O}$

III $C_6H_5-CH_2C(=O)CH_2CH_3$ + CH_3MgI $\xrightarrow[(2)\ H_3O^+]{(1)\ Et_2O}$

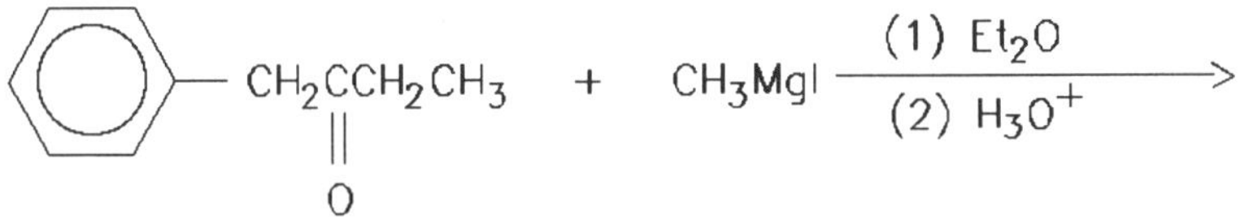

A) I
B) II
C) III
* D) All of the above
E) None of the above

22. What is the predominant product from the reaction of 2-hexanol with H_2CrO_4?

A) CH_3CO_2H
B) $CH_3(CH_2)_3CO_2H$
* C) $CH_3(CH_2)_3C(=O)CH_3$
D) $CH_3(CH_2)_4CO_2H$
E) A) and B)

23. What would be the product, O, of the following reaction sequence?

$$\mathrm{CH_3CH_2\underset{\substack{3\\ Br}}{C}HCH_3} \xrightarrow[\text{ether}]{\text{Mg}} \text{N} \xrightarrow{D_2O} \text{O}$$

A) $CH_3CH_2CH_2CH_3$

* B) $CH_3CH_2CHCH_3$
3
D

C) $CH_3CH_2CHCH_3$
3
OD

D) $CH_3CH_2CH_2CH_2OD$

E) $CH_3CH_2CH_2CH_2D$

24. Which of the following resonance structures is not a significant contributor to the hybrid for the carbonyl group?

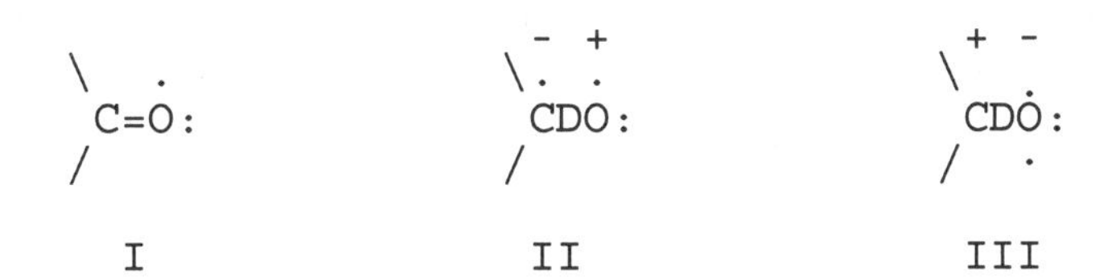

A) I
* B) II
C) III
D) Neither II nor III is important.
E) All are significant contributors.

25. Which of the following reactions would serve as a reasonable synthesis of the following racemic alcohol?

OH
3
$CH_3CC_6H_5$
3
C_2H_5

A) O
:
$CH_3CC_6H_5$ + C_2H_5MgBr (1) Et_2O DDDDDDDD4 (2) H_3O^+

B) O
:
$C_2H_5CC_6H_5$ + CH_3MgBr (1) Et_2O DDDDDDDD4 (2) H_3O^+

C) O
:
$CH_3CC_2H_5$ + C_6H_5MgBr (1) Et_2O DDDDDDDD4 (2) H_3O^+

D) Answers A) and B) only
* E) Answers A), B) and C)

26. The final product, D, in the following reaction sequence,

$C_6H_5CH_2OH$ PBr_3 DDDDDD4 A Mg DDDDDD4 B CH_2CH_2 \ / O DDDDDD4 C H_3O^+ DDDDDD4 D,

would be?

A) $C_6H_5CH_2OCH_2CH_3$
B) $C_6H_5CH_2CH_2OH$
* C) $C_6H_5CH_2CH_2CH_2OH$
D) $C_6H_5CHCH_2CH_2OH$
3
Br
E) $C_6H_5CH_2CH_2OCH_3$

27. What is the final product of the following reaction sequence?

$$CH_3I \xrightarrow[\text{ether}]{Mg} \text{organic product} \xrightarrow[(2)\ H_3O^+]{(1)\ CH_3\overset{CH_3}{\overset{|}{C}}HCH_2\overset{O}{\overset{\|}{C}}H} \text{organic product} \xrightarrow{H_2CrO_4} \text{final product}$$

A) $(CH_3)_2CHCH_2CHOHCH_3$

B) $(CH_3)_2CHCH_2\overset{O}{\overset{\|}{C}}OCH_3$

C) $(CH_3)_2CH\overset{O}{\overset{\|}{C}}CH_2CH_3$

D) $(CH_3)_2CHCHOHCH_2CH_3$

* E) $(CH_3)_2CHCH_2\overset{O}{\overset{\|}{C}}CH_3$

28. What would be the alkane C that is the product of the following reaction sequence?

$$CH_3\underset{CH_3}{\underset{|}{C}}HBr \xrightarrow{Li} A \xrightarrow{CuI} B \xrightarrow{CH_3\overset{CH_3}{\overset{|}{C}}HCH_2Br} C$$

A) $CH_3\overset{CH_3}{\overset{|}{C}}H\underset{CH_3}{\underset{|}{C}}HCH_2CH_3$

B) $CH_3\underset{CH_3}{\underset{|}{C}}H{-}\underset{CH_3}{\underset{|}{C}}HCH_2CH_3$

* C) $CH_3\underset{CH_3}{\underset{|}{C}}HCH_2\overset{CH_3}{\overset{|}{C}}HCH_3$

D) $CH_3\underset{CH_3}{\underset{|}{C}}H{-}\overset{CH_3}{\underset{CH_3}{\overset{|}{\underset{|}{C}}}}CH_3$

E) $CH_3CH_2CH_2CH_2\overset{CH_3}{\overset{|}{C}}HCH_3$

29. Which reaction is an oxidation?

A) $RCHO \longrightarrow RCO_2H$

B) $RCH_2OH \longrightarrow RCHO$

C) $RCH_2OH \longrightarrow RCO_2H$

D) Two of these

* E) All of these

30. Which reagent(s) would you use to convert $CH_3(CH_2)_6CO_2H$ to $CH_3(CH_2)_6CH_2OH$?

A) $NaBH_4/H_2O$
* B) $LiAlH_4$/ether, then H_3O^+
C) PCC/CH_2Cl_2
D) Zn, H^+
E) H_2, Pt

31. How could the following synthesis be accomplished?

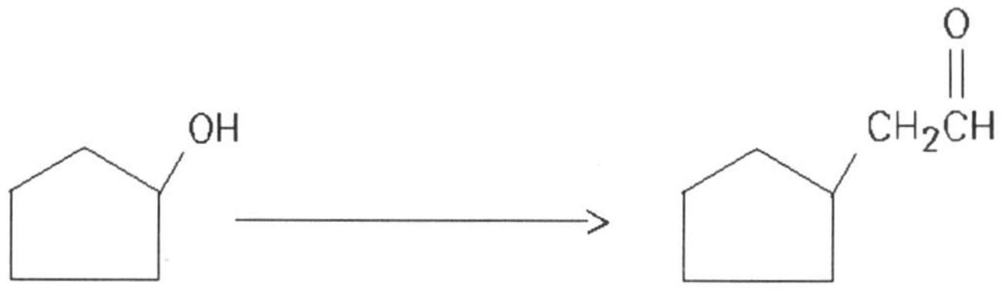

OH

O
||
CH_2CH

A)

O
:
(1) $SOCl_2$, (2) Mg, ether, (3) CH_3CH, then H^+

B) (1) $SOCl_2$, (2) Li, ether, (3) $(CH_3CH_2)_2CuLi$, (4) $KMnO_4$, OH^-

* C)

O
/ \
(1) PBr_3, (2) Mg, ether, (3) CH_2CH_2, then H_3O^+ (4) PCC, CH_2Cl_2

D) More than one of the above
E) None of the above

32. What compound(s) result(s) from the reaction of $CH_3CH_2CH_2MgBr$ with $CH_3CH_2CH_2CH_2CO_2H$?

A) $(CH_3CH_2CH_2)_2CCH_2CH_2CH_2CH_3$
3
OH

B)
O
:
$CH_3CH_2CH_2CCH_2CH_2CH_2CH_3$

C)
O
:
$CH_3CH_2CH_2CH_2COCH_2CH_2CH_3$

* D) $CH_3CH_2CH_3$ + $CH_3CH_2CH_2CH_2CO_2MgBr$

E)
O O
: :
$CH_3CH_2COCCH_2CH_2CH_2CH_3$

33. Which method would give $(CH_3)_2C(OH)C\equiv CH$?

```
                                  OH
                                  3
    Which method would give (CH3)2CCpCH ?
```

A)

```
              H2CrO4    HCpCNa    H3O+
  CH3CHCH3   DDDDDD4   DDDDDD4   DDDDD4
     3       acetic
     OH       acid
```

B)

```
             CH3CH2Li    H3O+    H2CrO4
  CH3CCH3   DDDDDDDD4   DDDDD4   DDDDDDD4
     :                           acetone
     O
```

C)

```
              2 CH3MgBr    H3O+
  CH3CCpCH   DDDDDDDDD4   DDDDD4
     :         ether
     O
```

* D) More than one of the above
E) None of the above

34. Which synthesis of a Grignard reagent would fail to occur as written?

A)

```
                    Mg, ether
  CH3OCH2CH2Br     DDDDDDDDD4   CH3OCH2CH2MgBr
```

B)

```
                 Mg, ether
  CH3CH2CH2I    DDDDDDDDD4   CH3CH2CH2MgI
```

* C)

```
                    Mg, ether
  HO2CCH2CH2Br     DDDDDDDDD4   HO2CCH2CH2MgBr
```

D)

```
              Mg, ether
  C6H5Br     DDDDDDDDD4   C6H5MgBr
```

E) All of the above will succeed.

35. What is the final product?

$$\mathrm{CH_3CpCH} \xrightarrow{\mathrm{NaNH_2}} \xrightarrow{\mathrm{CH_3CCH_3}\ (\overset{O}{:})} \xrightarrow{\mathrm{CH_3I}} \ ?$$

O
:
NaNH$_2$ CH$_3$CCH$_3$ CH$_3$I
CH$_3$CpCH DDDDDD4 DDDDDDD4 DDDDDD4 ?

A) OH
3
CH$_3$CCpCCH$_3$
3
CH$_3$

* B) OCH$_3$
3
CH$_3$CCpCCH$_3$
3
CH$_3$

C) OCH$_3$
3
CH$_3$CpCOCH$_2$CHCH$_3$

D) CH$_3$
3
CH$_3$CCpCONa
3
CH$_3$

E) None of these

36. Which Grignard synthesis will produce an optically active product or product mixture?

A) O
:
C$_6$H$_5$MgBr + CH$_3$CCH$_2$CH$_3$

B) O
:
CH$_3$MgI + C$_6$H$_5$CCH$_2$CH$_3$

C) O
:
CH$_3$CH$_2$MgBr + C$_6$H$_5$COCH$_3$

D) CH$_3$CHCH$_2$CH$_2$MgBr + H$_2$CMO
3
CH$_3$

* E) H H
3 3
CH$_3$CH$_2$CCH$_2$CMO + CH$_3$MgI
3
CH$_3$

37. CH$_3$
3
Consider the molecule CH$_3$CHCHMCHCH$_2$OH.

Which reagent will <u>not</u> give a positive test with this compound?

A) Cold concd. H_2SO_4
B) Br_2/CCl_4
* C) $Ag(NH_3)_2^+$
D) CrO_3/H_2SO_4
E) Dilute $KMnO_4/H_2O$

38. Which of these compounds will <u>not</u> be reduced by $LiAlH_4$?

* A) $CH_3CH_2CH_2CHMCH_2$

B) H
3
$CH_3CH_2CH_2CMO$

C) O
:
$CH_3CH_2CH_2COH$

D) O
:
$CH_3CH_2CH_2COCH_3$

E) O
:
$CH_3CH_2CH_2CCH_3$

39. When nucleophilic addition to a carbonyl group occurs, the carbon attacked undergoes this hybridization change:

* A) sp^2 DDDDDD4 sp^3

B) sp DDDDDD4 sp^2

C) sp DDDDDD4 sp^3

D) sp^3 DDDDDD4 sp^2

E) sp^2 DDDDDD4 sp

40. Which reaction leads to an optically active product?

A) O
:
$CH_3CH_2CCH_2CH_2CH_3$ + $NaBH_4$

B) O
:
$CH_3CC_6H_5$ + CH_3CH_2MgBr

* C) H
3
$CH_2MCHCH_2CCH_2CH_3$ + H_2, Pt
3
COOH

D) OH
3
$CH_3CH_2CCH(CH_3)_2$ + $KMnO_4$
3
H

E) CHO
3
$CH_3CCH_2CH_3$ + $LiAlH_4$ (xs)
3
$COOC_2H_5$

41. Which of these transformations cannot be classified as a reduction?

A) RCH_2Cl DDDDDD4 RCH_3

B) $RCHMCH_2$ DDDDDD4 RCH_2CH_3

C) RCOOH DDDDDD4 RCH_2OH

D) O
:
RCOR′ DDDDDD4 RCH_2OH + R′OH

* E) All of these are reductions.

42. Which of these compounds cannot be reduced by sodium borohydride?

A)
```
            H
            3
(CH3)2CHCMO
```

B)
```
        O
        :
CH3CH2CCH3
```

C)
```
    O
    :
C6H5COH
```

D)
```
          O
          :
CH3(CH2)4COCH3
```

* E) Neither C) nor D) can be reduced.

43. In the reaction of carbonyl compounds with $LiAlH_4$, the effective reducing species is:

A) Li^+ B) Al^{+3} C) AlH_4^- D) AlH_3 * E) H^-

44. Which of these is the least reactive type of organometallic compound?

A) RK * B) R_2Hg C) RLi D) R_2Zn E) R_3Al

45. If the role of the solvent is to assist in the preparation and stabilization of the Grignard reagent by coordination with the magnesium, which of these solvents should be least effective?

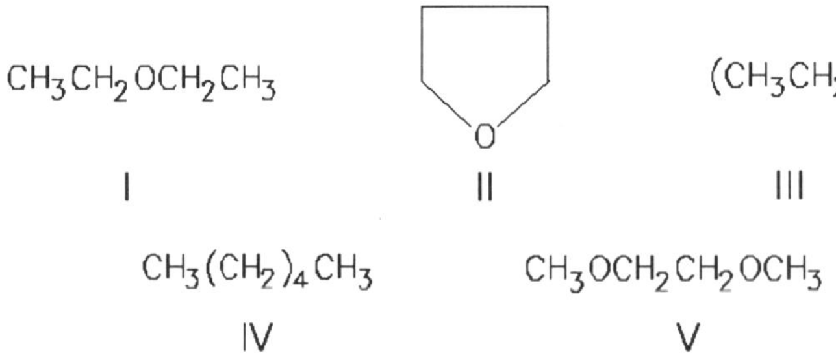

I: $CH_3CH_2OCH_2CH_3$

II: (five-membered ring containing O)

III: $(CH_3CH_2)_3N$

IV: $CH_3(CH_2)_4CH_3$

V: $CH_3OCH_2CH_2OCH_3$

A) I B) II C) III * D) IV E) V

46. The success in converting low molecular weight 1x alcohols to aldehydes by use of $K_2Cr_2O_7/H_2SO_4$ as oxidant can be attributed to the fact that:

A) dichromate is a relatively weak oxidizing agent.
B) the presence of H_2SO_4 limits the oxidation.
* C) the aldehyde can be separated, as formed, by distillation.
D) aldehydes are not oxidized by the $K_2Cr_2O_7/H_2SO_4$ mixture.
E) hydrogen bonding occurs between the alcohol and the acid present.

47. Fundamentally, <u>tert</u>-pentyl alcohol does not undergo oxidation by H_2CrO_4 because:

A) the intermediate chromate ester is not formed.
B) the oxidant isn't in a sufficiently high oxidation state.
C) the alcohol undergoes dehydration.
* D) the intermediate chromate ester cannot lose hydrogen.
E) Actually, this oxidation does occur.

48. Which of these is most likely to be a successful synthesis of an organometallic compound?

A) $CH_3CH_2CH_2MgBr$ + LiCl DDDDDD4 $CH_3CH_2CH_2Li$ + MgBrCl

* B) 2 $CH_3CH_2CH_2CH_2Li$ + $ZnCl_2$ DDDDDD4 $(CH_3CH_2CH_2CH_2)_2Zn$ + 2 LiCl

C) 3 $(CH_3CH_2)_2Hg$ + 2 $AlCl_3$ DDDDDD4 2 $(CH_3CH_2)_3Al$ + 3 $HgCl_2$

D) $(CH_3CH_2)_3Al$ + 3 NaCl DDDDDD4 3 CH_3CH_2Na + $AlCl_3$

E) $(CH_3)_2Cu$ + $MgBr_2$ DDDDDD4 $(CH_3)_2Mg$ + $CuBr_2$

49. Which of these reactions will not produce a 1x alcohol?

A) $CH_3CH_2COCH_2CH_2CH_3$ (C=O) $\xrightarrow[(2)\ H_3O^+]{(1)\ LiAlH_4,\ Et_2O}$

B) $(CH_3)_2CHCH_2CH_2MgBr$ $\xrightarrow[(2)\ H_3O^+]{(1)\ CH_2CH_2\ (\text{oxirane, } \backslash\,/\ O)}$

* C) $C_6H_5CCH_3$ (C=O) $\xrightarrow[(2)\ H_3O^+]{(1)\ NaBH_4}$

D) $CH_3CH_2CH_2CH_2Li$ $\xrightarrow[(2)\ H_3O^+]{(1)\ H_2CMO}$

E) $CH_3(CH_2)_5COOH$ $\xrightarrow[(2)\ H_3O^+]{(1)\ LiAlH_4,\ Et_2O}$

50. Grignard reagents react with oxirane (ethylene oxide) to form 1x alcohols but can be prepared in tetrahydrofuran solvent. Why is this difference in behavior observed?

A) Steric hindrance in the case of tetrahydrofuran precludes reaction with the Grignard.
B) There is a better leaving group in the oxirane molecule.
* C) The oxirane ring is the more highly strained.
D) It is easier to obtain tetrahydrofuran in anhydrous condition.
E) Oxirane is a cyclic ether, while tetrahydrofuran is a hydrocarbon.

51. What product(s) is/are formed in the following reaction:

$CH_3CH_2CH_2COCH_2CH_3$ (C=O) $\xrightarrow{LiAlD_4}$ $\xrightarrow{D_2O}$?

A) $CH_3CH_2CH_2CH_2OD$ + CH_3CH_2OD
* B) $CH_3CH_2CH_2CD_2OD$ + CH_3CH_2OD
C) $CH_3CH_2CH_2CD_2OH$ + CH_3CH_2OH
D) $CH_3CH_2CH_2CHDOD$ + CH_3CH_2OD
E) $CH_3CH_2CH_2CDOCH_2CH_3$ (with OD on C)

52. Which of these reduction reactions is unsuccessful?

A) O
:
$CH_3CH_2CH_2COCH_2CH_2CH_3$ + $LiAlH_4$, ether

B) H
3
$C_6H_5CH_2CMO$ + $NaBH_4$, CH_3OH

* C) O
:
$CH_3(CH_2)_{10}COH$ + H_2, Pt, high pressure

D) O
:
$C_6H_5CCH_2CH_2CH_2CH_3$ + $NaBH_4$, CH_3CH_2OH

E) All of these are successful reductions.

53. CrO_3 in H_2SO_4/H_2O will <u>fail</u> to give a positive test with which of these compounds?

A) $CH_3CH_2CH_2CH_2OH$

B) $CH_3CHCH_2CH_3$
3
OH

* C) $(CH_3)_3COH$

D) H
3
$CH_3CH_2CH_2CMO$

E) More than one of these

54. What product(s) is/are produced in the 1:1 reaction of <u>sec</u>-butylmagnesium bromide with
OH H
3 3
$CH_3CHCH_2CH_2CMO$?

A) OH OMgBr
3 3
$CH_3CHCH_2CH_2CHCHCH_2CH_3$
3
CH_3

* B) OMgBr H
3 3
$CH_3CHCH_2CH_2CMO$ + $CH_3CH_2CH_2CH_3$

C) OH OMgBr
3 3
$CH_3CHCH_2CH_2CHCH_2CH(CH_3)_2$

D) OMgBr H
3 3
$CH_3CHCH_2CH_2CMO$ + $(CH_3)_3CH$

E) H
3
$CH_3CHCH_2CH_2CMO$
3
$OCHCH_2CH_3$
3
CH_3

55. Which of these compounds cannot be used to prepare the corresponding Grignard reagent?

A) $CH_3OCH_2CH_2CH_2Br$

B) $(CH_3)_3CCl$

C) CH_2MCHCH_2Br

D) $(CH_3)_2NCH_2CH_2Br$

* E) H
3
$OMCCH_2CH_2I$

56. What is the principal product of the following reaction:

O
/ \
$CH_3CHDDCH_2$ + $CH_3CH_2CH_2CH_2MgBr$/ether; then H_2O DDDDD4 ?

A) $CH_3CHCH_2CH_2CH_2CH_3$
3
CH_2OH

* B) OH
3
$CH_3CHCH_2CH_2CH_2CH_2CH_3$

C) OH
3
$CH_3CHCH_2OCH_2CH_2CH_2CH_3$

D) $CH_3CHOCH_2CH_2CH_2CH_3$
3
CH_2OH

E) $CH_3CHCH_2OCH_2CH_2CH_2CH_3$
3
$OCH_2CH_2CH_2CH_3$

1. Which of the following dienes might react with bromine in CCl_4 to yield 2,5-dibromo-3-hexene?
   A) $CH_2MCHCH_2CH_2CHMCH_2$
   B) $CH_2MCHCHMCHCH_2CH_3$
   C) $CH_3CHMCMCHCH_2CH_3$
   * D) $CH_3CHMCHCHMCHCH_3$
   E) $CH_3CHMCHCH_2CHMCH_2$

2. Which of the following dienes would you expect to be the most stable?
   * A) $CH_3CHMCHCHMCHCH_3$
   B) $CH_3CHMCHCH_2CHMCH_2$
   C) $CH_2MCHCH_2CH_2CHMCH_2$
   D) $CH_2MCHCHCHMCH_2$
   3
   CH_3
   E) $CH_3CHMCMCHCH_2CH_3$

3. What product(s) would you expect from the following substitution reaction of ^{14}C-labelled propene?

N-bromosuccinimide
$^{14}CH_2MCHDCH_3$ DDDDDDDDDDDDDDDDDDD4 ?
ROOR, CCl_4

   A) $^{14}CH_2MCHDCH_2Br$ alone
   * B) $^{14}CH_2MCHDCH_2Br$ and $CH_2MCHD^{14}CH_2Br$ in equal amounts
   C) $CH_2MCHD^{14}CH_2Br$ alone
   D) More $^{14}CH_2MCHCH_2Br$ but a little $CH_2MCHD^{14}CH_2Br$
   E) More $CH_2MCH^{14}CH_2Br$ but a little $^{14}CH_2MCHCH_2Br$

4. Which carbon-carbon bond in the following compound would you expect to be shortest?

$HDCpCDCHMCHDCH_2DCH_3$
t t t t t
I II t IV V
III

* A) I B) II C) III D) IV E) V

5. Which of the following compounds would be the most stable?

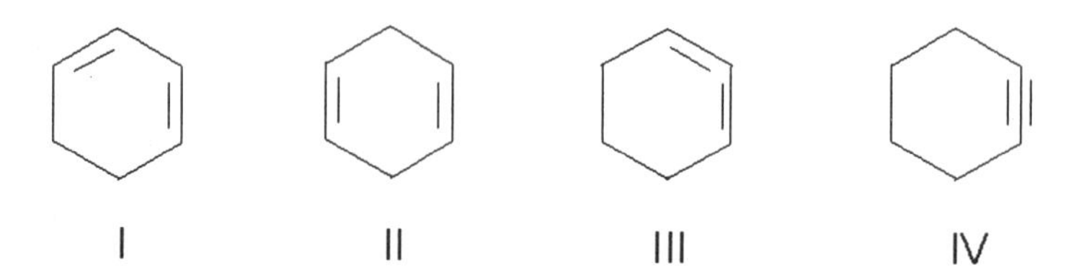

* A) I
B) II
C) III
D) IV
E) They are all of equal stability.

6. Treatment of 2-butene with Cl_2 at 400xC would yield mainly:

A) $CH_2ClCHClCH_2CH_3$
B) $CH_3CHClCH_2CH_3$
C) $CH_3CHMCClCH_3$
* D) $CH_3CHMCHCH_2Cl$ and $CH_3CHCHMCH_2$ 3 Cl
E) $CH_3CHClCHClCH_3$

7. Which carbocation would be most stable?

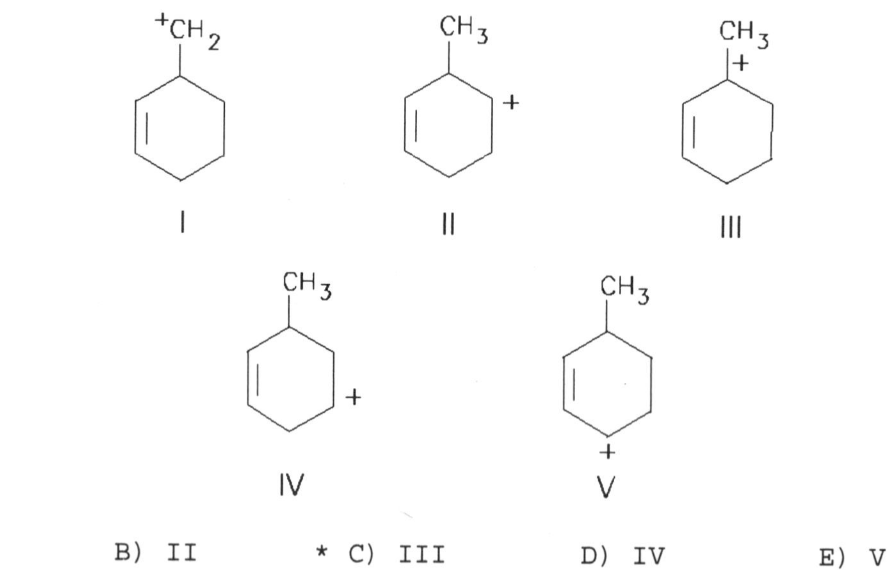

A) I
B) II
* C) III
D) IV
E) V

8. Which free radical would be most stable?

* A)

```
  CH3
  3
CH3CCHMCH2
  y
```

B)

```
   CH3
   3
CH2MCCH2CH2y
```

C)

```
  CH3
  3
CH3CCH2CH3
  y
```

D)

```
        CH3
        3
yCH2CHCHMCH2
```

E)

```
    CH3
    3
CH3CHCHCH3
     y
```

9. Which diene and dienophile would you choose to synthesize the following compound?

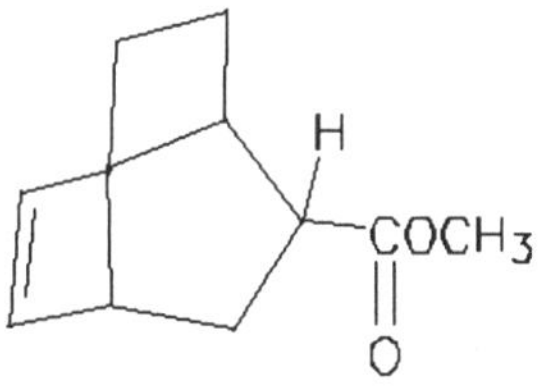

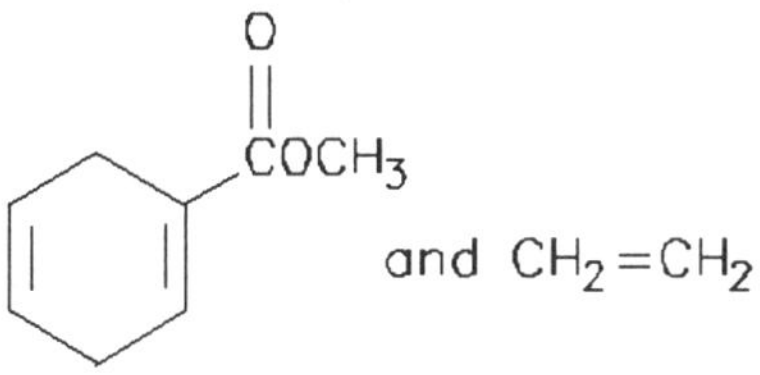

H

$-\overset{O}{\overset{\|}{C}}OCH_3$

I: $\overset{O}{\overset{\|}{C}}OCH_3$ and $CH_2=CH_2$

II: and $HC\equiv C\overset{O}{\overset{\|}{C}}OCH_3$

III: and $CH_2=CH\overset{O}{\overset{\|}{C}}OCH_3$

IV: and $CH_2=CH\overset{O}{\overset{\|}{C}}OCH_3$

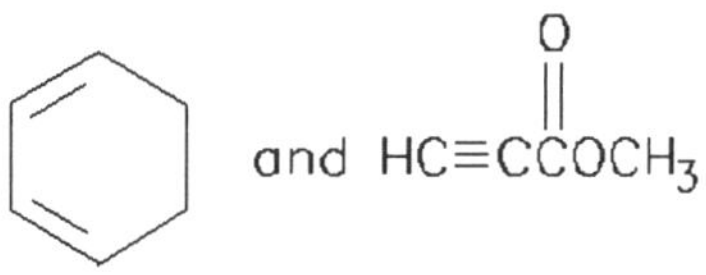

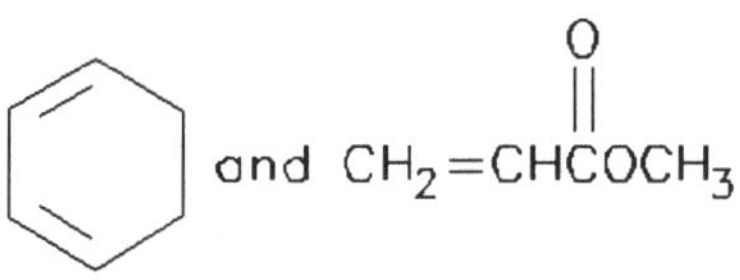

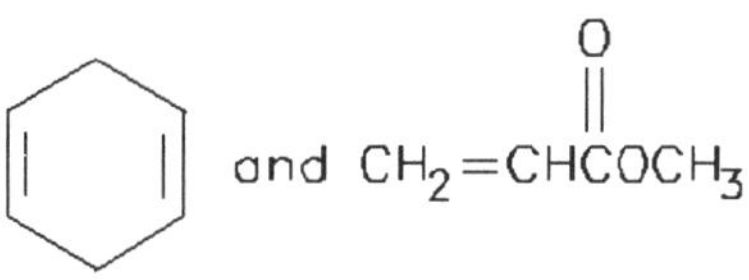

A) I
B) II
* C) III
D) IV
E) None of these

10. Which alkene would you expect to have the lowest heat of hydrogenation?

A) $CH_2MCHCH_2CH_2CHMCH_2$

B)
```
CH2MCHCH2     H
         \   /
          CMC
         /   \
        H     CH3
```

C)
```
CH3      CH2CHMCH2
   \    /
    CMC
   /   \
  H     H
```

D)
```
         H     H
          \   /
CH3        CMC
   \      /   \
    CMC        CH3
   /   \
  H     H
```

* E)
```
CH3       H
   \     /
    CMC         H
   /   \       /
  H     CMC
       /   \
      H     CH3
```

11. Which compound would have the shortest carbon-carbon single bond?

A) CH_3DCH_3

B) $CH_2MCHDCH_3$

* C) HCpCDCpCH

D) $CH_2MCHDCpCH$

E) $CH_2MCHDCHMCH_2$

12. Indicate which products would be obtained from the 1:1 chlorination of 1,5-hexadiene at high temperature (500xC)?

```
ClCH2CHCH2CH2CHMCH2
     3
     Cl

        I
```

```
ClCH2CHMCHCHMCHCH2Cl

          II
```

```
ClCH2CHMCHCH2CHMCH2

        III
```

```
CH2MCHCH2CHCHMCH2
         3
         Cl

        IV
```

```
ClCH2CHCH2CH2CHCH2Cl
     3        3
     Cl       Cl

          V
```

A) I and II

B) II and III

* C) III and IV

D) IV and V

E) V and I

13. Which reagent would serve as the basis for a simple chemical test to distinguish between $CH_3CH=CHBr$ and $CH_2=CHCH_2Br$?

* A) $AgNO_3/C_2H_5OH$
B) $KMnO_4/H_2O$
C) Br_2/CCl_4
D) Concd. H_2SO_4
E) CrO_3 in H_2SO_4

14. Which reagent would convert 1,3-pentadiene into the alcohol shown below?

$CH_3CHMCHCHMCH_2$ DDDDDD4 (?) $CH_3CHMCHCHCH_3$
3
OH

A) $KMnO_4/^{-}OH$
B) OsO_4
C) H_2O_2, then H^+
D) Cl_2/H_2O
* E) H^+/H_2O

15. What would be the product of the following reaction?

+ $CH_2=CHCCH_3$ (O) —heat→

I: CH_3, C, $CH=CH_2$

II: O, C–CH_3

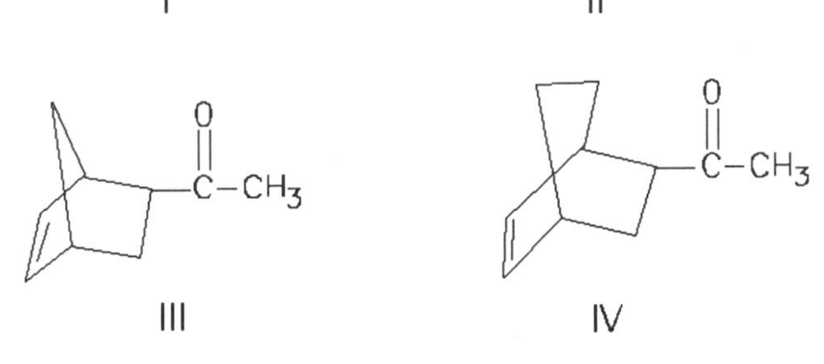

A) I
B) II
* C) III
D) IV
E) All of these

16. How would you synthesize:

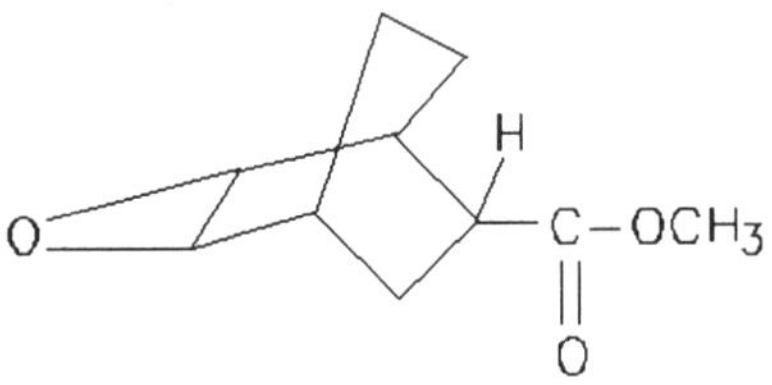

I 1,3–Cyclohexadiene + $CH_2=CHCOCH_3$ (C=O), then RCOOH (C=O)

II 1,3-Cyclohexadiene + RCOOH (C=O), then $CH_2=CHCOCH_3$ (C=O)

III
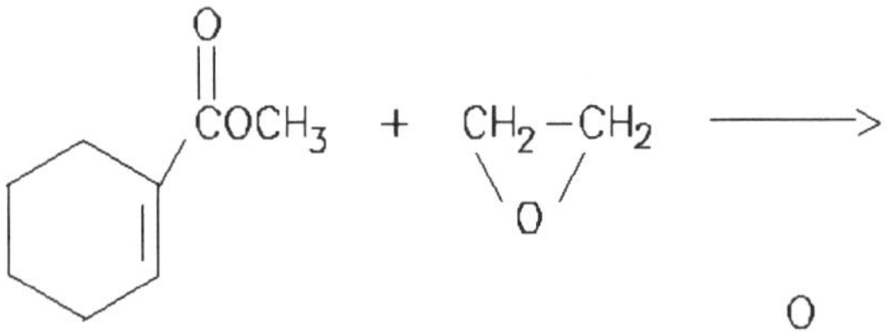

IV 1,4–Cyclohexadiene + $CH_2=CHCOCH_3$ (C=O), then CH_2-CH_2 (epoxide, O)

V
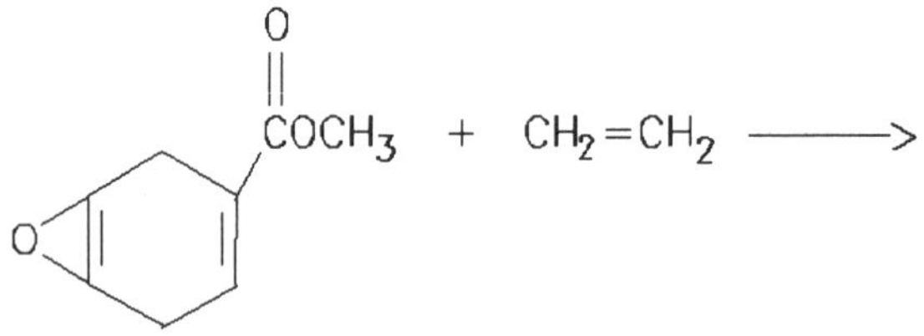

* A) I B) II C) III D) IV E) V

17. Which of the following would afford a synthesis of the following compound?

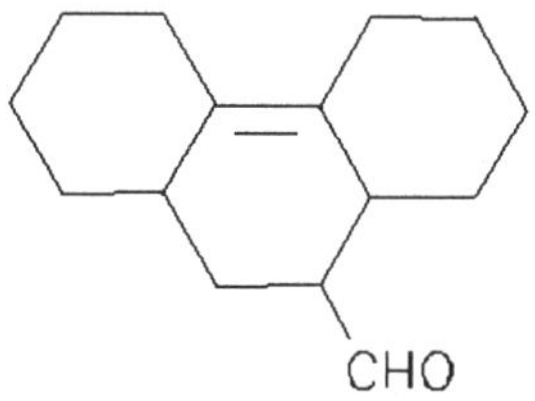

CHO

I $2\ CH_3CH{=}CHCH_3$ + (CHO)

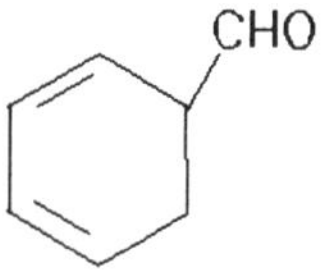

II $CH_3CH{=}CH_2$ + (CHO)

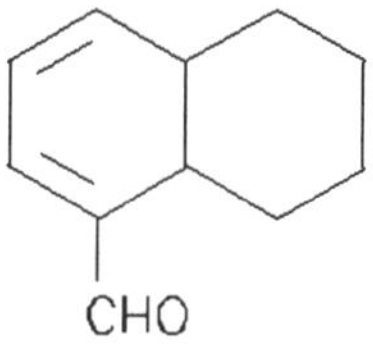

III 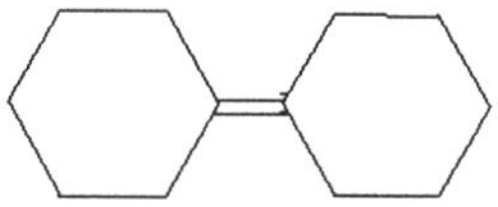 + $CH_2{=}CHCHO$

IV + $CH_2{=}CHCHO$

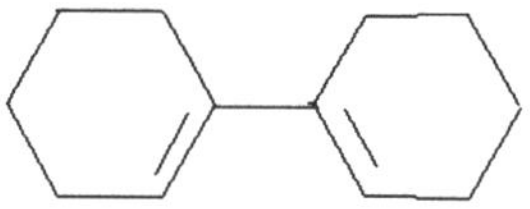

A) I
B) II
C) III
* D) IV
E) None of these

18. Which diene would be least reactive toward Diels-Alder addition of maleic anhydride?

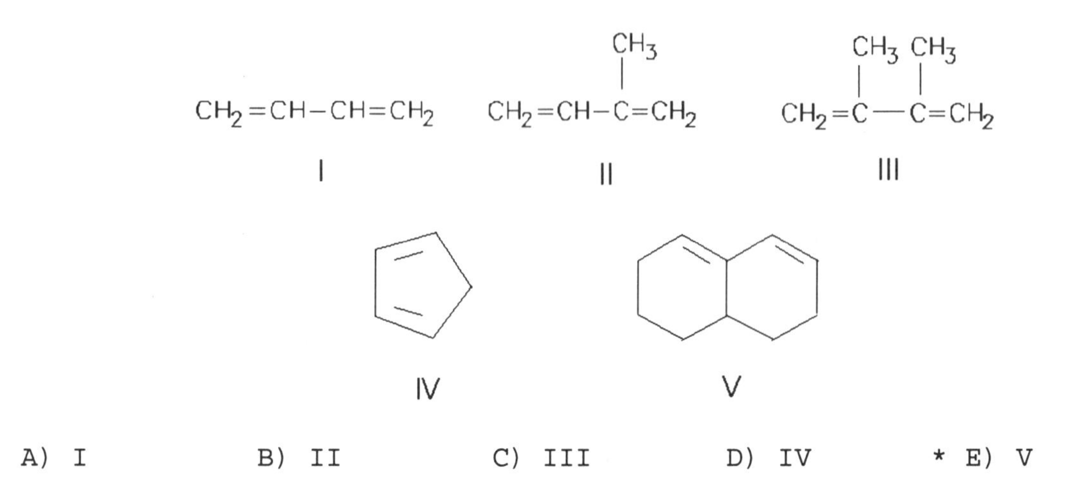

A) I B) II C) III D) IV * E) V

19. Which diene would you expect to react most rapidly with maleic anhydride?

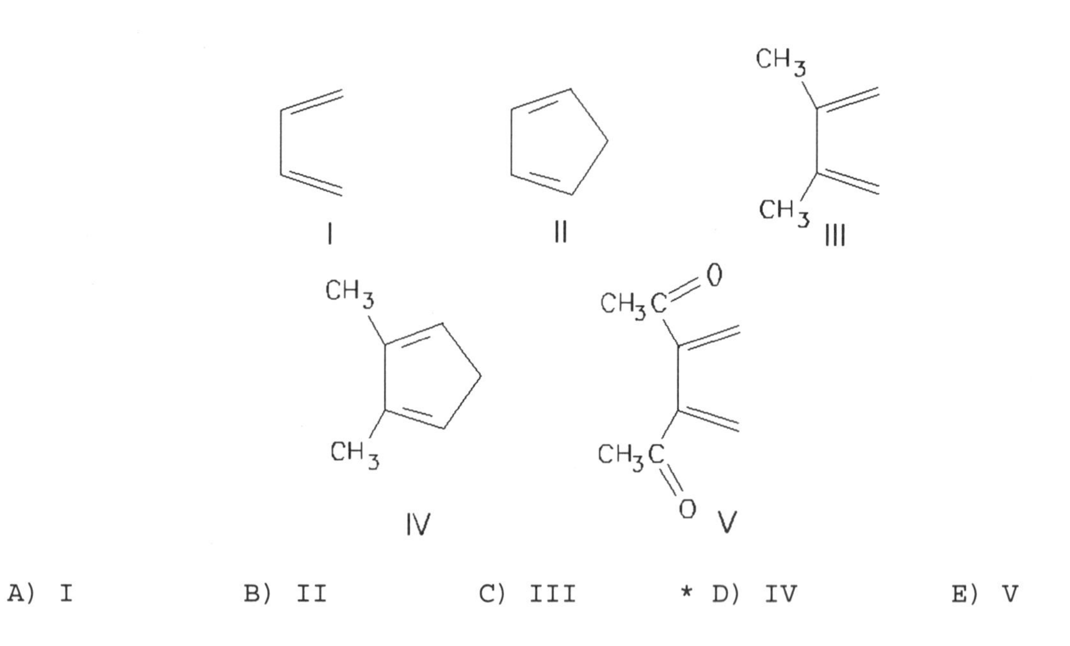

A) I B) II C) III * D) IV E) V

20. Select the structure of the conjugated diene.

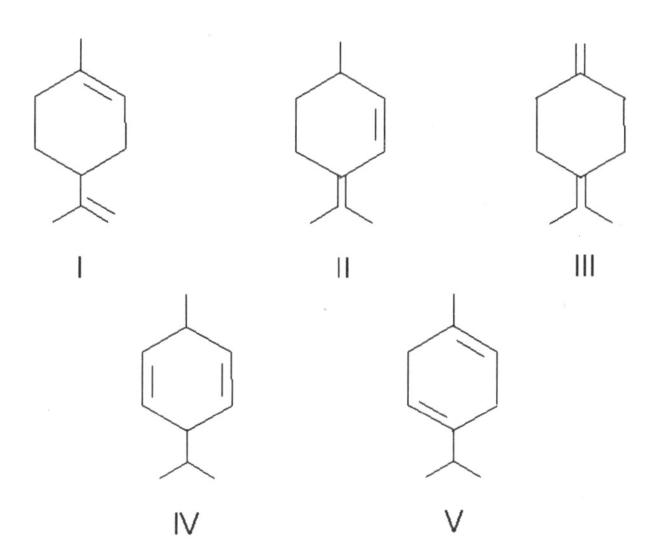

A) I * B) II C) III D) IV E) V

21. The allyl radical has ___________ bonding c molecular orbitals.
* A) 1 B) 2 C) 3 D) 4 E) 5

22. Which is <u>not</u> an example of resonance?

I $CH_2=CH-CH_2\cdot \longleftrightarrow \cdot CH_2-CH=CH_2$

II (cyclopentene with $\overset{+}{C}H_2$ substituent) $\longleftrightarrow$ (methylenecyclopentane, $=CH_2$, with + on ring carbon)

III (methylcyclohexane with + on ring carbon, CH_3) $\longleftrightarrow$ (cyclohexane with $\overset{+}{C}H_2$)

IV $CH_2=CH-\dot{C}H-CH_2CCl_3 \longleftrightarrow \dot{C}H_2-CH=CH-CH_2CCl_3$

A) I
B) II
* C) III
D) IV
E) None of these.

23. Estimate the stabilization energy for 1,3-butadiene using the heats of hydrogenation in Table 1.

Table 1. Heats of Hydrogenation for Selected Compounds

Compound	Moles H_2	(H(kcal mol^{-1})
1-Butene	1	-30.3
1-Pentene	1	-30.1
1,3-Butadiene	2	-57.1
1,3-Pentadiene	2	-54.1

A) 3.0 kcal mol^{-1}
* B) 3.5 kcal mol^{-1}
C) 6.7 kcal mol^{-1}
D) 57.1 kcal mol^{-1}
E) 26.8 kcal mol^{-1}

24. Considering both configurational and conformational factors, select the most stable form of 2,4-hexadiene.

I　　II　　III　　IV　　V

* A) I　　B) II　　C) III　　D) IV　　E) V

25. The correct IUPAC name of the compound on the right is:

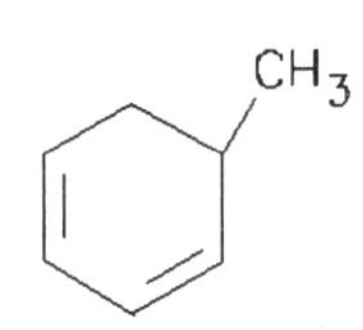

A) 1-Methyl-2,4-cyclohexadiene
B) 3-Methyl-1,3-cyclohexadiene
* C) 5-Methyl-1,3-cyclohexadiene
D) 6-Methyl-1,3-cyclohexadiene
E) None of these

26. The allyl radical has ______ electrons in bonding c molecular orbitals.
A) 1　　* B) 2　　C) 3　　D) 4　　E) 5

27. How could the following synthesis be carried out?

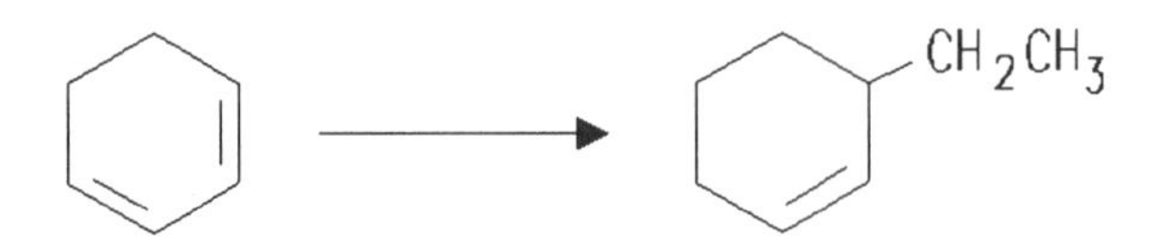

A) (1) Br_2/CCl_4; (2) CH_3CH_2MgCl; (3) CH_3ONa/CH_3OH
* B) (1) HBr, 80xC; (2) $(CH_3CH_2)_2CuLi$, ether
C) (1) HBr, 80xC; (2) Mg, ether; (3) CH_3CH_2OH, then H_3O^+
D) More than one of the above
E) All of the above

28. Which of these conjugated dienes can undergo a Diels-Alder reaction?

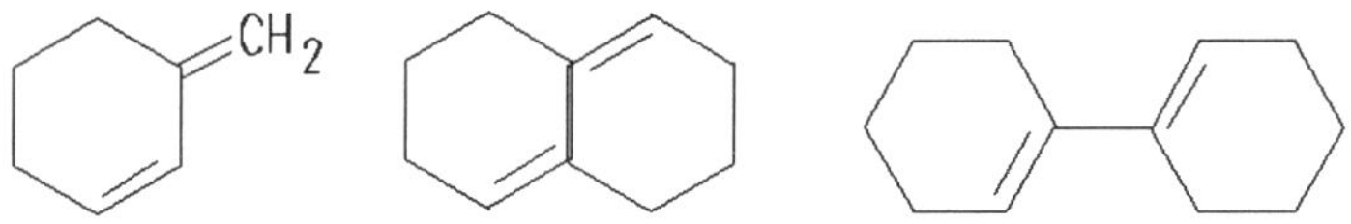

A) I B) II * C) III D) IV E) V

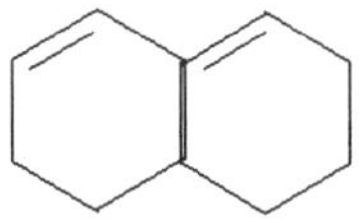

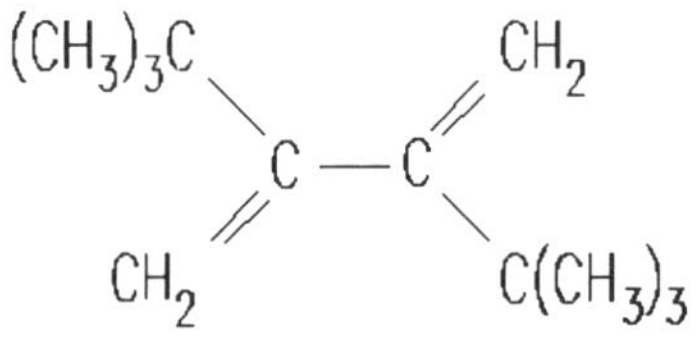

29. Which of the following dienes is a cumulated diene?
A) $CH_2MCHCH_2CH_2CHMCH_2$
B) $CH_2MCHCHMCHCH_2CH_3$
* C) $CH_3CHMCMCHCH_2CH_3$
D) $CH_3CHMCHCHMCHCH_3$
E) $CH_3CHMCHCH_2CHMCH_2$

30. Which carbon-carbon bond in the following compound would you expect to be longest?

H-CpC-CH=CH-CH_2-CH_3
t t t t t
I IIIII IV V

A) I B) II C) III D) IV * E) V

31. What product(s) would you expect from the following substitution reaction of ^{14}C-labelled propene?

```
                    Cl2
14CH2=CH-CH3   DDDDDD4   ?
                  500xC
```

A) $^{14}CH_2MCHDCH_2Cl$ alone
* B) $^{14}CH_2MCHDCH_2Cl$ and $CH_2MCHD^{14}CH_2Cl$, in equal amounts
C) $CH_2MCHD^{14}CH_2Cl$ alone
D) more $^{14}CH_2MCHCH_2Cl$, but a little $CH_2MCHD^{14}CH_2Cl$
E) more $CH_2MCHD^{14}CH_2Cl$, but a little $^{14}CH_2MCHCH_2Cl$

32. Which alkene would you expect to be most stable?

A) $CH_2MCHCH_2CH_2CHMCH_2$

```
B) CH2MCHCH2      H
            \    /
             CMC
            /    \
           H      CH3

C) CH3    CH2CHMCH2
     \    /
      CMC
     /    \
    H      H

D)      H      H
         \    /
   CH3    CMC
     \    /   \
      CMC      CH3
     /    \
    H      H

* E) CH3    H
       \    /
        CMC      H
       /    \   /
      H      CMC
            /    \
           H      CH3
```

33. Which of the following could be used to synthesize 3-bromopropene?

```
A)
                          25xC
   CH3CH=CH2   +   Br2   DDDDDD4
                          CCl4

B)
                                        ROOR
   CH3CHMCH2   +   N-bromosuccinimide  DDDDDD4
                                        CCl4

C) CH2MCHCH2OH   +   PBr3   DDDDDD4
```

* D) More than one of these
E) None of these

34. Which carbocation would be most stable?

A) $CH_3DCHDCH_2DCHMCH_2$
      +

B) $CH_2DCH_2DCHMCHDCH_3$
  +

* C)
      CH_3
      3
  $CH_3DCDCHMCH_2$
      +

D)
      CH_3
      3
  $CH_3DCHDCH_2DCH_2{}^+$

E)
      CH_3
      3
  $CH_2MCDCH_2DCH_2{}^+$

35. Which is <u>not</u> a proper resonance structure for 1,3-butadiene?

A) $CH_2MCHDCHMCH_2$

* B)
  .   .
  $CH_2DCHMCHDCH_2$

C) -   +
  $:CH_2DCHMCHDCH_2$

D) +   .-
  $CH_2DCHMCHDCH_2$

E) All are correct.

36. Treatment of 2-butene with N-bromosuccinimide in CCl_4 would yield mainly:

A) $CH_3CHBrCHBrCH_3$ alone
B) $CH_3CHMCBrCH_3$ alone
C) $CH_3CHMCHCH_2Br$ alone
D) $CH_2BrCH_2CHMCH_2$ and $CH_3CHMCHCH_2Br$
* E) $CH_3CHBrCHMCH_2$ and $CH_3CHMCHCH_2Br$

37. Which of the following compounds is <u>not</u> formed as a result of a chain-termination step in the free radical chlorination of propene?

A) CH_2MCHCH_2Cl
B) Cl_2
C) $CH_2MCHCH_2CH_2CHMCH_2$
* D) HCl
E) All can be formed in chain-termination steps.

38. A thermodynamically-controlled reaction will yield predominantly:

* A) the more/most stable product.
B) the product whose formation requires the smallest free energy of activation.
C) the product that can be formed in the fewest steps.
D) the product that is formed at the fastest rate.
E) the product which possesses the greatest potential energy.

39. What would be the best synthesis?

$$CH_2MCHCH_3 \xrightarrow{?} \underset{Br}{CH_2}\underset{Br}{CH}CH_2Cl$$

A) Propene $\xrightarrow{Br_2/CCl_4}$ $\xrightarrow{Cl_2,\ h0}$

* B) Propene $\xrightarrow{Cl_2,\ 400xC}$ $\xrightarrow{Br_2/CCl_4}$

C) Propene $\xrightarrow{HCl}$ $\xrightarrow{Br_2,\ h0}$

D) Propene $\xrightarrow{NBS/CCl_4}$ $\xrightarrow{Cl_2/CCl_4}$

E) Propene $\xrightarrow{Cl_2/CCl_4}$ $\xrightarrow{NBS/CCl_4}$

40. Which is the diene that yields on ozonolysis (O_3, followed by H_2O/Zn) an equimolar mixture of CH_2O, CH_3CHO, and $OHC-CH_2-CHO$?

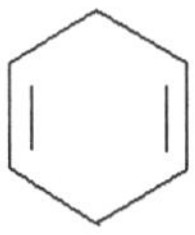

I: cyclohexa-1,4-diene (ring structure)

II: 5-methylcyclopentadiene (ring with CH_3)

III: $CH_2=CH-CH=CH-CH_2-CH_3$

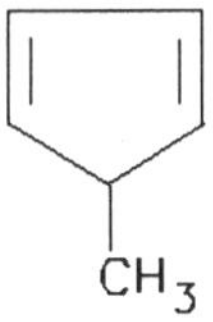

IV: $CH_2=CH-CH_2-CH=CH-CH_3$

V: $CH_2=CH-CH_2-CH_2-CH=CH_2$

A) I   B) II   C) III   * D) IV   E) V

41. Which diene and dienophile would you choose to synthesize the following compound?

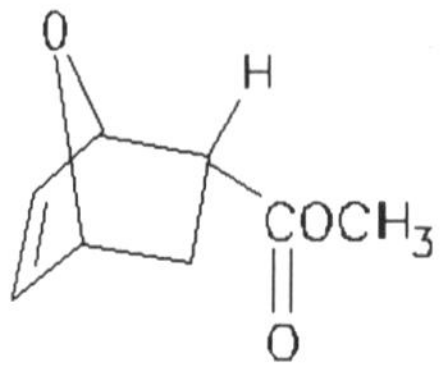

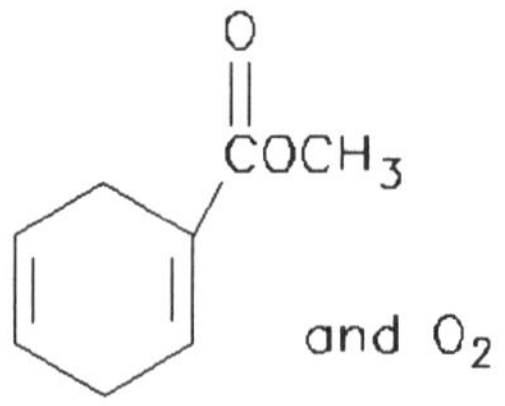

$$\text{[7-oxabicyclo[2.2.1]hept-5-ene with } -\overset{O}{\overset{\|}{C}}OCH_3 \text{ and H substituents]}$$

I: 1,3-cyclohexadiene bearing $\overset{O}{\overset{\|}{C}}OCH_3$ and O_2

II: furan and $HC{\equiv}C{-}\overset{O}{\overset{\|}{C}}OCH_3$

III: furan and $CH_2{=}CH{-}\overset{O}{\overset{\|}{C}}OCH_3$

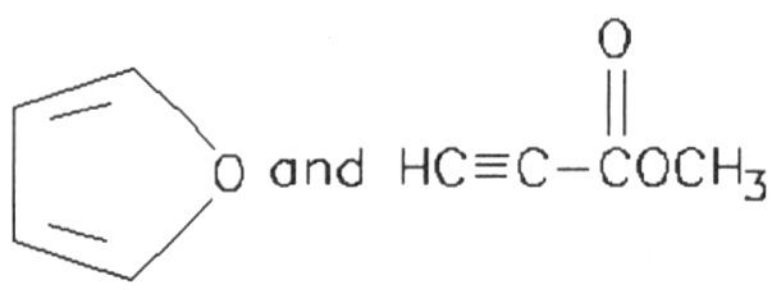

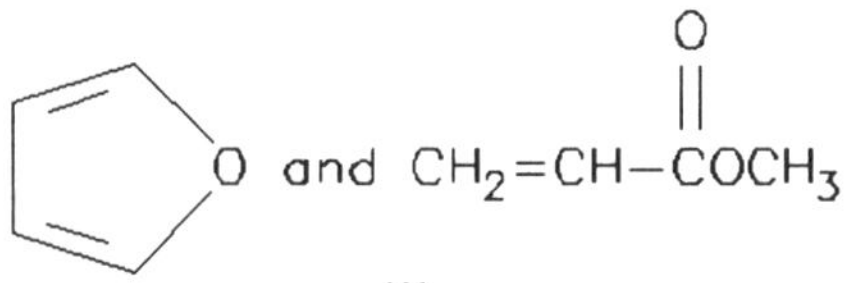

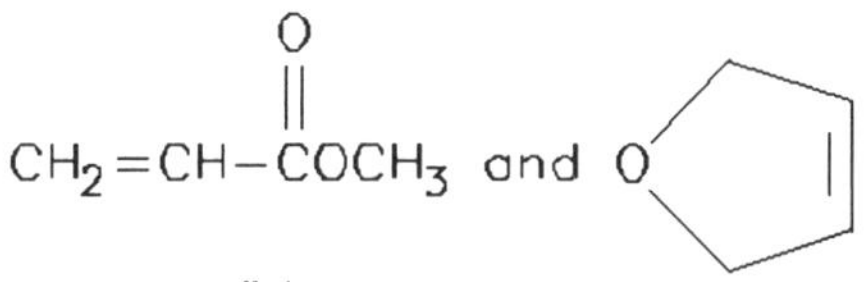

IV: $CH_2{=}CH{-}\overset{O}{\overset{\|}{C}}OCH_3$ and 2,3-dihydrofuran

A) I
B) II
* C) III
D) IV
E) None of these

42. Which compounds could be used in a Diels-Alder synthesis of

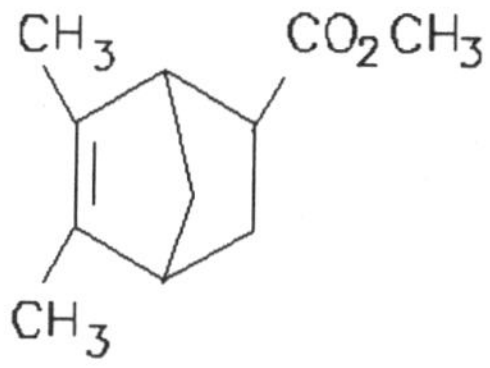

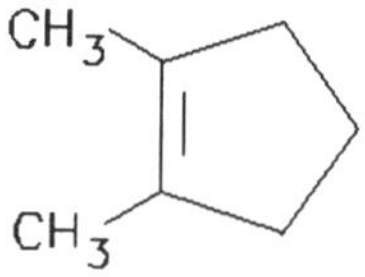

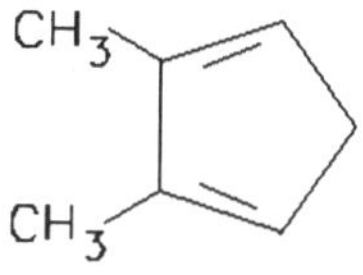

[Structure: bicyclo[2.2.1]heptene with two CH_3 groups on the double bond and a CO_2CH_3 substituent] ?

I: 1,2-dimethylcyclopentene (CH_3, CH_3)

II: 1,2-dimethylcyclopentadiene (CH_3, CH_3)

III: $CH_2{=}CHCO_2CH_3$

IV: $HC{\equiv}CCO_2CH_3$

A) I and III B) I and IV * C) II and III D) II and IV

43. Which is the major product of the following reaction?

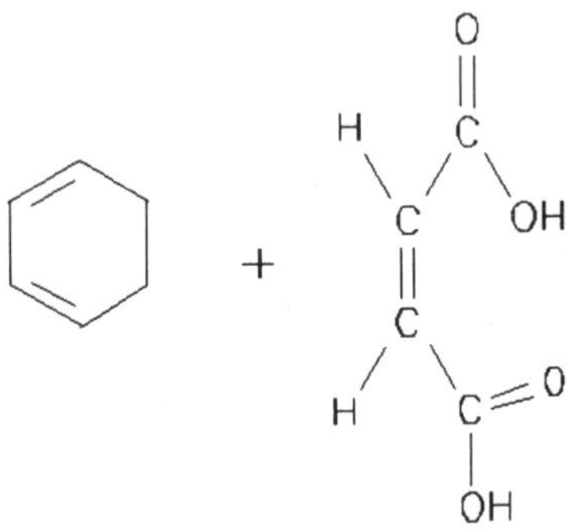

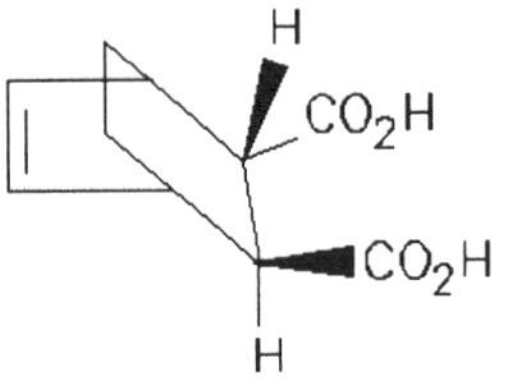

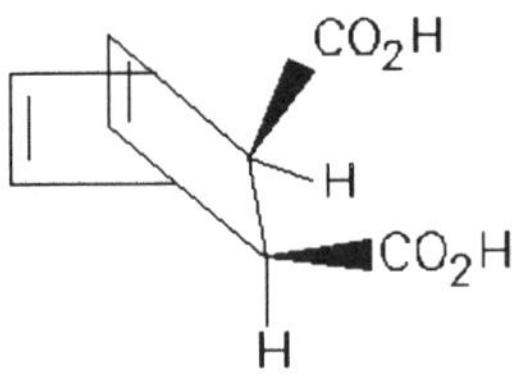

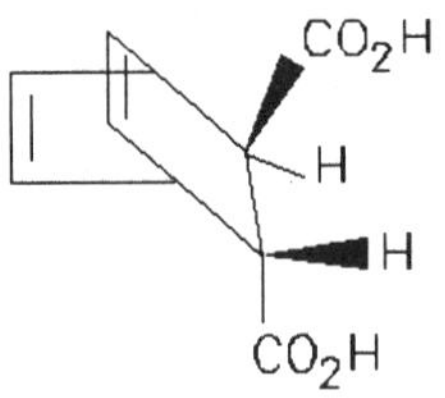

A) I
B) II
C) III

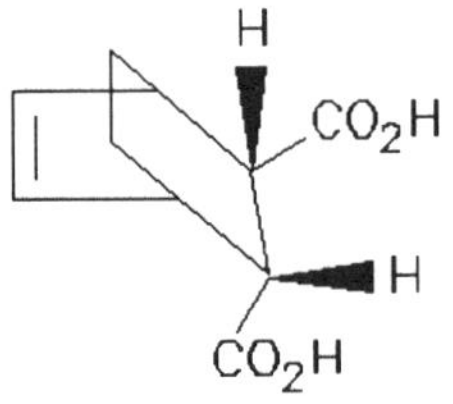

* D) IV
E) None of these

44. Which diene and dienophile would you choose to synthesize the following compound?

(structure: bicyclic compound with O bridge and $\overset{O}{\|}$ COCH$_3$ substituent)

I: (cyclohexadiene with $\overset{O}{\|}$COCH$_3$) and O_2

II: (furan) and $HC{\equiv}C{-}\overset{O}{\overset{\|}{C}}OCH_3$

III: (butadiene) and $CH_2{=}CH{-}\overset{O}{\overset{\|}{C}}OCH_3$

IV: $CH_2{=}CH{-}\overset{O}{\overset{\|}{C}}OCH_3$ and (dihydrofuran)

A) I
* B) II
C) III
D) IV
E) None of these

45. From the standpoint of reactivity, which is the poorest choice of dienophile to react with

H3C CH3
 3 3
$CH_2{=}C{-}C{=}CH_2$ in a Diels-Alder reaction.

A) H
 3
$CH_2{=}CHC{=}O$

B) H CO2CH3
 \ /
 C=C
 / \
CH_3O_2C H

* C) CH_3 CH_3
 \ /
 C=C
 / \
 H H

D) $CH_3O_2C{-}CpC{-}CO_2CH_3$

E) 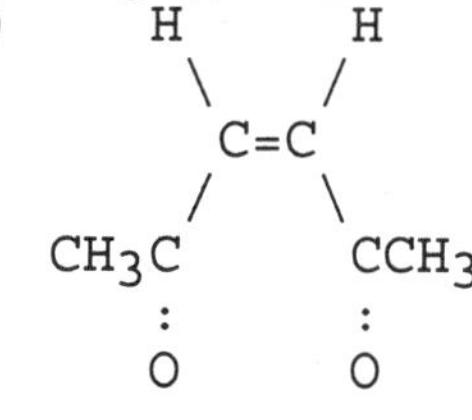

46. The allyl radical has how many electrons in nonbonding c molecular orbitals?
* A) 1 B) 2 C) 3 D) 4 E) 0

47. The following reaction

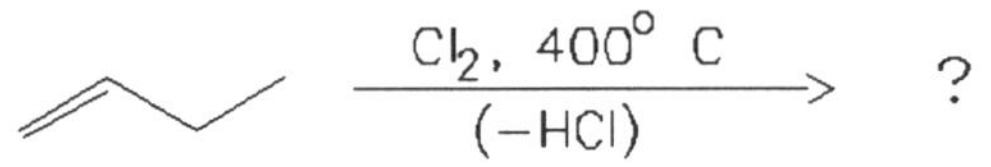

Cl_2, 400° C
(–HCl)
?

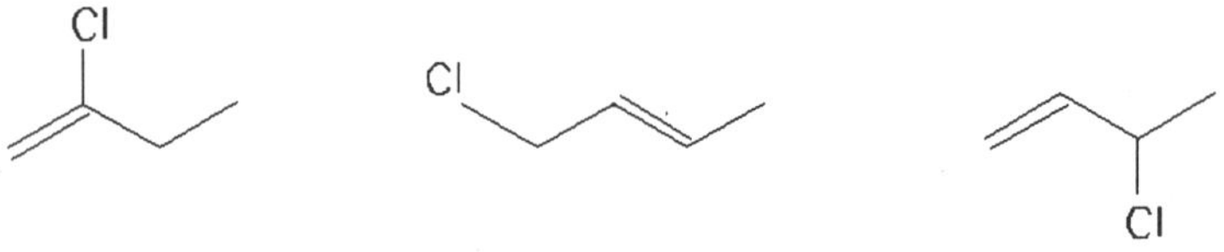

Cl
I

Cl
II

Cl
III

would yield:
A) I
B) II
C) III
D) a mixture of I and II
* E) a mixture of II and III

48. Which diene would be <u>least</u> stable?

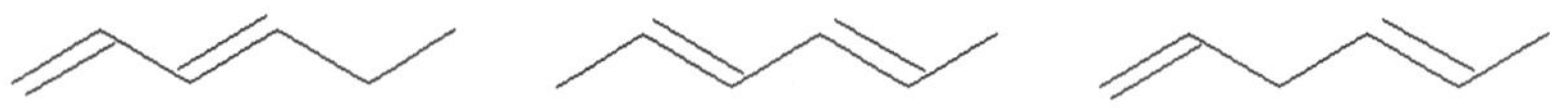

I II III

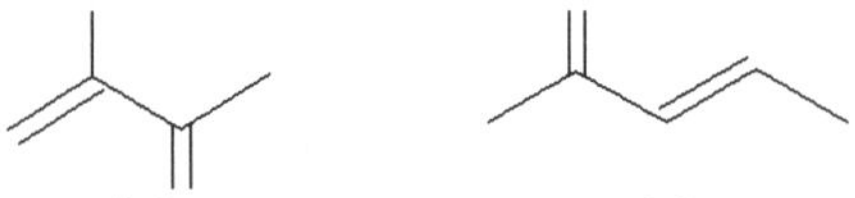

IV V

A) I B) II * C) III D) IV E) V

49. What is the product of the following reaction?

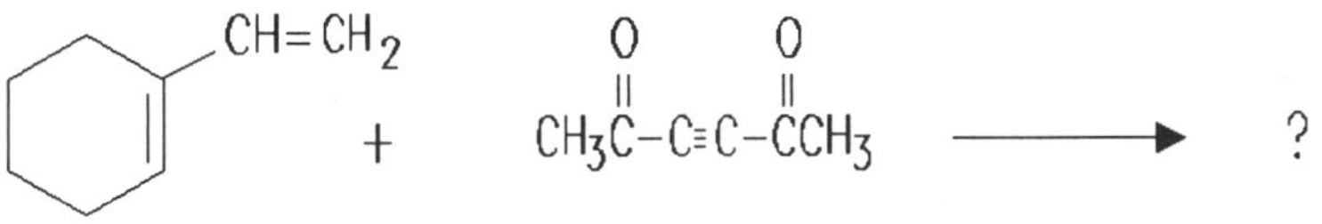

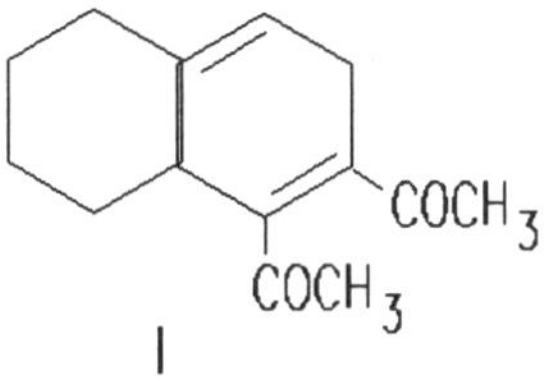

$$\text{1-vinylcyclohexene} + CH_3\overset{O}{\overset{\|}{C}}-C\equiv C-\overset{O}{\overset{\|}{C}}CH_3 \longrightarrow ?$$

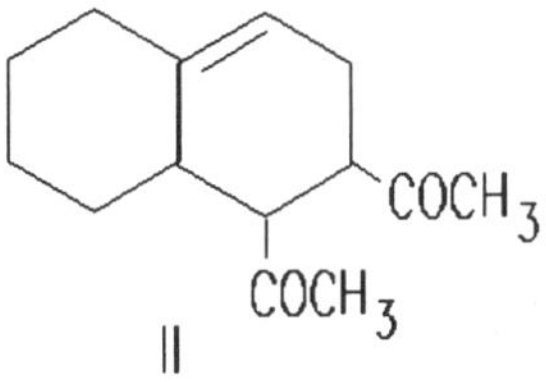

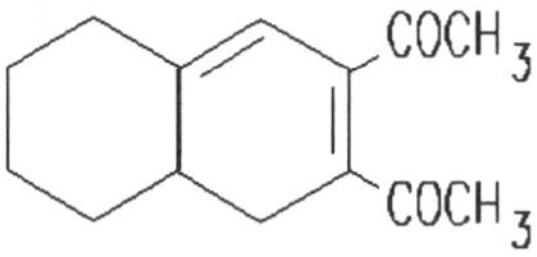

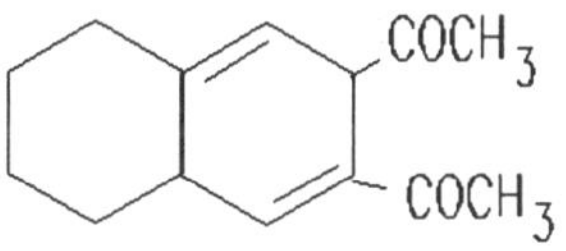

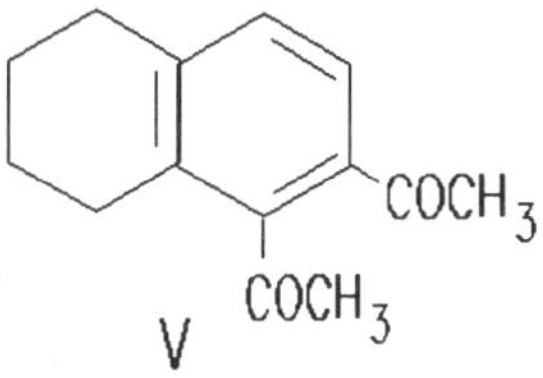

* A) I B) II C) III D) IV E) V

50. Which would be the best synthesis of the following compound?

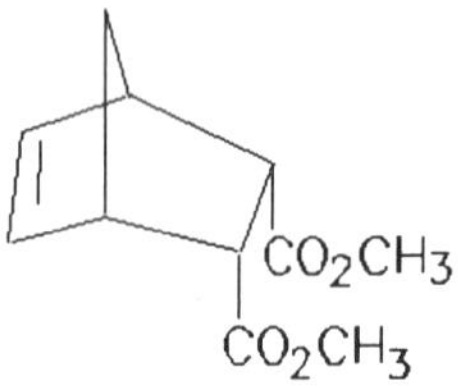

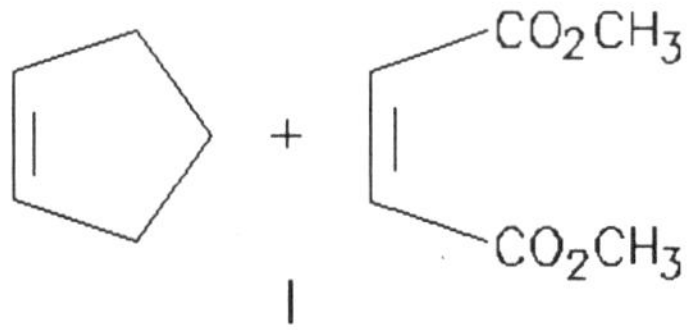

CO_2CH_3
CO_2CH_3

+ CO_2CH_3 CO_2CH_3
I

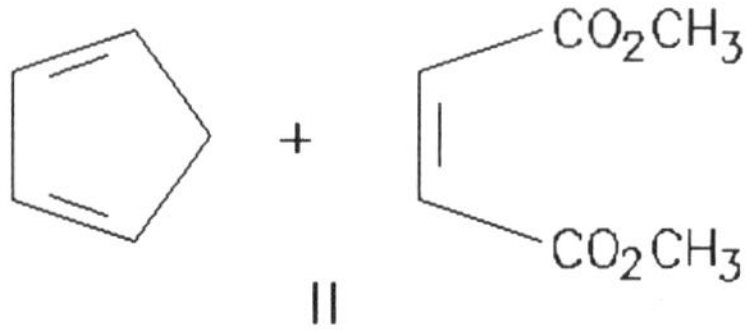

+ CO_2CH_3 CO_2CH_3
II

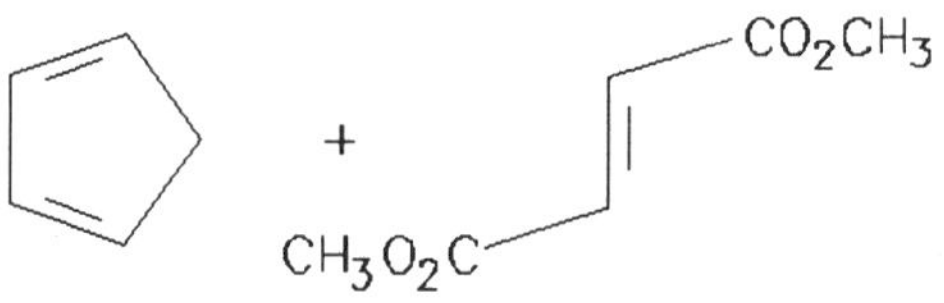

+ CO_2CH_3 CH_3O_2C
III

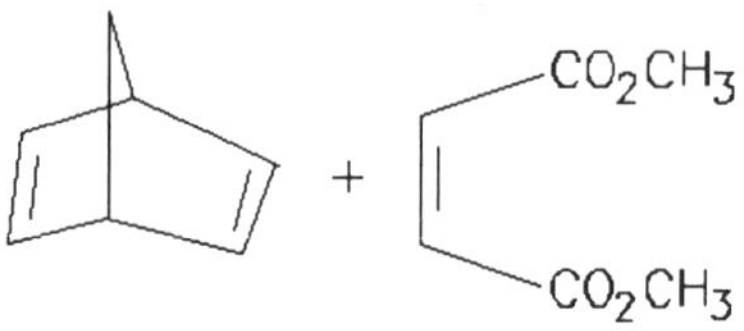

+ CO_2CH_3 CO_2CH_3
IV

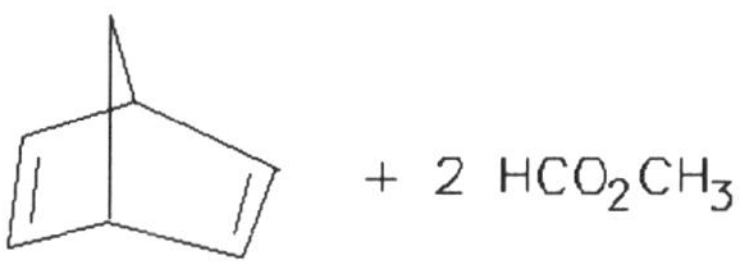

+ 2 HCO_2CH_3
V

A) I * B) II C) III D) IV E) V

51. Which pair does not represent a pair of resonance structures?

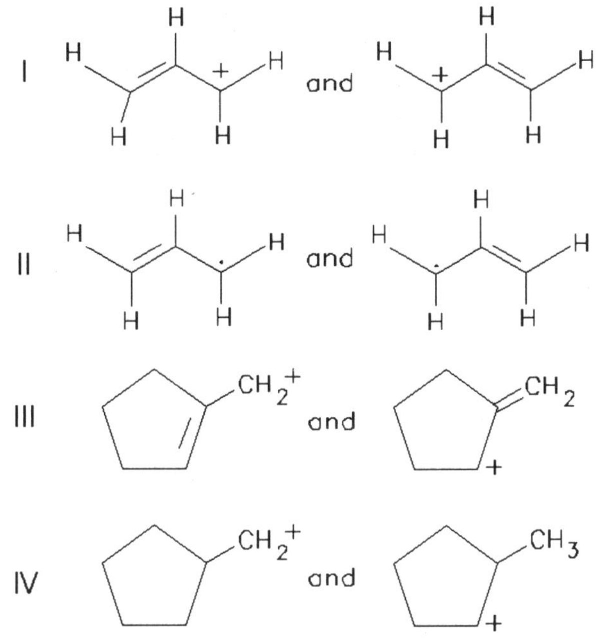

I
and
II
and
III
CH_2^+
and
CH_2
IV
CH_2^+
and
CH_3

A) I
B) II
C) III
* D) IV
E) All of these represent pairs of resonance structures.

52. Which reaction would produce the following compound?

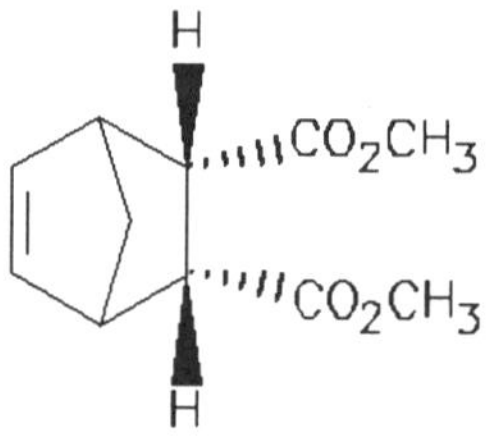

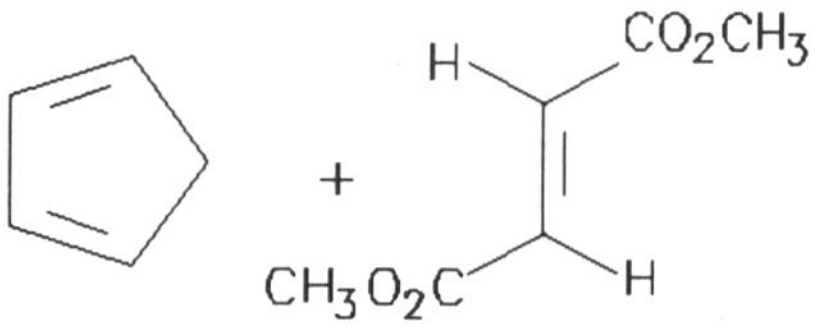

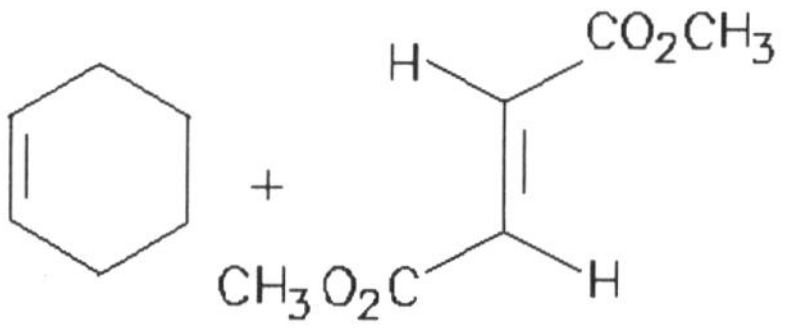

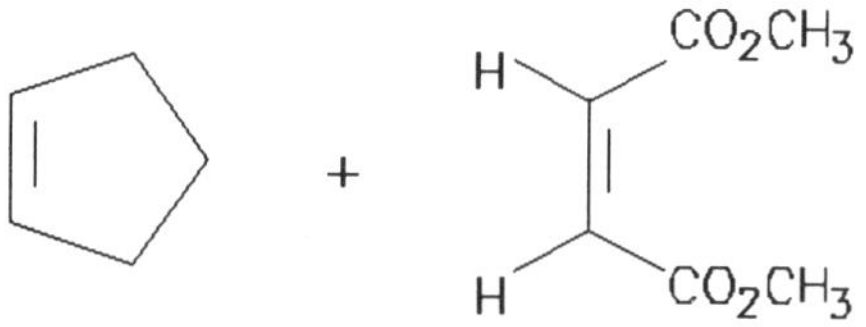

A) I
B) II
C) III
* D) IV
E) None of the above

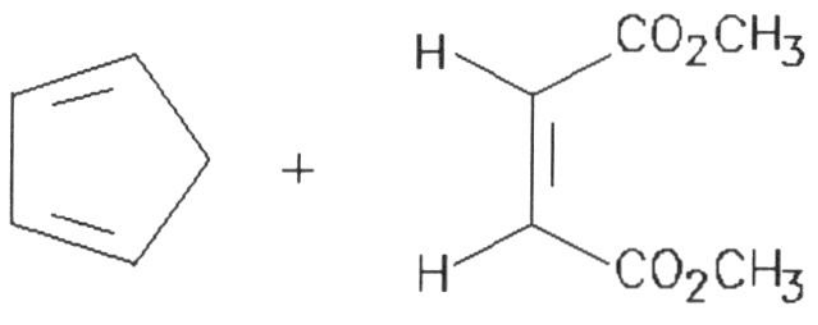

53. Which of the following pairs of compounds could be used as the basis for a Diels-Alder synthesis of the compound shown below?

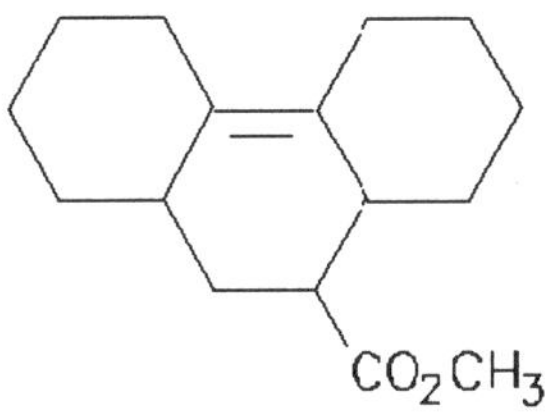

I $2\ CH_3CH{=}CHCH_3$ + 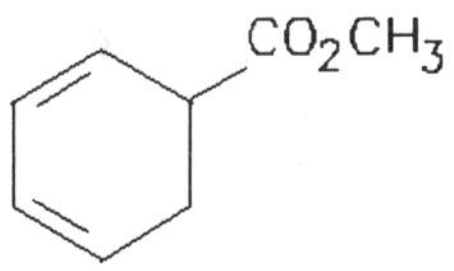

II 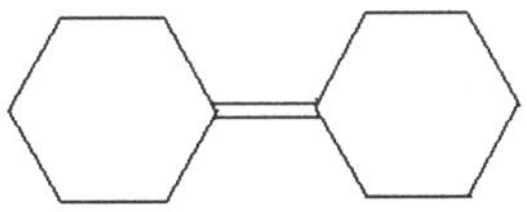 + $CH_2{=}CHCO_2CH_3$

III $CH_3CH{=}CH_2$ + 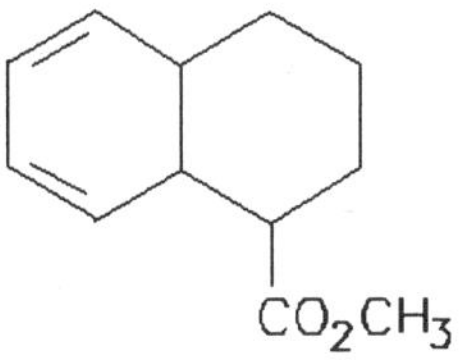

IV 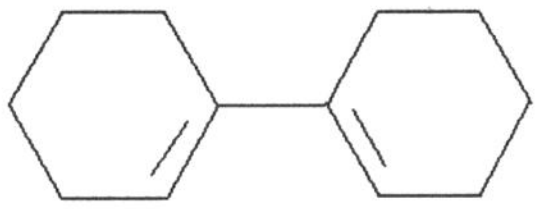 + $CH_2{=}CHCO_2CH_3$

A) I
B) II
C) III
* D) IV
E) More than one of the above

54. A reaction under kinetic (or rate) control will yield predominantly:
A) the most stable product.
B) the product that can be formed in the fewest steps.
* C) the product whose formation requires the smallest free energy of activation.
D) the product with the greatest potential energy.
E) the product with the least potential energy.

55. Which set of conditions does <u>not</u> result in allylic halogenation of an alkene?

A) Cl_2 at 400xC
* B) Cl_2 in CCl_4 at 25xC
C) Cl_2, ROOR,h0
D) Br_2 at low concentration in CCl_4
E) N-Bromosuccinimide in CCl_4, ROOR

56.

CH_3
3
Which carbon of $CH_3DCHDCH_2DCHMCH_2$ is predicted to be the major
t t t t t
V IV III II I

site of substitution when this alkene reacts with chlorine at 400xC?

A) I B) II * C) III D) IV E) V

57. Which of these is <u>not</u> a useful method for the synthesis of 1,3-pentadiene?

A)
OH
3
$HOCH_2CH_2CH_2CHCH_3$ + H_2SO_4 at 180xC

* B)
Br Br
3 3
$CH_3CHCH_2CHCH_3$ + $(CH_3)_3COK/(CH_3)_3COH$ at 75xC

C)
OH OH
3 3
$CH_3CHCH_2CHCH_3$ + H_2SO_4 at 180xC

D) $HCpCCHMCHCH_3$ + H_2, Ni_2B (P-2)

E)
Br
3
$CH_2MCHCHCHCH_3$ + Zn, CH_3CO_2H
3
Br

58.

V
u
CH_3
3
Which hydrogen atom(s) of $CH_3DCHMCHDCHDCHMCHDCH_3$ is/are most
t t t t
I II III IV

susceptible to abstraction by free radicals?

A) I B) II C) III * D) IV E) V

59. Ignoring stereochemistry, the 1:1 reaction of bromine with 1,3-cyclohexadiene at 25xC in the dark and in the absence of peroxide forms which of these?

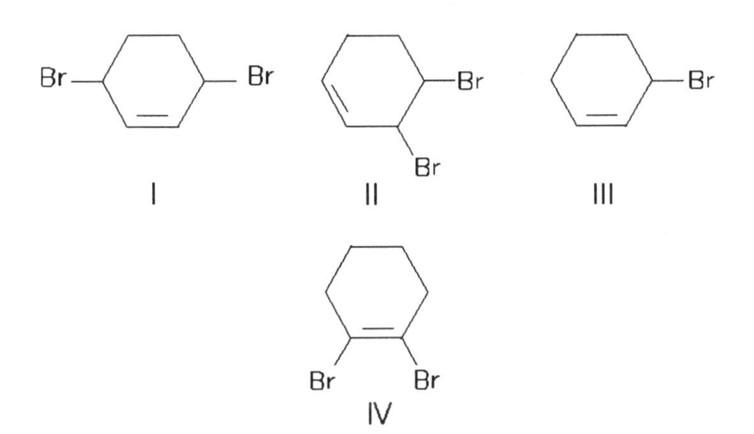

A) I
B) II
C) III
D) IV
* E) Both I and II

60. Which is the only compound which can be <u>completely</u> ruled out as a product of the reaction of 1,3-butadiene with HCl?

A)
H
3
$CH_3CCHMCH_2$
3
Cl

B)
Cl
3
$CH_3CCHMCH_2$
3
H

C)
H
3
CH_3CMCCH_2Cl
3
H

D)
H H
3 3
CH_3CMCCH_2Cl

* E)
Cl
3
CH_3CMCCH_3
3
H

61. When 3-methyl-2,3-pentanediol vapor is passed over hot alumina (Al_2O_3), the chief product to be expected is which of these?

A) 2-Ethyl-1,3-butadiene
* B) 3-Methyl-1,3-pentadiene
C) 3-Methyl-1,2-pentadiene
D) 3-Methyl-2,3-pentadiene
E) 3-Methyl-1,4-pentadiene

62. Which is an <u>untrue</u> statement concerning the Diels-Alder reaction?
A) The reaction is a *syn* addition.
B) The diene must be in the *s*-cis conformation to react.
C) Most Diels-Alder reactions are reversible.
* D) Generally, the adduct formed most rapidly is the *exo* product.
E) Depending on the nature of the dienophile, both electron-releasing and electron-withdrawing groups in the diene can favor adduct formation.

63.

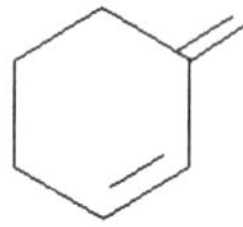

does not undergo the Diels-Alder reaction because:
A) ring systems cannot function as the diene component.
* B) it cannot adopt the *s*-cis conformation.
C) it lacks electron-withdrawing groups.
D) it lacks strong electron-releasing groups.
E) the two double bonds are further apart than in a non-cyclic conjugated system.

64. The accompanying diagram which describes the fate of the intermediate in a reversible reaction implies that:

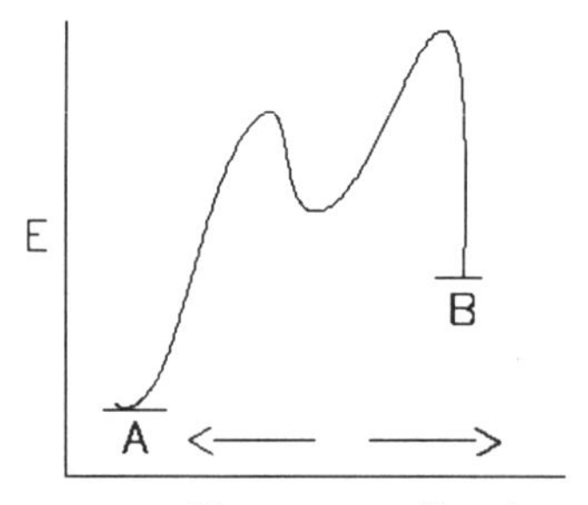

A) the less stable product forms the more rapidly.
* B) the more stable product forms the more rapidly.
C) product B will predominate at equilibrium.
D) the intermediate has a short lifetime.
E) No conclusions can be drawn as to either reaction rate or product stability.

65. An unsaturated product results from the reaction of cyclohexene with which of these?

A) Br_2/CCl_4 at 25xC
* B) NBS/CCl_4, ROOR
C) HCl, ROOR
D) HCl, no peroxides
E) More than one of these

66. What is an IUPAC name for

```
CH3DCpCDCHDCHMCH2     ?
          3
          CH
         / \
       CH3 CH3
```

* A) 3-Isopropyl-1-hexen-4-yne
B) 3-Isopropyl-4-hexyn-1-ene
C) 4-Isopropyl-5-hexen-2-yne
D) 4-Isopropyl-2-hexyn-5-ene
E) 5-Methyl-4-vinyl-2-hexyne

67. Which of these dienes is the most reactive in the Diels-Alder reaction?

A) 1,3-Butadiene
B) 1,4-Pentadiene
* C) Cyclopentadiene
D) 1,2-Butadiene
E) 1,4-Cyclohexadiene

68. Arrange these hexadienes in order of expected <u>decreasing</u> stability.

```
            H                       H H                  H H H
            3                       3 3                  3 3 3
CH2MCHDCH2DCMCDCH3      CH2MCHDCH2DCMCDCH3      CH3DCMCDCMCDCH3
              3                                               3
              H                                               H

         I                       II                      III


     H   H
     3   3
CH3DCMCDCMCDCH3            CH2MCHDCH2DCH2DCHMCH2
       3   3
       H   H

      IV                              V
```

A) V > II > I > III > IV
B) III > IV > II > I > V
C) IV > III > II > V > I
* D) IV > III > I > II > V
E) I > II > IV > III > V

69. What is the IUPAC name for this triene?

```
          H      H H      CH3
           \    /   \    /
            CMC      CMC
           /   \    /   \
    CH3CH2      CMC      H
               /   \
              H     H
```

A) (2E,4Z,6E)D2,4,6DNonatriene

B) (2Z,4E,6Z)D2,4,6DNonatriene

* C) (2E,4Z,6Z)D2,4,6DNonatriene

D) (3Z,5Z,7E)D3,5,7DNonatriene

E) (3Z,5E,7E)D3,5,7DNonatriene

1. If all the protons of 1-fluoropentane could be discerned, which would you expect to be at the lowest field in the ^1H NMR spectrum of this compound?

$CH_3CH_2CH_2CH_2CH_2F$
t t t t t
V IV III II I

* A) Protons on carbon I
B) Protons on carbon II
C) Protons on carbon III
D) Protons on carbon IV
E) Protons on carbon V

2. Which proton(s) of the compound below would appear as a septet in the ^1H NMR spectrum?

IDD4CH_3
/
$CH_3CH_2CH_2$-O-CH
t t t t\
3 3 3 3 CH_3
V IV III II t
3
I

A) The protons on carbon I
* B) The proton on carbon II
C) The protons on carbon III
D) The protons on carbon IV
E) The protons on carbon V

3. A compound with the molecular formula $C_4H_{10}O$ gives a ^1H NMR spectrum consisting only of a quartet centered at k3.5 and a triplet at k1.1. The most likely structure for the compound is:

A) CH_3
3
CH_3COH
3
CH_3

B) CH_3
3
CH_3OCHCH_3

C) $CH_3CH_2CH_2CH_2OH$

* D) $CH_3CH_2OCH_2CH_3$

E) CH_3CHCH_2OH
3
CH_3

4. How many signals would you expect to find in the ^1H NMR spectrum of $CH_3OCH_2CH_2OCH_3$?

A) 1 * B) 2 C) 3 D) 4 E) 5

5. A compound with the molecular formula C_9H_{12} gave a ^1H NMR spectrum consisting of:

a doublet at k1.25
a septet at k2.90
and a multiplet at k7.25

The most likely structure is:

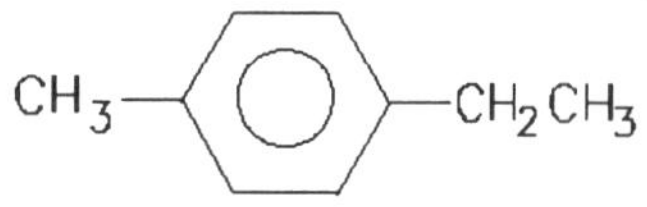

CH_3— —CH_2CH_3
I

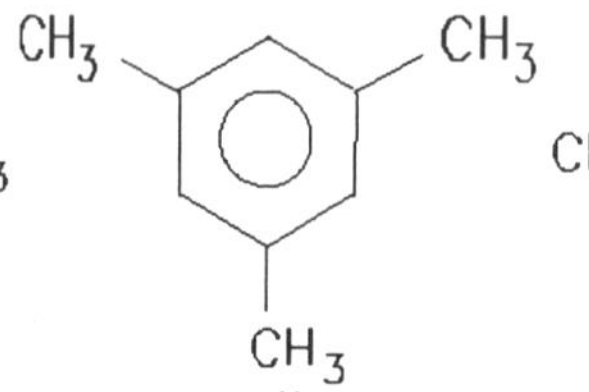

CH_3 CH_3 CH_3
II

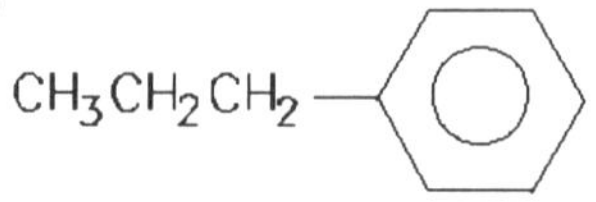

$CH_3CH_2CH_2$—
III

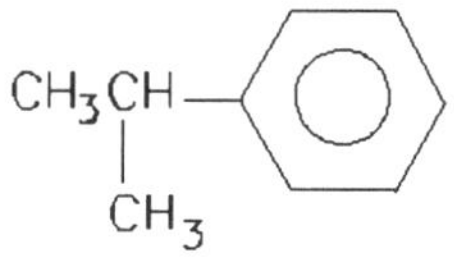

CH_3CH— CH_3
IV

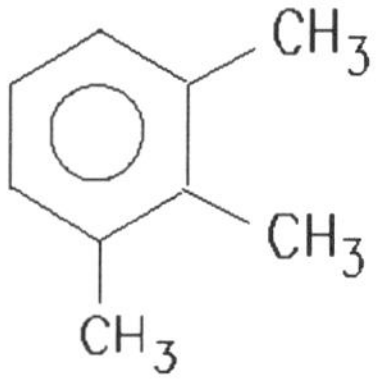

CH_3 CH_3 CH_3
V

A) I   B) II   C) III   * D) IV   E) V

6. A compound with the molecular formula $C_{10}H_{13}Cl$ gave the following ^1H NMR spectrum:

singlet, k1.6
singlet, k3.1
multiplet, k7.2 (5H)

The most likely structure for the compound is:

A)
```
        CH3
        3
C6H5CH2CHCH2Cl
```

* B)
```
        CH3
        3
C6H5CH2CCH3
        3
        Cl
```

C)
```
p-CH3C6H4CH2CHCH3
              3
              Cl
```

D)
```
C6H5CH2CH2CHCH3
            3
            Cl
```

E)
```
      Cl
      3
C6H5CCH2CH3
      3
      CH3
```

7. A compound with the molecular formula C_8H_9BrO gave the following ^1H NMR spectrum:

triplet, k1.4
quartet, k3.9
multiplet, k7.0 (4H)

There was no evidence of an -OH band in the IR spectrum. A possible structure for the compound is:

A) $C_6H_5OCH_2CH_2Br$

B) p-$CH_3C_6H_4OCH_2Br$

* C) p-$BrC_6H_4OCH_2CH_3$

D) $C_6H_5OCHCH_3$
3
Br

E) p-$CH_3OC_6H_4CH_2Br$

8. A compound with the molecular formula $C_6H_{15}N$ gave the following ^1H NMR spectrum:

triplet, k0.90
quartet, k2.4

There were no other signals. The most likely structure for the compound is:

A) $CH_3NCH_2CH_3$
3
$CH_2CH_2CH_3$

B) $CH_3NCH_2CH_2CH_2CH_3$
3
CH_3

C) $CH_3CH_2CH_2CH_2CH_2CH_2NH_2$

* D) $CH_3CH_2NCH_2CH_3$
3
CH_2CH_3

E) $CH_3CH_2CH_2NCH_2CH_2CH_3$
3
H

9. A compound with the molecular formula C_8H_9ClO gave the following ^1H NMR spectrum:

triplet, k3.7
triplet, k4.2
multiplet, k7.1

There was no evidence of an -OH band in the IR spectrum. The most likely structure for the compound is:

* A) $C_6H_5OCH_2CH_2Cl$

B) $C_6H_5OCHCH_3$
3
Cl

C) p-$ClC_6H_4OCH_2CH_3$

D) o-$ClC_6H_4OCH_2CH_3$

E) p-$CH_3OC_6H_4CH_2Cl$

10. A compound with the molecular formula $C_3H_6Cl_2$ gave a ^1H NMR spectrum consisting only of a triplet centered at k3.7 and a quintet centered at k2.2. The most likely structure for the compound is:

A) $CH_3CH_2CHCl_2$
B) $CH_3CHClCH_2Cl$
C) $ClCH_2CHClCH_3$
* D) $ClCH_2CH_2CH_2Cl$
E) $CH_3CCl_2CH_3$

11. A compound with the molecular formula $C_{10}H_{14}$ gave the following ^1H NMR spectrum:

doublet, k1.2
singlet, k2.3
septet, k2.8
multiplet, k7.1

A possible structure for the compound is:

A)
CH_3
3
$C_6H_5CCH_3$
3
CH_3

* B)
CH_3
3
p-$CH_3C_6H_4CHCH_3$

C) p-$CH_3C_6H_4CH_2CH_2CH_3$

D) $C_6H_5CH_2CH_2CH_2CH_3$

E)
CH_3
3
$C_6H_5CH_2CHCH_3$

12. The broad band proton-decoupled ^{13}C NMR spectrum of a hexyl chloride exhibits five signals. Which of these structures could be the correct one for the compound?

A)
CH_3
3
$CH_3CCH_2CH_2Cl$
3
CH_3

B)
Cl
3
$CH_3CH_2CCH_2CH_3$
3
CH_3

C)
H_3C CH_3
3 3
CH_3C-$CHCH_3$
3
Cl

* D)
CH_3
3
$CH_3CHCHCH_2CH_3$
3
Cl

E)
CH_3
3
CH_3C-$CHCH_3$
3 3
H_3C Cl

13. Predict the splitting you would observe for the proton at C-3 of 2,3-dimethyl-2-phenylbutane.

A) Doublet B) Singlet C) Quartet * D) Septet E) Octet

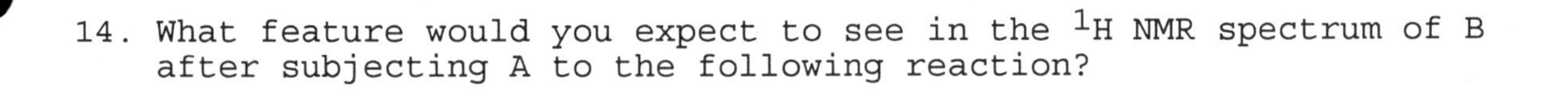

14. What feature would you expect to see in the ^{1}H NMR spectrum of B after subjecting A to the following reaction?

$$\text{A (C}_6\text{H}_5\text{CH}_2\text{CH}_3) \xrightarrow[\text{heat or light}]{Cl_2} \underset{\text{(chief product) B}}{C_8H_9Cl} + HCl$$

A) There would be only 4 aromatic protons at low field.
B) The signal for the protons on the benzyl carbon would be a doublet.
C) The signal for the methyl protons would be a triplet.
* D) The signal for the methyl protons would be a doublet.
E) The signal for the methyl protons would integrate for only 2 hydrogens.

15. How many ^{1}H NMR signals would trans-1,2-dichlorocyclopropane give?
A) 1 * B) 2 C) 3 D) 4 E) 5

16. How many signals will be recorded in the broad band proton-decoupled ^{13}C spectrum of 4-chloro-1-ethylbenzene?
A) 2 B) 3 C) 4 * D) 6 E) 7

17. How will the methyl carbon appear in the proton off-resonance decoupled ^{13}C spectrum of toluene?
A) Singlet B) Doublet C) Triplet * D) Quartet E) Quintet

18. How many ^{13}C signals would 1,4-dimethylbenzene give?
A) 1 B) 2 * C) 3 D) 4 E) 5

19. How many ^{13}C signals would 1,3-dichlorobenzene give?
A) 1 B) 2 C) 3 * D) 4 E) 5

20. How many ^{13}C signals would 1,2-dimethylbenzene give?
A) 1 B) 2 C) 3 * D) 4 E) 5

21. How many ^{1}H NMR signals would the following compound give?

$ClCH_2CHCH_3$
3
Br

A) 1 B) 2 C) 3 * D) 4 E) 5

22. Which compound would have an UV absorption band at longest wavelength?

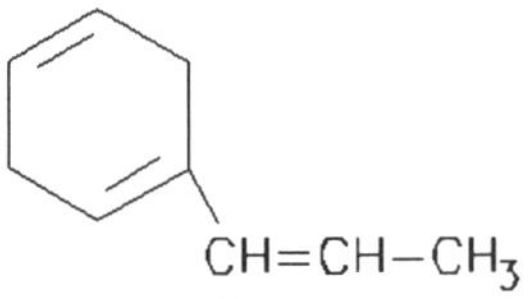

CH=CH–CH$_3$ I

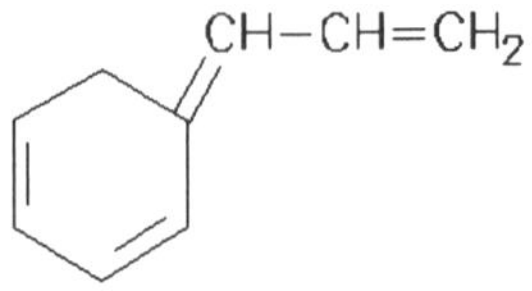

CH–CH=CH$_2$ II

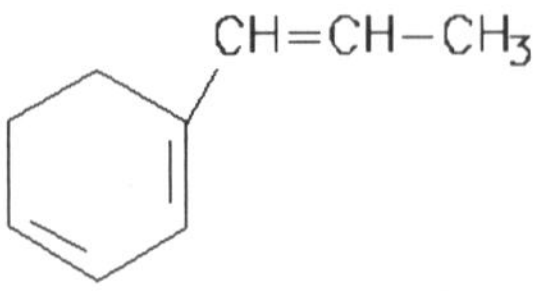

CH=CH–CH$_3$ III

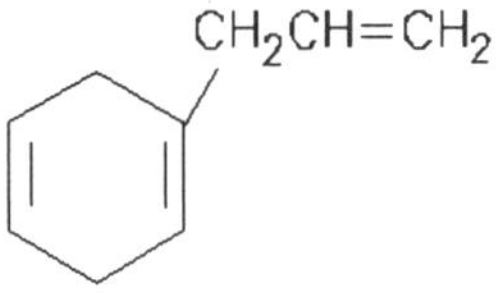

CH$_2$CH=CH$_2$ IV

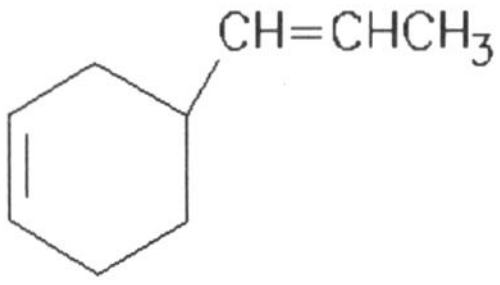

CH=CHCH$_3$ V

A) I   * B) II   C) III   D) IV   E) V

23. Select the most energetically favorable UV transition for 1,3-butadiene.

A) n DDDDDD4 e^*

B) n DDDDDD4 c^*

* C) c_2 DDDDDD4 c_3^*

D) e DDDDDD4 e^*

E) c_1 DDDDDD4 c_4^*

24. Which compound would have an UV absorption band at longest wavelength?

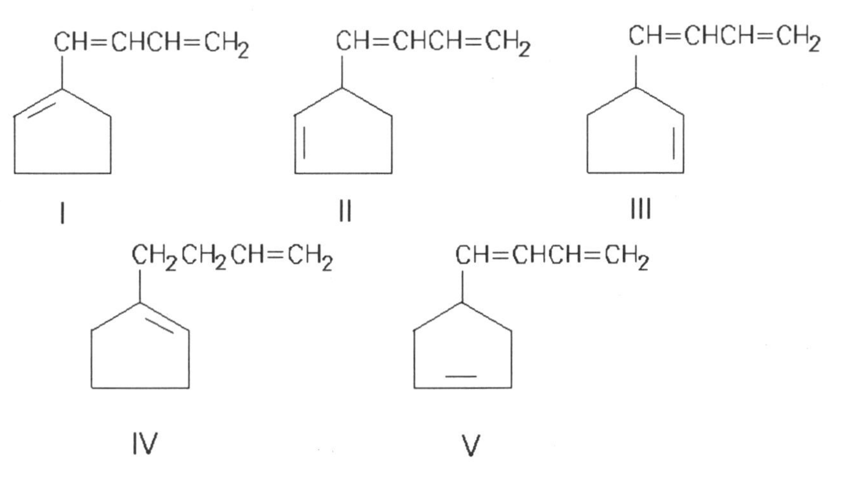

* A) I B) II C) III D) IV E) V

25. How many ^{13}C signals would you expect from $C_6H_5OCH_3$?

A) 1 B) 2 C) 3 D) 4 * E) 5

26. A compound $C_5H_{10}O$ gave the following spectral data:

^1H NMR spectrum	IR spectrum,
doublet, k 1.10	strong peak
singlet, k 2.10	near 1720 cm^{-1}
septet, k 2.50	

Which is a reasonable structure for the compound?

A)
O
:
$CH_3CCH_2CH_2CH_3$

B)
O
:
$CH_3(CH_2)_3CH$

C)
CH_3
3
CH_3CCHO
3
CH_3

D)
O
:
$CH_3CH_2CCH_2CH_3$

* E)
O
:
CH_3CCHCH_3
3
CH_3

27. A compound C_4H_9Br gave the following ^1H NMR spectrum:
multiplet, k4.1 (1H); multiplet, k1.8; doublet, k1.7;
triplet, k1.0 (3H)
Which is a reasonable structure for the compound?

* A) $CH_3CH_2CHBrCH_3$

B) $CH_3CH_2CH_2CH_2Br$

C) CH_3CHCH_2Br
3
CH_3

D) CH_3
3
CH_3CBr
3
CH_3

28. How many ^1H NMR signals would cis-1,2-dichlorocyclopropane give?
A) 1 B) 2 * C) 3 D) 4 E) 5

29. How many ^1H NMR signals would you expect from this compound?

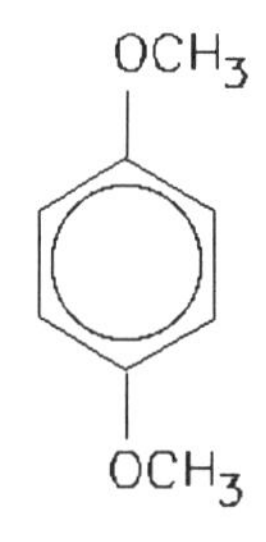

A) 1 * B) 2 C) 3 D) 4 E) 5

30. A compound has the molecular formula of $C_6H_{10}O$. It decolorizes bromine in carbon tetrachloride and yields a precipitate when treated with ammoniacal silver nitrate. The IR spectrum of the compound has a broad peak at 3500 cm^{-1} and a sharp peak at 3300 cm^{-1}. The ^1H NMR spectrum consists of:

triplet, k1.0	singlet, k2.4 (1H)
singlet, k1.4	singlet, k3.4 (1H)
quartet, k1.6	

A likely structure for the compound is which of these?

```
A) CH2MCCHMCHCH2OH                     D) HOCH2CH2CHCpCH
        3                                          3
        CH3                                        CH3

* B)      CH3                          E) CH3CHCH2CpCCH3
          3                                   3
   CH3CH2CCpCH                                OH
          3
          OH

C)       CH3
         3
   CH2MCCHCHMCH2
          3
          OH
```

31. The ^1H NMR spectrum of which of the compounds below, all of formula $C_7H_{12}O_2$, would consist of two singlets only?

```
A)     O H3C H                          D)  H3C O    H
       :   3 3                                3 :    3
   CH3CCH2C-C=O                         CH3C-CCH2C=O
           3                                3
           CH3                              CH3
B)     CH3 O H                        * E)     O
       3   : 3                                 :
   CH3CCH2C-C=O                            CH3C   CH3
       3                                        \ /
       CH3                                       C
C)         O CH3 H                              / \
           : 3  /                          CH3C   CH3
   CH3CH2C-C-C                                 :
             3  [                              O
             CH3 O
```

32. An oxygen-containing compound which shows no IR absorption at 1630-1780 cm^{-1} or at 3200-3550 cm^{-1} is likely to be what type of compound?

A) An alcohol
B) A carboxylic acid
* C) An ether
D) A ketone
E) An aldehyde

33. The C_7 compound which gives 3 signals in the broad band proton-decoupled ^{13}C spectrum could be:

A) Heptane
B) 2-Methylhexane
C) 3,3-Dimethylpentane
* D) 2,4-Dimethylpentane
E) 2,2,3-Trimethylbutane

34. Which of these compounds will not be represented by a singlet only in the ^1H NMR spectrum?

A) Neopentane
B) Hexamethylbenzene
* C) Isobutane
D) (Z)-1,2-Dichloroethene
E) (E)-1,2-Dichloroethene

35. For the C-2 methylene group in 1-bromopropane, $CH_3DCH_2DCH_2DBr$, the

c b a

theoretical multiplicity in the ^1H NMR spectrum, presuming that J_{ab} is sufficiently different from J_{bc} and that the instrument has sufficient resolving power, is which of these?

A) 2 B) 5 C) 6 D) 8 * E) 12

36. A compound $C_5H_{11}Cl$ which exhibits only two singlets in the ^1H NMR spectrum must be:

A) 1-Chloropentane
* B) 1-Chloro-2,2-dimethylpropane
C) 1-Chloro-2-methylbutane
D) 3-Chloropentane
E) 1-Chloro-3-methylbutane

37. A downfield (k 9-10) singlet is observed in the ^1H NMR spectrum of:

A) O
:
$CH_3CCH_2CH_3$

B) H
3
$C_6H_5CH_2CMO$

C) H
3
$(CH_3)_2CHCMO$

* D) H
3
$(CH_3)_3CCMO$

E) O
:
$C_6H_5CH_2CCH_3$

38. Which of these forms of electromagnetic radiation possesses the least energy?

A) Visible light
* B) Radio frequency radiation
C) X-rays
D) Infrared radiation
E) Ultraviolet light

39. The absorption band for the O-H stretch in the IR spectrum of an alcohol is sharp and narrow in the case of:

A) a Nujol mull of the alcohol.
B) a concentrated solution of the alcohol.
* C) a gas phase spectrum of the alcohol.
D) the spectrum of the neat liquid.
E) none of these.

40. A bromodichlorobenzene which gives four signals in the broad band proton-decoupled ^{13}C spectrum could be:

* A) 2-Bromo-1,3-dichlorobenzene
B) 2-Bromo-1,4-dichlorobenzene
C) 3-Bromo-1,2-dichlorobenzene
D) 4-Bromo-1,2-dichlorobenzene
E) 4-Bromo-1,3-dichlorobenzene

41. A split peak for the IR absorption due to bond stretching is observed for the carbonyl group in which of these compounds?

A) $CH_3C(=O)OH$
B) $CH_3C(=O)Cl$
C) $CH_3C(=O)NH_2$
D) $CH_3C(=O)OCH_2CH_3$
* E) $CH_3C(=O)OC(=O)CH_3$

42. In the structure shown, H_b and H_c are classified as:

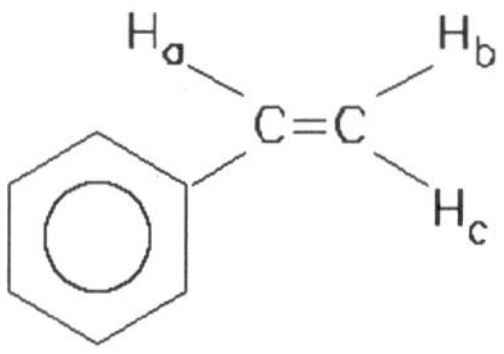

A) homotopic protons.
B) vicinal protons.
C) enantiotopic protons.
* D) diastereotopic protons.
E) isomeric protons.

43. Distinction between the methine protons in the compounds

```
      H                         Cl
      3                         3
CH3DCDCH2CH3       and    CH3DCDCH2CH3
      3                         3
      Cl                        H
```

should be possible in the ^{1}H NMR spectra if:

A) a very high field instrument is used.
* B) the spectra are determined in a chiral solvent.
C) a long scan time is used for each compound.
D) a high amplitude setting is employed.
E) Distinction between the enantiomers is impossible.

44. The IR stretching frequency occurs at the lowest frequency for which of these bonds?

A) CDH B) CDO * C) CDBr D) CDN E) CDF

45. Which of these symbols is used in connection with the intensity of absorption in the UV-visible region?

A) ! B) 0 * C) n D) $ E) i

46. An anticipated IR absorption band may not be observed because:

A) it occurs outside the range of the instrument used.
B) no change occurs in the dipole moment during the vibration.
C) the absorption band is eclipsed by another.
D) the intensity is so weak that it cannot be differentiated from instrument noise.
* E) All of these

47. In ^{13}C NMR spectroscopy, the signal due to this type of carbon occurs furthest downfield.

```
A) DCpN    * B) \         C) \          D) \         E) DCpCD
                CMO          DCDOD         DCDX
               /            /             /
```

48. IR evidence for the presence of the C=C would be most difficult to detect in the case of which of these alkenes?

A)
```
          H H
          3 3
CH3CH2CH2CMCCH2CH3
```

B)
```
      H
      3
CH3CH2CMCCH2CH2CH2CH3
        3
        H
```

C)
```
CH3CH2      CH2CH3
      \    /
       CMC
      /   \
    CH3    H
```

* D)
```
CH3    CH3
   \  /
   CMC
   /  \
CH3    CH3
```

E)
```
H       CH3
  \   /
   CMC
  /   \
CH3    CH3
```

49. The ^{1}H NMR signal for which of the indicated protons occurs furthest downfield?

```
            Cl
            |
HC≡C–CH–CH–CH3
I       II | III  IV
           |
        (benzene) } V
```

A) I B) II C) III D) IV * E) V

50. In NMR terminology, protons H_a and H_b are said to be:

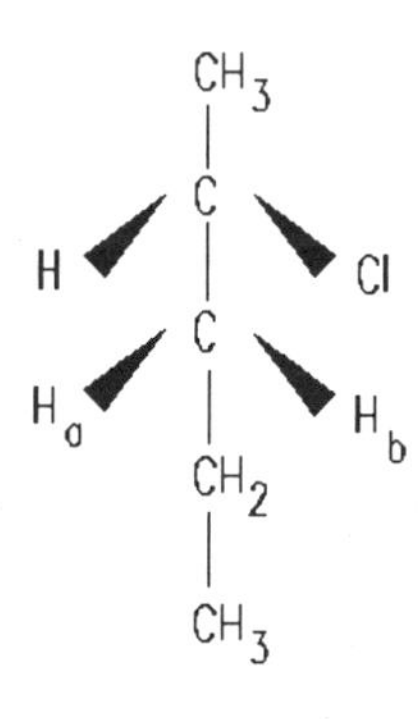

A) identical
B) enantiotopic
* C) diastereotopic
D) homotopic
E) mesotopic

51. IR absorption due to the stretching of which of these carbon-hydrogen bonds occurs at the highest frequency?

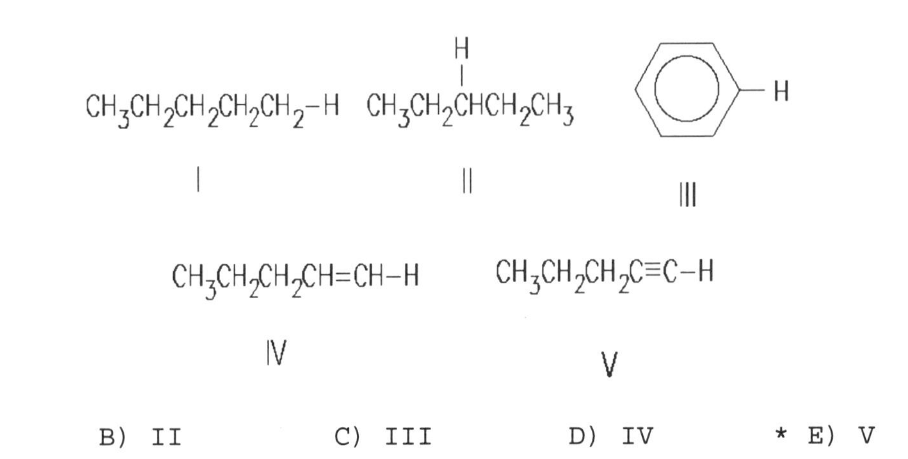

A) I B) II C) III D) IV * E) V

52. The IR spectrum of which type of compound will not show evidence of hydrogen bonding?

* A) Aldehyde
B) Alcohol
C) Carboxylic acid
D) Phenol
E) Primary amine

53. Consider the ^{1}H NMR spectrum of very pure 1-propanol, CH_3-CH_2-CH_2-OH. Assuming the maximum multiplicity of signals and
a b c d

non-superposition of peaks, what is the number of peaks to be observed theoretically, in the order: a, b, c, d.

A) 3, 6, 4, 1
B) 3, 6, 4, 3
C) 3, 12, 3, 1
D) 3, 12, 3, 3
* E) 3, 12, 6, 3

54. The ^{1}H NMR spectrum of which of these compounds would consist of a triplet, singlet and quartet <u>only</u>?

A) Cl
3
$CH_3CHCH_2CH(CH_3)_2$

B) Cl
3
$CH_3CH_2CHCH(CH_3)_2$

C) Cl
3
$CH_3CH_2CHCH_2CH_3$

D) CH_3
3
$CH_3CH_2CCH_2Cl$
3
CH_3

* E) Cl
3
$CH_3CH_2CCH_2CH_3$
3
CH_3

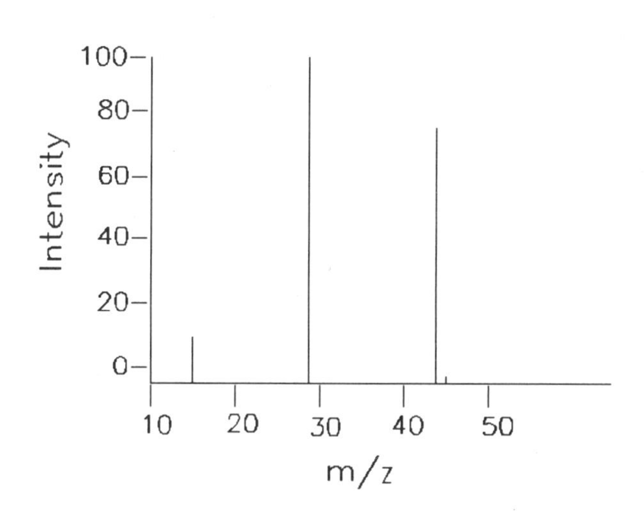

55. Which is the base peak?

A) 15 * B) 29 C) 44 D) 45 E) 100

56. Which is the likely molecular ion (MB)?

A) 15 B) 29 * C) 44 D) 45 E) 100

57. What is the molecular formula of this compound?

m/z	intensity
84 MB	10.00
85	0.56
86	0.04

A) $C_5H_{10}O$ * B) C_5H_8O C) C_5H_{24} D) C_6H_{12} E) $C_4H_6O_2$

58. What is the molecular formula of this compound?

m/z	intensity
78 MB	10.00
79	1
80	3.3
81	0.3

A) C_6H_6 B) C_3H_5Cl C) C_6H_8 D) C_6H_9 * E) C_3H_7Cl

59. The mass spectra of alkyl halides are characterized by an unusually intense __________.

A) base peak
B) parent peak
C) MB +1 peak
* D) MB +2 peak
E) None of these

60. Predict the base peak for

```
     CH3
     3
CH3-C-Cl
     3
     CH3
```

A) m/z 15 B) m/z 92 C) m/z 43 * D) m/z 57 E) m/z 77

61. Select the structure of a compound C_6H_{14} with a base peak at m/z 43.

A) $CH_3CH_2CH_2CH_2CH_2CH_3$
B) $(CH_3CH_2)_2CHCH_3$
C) $(CH_3)_3CCH_2CH_3$
* D) $(CH_3)_2CHCH(CH_3)_2$
E) None of these

62. The data below from the molecular ion region of the mass spectrum of a halogen-containing compound are consistent with the presence of what halogen(s) in the original compound?

	intensity
MB	51.0
MB +2	100.0
MB +4	49.0

A) One Br
B) One Cl
C) One Br and one Cl
* D) Two Br
E) Two Cl

63. A prominent MB -18 peak suggests that the compound might be a(n):

A) alkane
* B) alcohol
C) ether
D) ketone
E) primary amine

64. An organic compound absorbs strongly in the IR at 1687 cm^{-1}. On oxidation it is converted to benzoic acid. Its mass spectrum shows significant peaks at m/z 120, m/z 105 and m/z 77. This information is consistent with which of the following structures?

* A)
```
   O
   :
C6H5CCH3
```
B)
```
        H
        3
C6H5CH2C=O
```
C)
```
         H
         3
p-CH3C6H4C=O
```
D) $C_6H_5CH_2CH_2CH_3$
E) $C_6H_5CH(CH_3)_2$

Chapter 14: Aromatic Compounds

1. Which of the following is NOT true of benzene?
   A) Benzene tends to undergo substitution rather than addition reactions, even though it has a high index of hydrogen deficiency.
   B) All of the hydrogen atoms of benzene are equivalent.
 * C) The carbon-carbon bonds of benzene are alternately short and long around the ring.
   D) Only one o-dichlorobenzene has ever been found.
   E) Benzene is more stable than the hypothetical compound 1,3,5-cyclohexatriene.

2. There are three known dimethylbenzenes, $C_6H_4(CH_3)_2$. One of these dimethylbenzenes reacts with nitric acid (in the presence of sulfuric acid) to yield only two different nitrodimethylbenzenes, $C_6H_3NO_2(CH_3)_2$. The structure of this dimethylbenzene is:

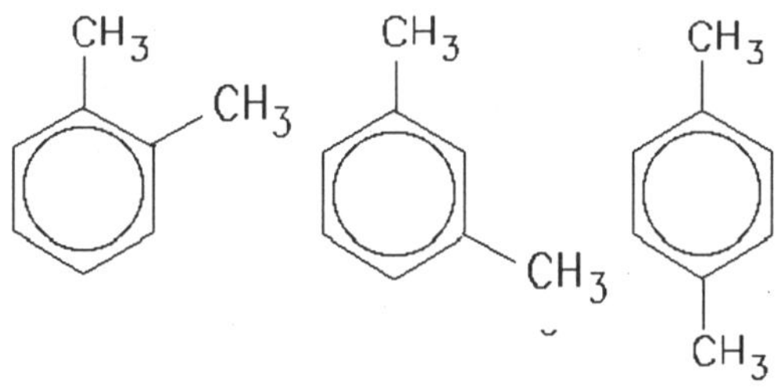
CH3
CH3
CH3
CH3
CH3
CH3

I II III

 * A) I
   B) II
   C) III
   D) All of the above would give two different nitrodimethylbenzenes.
   E) None of the above would give two different nitrodimethylbenzenes.

3. In which of the following compounds would the carbon-carbon bond(s) be shortest?

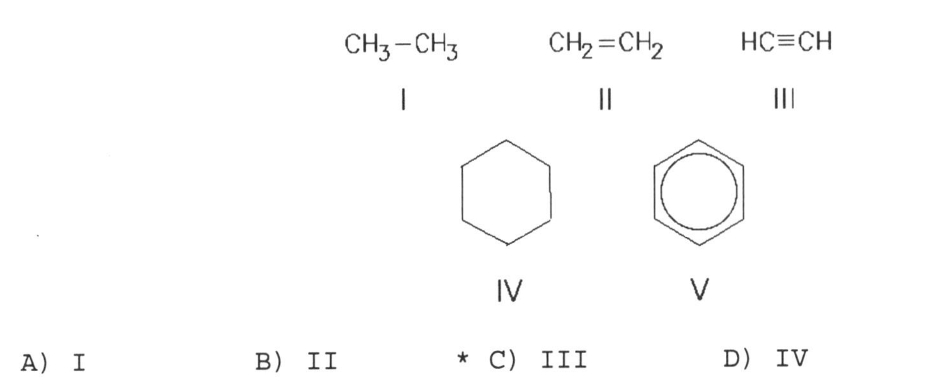

A) I   B) II   * C) III   D) IV   E) V

4. Which of the following would you expect to be aromatic?

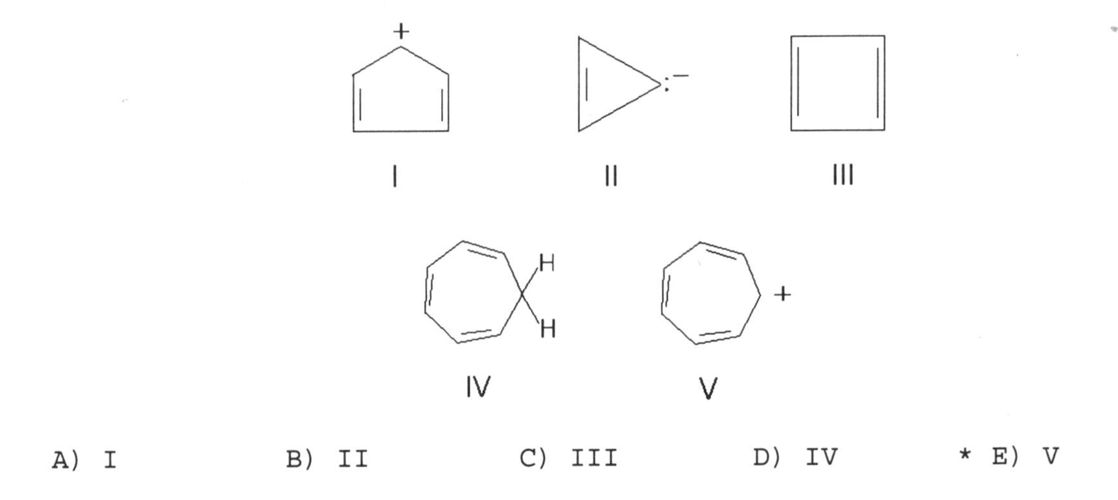

A) I   B) II   C) III   D) IV   * E) V

5. Which of these would you expect to have a significant resonance energy?

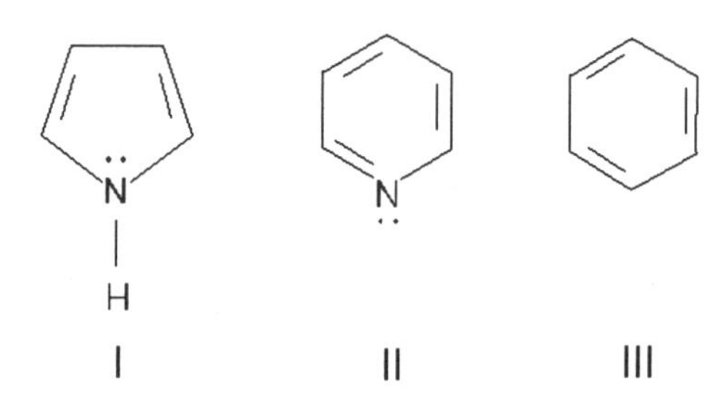

A) I
B) II
C) III
* D) All of the above
E) None of the above

6. Which of the following would you expect to be aromatic?

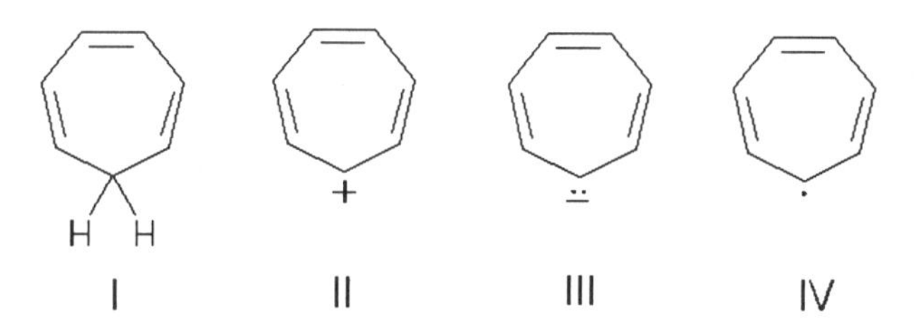

A) I
* B) II
C) III
D) IV
E) None of these

7. Which reagent(s) would serve as the basis for a simple chemical test that would distinguish between benzene and 1-hexene?

A) $Ag(NH_3)_2OH$
* B) Br_2 in CCl_4
C) $AgNO_3$ in C_2H_5OH
D) CuCl in NH_3/H_2O
E) None of these

8. Which dibromobenzene can, in theory, yield three mononitro derivatives?

A) o-Dibromobenzene
* B) m-Dibromobenzene
C) p-Dibromobenzene
D) All of these
E) None of these

9. Cyclopentadiene is unusually acidic for a hydrocarbon. An explanation for this is the following statement.

A) The carbon atoms of cyclopentadiene are all sp^2-hybridized.
B) Cyclopentadiene is aromatic.
* C) Removal of a proton from cyclopentadiene yields an aromatic anion.
D) Removal of a hydrogen atom from cyclopentadiene yields a highly stable free radical.
E) Removal of a hydride ion from cyclopentadiene produces an aromatic cation.

10. In the molecular orbital model of benzene, the six p-orbitals combine to form how many molecular orbitals?

* A) 6 B) 5 C) 4 D) 3 E) 2

11. Consider the molecular orbital model of benzene. In the ground state how many molecular orbitals are filled with electrons?

A) 1 B) 2 * C) 3 D) 4 E) 5

12. In the molecular orbital model of benzene, how many pi-electrons are delocalized about the ring?

A) 2 B) 3 C) 4 D) 5 * E) 6

13. The carbon-carbon bonds in benzene are:

A) of equal length and are shorter than the double bond of ethene.
* B) of equal length and are intermediate between a double bond and a single bond.
C) of unequal length and are alternately short and long around the ring.
D) due only to p-orbital overlap.
E) of equal length and intermediate between the carbon-carbon bond lengths in ethene and ethyne.

14. We now know that the two Kekule structures for benzene are related in the following way:

A) They are each equally correct as a structure for benzene.
B) Benzene is sometimes one structure and sometimes the other.
C) The two structures are in a state of rapid equilibrium.
* D) Neither of the two structures adequately describes benzene; benzene is a resonance hybrid of the two.
E) None of the above

15. Which compound would you NOT expect to be aromatic?

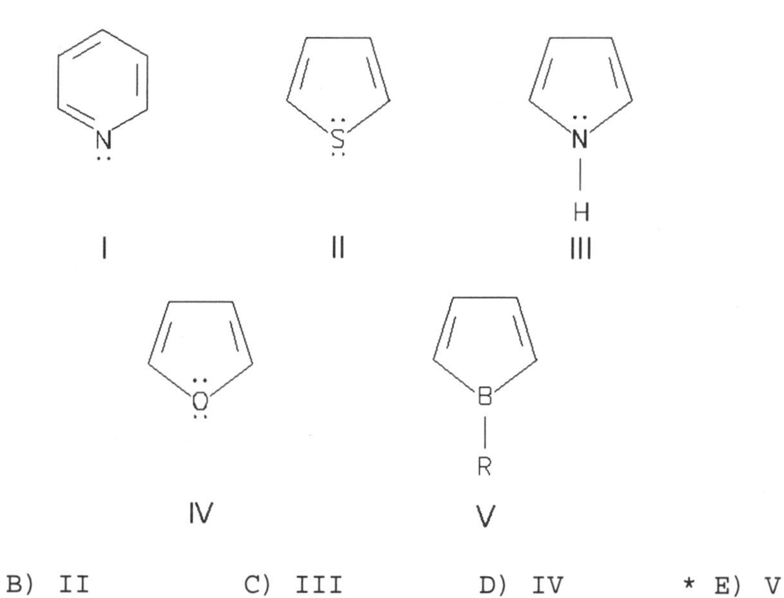

A) I   B) II   C) III   D) IV   * E) V

16. Which annulene would you NOT expect to be aromatic?
A) [6]-Annulene
B) [14]-Annulene
* C) [16]-Annulene
D) [18]-Annulene
E) [22]-Annulene

17. Which cyclization(s) should occur with a decrease in pi-electron energy?

I $CH_2{=}CH{-}CH{=}CH_2 \longrightarrow$ (cyclobutadiene) $+ H_2$

II $CH_2{=}CH{-}CH_2^+ \longrightarrow$ (cyclopropenyl cation) $+ H_2$

III $CH_2{=}CH{-}CH_2\cdot \longrightarrow$ (cyclopropenyl radical) $+ H_2$

IV $CH_2{=}CH{-}CH_2{:}^- \longrightarrow$ (cyclopropenyl anion) $+ H_2$

A) I
* B) II
C) III
D) IV
E) All of the above

18. Which of the following statements regarding the cyclopentadienyl radical is correct?
A) It is aromatic.
* B) It is not aromatic.
C) It obeys Huckel's rule.
D) It undergoes reactions characteristic of benzene.
E) It has a closed shell of 6 pi-electrons.

19. If thiophene is an aromatic molecule and reacts similarly to benzene, how many (neutral) monobromothiophenes could be obtained in the following reaction?

S $\xrightarrow[FeBr_3]{Br_2}$

A) 1 * B) 2 C) 3 D) 4 E) 5

20. 2-Bromo-4-nitroaniline is:

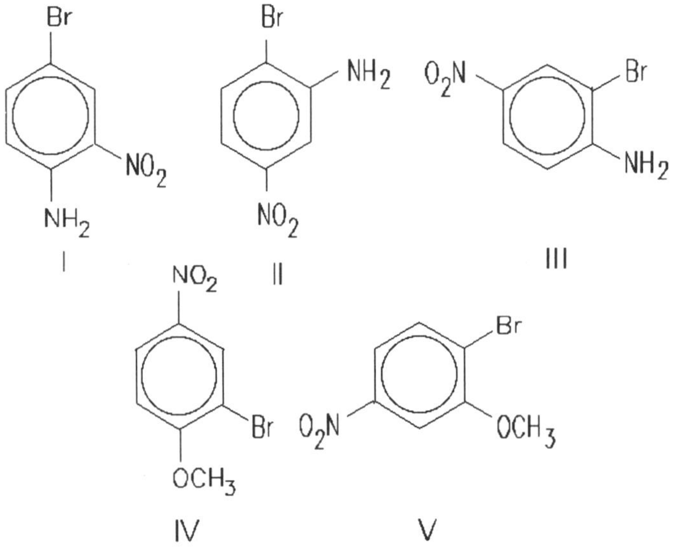

Br, NO_2, NH_2 — I
Br, NH_2, NO_2 — II
O_2N, Br, NH_2 — III
NO_2, Br, OCH_3 — IV
Br, O_2N, OCH_3 — V

A) I B) II * C) III D) IV E) V

21. How many different dibromophenols are possible?
A) 8 B) 7 * C) 6 D) 5 E) 4

22. Why would 1,3-cyclohexadiene undergo dehydrogenation readily?
  A) It is easily reduced.
  B) Hydrogen is a small molecule.
  C) 1,3-Cyclohexadiene has no resonance energy.
* D) It would gain considerable stability by becoming benzene.
  E) It would not undergo dehydrogenation.

23. The correct name for the compound shown below is:

  A) 3,4-Dibromoaniline
  B) 2,4-Dibromoaniline
* C) 2,5-Dibromoaniline
  D) 3,6-Dibromoaniline
  E) 2,6-Dibromoaniline

24. Which of the following would you expect to be aromatic?

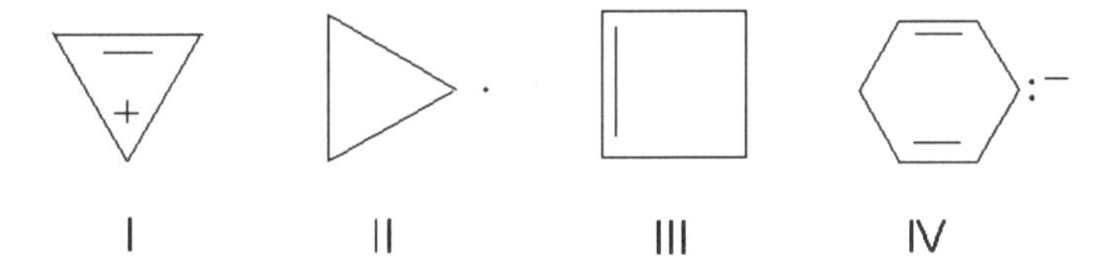

* A) I
  B) II
  C) III
  D) IV
  E) All of these

25. Which of the following is true of benzene?
  A) Benzene tends to undergo addition rather than substitution reactions.
* B) All of the hydrogen atoms of benzene are equivalent.
  C) The carbon-carbon bonds of benzene are alternately short and long around the ring.
  D) The benzene ring is a distorted hexagon.
  E) Benzene has the stability expected for cyclohexatriene.

26. Which of these species is aromatic?

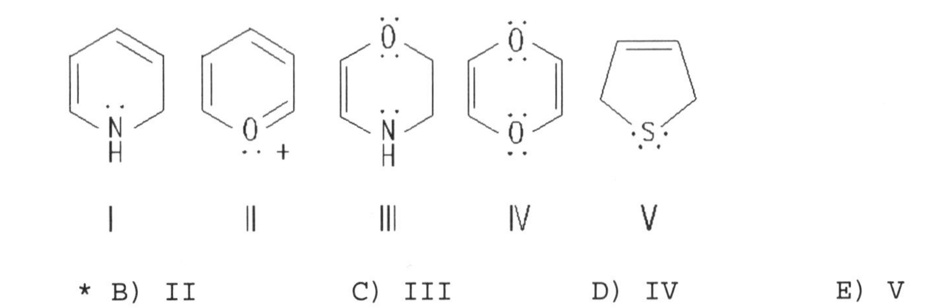

A) I * B) II C) III D) IV E) V

27. In the molecular orbital model of benzene, how many pi-electrons are in bonding molecular orbitals?

* A) 6 B) 5 C) 4 D) 3 E) 2

28. Which of the following structures would be aromatic?

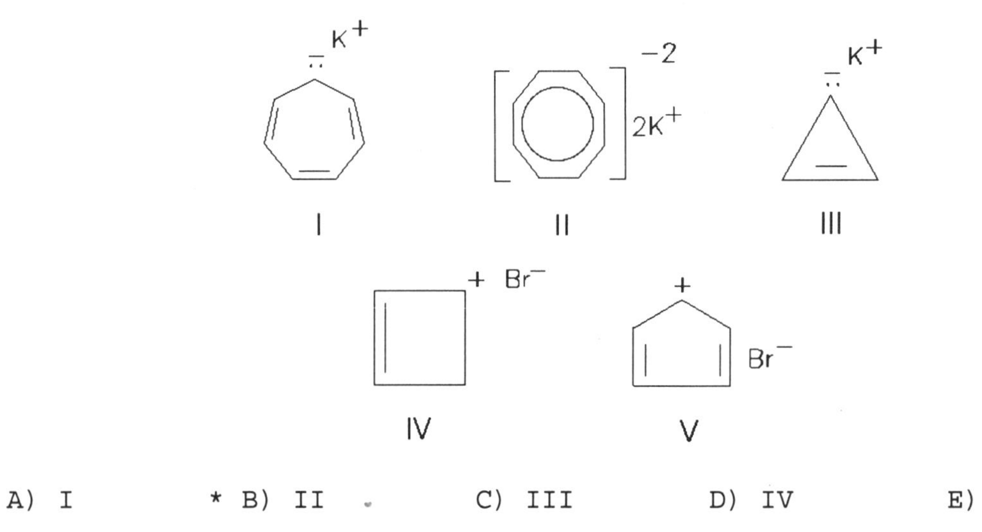

A) I * B) II C) III D) IV E) V

29. Which dibromobenzene can yield only one mononitro derivative?

A) o-Dibromobenzene
B) m-Dibromobenzene
* C) p-Dibromobenzene
D) More than one of these
E) None of these

30. Which of the following would you expect to be antiaromatic?

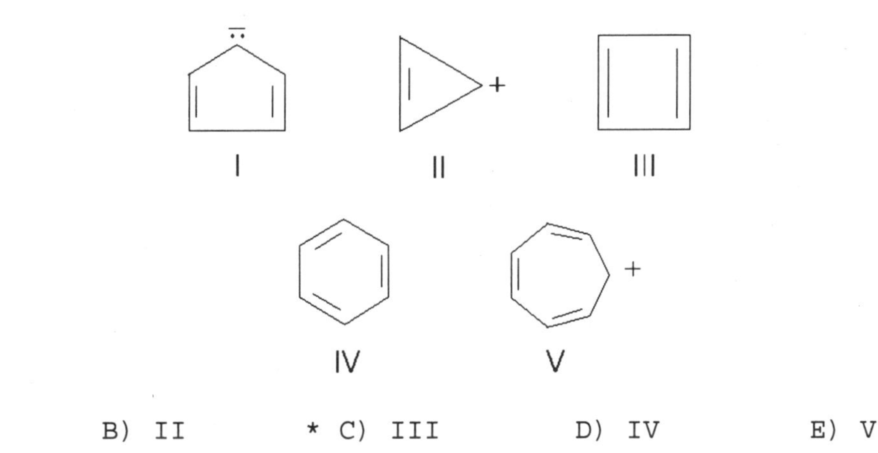

A) I B) II * C) III D) IV E) V

31. Which of the following would you expect to be aromatic?

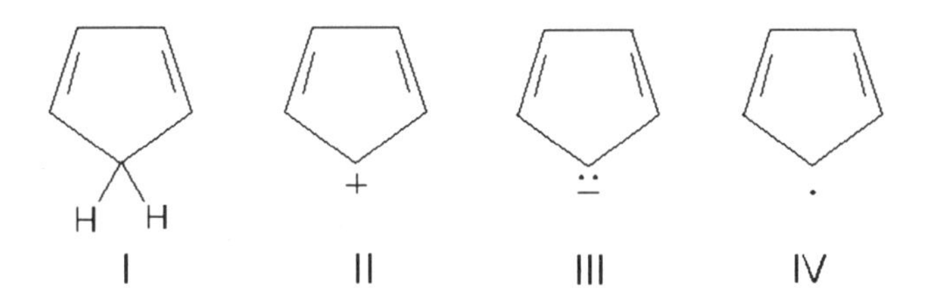

A) I
B) II
* C) III
D) IV
E) None of the above

32. On the basis of molecular orbital theory and Huckel's rule, which molecules and/or ions should be aromatic?

I II III IV V

VI VII VIII IX

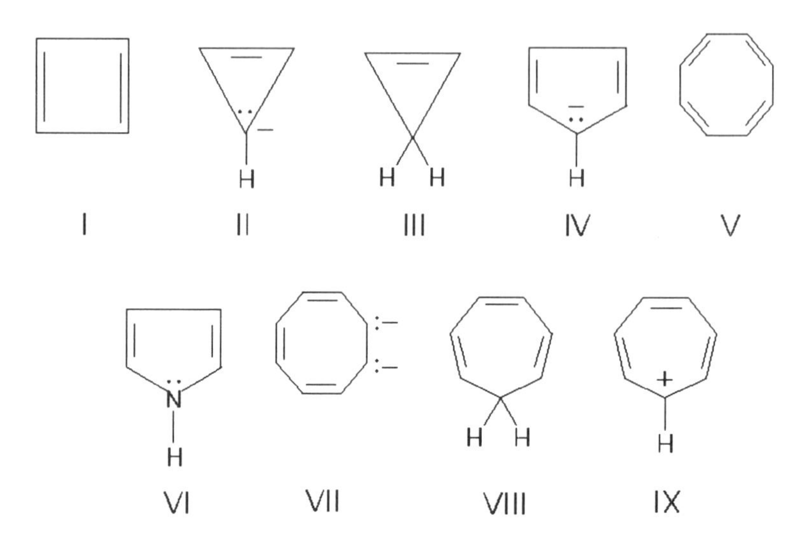

A) I and V
B) III and VIII
C) IV, VII and IX
* D) IV, VI, VII and IX
E) All of the structures, I-IX

33. A dichlorobenzene that on reaction with nitric acid and sulfuric acid might theoretically yield three mononitro products would be:
A) o-Dichlorobenzene
* B) m-Dichlorobenzene
C) p-Dichlorobenzene
D) None of these
E) All of these

34. Which of the following would have the longest carbon-carbon bond?

CH_3-CH_3 (I) $CH_2=CH_2$ (II) $HC\equiv CH$ (III)

$CH_2=CH-CH=CH_2$ (IV)

V

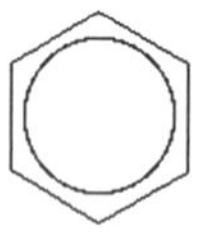

* A) I
B) II
C) III
D) IV
E) V

35. On the basis of molecular orbital theory and Huckel's rule, which of these compounds should be aromatic?

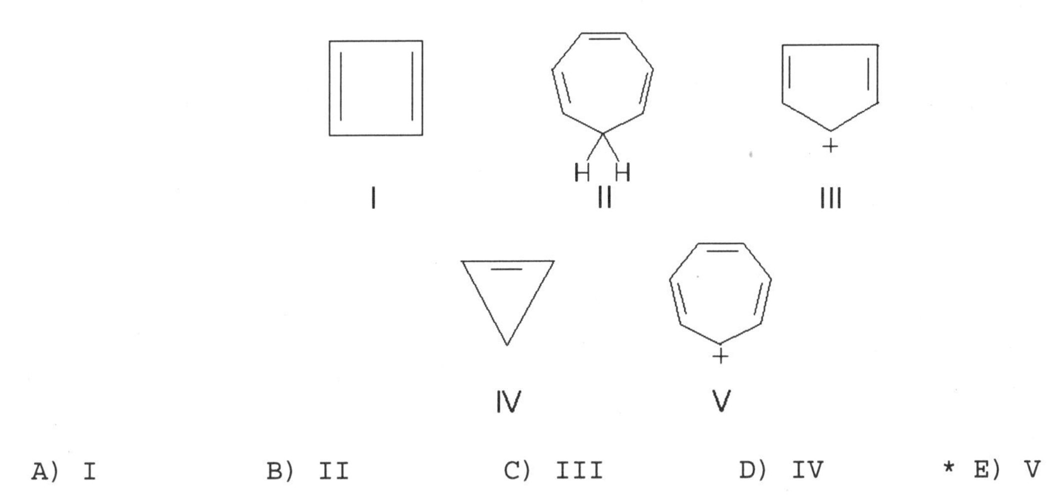

A) I B) II C) III D) IV * E) V

36. In theory, a single molecule of this compound will rotate plane-polarized light.

A) Butylbenzene
B) Isobutylbenzene
* C) <u>sec</u>-Butylbenzene
D) <u>tert</u>-Butylbenzene
E) None of these

37. Which of the following is NOT 2-bromo-5-nitrobenzoic acid?

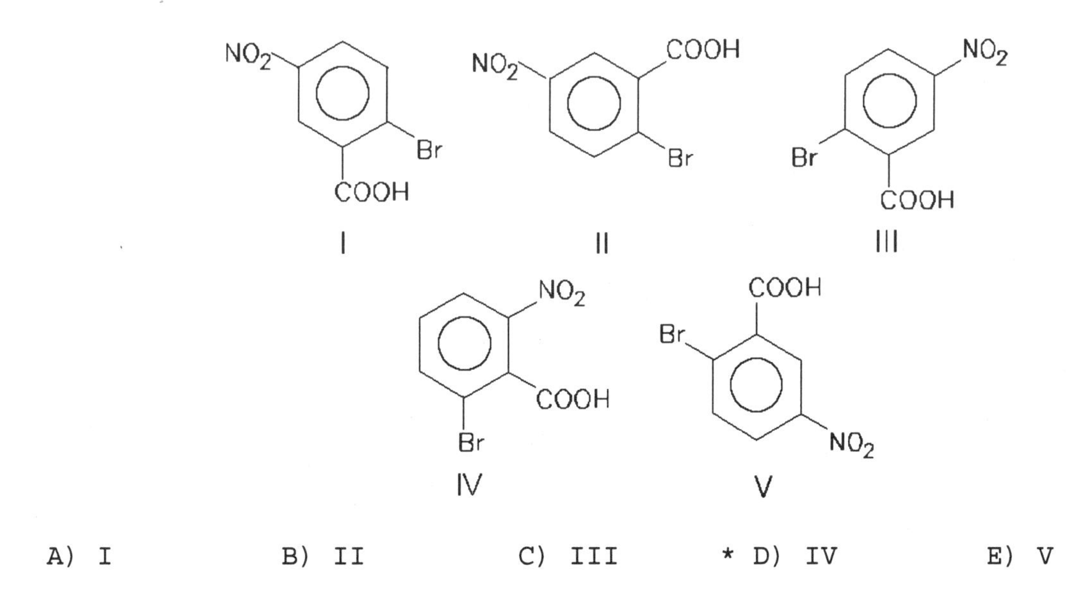

A) I B) II C) III * D) IV E) V

38. Toluene is the name commonly assigned to:
A) Hydroxybenzene
B) Aminobenzene
* C) Methylbenzene
D) Ethylbenzene
E) Methoxybenzene

39. This compound

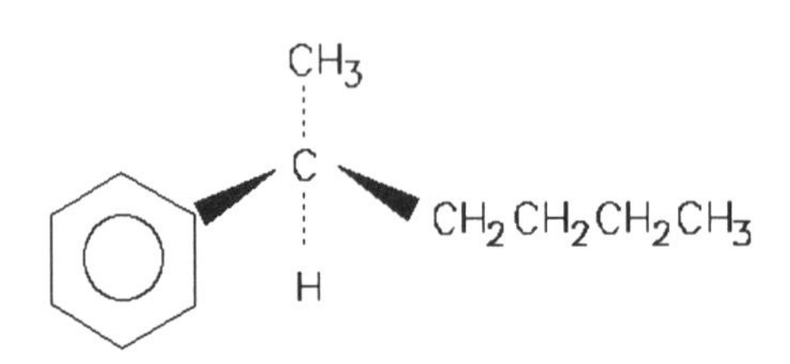

is most precisely named:
A) <u>sec</u>-Hexylbenzene
B) 2-Phenylhexane
C) (R)-2-Phenylhexane
* D) (S)-2-Phenylhexane
E) Butylmethylphenylmethane

40. How many dichloronaphthalenes are possible?
A) 7
B) 8
C) 9
* D) 10
E) 12

41. Of Huckel's requirements for aromatic character, only this one is waived in the case of certain compounds considered to be aromatic.
A) The ring system must be planar.
* B) The system must be monocyclic.
C) There must be (4n + 2)c electrons.
D) The Huckel number of electrons must be completely delocalized.
E) None. All of these rules must apply in every case.

42. Which of these is an aromatic molecule?

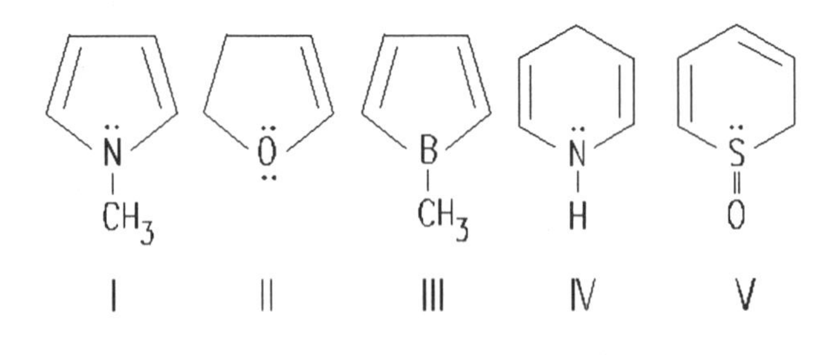

* A) I
B) II
C) III
D) IV
E) V

43. Which of these is the single best representation for naphthalene?

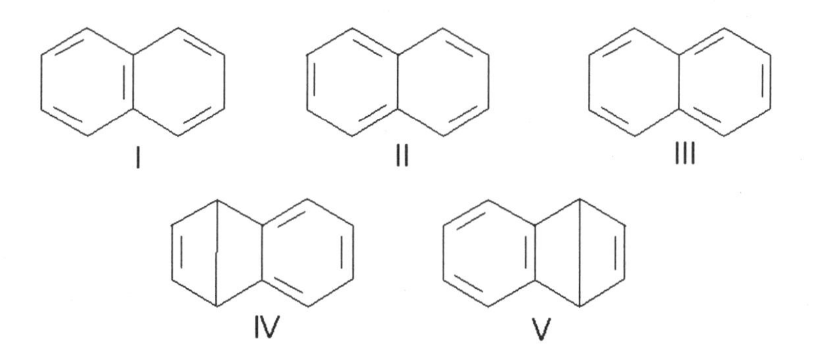

* A) I   B) II   C) III   D) IV   E) V

44. Of the following C-10 compounds, which is expected to possess the greatest resonance (delocalization) energy?

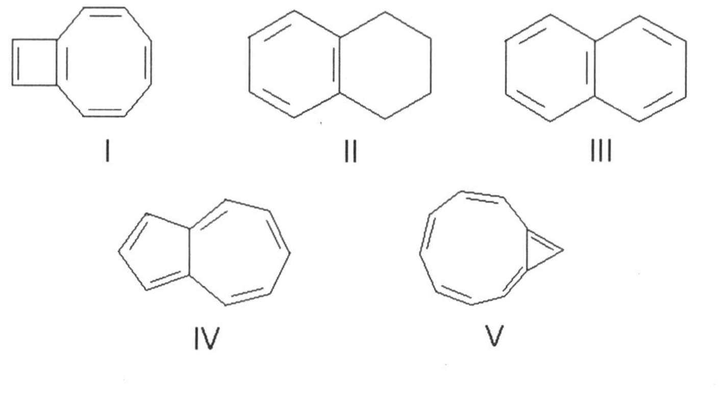

A) I   B) II   * C) III   D) IV   E) V

45. In which case is the indicated unshared pair of electrons NOT a contributor to the c aromatic system?

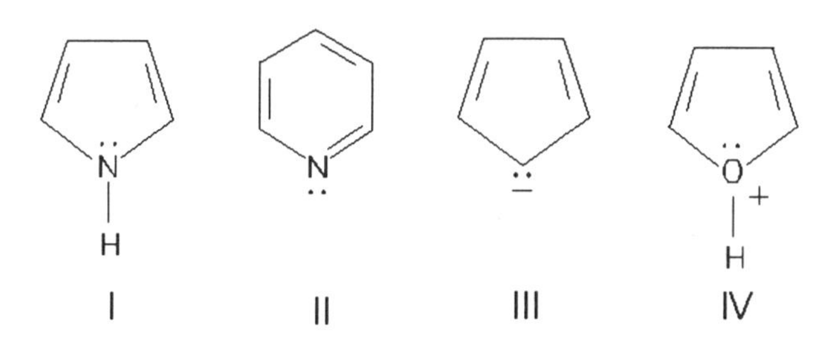

A) I
* B) II
C) III
D) IV
E) None of these

46. Which is the only one of these reagents which will react with benzene under the specified conditions?
* A) Cl_2, $FeCl_3$, heat
  B) H_2, 25xC
  C) Br_2/CCl_4, 25xC, dark
  D) $KMnO_4/H_2O$, 25xC
  E) H_3O^+/H_2O, heat

47. Which of the following statements about cyclooctatetraene is NOT true?
  A) The compound rapidly decolorizes Br_2/CCl_4 solutions.
  B) The compound rapidly decolorizes aqueous solutions of $KMnO_4$.
  C) The compound readily adds hydrogen.
  D) The compound is nonplanar.
* E) The compound is comparable to benzene in stability.

48. Application of the polygon-and-circle technique reveals that single electrons occupy each of the two nonbonding orbitals in the molecular orbital diagram of:
* A) Cyclobutadiene
  B) Benzene
  C) Cyclopropenyl cation
  D) Cyclopentadienyl anion
  E) Cycloheptatrienyl cation

49. Recalling that benzene has a resonance energy of 36 kcal mol^{-1} and naphthalene has a resonance energy of 61 kcal mol^{-1}, predict the positions which would be occupied by bromine when phenanthrene (below) undergoes addition of Br_2.

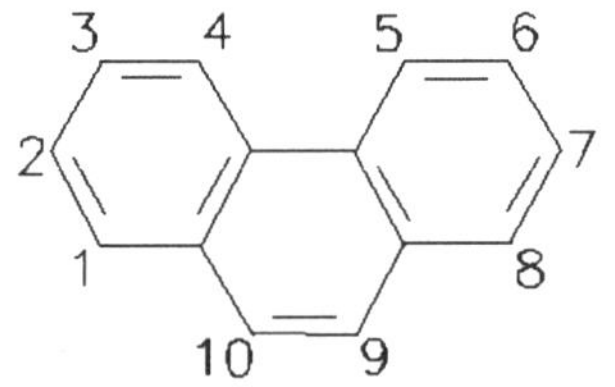

3 4 5 6
2 7
1 8
10 9

A) 1, 2 B) 1, 4 C) 3, 4 D) 7, 8 * E) 9, 10

50. Which isomer of C_7H_7Cl exhibits strong IR absorbances at 690 and 750 cm^{-1}?

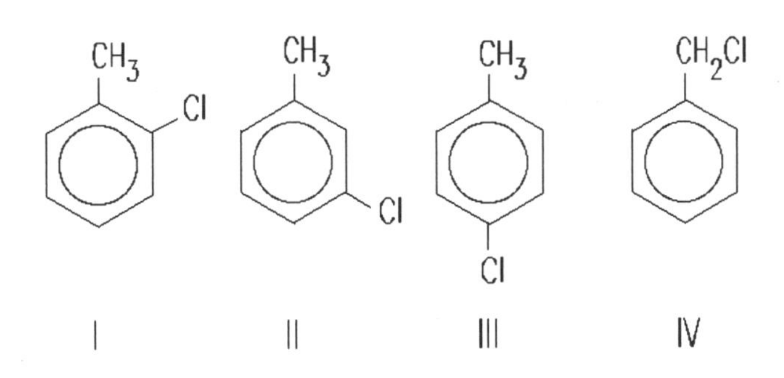

A) I B) II C) III D) IV * E) II and IV

51. Which of these compounds absorbs at the longest wavelength in the UV-visible region?

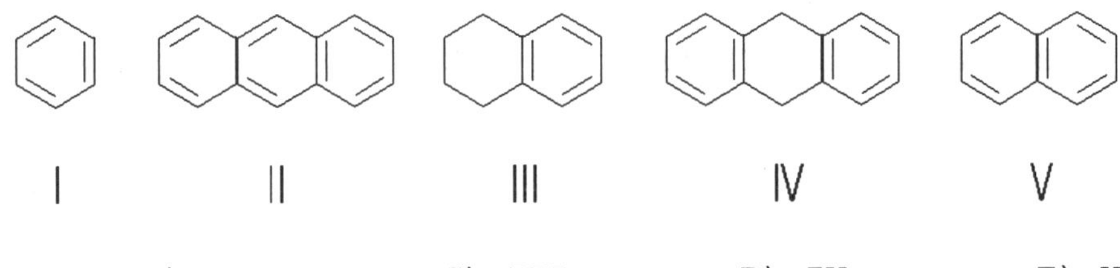

A) I * B) II C) III D) IV E) V

52. A compound has the formula C_8H_9Br. Its ¹H NMR spectrum consists of:

doublet, k 2.0
quartet, k 5.15
multiplet, k 7.35

The IR spectrum shows two peaks in the 680-840 cm^{-1} region; one is between 690 and 710 cm^{-1} and the other is between 730 and 770 cm^{-1}. Which is a possible structure for the compound?

I: p-$BrC_6H_4CH_2CH_3$

II: p-$CH_3C_6H_4CH_2Br$

III: $C_6H_5CH_2CH_2Br$

IV: $C_6H_5CHBrCH_3$

V: Br–(benzene ring)–CH_3, with a second CH_3 on the ring

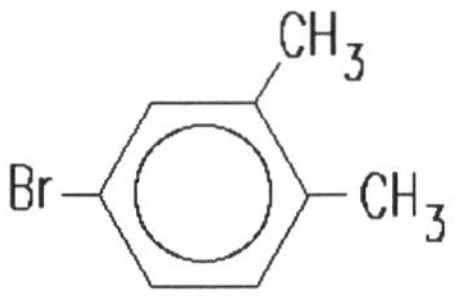

A) I B) II C) III * D) IV E) V

1. Which of the following compounds would be most reactive toward ring bromination?

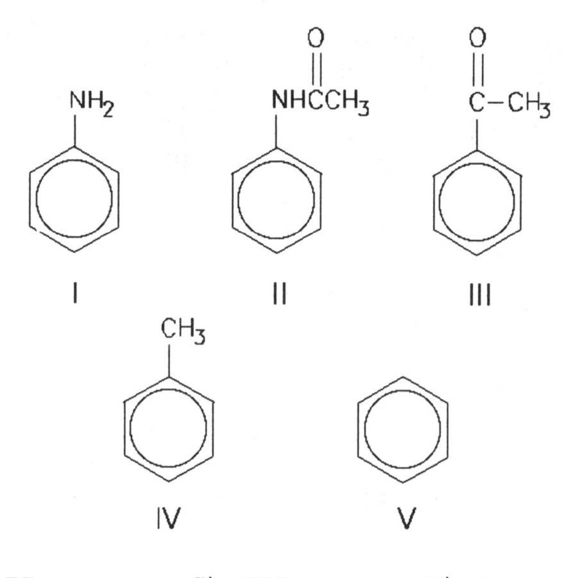

* A) I B) II C) III D) IV E) V

2. Which of the compounds listed below would you expect to give the greatest amount of meta-product when subjected to ring nitration?

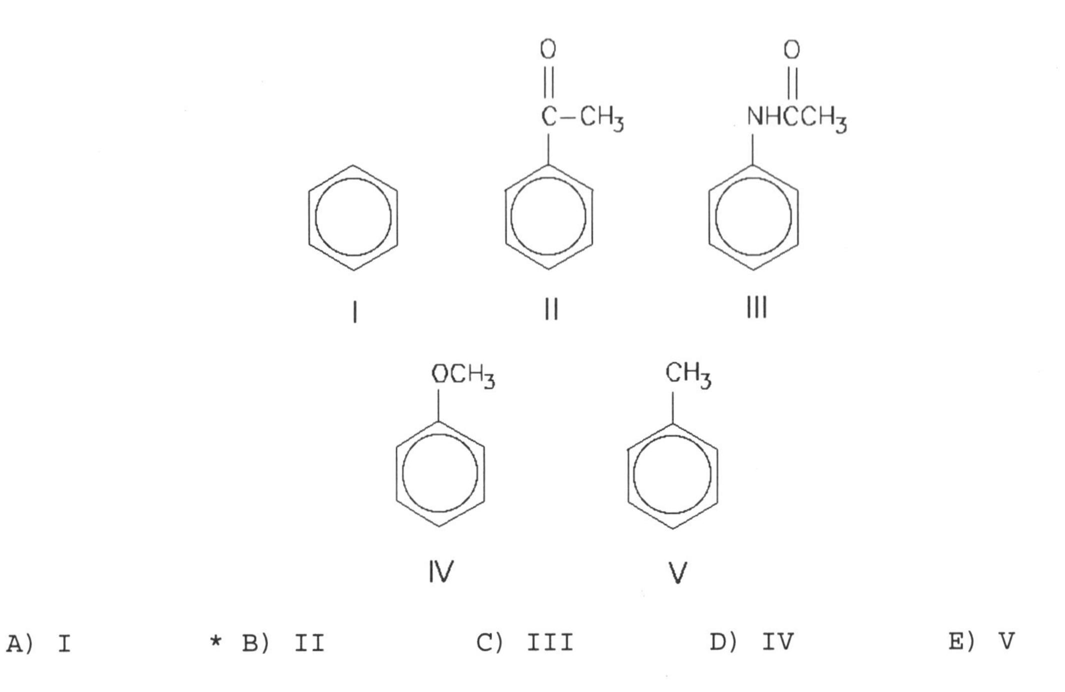

A) I * B) II C) III D) IV E) V

3. The major product(s), A, of the following reaction,

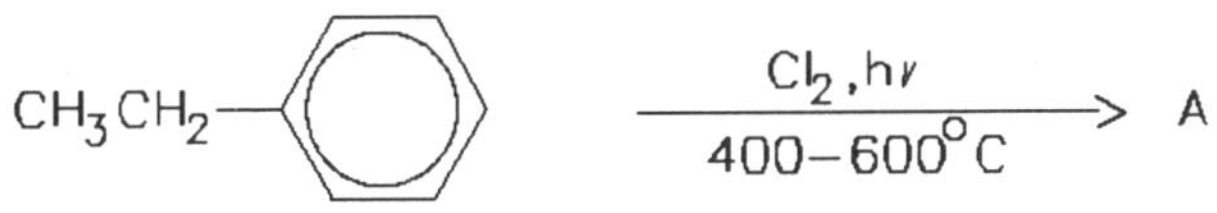

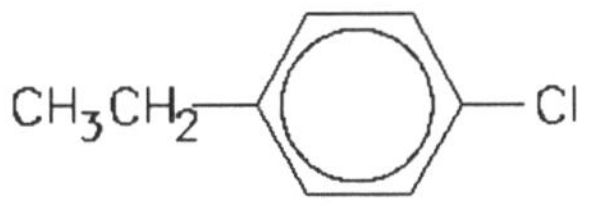

$$CH_3CH_2-C_6H_5 \xrightarrow[400-600^{\circ}C]{Cl_2,\ h\nu} A$$

I: $CH_3CH_2-C_6H_4-Cl$ (para)

II: $CH_3CH_2-C_6H_4-Cl$ (ortho)

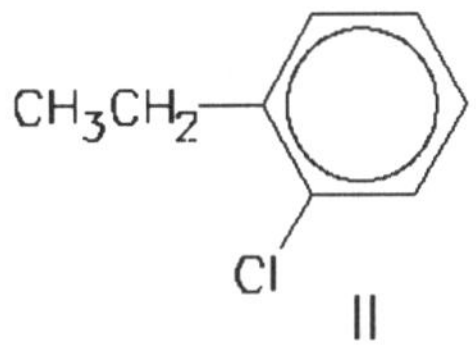

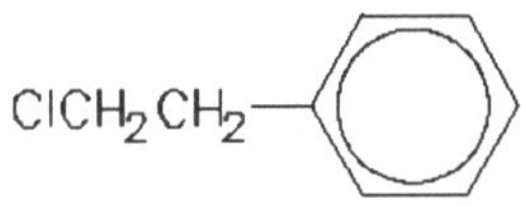

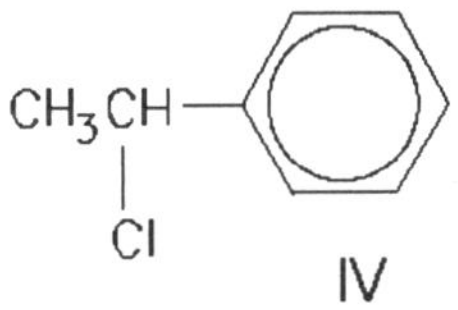

III: $ClCH_2CH_2-C_6H_5$

IV: $CH_3CH(Cl)-C_6H_5$

would be:

A) I
B) II
C) A mixture of I and II
D) III
* E) IV

4. The major product(s), B, of the following reaction,

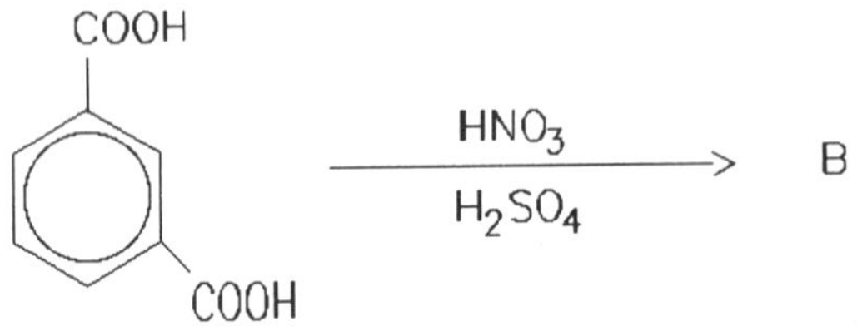

COOH

HNO_3

H_2SO_4

B

COOH

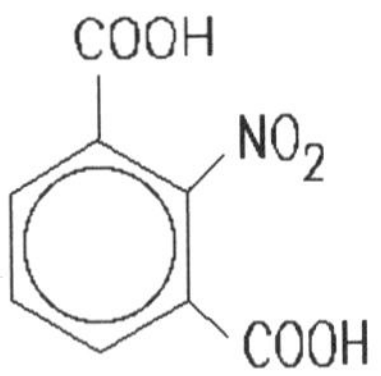

COOH	COOH	COOH
NO_2	
COOH	COOH	O_2N COOH
	NO_2
I 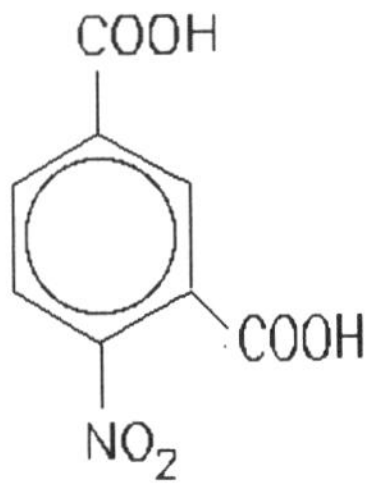	II 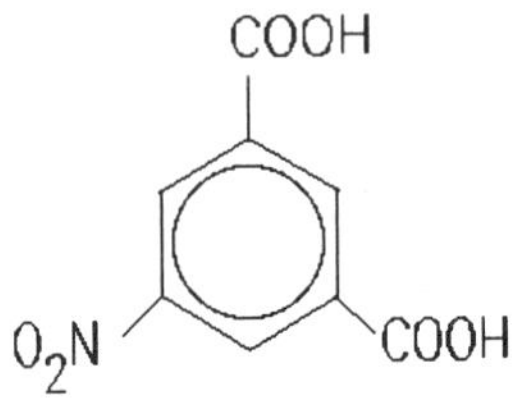	III

would be:

A) I
B) II
* C) III
D) Equal amounts of I and II
E) Equal amounts of I, II and III

5. The major product(s), C, of the following reaction,

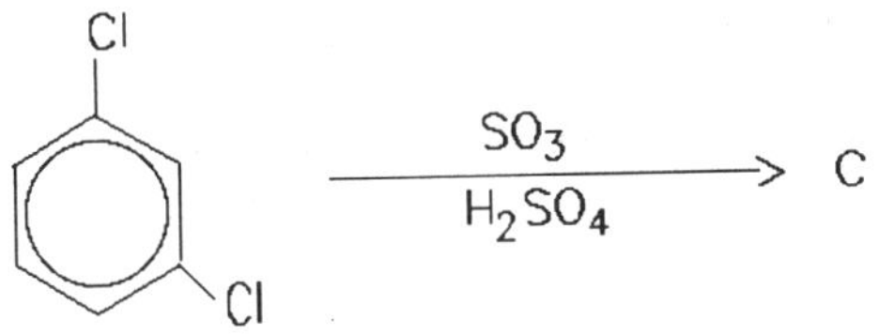

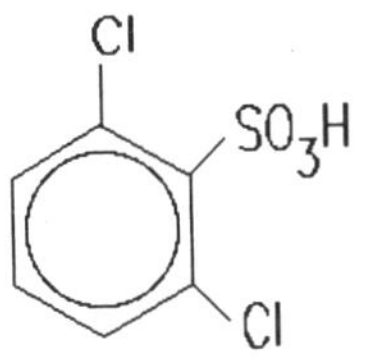

$$\text{1,3-dichlorobenzene} \xrightarrow[H_2SO_4]{SO_3} C$$

I: 2,6-dichlorobenzenesulfonic acid (Cl, SO_3H, Cl)

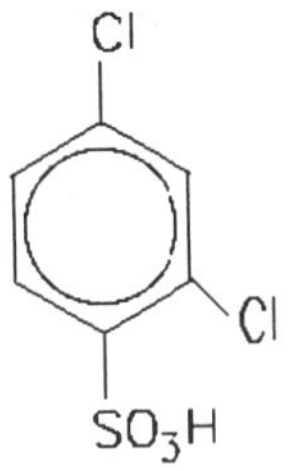

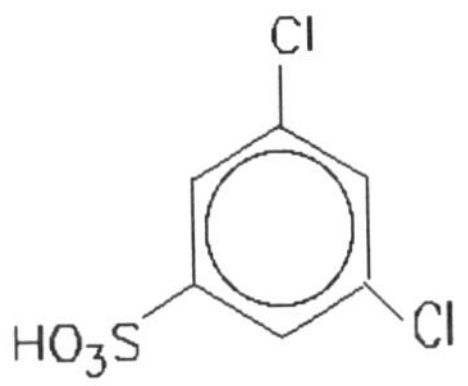

II: 2,4-dichlorobenzenesulfonic acid (Cl, Cl, SO_3H)

III: 3,5-dichlorobenzenesulfonic acid (HO_3S, Cl, Cl)

would be:

A) I
* B) II
C) III
D) Equal amounts of I and II
E) Equal amounts of I and III

6. The major product(s), D, of the following reaction,

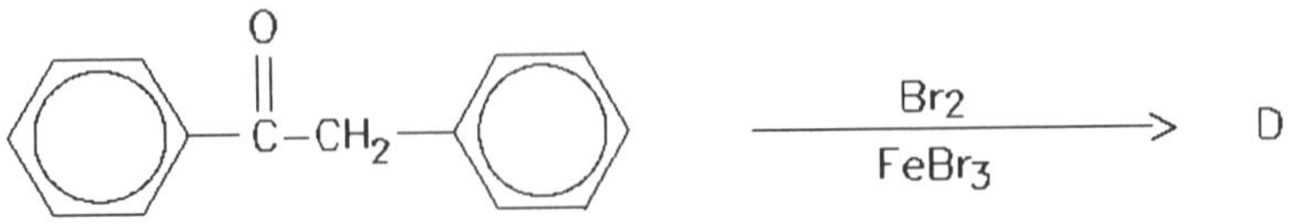

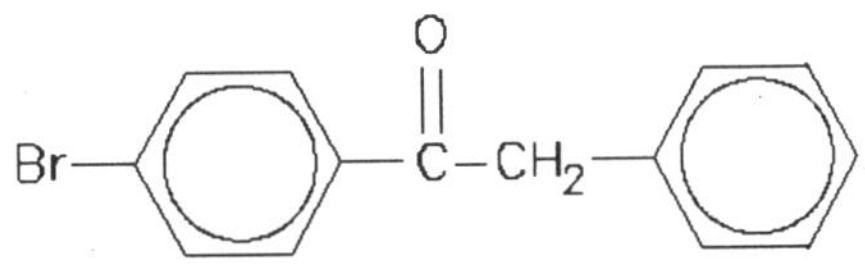

$$C_6H_5-\overset{O}{\overset{\|}{C}}-CH_2-C_6H_5 \xrightarrow[FeBr_3]{Br_2} D$$

I, 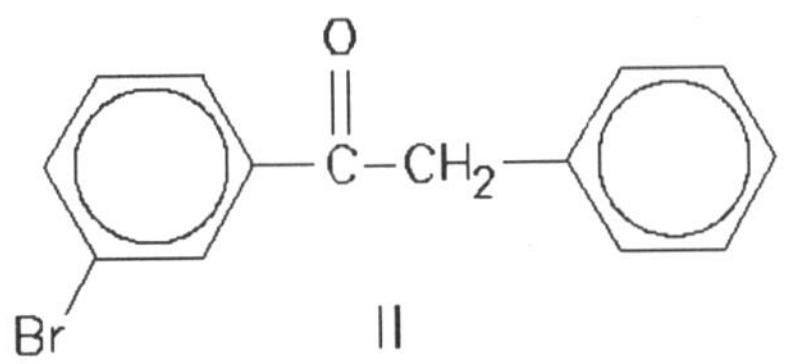 II, 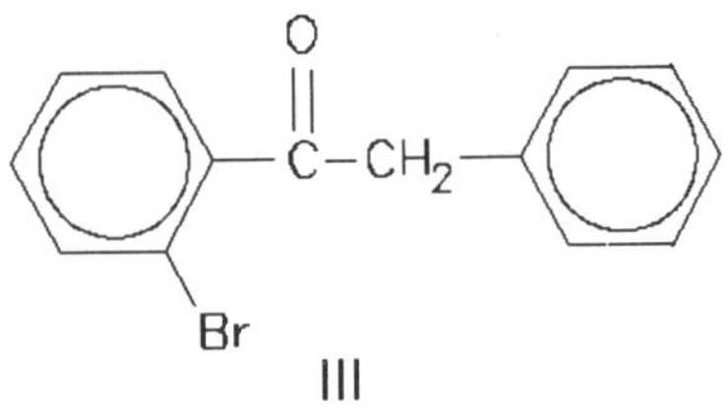 III, 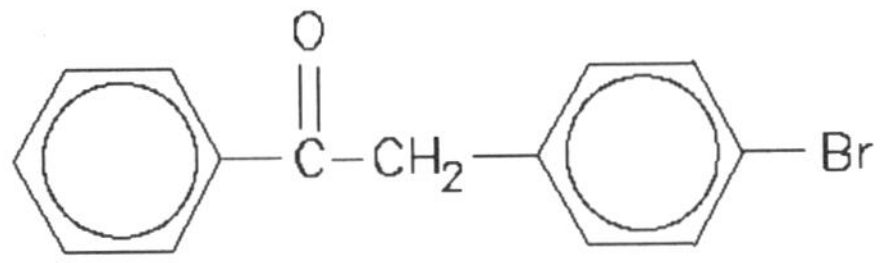 IV

would be:

A) I
B) II
C) III
* D) IV
E) Equal amounts of I and II

7. The product, F, of the following reaction sequence,

$$CH_3CH(CH_3)COH(=O) \xrightarrow{SOCl_2} \text{organic product} \xrightarrow[AlCl_3]{C_6H_6} \text{organic product} \xrightarrow[\text{HCl, heat}]{Zn(Hg)} F$$

would be:

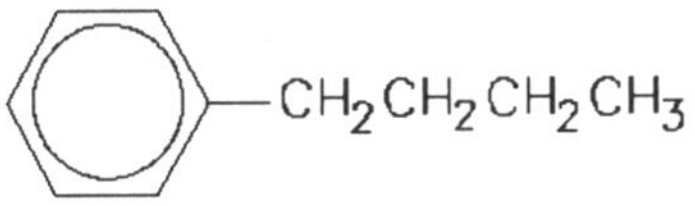

C_6H_5–$CH_2CH_2CH_2CH_3$

I

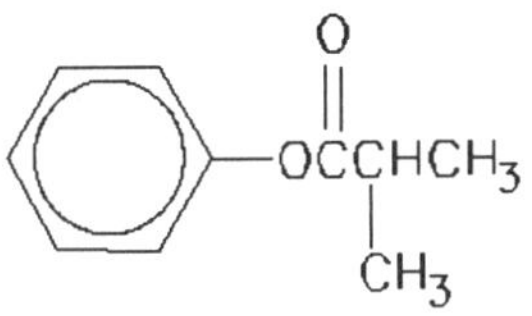

C_6H_5–$OC(=O)CH(CH_3)CH_3$

II

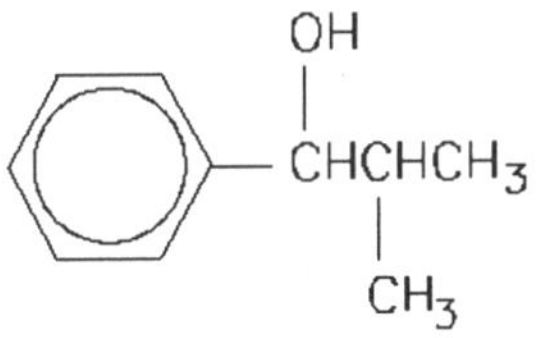

C_6H_5–$CH(OH)CH(CH_3)CH_3$

III

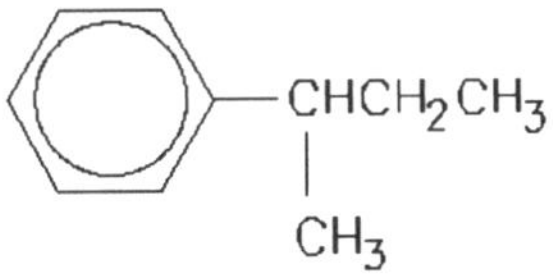

C_6H_5–$CH(CH_3)CH_2CH_3$

IV

C_6H_5–$CH_2CH(CH_3)CH_3$

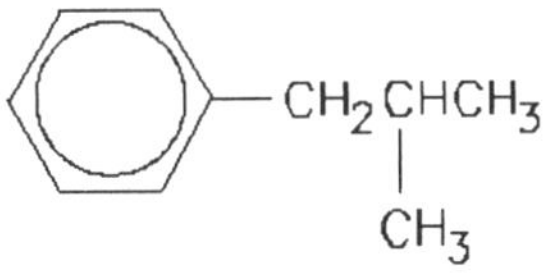

V

A) I B) II C) III D) IV * E) V

8. Which reagent(s) would you use to carry out the following transformation?

toluene DDDDDD4 benzoic acid

A) Br_2, heat, and light
B) Cl_2, $FeCl_3$
* C) $KMnO_4$, OH^-, heat (then H_3O^+)
D) HNO_3/H_2SO_4
E) SO_3/H_2SO_4

9. Which reagent(s) would you use to carry out the following transformation?

ethylbenzene DDDDDD4 1-chloro-1-phenylethane

* A) Cl_2, light, and heat
B) Cl_2, $FeCl_3$
C) $SOCl_2$
D) C_2H_5Cl, $AlCl_3$
E) HCl, O_2

10. Which reagent(s) would you use to carry out the following transformation?

isopropylbenzene DDDDDD4 2- and 4-chloro-1-isopropylbenzene

A) Cl_2, light
* B) Cl_2, $FeCl_3$
C) $SOCl_2$
D) C_2H_5Cl, $AlCl_3$
E) HCl, peroxides

11. Which reagent could you use as the basis for a simple chemical test to distinguish between:

C_6H_5–C≡CH and C_6H_5–CH=CH_2 ?

A) $NaOH/H_2O$
B) Br_2/CCl_4
* C) $Ag(NH_3)_2OH$
D) CrO_3/H_2SO_4
E) Concd. H_2SO_4

12. What would you expect to be the major product obtained from the following reaction?

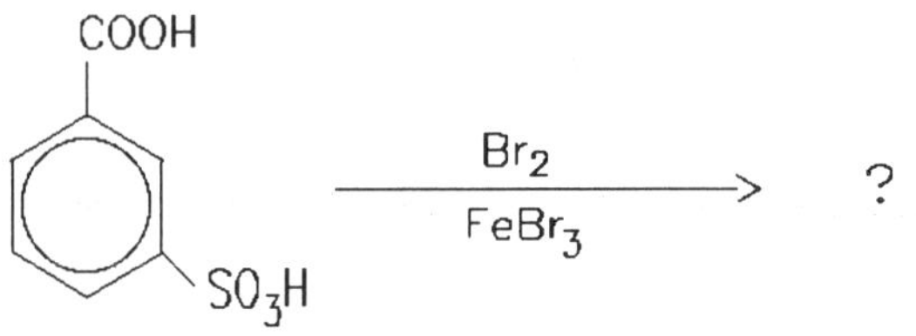

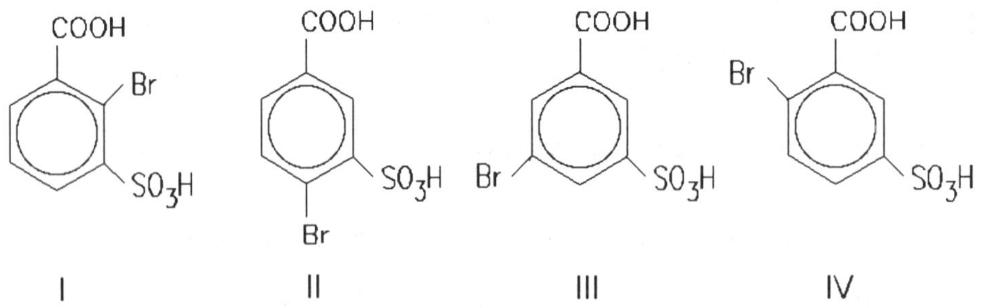

m-HO_3S–C_6H_4–COOH $\xrightarrow[FeBr_3]{Br_2}$?

I: COOH, Br, SO_3H (Br between COOH and SO_3H)
II: COOH, SO_3H, Br (Br para to COOH)
III: COOH, Br, SO_3H (Br meta to both)
IV: COOH, Br, SO_3H (Br ortho to COOH)

A) I
B) II
* C) III
D) IV
E) Equal amounts of II and IV

13. Which of the following reactions would yield isopropylbenzene as the major product?

A)
CH_3CHMCH_2
Benzene DDDDDDDDDD4
H_2SO_4

B)
CH_3CHCH_3
3
OH
Benzene DDDDDD4
H_2SO_4

C)
CH_3CHCH_3
3
Cl
Benzene DDDDDD4
$AlCl_3$

D)
$CH_3CH_2CH_2Cl$
Benzene DDDDDDDDDDD4
$AlCl_3$

* E) All of these

14. Which of the following reactions could be used to synthesize <u>tert</u>-butylbenzene?

A) C_6H_6 + $CH_2MC(CH_3)_2$ $\overset{H_2SO_4}{\text{DDDDDD4}}$

B) C_6H_6 + $(CH_3)_3COH$ $\overset{H_2SO_4}{\text{DDDDDD4}}$

C) C_6H_6 + $(CH_3)_3CCl$ $\overset{AlCl_3}{\text{DDDDDD4}}$

* D) All of the above

E) None of the above

15. Which of the following compounds would be most reactive toward electrophilic substitution?

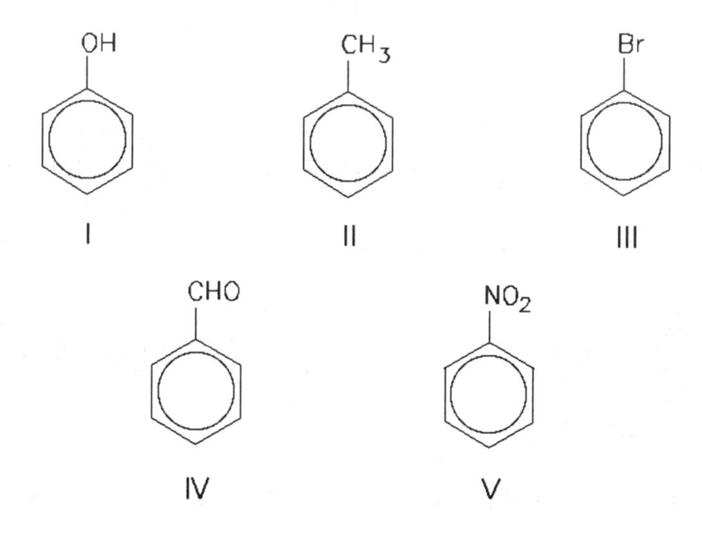

* A) I B) II C) III D) IV E) V

16. Which of these liquids would be unsuitable as an inert solvent for a Friedel-Crafts reaction?

* A) Chlorobenzene
  B) Nitrobenzene
  C) Acetophenone
  D) (Trifluoromethyl)benzene
  E) All could be used.

17. Which of the following contributors to the resonance stabilized hybrid formed when aniline undergoes para-chlorination would be exceptionally stable?

I II III IV

  A) I
  B) II
* C) III
  D) IV
  E) None of these

18. What would be the major product of the following reaction?

$$C_6H_5\text{—}CH_2CH_2CH_3 \xrightarrow[400^\circ C]{Cl_2} ?$$

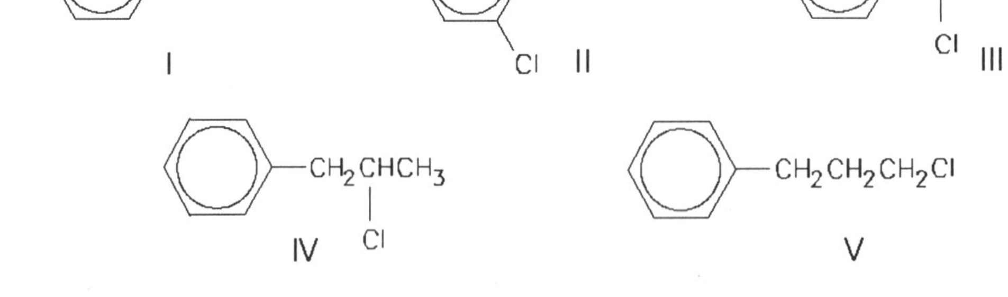

  A) I   B) II   * C) III   D) IV   E) V

19. The ortho/para product ratio is expected to be the smallest for the bromination of which of these?

A) Toluene
B) Isopropylbenzene
C) Butylbenzene
D) sec-Butylbenzene
* E) tert-Butylbenzene

20. Which of the following compounds would be most reactive to ring bromination?

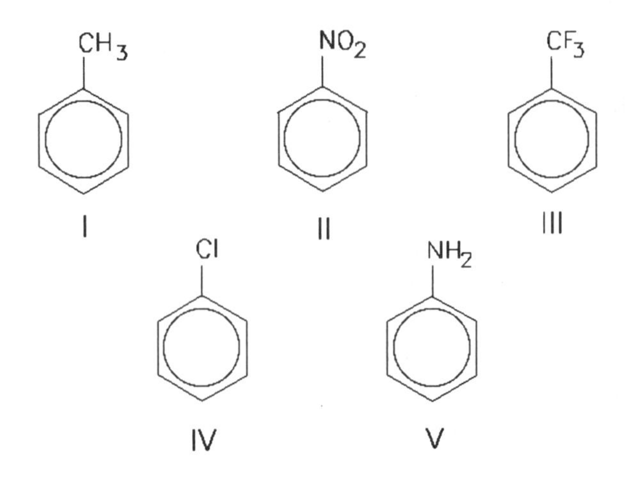

A) I B) II C) III D) IV * E) V

21. What would you expect to be the major product obtained from the following reaction?

NO_2, SO_3H $\xrightarrow[FeBr_3]{Br_2}$?

I: NO_2, Br, SO_3H

II: NO_2, SO_3H, Br

III: NO_2, Br, SO_3H

IV: NO_2, Br, SO_3H

A) I
B) II
* C) III
D) IV
E) Equal amounts of II and IV

22. Consider the structures given below. Which of them would be a relatively stable contributor to the hybrid formed when toluene undergoes para bromination?

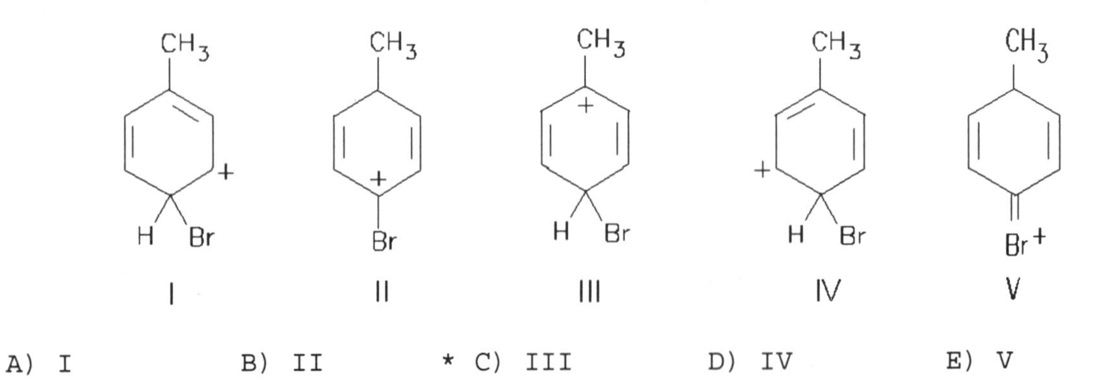

A) I   B) II   * C) III   D) IV   E) V

23. Which reagent would you use to carry out the following transformation?

<u>tert</u>-butylbenzene DDDDDD4 <u>p</u>-<u>tert</u>-butylbenzenesulfonic acid + <u>o</u>-<u>tert</u>-butylbenzenesulfonic acid

A) HNO_3/H_2SO_4
B) <u>tert</u>-$C_4H_9Cl/AlCl_3$
C) H_2SO_3/peroxides
* D) SO_3/H_2SO_4
E) SO_2/H_2SO_3

24. Which reagent would you use as the basis for a simple chemical test that would distinguish between toluene and

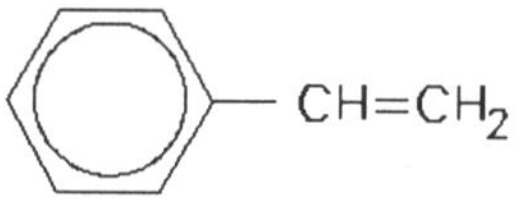

C_6H_5—CH=CH_2 ?

A) NaOH/H_2O
* B) Br_2/CCl_4
C) $Ag(NH_3)_2OH$
D) HCl/H_2O
E) CuCl in NH_3/H_2O

25. What would you expect to be the major product obtained from the following reaction?

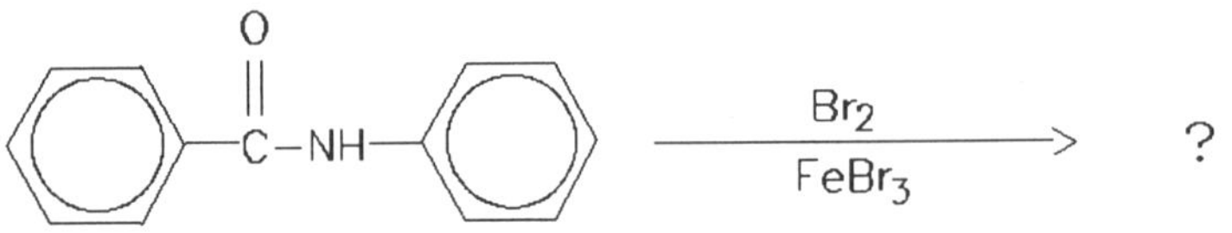

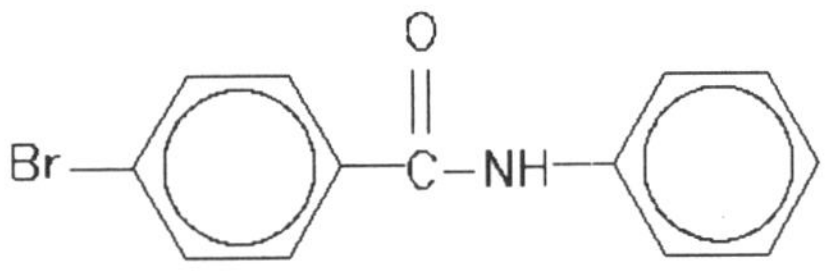

$\xrightarrow[FeBr_3]{Br_2}$?

I 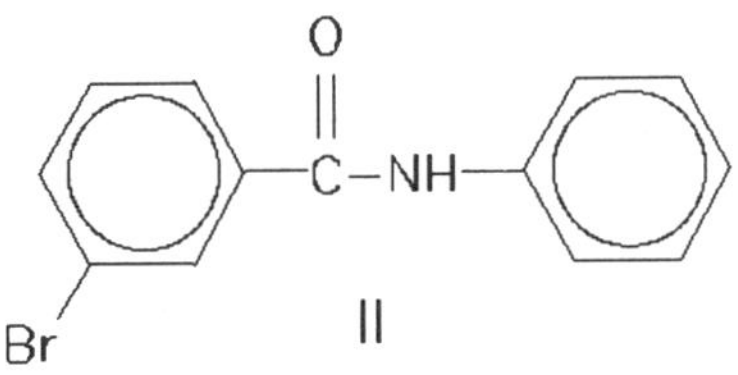 II 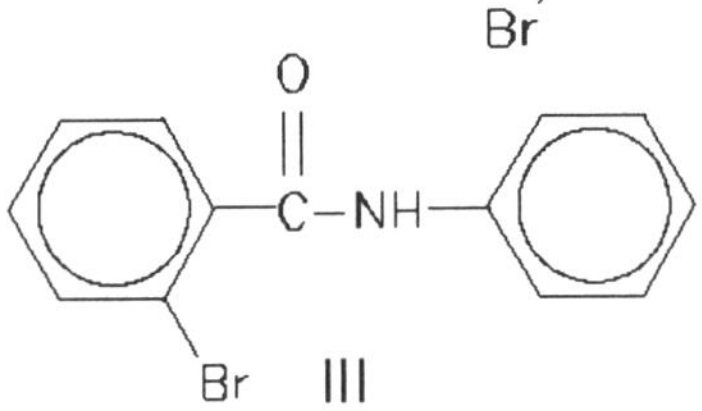 III 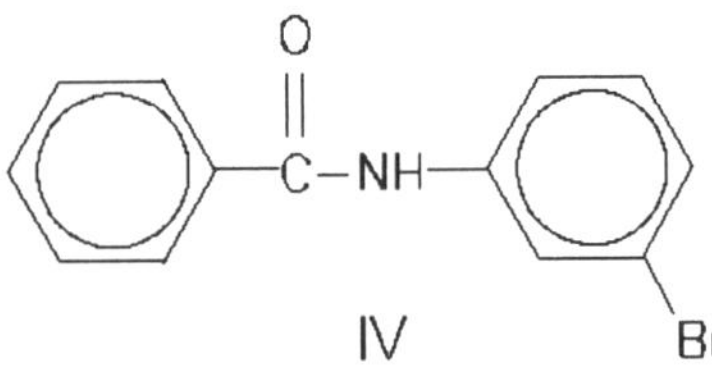 IV 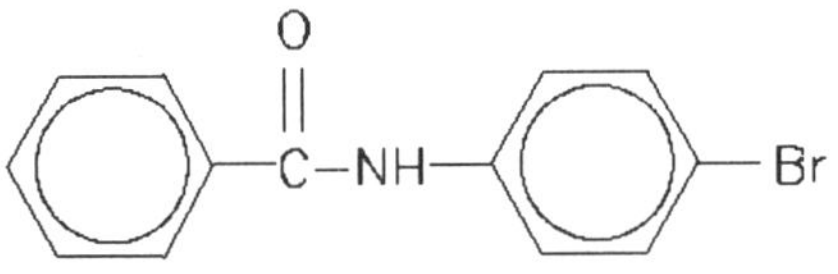 V

A) I B) II C) III D) IV * E) V

26. Which of the following structures does not contribute to the resonance hybrid of the intermediate formed when bromobenzene undergoes para-chlorination?

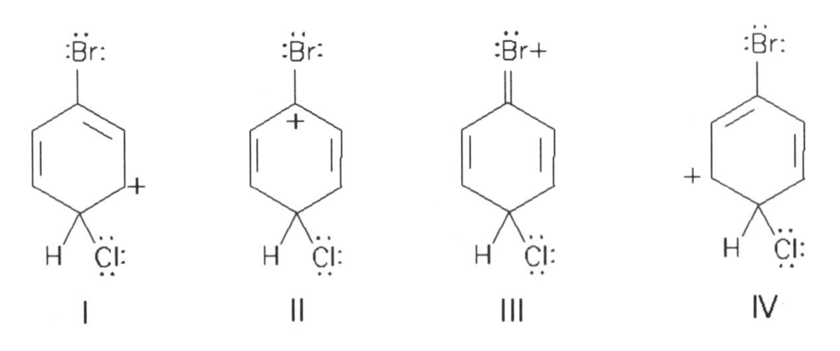

A) I
B) II
C) III
D) IV
* E) All of these contribute.

27. Which of the following reactions would give the product(s) indicated in substantial amounts (i.e., in greater than 50% yield)?

I $C_6H_5NH_2 + CH_3Cl \xrightarrow{AlCl_3}$ p-$CH_3C_6H_4NH_2$ and o-$CH_3C_6H_4NH_2$

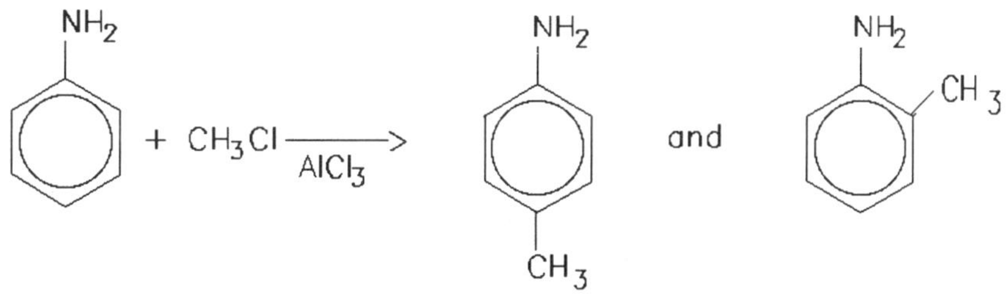

II $C_6H_5NO_2 + CH_3Cl \xrightarrow{AlCl_3}$ m-$CH_3C_6H_4NO_2$

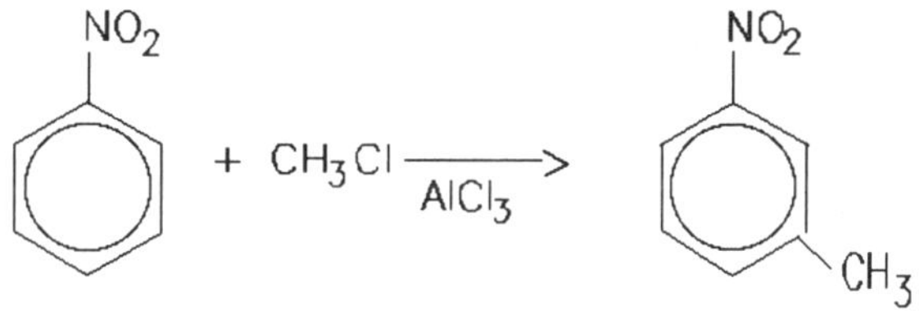

III $C_6H_6 + CH_3CH_2CH_2Cl \xrightarrow{AlCl_3} C_6H_5CH_2CH_2CH_3$

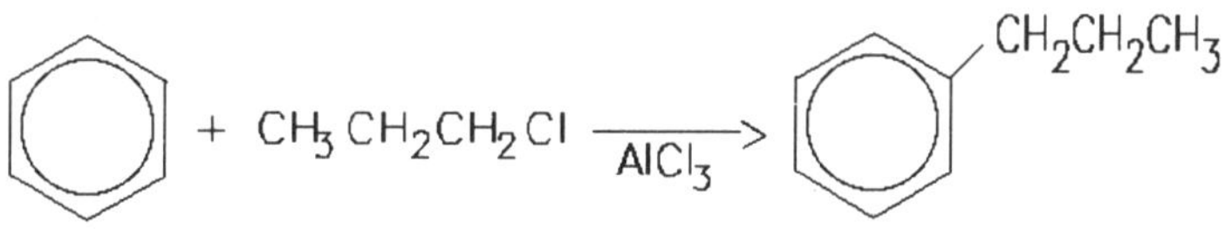

A) I
B) II
C) III
D) All of these
* E) None of these

28. Which of the following is not a meta-directing substituent when present on the benzene ring?

* A) $-NHCOCH_3$
B) $-NO_2$
C) $-N(CH_3)_3^+$
D) -CpN
E) $-CO_2H$

29. The best synthesis of

CH_3
NO_2
CH_3

would be:

A) HNO_3 / $2\ CH_3Cl$
Benzene DDDDDD4 product DDDDDDD4
H_2SO_4 / $2\ AlCl_3$

B) HNO_3 / CH_3Cl
Toluene DDDDDD4 product DDDDDD4
H_2SO_4 / $AlCl_3$

* C) HNO_3
p-Xylene DDDDDD4
H_2SO_4

D) CH_3Cl
m-Nitrotoluene DDDDDD4
$AlCl_3$

E) All of these are equally good.

30. Starting with benzene, the best method for preparing p-nitrobenzoic acid is:

A) HNO_3/H_2SO_4; then $CH_3Cl/AlCl_3$; then separation of isomers; then $KMnO_4/^-OH$, followed by H_3O^+.
* B) $CH_3Cl/AlCl_3$; then HNO_3/H_2SO_4; then separation of isomers; then $KMnO_4/^-OH$, followed by H_3O^+.
C) $CH_3Cl/AlCl_3$; then $KMnO_4/^-OH$, followed by H_3O^+; then HNO_3/H_2SO_4.
D) HNO_3/H_2SO_4; then $KMnO_4/^-OH$, followed by H_3O^+; then $CH_3Cl/AlCl_3$.
E) HNO_3/H_2SO_4; then CO_2, followed by H_3O^+.

31. A good synthesis of

O

$(CH_3)_3C$ — — CCH_3

would be:

A)

Benzene $\overset{O}{\underset{}{:}}$ CH_3CCl DDDDDD4 ($AlCl_3$) $(CH_3)_3CCl$ DDDDDDDDD4 ($AlCl_3$)

* B)

Benzene $(CH_3)_3CCl$ DDDDDDDDD4 ($AlCl_3$) $\overset{O}{\underset{}{:}}$ CH_3CCl DDDDDD4 ($AlCl_3$)

C)

Benzene $\overset{O}{\underset{}{:}}$ CH_3CCl DDDDDD4 ($AlCl_3$) $(CH_3)_2C{=}CH_2$ DDDDDDDDDDD4 (HF)

D) More than one of these

E) None of these

32. Which would be a good synthesis of <u>m</u>-nitrobenzoic acid?

A) Benzene HNO_3/H_2SO_4 DDDDDDDDDD4 (heat) CH_3Cl DDDDDD4 ($AlCl_3$) (1) $KMnO_4, OH^-$, heat DDDDDDDDDDDDDDDDDDD4 (2) H_3O^+

B) Toluene HNO_3/H_2SO_4 DDDDDDDDDD4 (heat) (1) $KMnO_4, OH^-$, heat DDDDDDDDDDDDDDDDDDD4 (2) H_3O^+

* C) Toluene (1) $KMnO_4, OH^-$, heat DDDDDDDDDDDDDDDDDDD4 (2) H_3O^+ HNO_3/H_2SO_4 DDDDDDDDDD4 (heat)

D) More than one of the above

E) None of the above

33. Which reagent(s) would you use to carry out the following transformation?

ethylbenzene DDDDDD4 benzoic acid

A) Cl_2, heat, and light

B) Cl_2, $FeCl_3$

* C) $KMnO_4$, OH^-, heat (then H_3O^+)

D) HNO_3/H_2SO_4

E) SO_3/H_2SO_4

34. The major product(s), B, of the following reaction,

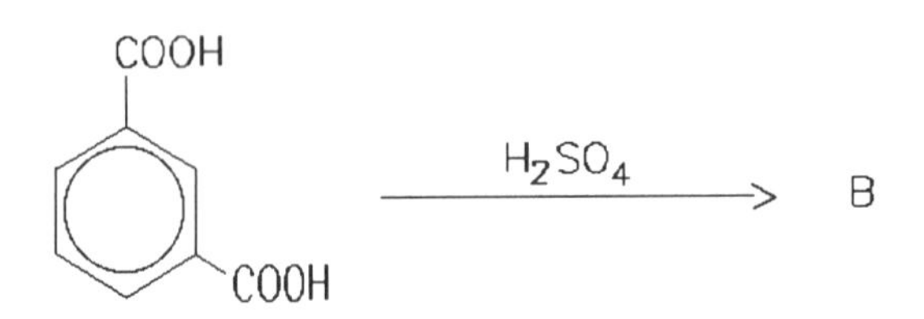

COOH, SO_3H, COOH — I

COOH, COOH, SO_3H — II

COOH, HO_3S, COOH — III

would be:

A) I

B) II

* C) III

D) Equal amounts of I and II

E) Equal amounts of I, II and III

35. Which of the following compounds would you expect to be most reactive toward ring nitration?

A) Benzene

B) Toluene

* C) _m_-Xylene

D) _p_-Xylene

E) Benzoic acid

36. Benzoic acid can be prepared by the oxidation of all of the following compounds <u>except</u> this one:

A) $C_6H_5CH{=}CH_2$

B) C_6H_5CpCH

C) $C_6H_5CCH_3$ (with O : above the C)

D) $C_6H_5CH_2CH_2CH_3$

* E) $C_6H_5C(CH_3)_3$

37. Each of the five disubstituted benzenes is nitrated. In which of these cases does the arrow not indicate the chief position of nitration.

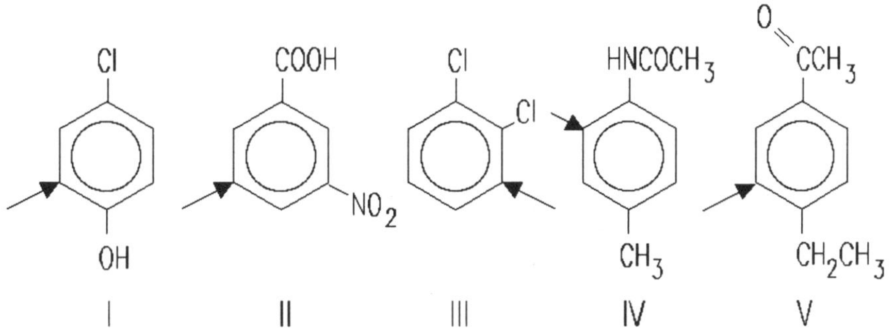

Cl
OH
I
COOH
NO_2
II
Cl
Cl
III
$HNCOCH_3$
CH_3
IV
O
CCH_3
CH_2CH_3
V

A) I B) II * C) III D) IV E) V

38. Which of the following compounds would be least reactive toward electrophilic substitution?

NH_2
I
CH_3
II
Br
III
CHO
IV
NO_2
V

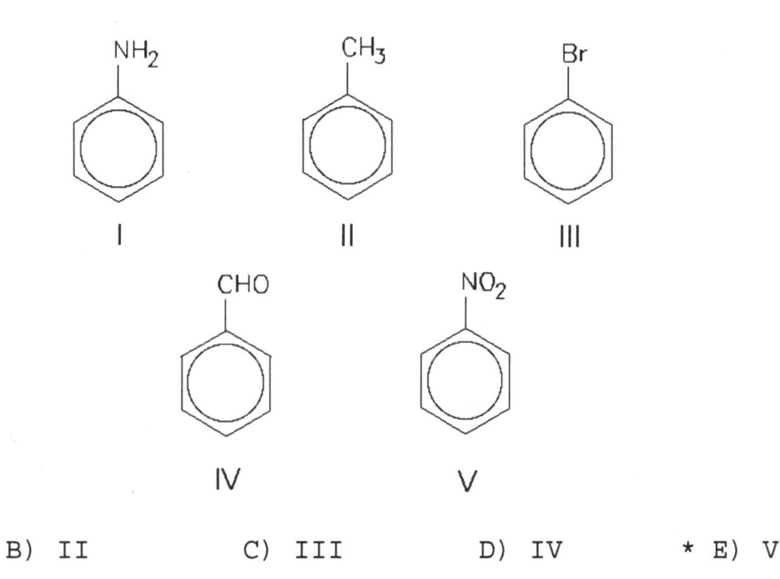

A) I B) II C) III D) IV * E) V

39. Which of these is a satisfactory synthesis of 1-chloro-2-phenylethane?

A) $C_6H_5CH_2CH_3$ + Cl_2, Fe DDDDD4
B) $C_6H_5CH_2CH_3$ + Cl_2, 400xC DDDDD4
* C) $C_6H_5CH_2CH_2OH$ + $SOCl_2$ DDDDD4
D) $C_6H_5CH{=}CH_2$ + HCl, peroxide DDDDD4
E) C_6H_6 + CH_3CH_2Cl, $AlCl_3$ DDDDD4

40. What would you expect to be the major product obtained from the mononitration of m-dichlorobenzene?

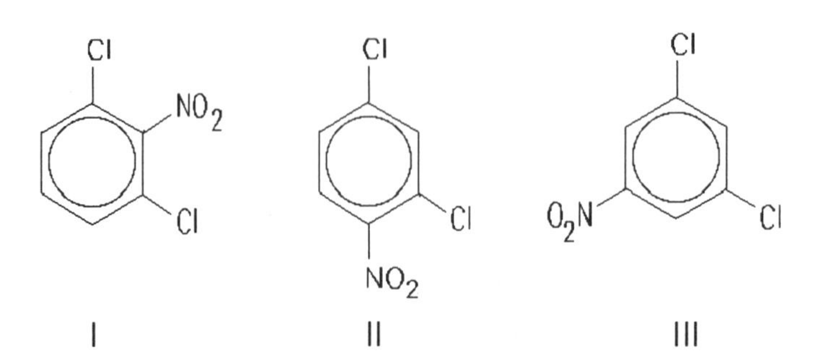

A) I
* B) II
C) III
D) Equal amounts of I and II
E) Equal amounts of I, II and III

41. Which of the following reactions would produce ethylbenzene?

A) Benzene $\xrightarrow[AlCl_3]{CH_3CH_2Cl}$

B) Benzene $\xrightarrow[HF]{CH_2MCH_2}$

C) Benzene $\xrightarrow[H_2SO_4]{CH_3CH_2OH}$

D) Acetophenone $\xrightarrow[\text{concd HCl}]{Zn(Hg)}$

* E) All of these

42. Which of the following compounds would yield the greatest amount of meta product when subjected to ring nitration?

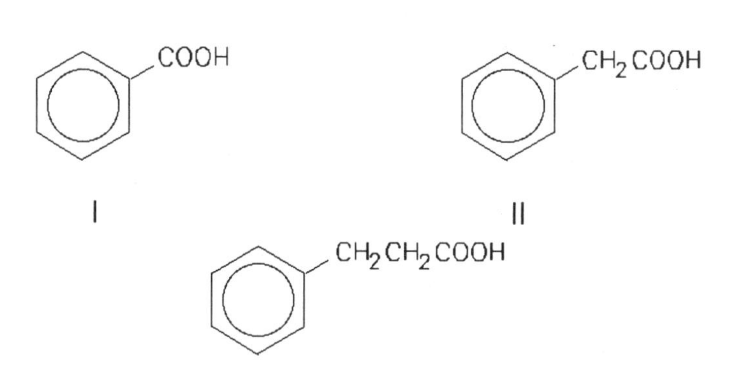

$CH_2CH_2CH_2COOH$

IV

$CH_2CH_2CH_2CH_2COOH$

V

* A) I B) II C) III D) IV E) V

43. Which reagent(s) would you use to carry out the following transformation?

ethylbenzene DDDDDD4 2- and 4-chloro-1-ethylbenzene

A) Cl_2, light
* B) Cl_2, $FeCl_3$
C) $SOCl_2$
D) C_2H_5Cl, $AlCl_3$
E) None of these

44. Which of the following compounds would be most reactive toward ring nitration?

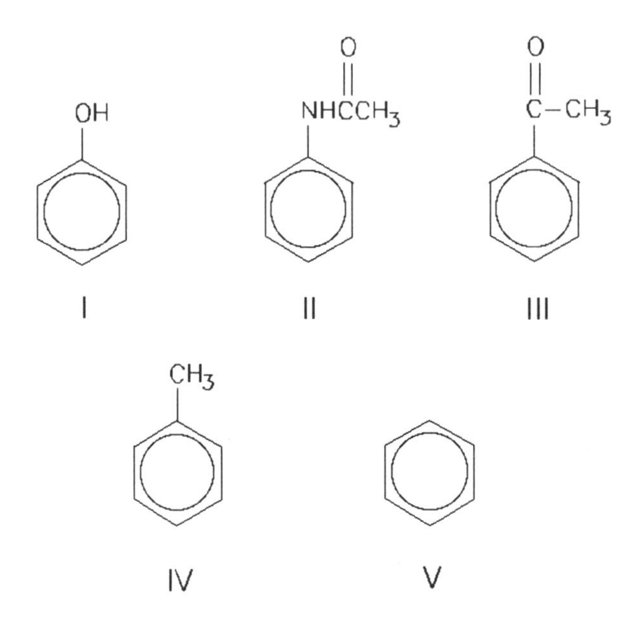

* A) I B) II C) III D) IV E) V

45. What would you expect to be the major product obtained from the following reaction?

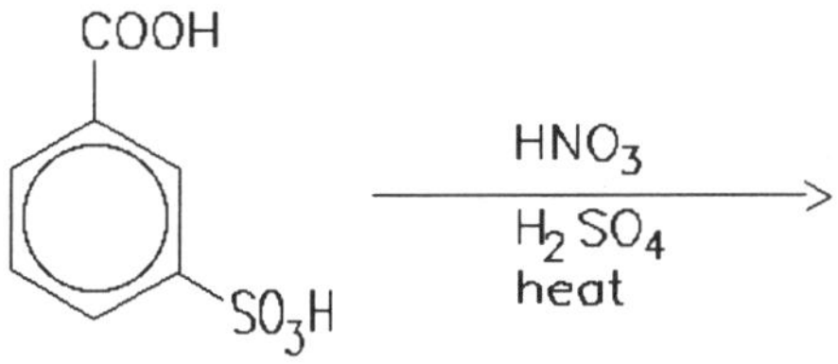

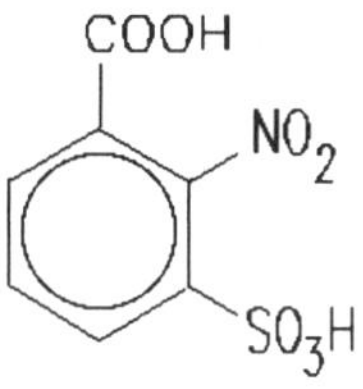

COOH, SO_3H $\xrightarrow[\text{heat}]{HNO_3 / H_2SO_4}$

I: COOH, NO_2, SO_3H

II: COOH, SO_3H, NO_2

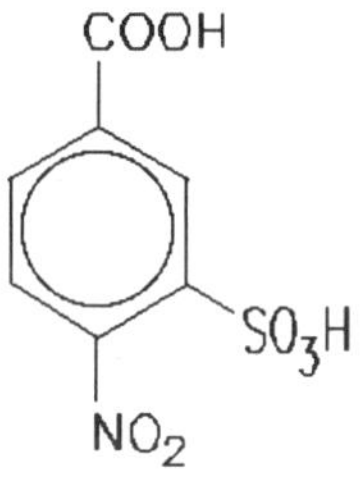

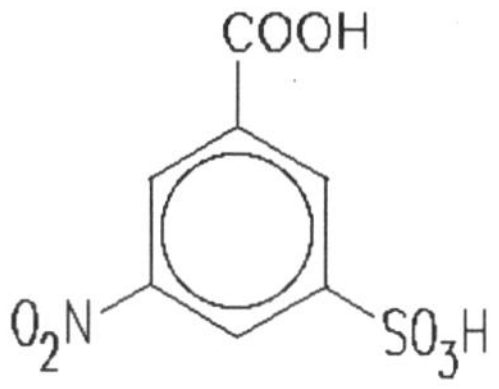

III: COOH, O_2N, SO_3H

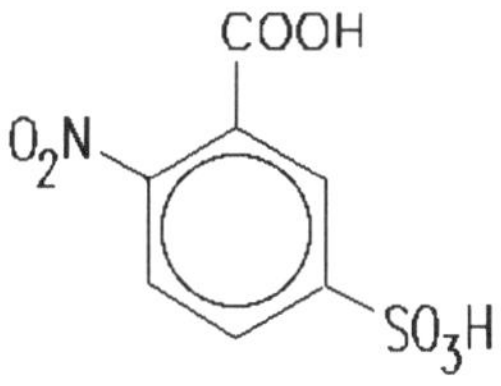

IV: COOH, O_2N, SO_3H

A) I
B) II
* C) III
D) IV
E) Equal amounts of I and II

46. The product, C, that would result from the following series of reactions,

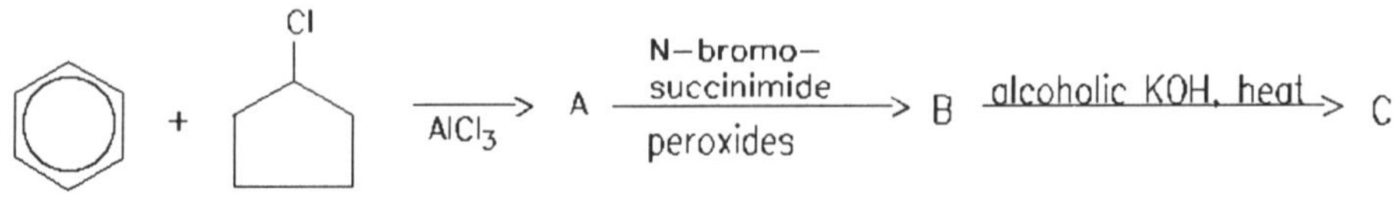

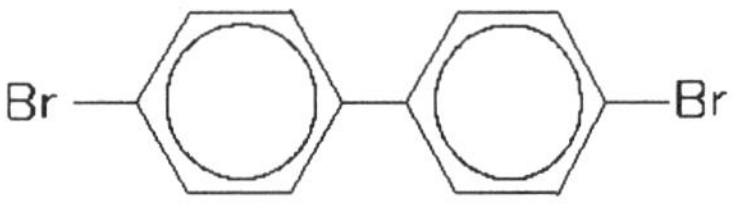

Cl

+ $\xrightarrow{AlCl_3}$ A $\xrightarrow[\text{peroxides}]{\text{N-bromo-succinimide}}$ B $\xrightarrow{\text{alcoholic KOH, heat}}$ C

Br — — Br

I

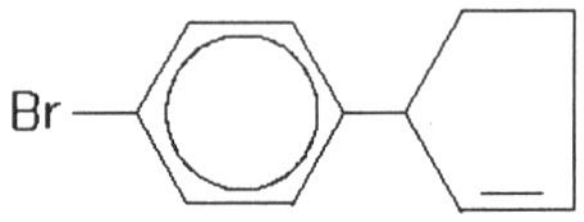

Br —

II

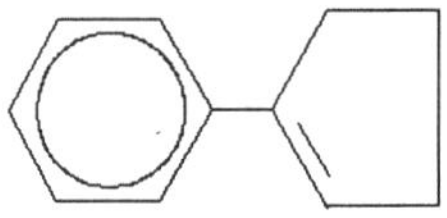

III

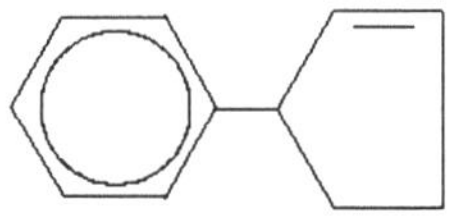

IV

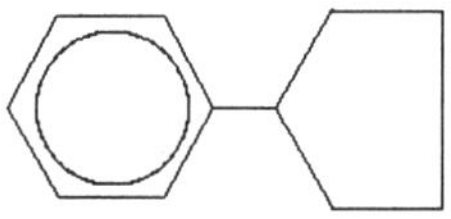

V

would be:

A) I B) II * C) III D) IV E) V

47. The major product(s), D, of the following reaction

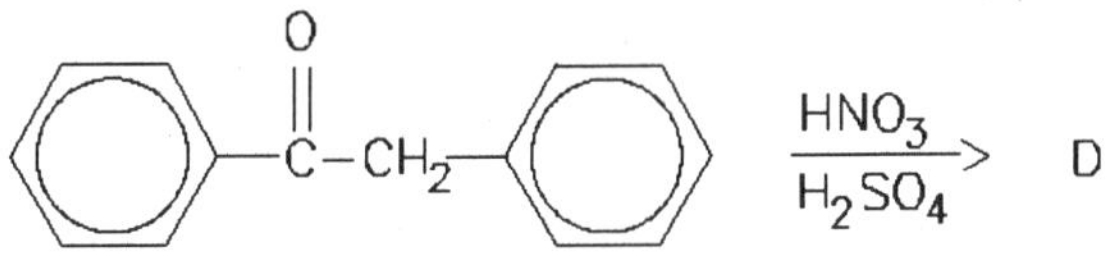

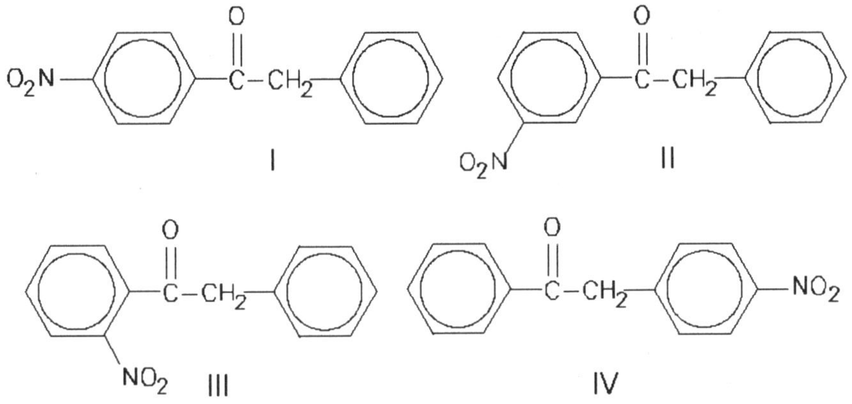

$$C_6H_5-C(=O)-CH_2-C_6H_5 \xrightarrow[H_2SO_4]{HNO_3} D$$

O_2N—C(=O)—CH_2— (I)

C(=O)—CH_2—, O_2N (II)

C(=O)—CH_2—, NO_2 (III)

C(=O)—CH_2—NO_2 (IV)

would be:

A) I
B) II
C) III
* D) IV
E) Equal amounts of I and II

48. What would be the major product of the following reaction?

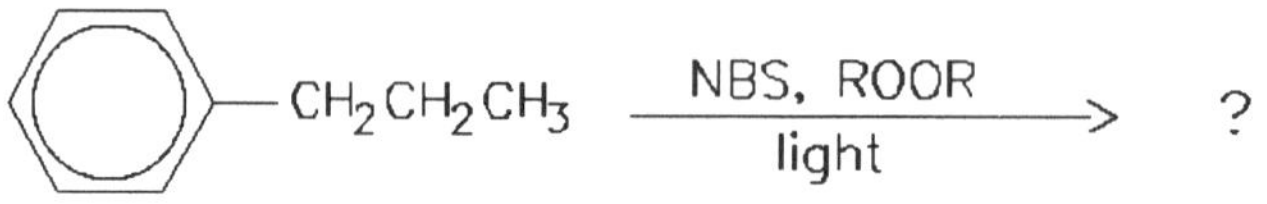

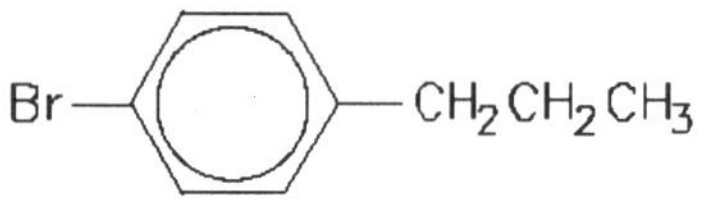

$C_6H_5-CH_2CH_2CH_3 \xrightarrow[\text{light}]{\text{NBS, ROOR}} ?$

I: $Br-C_6H_4-CH_2CH_2CH_3$ (para)

II: $C_6H_4(Br)-CH_2CH_2CH_3$ (ortho)

III: $C_6H_5-CH(Br)CH_2CH_3$

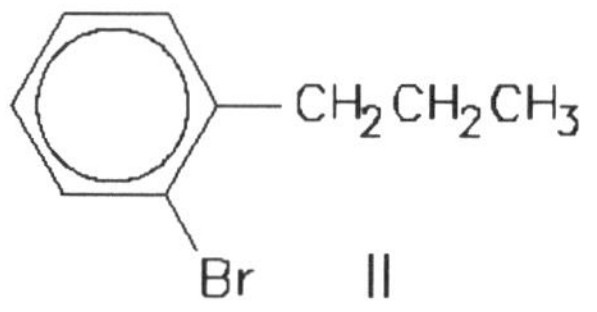

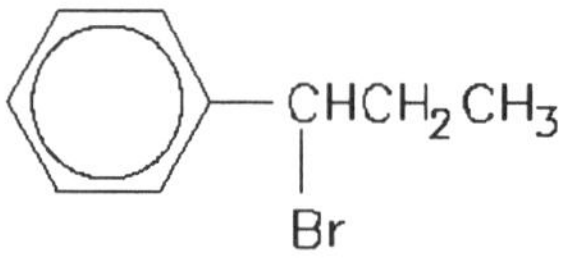

IV: $C_6H_5-CH_2CH(Br)CH_3$

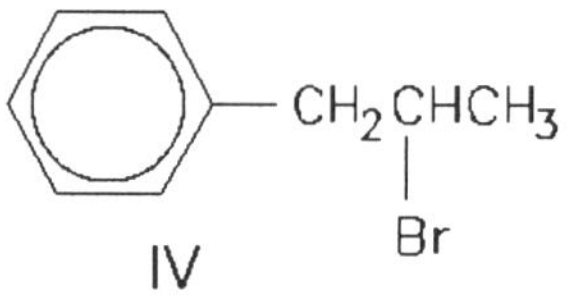

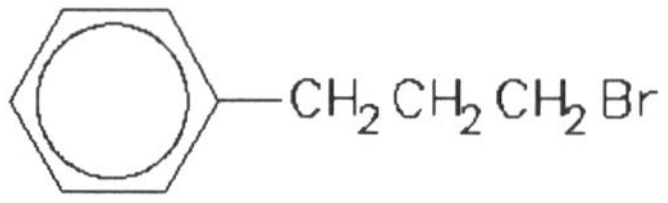

V: $C_6H_5-CH_2CH_2CH_2Br$

A) I B) II * C) III D) IV E) V

49. Which of the compounds listed below would you expect to give the greatest amount of meta-product when subjected to ring bromination?

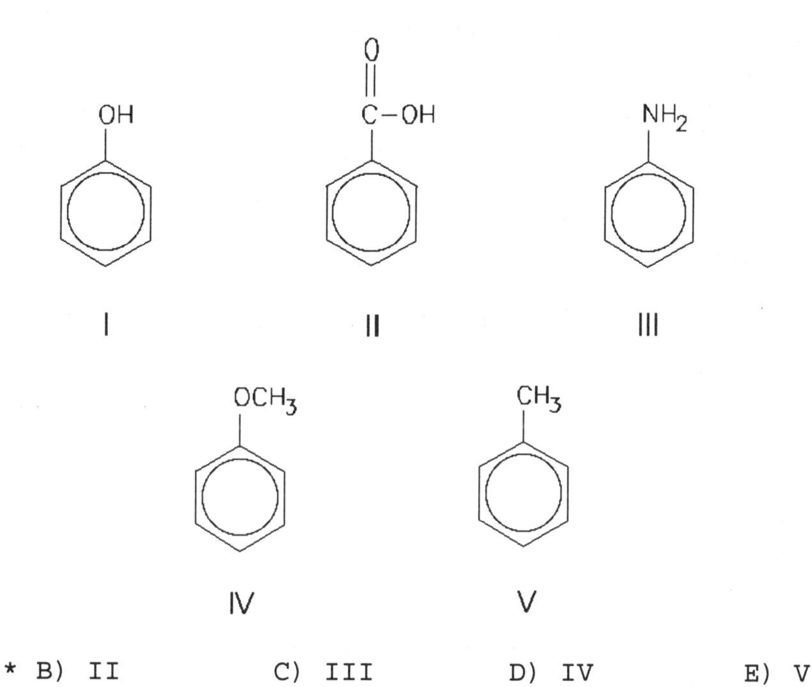

A) I
* B) II
C) III
D) IV
E) V

50. Which of the following structures contribute(s) to the resonance hybrid of the intermediate formed when bromobenzene undergoes para-chlorination?

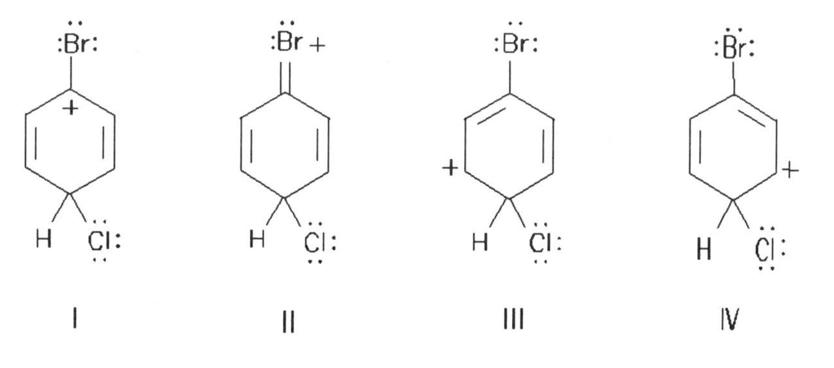

A) I
B) II
C) III
D) IV
* E) All of the above

51. Which reagent(s) would you use to carry out the following transformation?

toluene DDDDDD4 o- and p-chlorotoluene

A) Cl_2, light
* B) Cl_2, $FeCl_3$
C) $SOCl_2$
D) C_2H_5Cl, $AlCl_3$
E) HCl, peroxide

52. Which reagent(s) would you use to carry out the following transformation?

toluene DDDDDD4 benzyl bromide

A) Br_2, $FeBr_3$
* B) N-Bromosuccinimide, ROOR, h0
C) HBr
D) Br_2/CCl_4
E) NaBr, H_2SO_4

53. Which of the following is not an ortho-para director in electrophilic aromatic substitution?

* A) $-CF_3$ B) $-OCH_3$ C) $-CH_3$ D) -F E) $-NH_2$

54. Which would be the major product(s) of the following reaction?

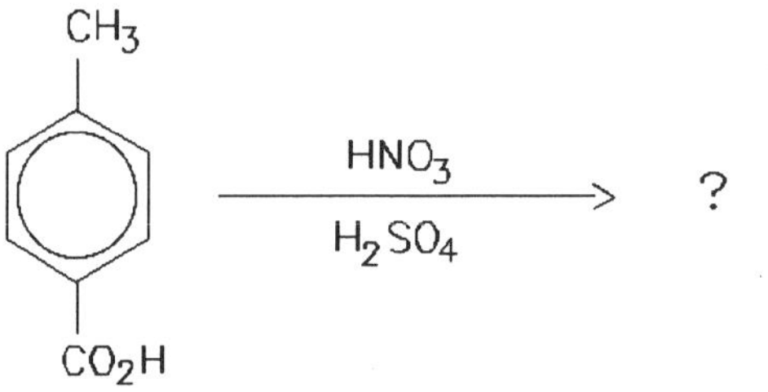

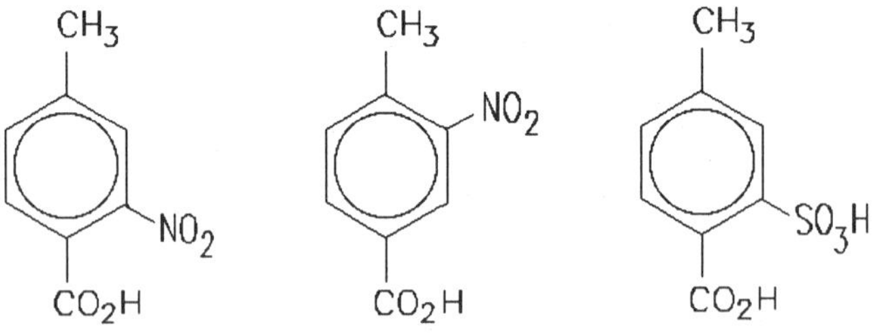

CH3 / CO2H (benzene ring) —HNO_3 / H_2SO_4→ ?

I: CH3, NO2, CO2H
II: CH3, NO2, CO2H
III: CH3, SO3H, CO2H

A) I
* B) II
C) III
D) I and II in roughly equal amounts
E) I and III in roughly equal amounts

55. Which would be the product, X, of the following reaction sequence?

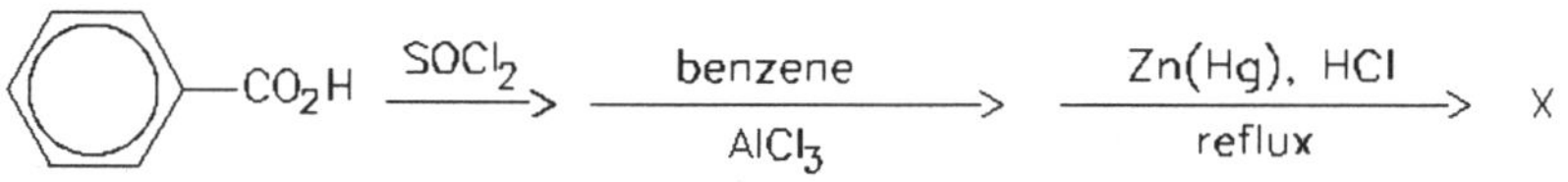

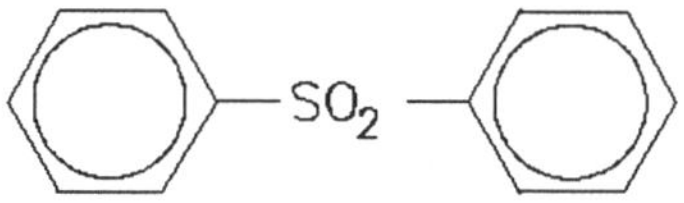

$C_6H_5CO_2H \xrightarrow{SOCl_2} \xrightarrow[AlCl_3]{benzene} \xrightarrow[reflux]{Zn(Hg),\ HCl} X$

I: $C_6H_5-SO_2-C_6H_5$

II: $C_6H_5-CO_2-C_6H_5$

III: $C_6H_5-CH_2-C_6H_5$

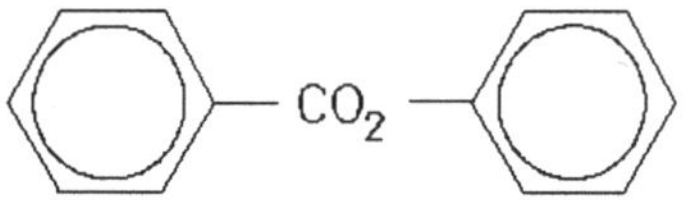

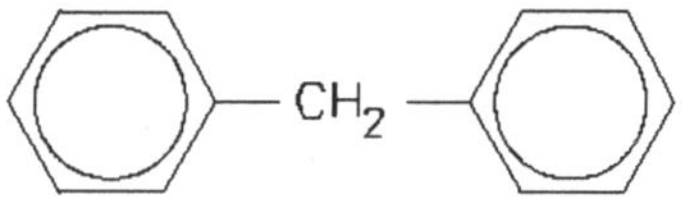

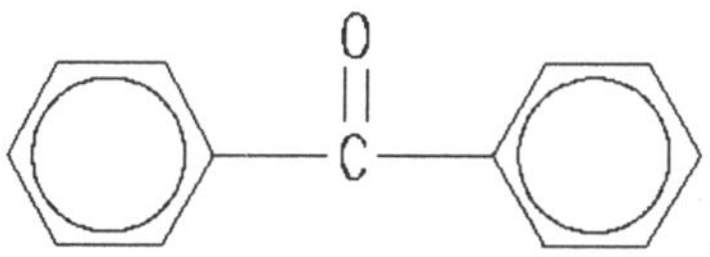

IV: $C_6H_5-C(=O)-C_6H_5$

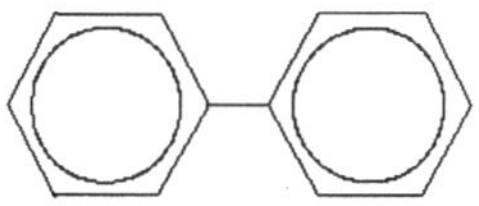

V: $C_6H_5-C_6H_5$

A) I B) II * C) III D) IV E) V

56. Which of these compounds gives essentially a single product on electrophilic substitution of a third group?

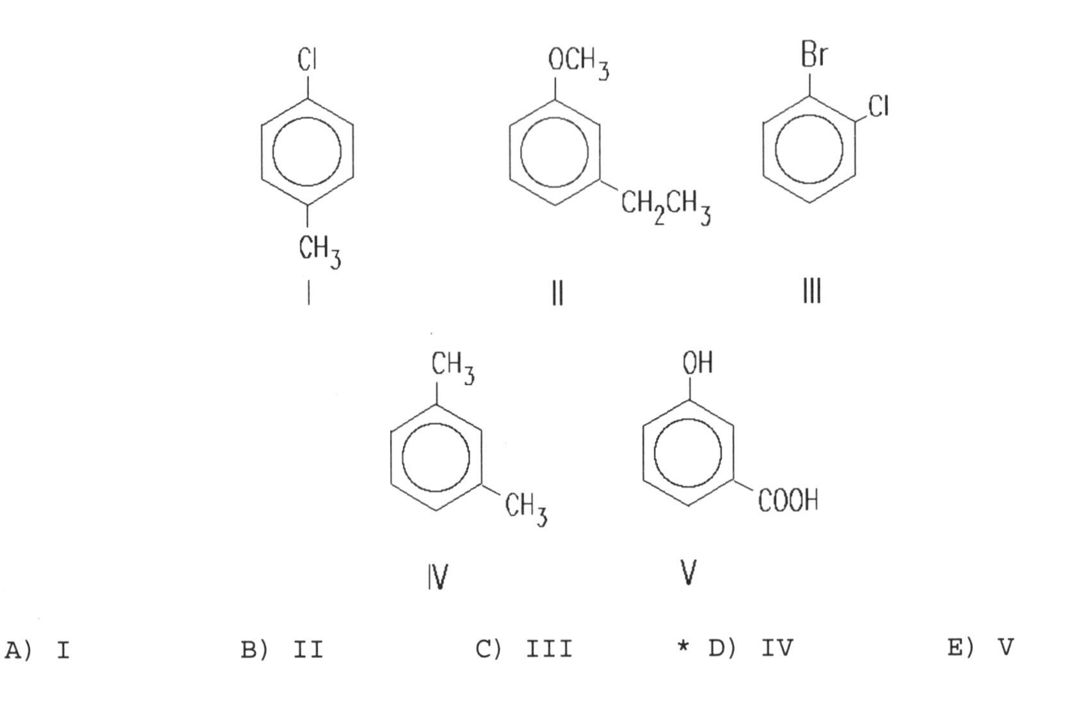

A) I B) II C) III * D) IV E) V

57. Which reagent(s) would you use to synthesize 2- and 4-bromo-1-ethylbenzene from ethylbenzene?
A) N-Bromosuccinimide (NBS), CCl_4, light
B) PBr_3
* C) Br_2, $FeBr_3$
D) CH_3CH_2Br, $AlBr_3$
E) HBr, ROOR

58. Undesired polysubstitution of an aromatic nucleus is most likely to be encountered in the case of:
* A) Friedel-Crafts alkylation
B) Friedel-Crafts acylation
C) Nitration
D) Sulfonation
E) Chlorination

59. What is a feature found in all ortho-para directing groups?
A) An oxygen atom is directly attached to the aromatic ring.
B) The atom attached to the aromatic ring possesses an unshared pair of electrons.
* C) The group has the ability to delocalize the positive charge of the arenium ion.
D) The atom directly attached to the aromatic ring is more electronegative than carbon.
E) The group contains a multiple bond.

60. This molecule <u>cannot</u> participate as a reactant in a Friedel-Crafts reaction.

A) Benzene
B) Chlorobenzene
* C) Nitrobenzene
D) Toluene
E) *tert*-Butylbenzene

61. Which one of these molecules <u>can be</u> a reactant in a Friedel-Crafts reaction?

A) Aniline
B) Nitrobenzene
C) Chloroethene
* D) Bromobenzene
E) *p*-Bromonitrobenzene

62. This reaction can be carried out in such a way that the outcome is determined by kinetics or in another way when thermodynamics is the determining factor:

A) Nitration of benzene
B) Chlorination of benzoic acid
C) Acetylation of bromobenzene
* D) Sulfonation of toluene
E) Bromination of ethyl benzoate

63. This substituent deactivates the benzene ring towards electrophilic substitution but directs the incoming group chiefly to the ortho and para positions.

A) $-OCH_2CH_3$ B) $-NO_2$ * C) -F D) $-CF_3$ E) $-NHCOCH_3$

64. Which of these electrophilic aromatic substitution reactions is reversible?

A) Nitration
B) Bromination
C) Sulfonation
D) Friedel-Crafts alkylation
* E) Both C) and D)

65. In electrophilic aromatic substitution, the attacking species (the electrophile) <u>necessarily</u> is a:

A) Neutral species.
B) Positively charged species.
* C) Lewis acid.
D) Proton.
E) Carbocation.

66. When toluene is reacted in turn with 1) Cl_2 (large excess), heat, and light and 2) Br_2,$FeBr_3$, the chief product is:

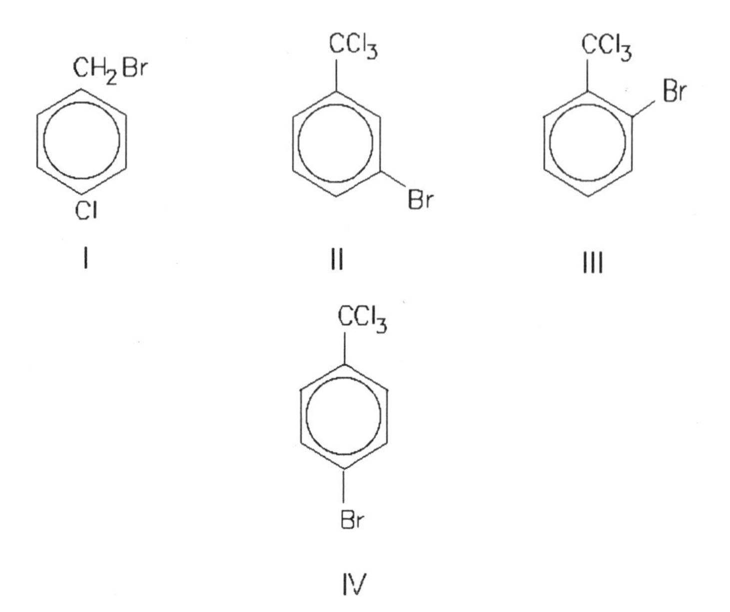

A) I
* B) II
C) III
D) IV
E) A mixture of III and IV

67. The electrophilic bromination or chlorination of benzene requires, in addition to the halogen:

A) Hydroxide ion.
B) A Lewis base.
* C) A Lewis acid.
D) Peroxide.
E) Ultraviolet light.

68. The reaction of benzene with $(CH_3)_3CCH_2Cl$ in the presence of anhydrous aluminum chloride produces principally which of these?

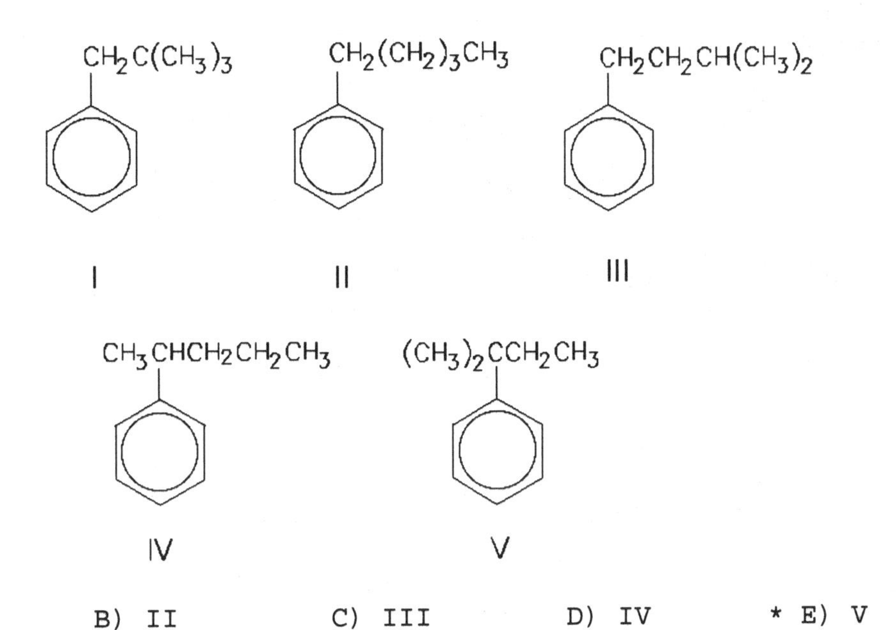

A) I B) II C) III D) IV * E) V

69. Toluene is subjected to the action of the following reagents in the order given: (1) $KMnO_4, OH^-$, heat; then H_3O^+ (2) HNO_3, H_2SO_4 (3) Br_2, $FeBr_3$
What is the final product of this sequence?

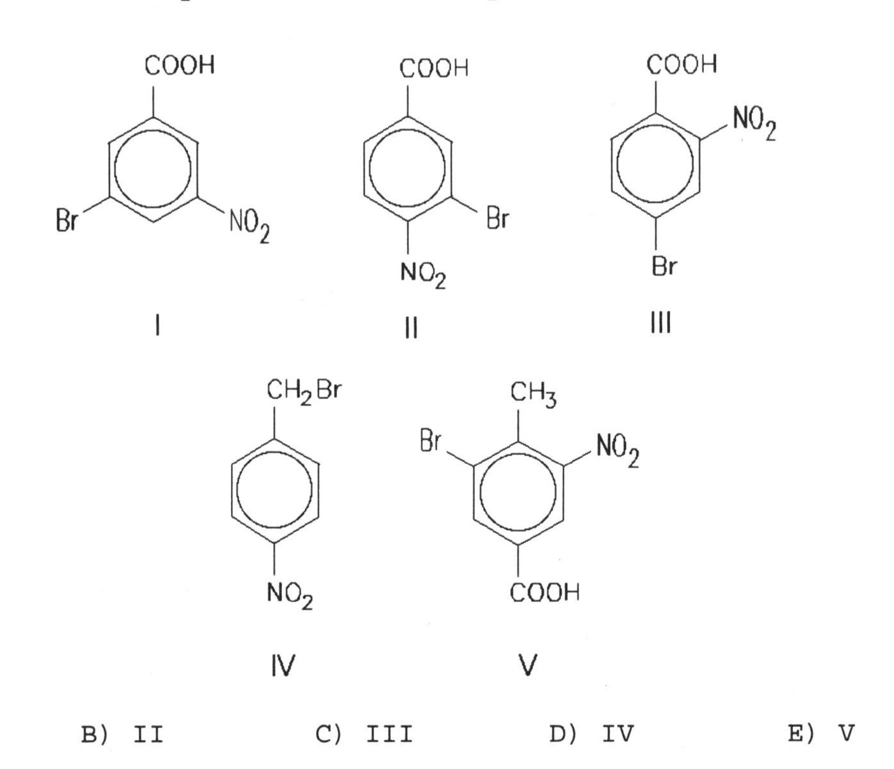

* A) I B) II C) III D) IV E) V

70. Which of these is the best model for the transition state of the rate-determining step in the nitration of benzene?

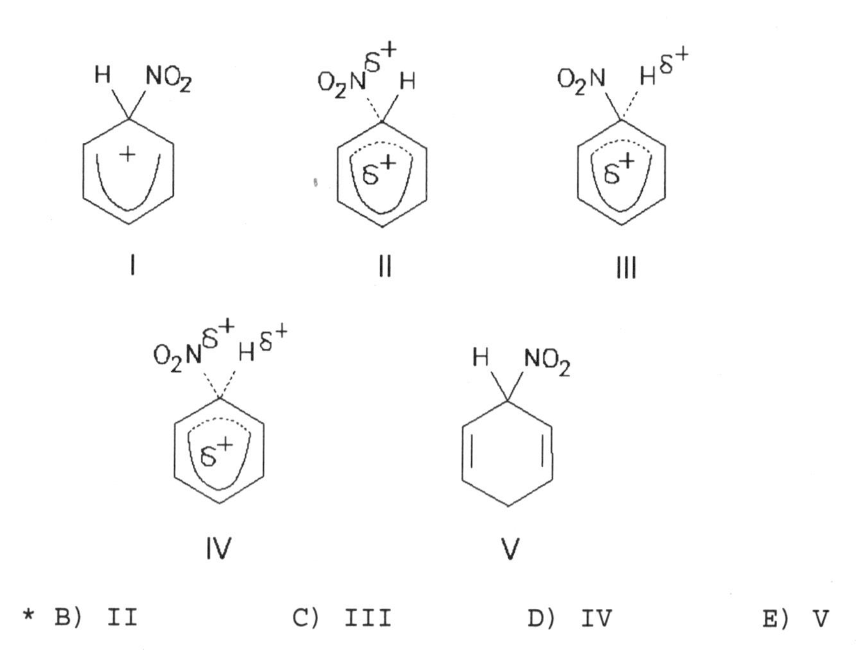

A) I * B) II C) III D) IV E) V

71. When "mixed acid" is used for the nitration of an aromatic compound, the role of the sulfuric acid is which of these?

A) Sulfuric acid removes the water formed in the nitration process.
* B) Sulfuric acid participates in the formation of the actual electrophile.
C) The reaction is catalyzed by sulfuric acid.
D) The sulfonic acid group must be introduced initially; then it is replaced by the nitro group.
E) Sulfuric acid provides the necessary low pH for the reaction.

72. When

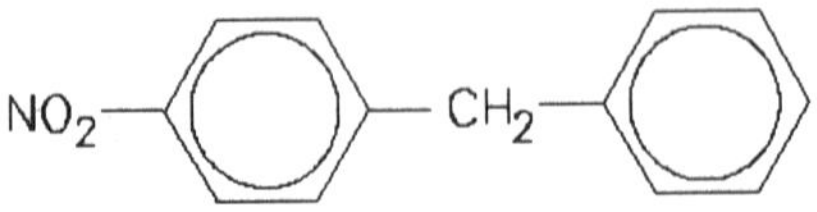

NO_2 — CH_2 —

undergoes Friedel-Crafts acylation, the chief product is which of these?

I NO_2 — CH_2 — C–R (O)

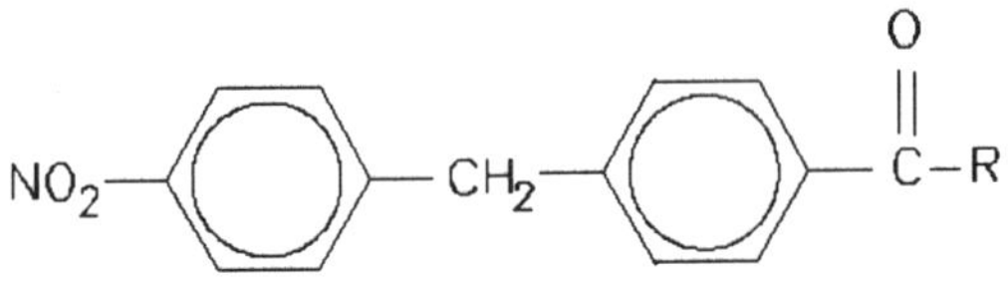

II NO_2 — CH_2 — R–C (O)

III NO_2 — CH_2 — C–R (O)

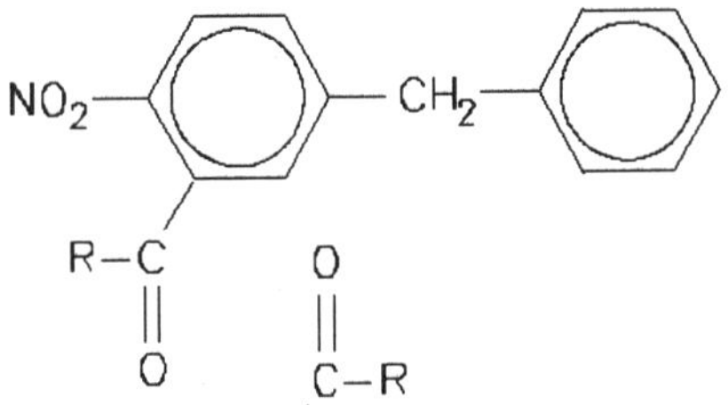

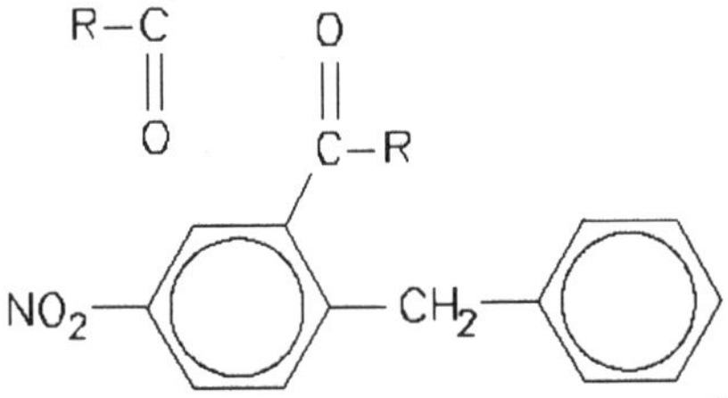

IV NO_2 — CH_2 — C–R (O)

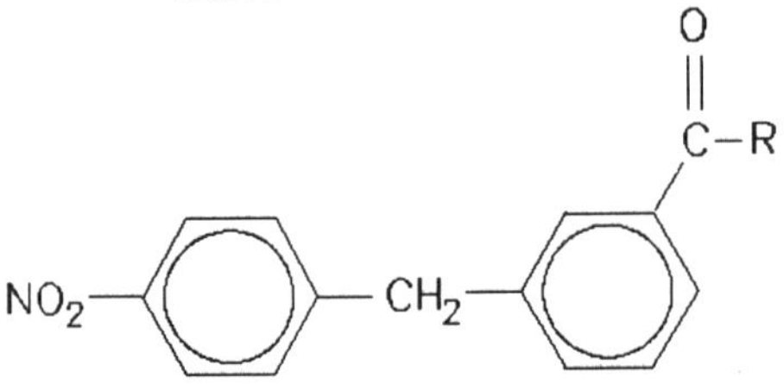

* A) I
B) II
C) III
D) IV
E) None of the above. No reaction will occur.

73. Which is the best prediction of the site(s) of substitution when this compound is nitrated?

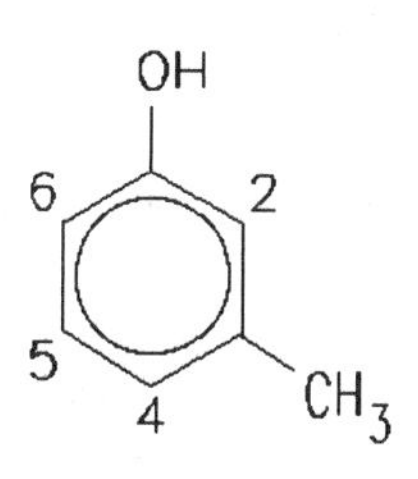

A) 2 B) 4 C) 5 D) 6 * E) 4 and 6

74. This compound

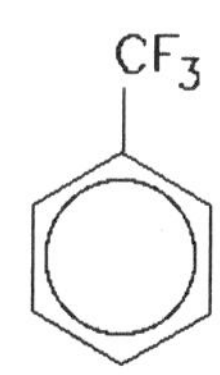

will:
A) nitrate rapidly in the ortho-para positions.
B) nitrate slowly in the ortho-para positions.
C) nitrate rapidly in the meta position.
* D) nitrate slowly in the meta position.
E) fail to nitrate under any conditions.

75. The compound 4-bromo-1-propylbenzene is best made from benzene by the application of these reagents in the order shown:
A) (1) Br_2,Fe (2) $CH_3CH_2CH_2Cl$, $AlCl_3$

B) (1) $CH_3CH_2CH_2Cl$, $AlCl_3$ (2) Br_2,Fe

C) (1) $CH_3CH_2C(=O)Cl$, $AlCl_3$ (2) Br_2,Fe (3) Zn(Hg), HCl

* D) (1) $CH_3CH_2C(=O)Cl$, $AlCl_3$ (2) Zn(Hg), HCl (3) Br_2,Fe

E) (1) $(CH_3)_2CHCl$, $AlCl_3$ (2) Br_2,Fe

76. What feature is common to all meta-directing groups?
* A) The atom directly attached to the ring has a full or well-developed partial positive charge.
  B) The atom directly attached to the ring is doubly bonded to oxygen.
  C) One or more halogen atoms are present in the group.
  D) One or more oxygen atoms are present in the group.
  E) The group is attached to the ring through a carbon atom.

77. Arrange the following compounds in order of decreasing reactivity in electrophilic substitution:

I: $C_6H_5-C(=O)CH_3$ II: $C_6H_5-OCH_3$ III: C_6H_5-Br

IV: $C_6H_5-NO_2$ V: $C_6H_5-CH_3$

  A) V > II > I > III > IV
* B) II > V > III > I > IV
  C) IV > I > III > V > II
  D) III > II > I > IV > V
  E) IV > V > II > I > III

78. An equimolecular mixture of chlorobenzene and acetanilide reacts with a deficiency of Br_2 in the presence of $FeBr_3$. What is the principal product of the competing reactions?

I: p-Br–C_6H_4–Cl   II: o-Br–C_6H_4–Cl   III: m-Br–C_6H_4–Cl

IV: p-Br–C_6H_4–NH–C(=O)CH$_3$ V: m-Br–C_6H_4–NH–C(=O)CH$_3$

A) I   B) II   C) III   * D) IV   E) V

79. How might the following synthesis be carried out:

benzene $\xrightarrow[\text{steps}]{\text{several}}$ (benzene ring with CH_2CH_3 and Cl in meta positions) ?

A)
C_6H_6 DDDDDDDDDD4 (CH_3CH_2Cl / $AlCl_3$) DDDDDDD4 (Cl_2 / $FeCl_3$) product

B)
C_6H_6 DDDDDDD4 (Cl_2 / $FeCl_3$) DDDDDDDDDD4 (CH_3CH_2Cl / $AlCl_3$) product

* C)
C_6H_6 DDDDDDD4 (CH_3CCl with O : above C / $AlCl_3$) DDDDDDD4 (Cl_2 / $FeCl_3$) DDDDDDDD4 (Zn(Hg) / HCl) product

D)
C_6H_6 DDDDDDDD4 (CH_3CCl with O : above C / $AlCl_3$) DDDDDDDD4 (Zn(Hg) / HCl) DDDDDDD4 (Cl_2 / $FeCl_3$) product

E) None of these syntheses is satisfactory.

80. What is the chief product of the Friedel-Crafts alkylation of benzene with 1-butene and $AlCl_3$?

A) $C_6H_5CH_2CH_2CH_2CH_3$

* B) $C_6H_5CHCH_2CH_3$
3
CH_3

C) $C_6H_5CH_2CH(CH_3)_2$

D) $C_6H_5C(CH_3)_3$

E) CH_3
3
$C_6H_5CC_6H_5$
3
CH_2
3
CH_3

81. Which is the best sequence of reactions for the preparation of p-bromostyrene from ethylbenzene?

A) ethylbenzene DDDDDDDDD4 (NBS, h0 / CCl_4) DDDDD4 (Br_2 / Fe) DDDDDDDDDD4 (KOH / CH_3CH_2OH) product

* B) ethylbenzene DDDDD4 (Br_2 / Fe) DDDDDDDDD4 (NBS, h0 / CCl_4) DDDDDDDDDD4 (KOH / CH_3CH_2OH) product

C) ethylbenzene DDDDDDDDD4 (NBS, h0 / CCl_4) DDDDDDDDDD4 (KOH / CH_3CH_2OH) DDDDD4 (Br_2 / Fe) product

D) ethylbenzene DDDDDDD4 (ZnO / 630xC) DDDDD4 (Br_2 / Fe) product

E) None of these syntheses is satisfactory.

82. Which of these species is/are proposed as intermediate(s) in the mechanism for the Birch reduction?

A) Free radical
B) Carbanion
C) Radical anion
D) Both A) and B)
* E) All of A), B) and C)

83. What is the Birch reduction product of the following reaction:

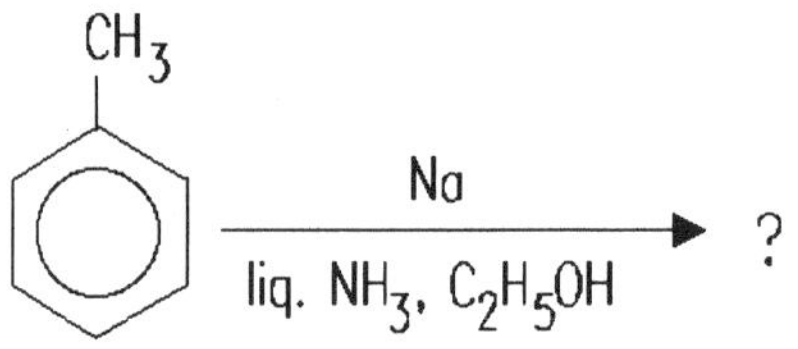

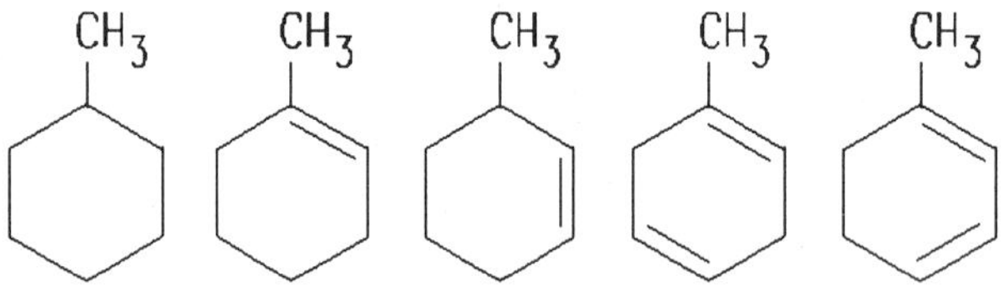

CH_3 (toluene) —Na / liq. NH_3, C_2H_5OH→ ?

CH_3 CH_3 CH_3 CH_3 CH_3

I II III IV V

A) I B) II C) III * D) IV E) V

84. S_N1 solvolysis of $C_6H_5CH{=}CHCH_2Cl$ in water produces:

A) $C_6H_5CH_2C{=}CH_2$
3
OH
B) $C_6H_5CH{=}CHCH_2OH$
C) $C_6H_5CHCH{=}CH_2$
3
OH
* D) A mixture of B) and C)
E) A mixture of A), B) and C)

85. The rate of solvolysis in ethanol is <u>least</u> for which of these compounds?

* A) $C_6H_5CH_2Cl$
  B) $C_6H_5C(CH_3)_2$ (with Cl on the central C)
  C) $(C_6H_5)_2CHCl$
  D) C_6H_5CHCl (with CH_3 on the central C)
  E) $(C_6H_5)_3CCl$

86. Which of the following structures <u>would</u> <u>not</u> be a contributor to the resonance hybrid of the benzyl cation?

I II III IV V

* A) I B) II C) III D) IV E) V

87. Which compound is capable of undergoing both S_N1 and S_N2 reactions in ordinary nonacidic solvents?

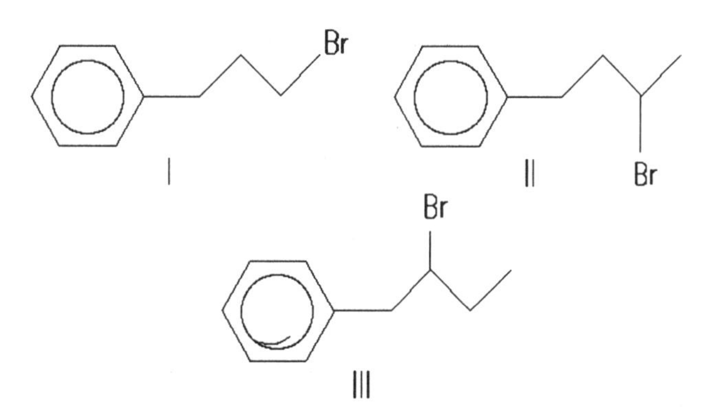

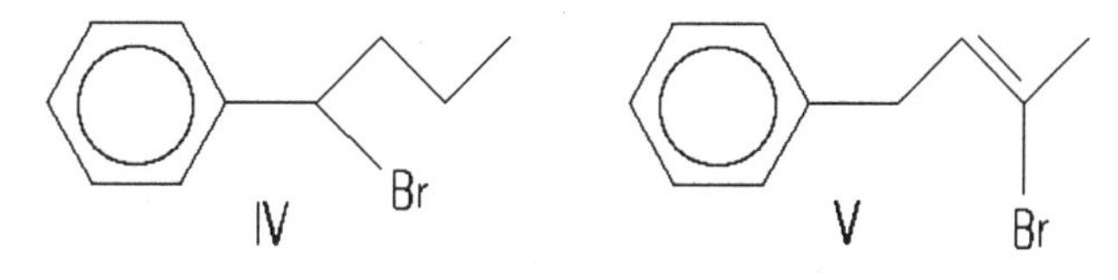

A) I B) II C) III * D) IV E) V

88. Which of the following carbocations would be most stable?

A) CH_3
3
$C_6H_5CH_2CH_2CHCH_2^+$

B) CH_2^+
3
$C_6H_5CH_2CH_2CHCH_3$

C) CH_3
3
$C_6H_5CH_2CH_2CCH_3$
+

D) CH_3
+ 3
$C_6H_5CH_2CHCHCH_3$

* E) CH_3
+ 3
$C_6H_5CHCH_2CHCH_3$

89. Which alkyl halide would be most reactive in an S_N2 reaction?

* A) $C_6H_5CH_2CH_2CH_2Br$

B) $C_6H_5CHCH_2Br$
3
CH_3

C) $C_6H_5CH_2CHCH_3$
3
Br

D) CH_3
3
$C_6H_5CCH_2Br$
3
CH_3

E) Br
3
$C_6H_5CCH_3$
3
CH_3

90. Which alkyl halide would be most reactive in an S_N1 reaction?

A) $C_6H_5CH_2CH_2CH_2Br$

B) $C_6H_5CH_2CHCH_3$
3
Br

C) CH_3
3
$C_6H_5CH_2CBr$
3
CH_3

* D) Br
3
$C_6H_5CCH_3$
3
CH_3

E) CH_3
3
$C_6H_5CCH_2Br$
3
CH_3

91. Which alkyl halide can undergo both S_N1 and S_N2 reactions in nonacidic solvents?

A) CH_3Br

B) $CH_3CH_2CH_2Br$

* C) $CH_2{=}CHCHCH_3$
3
Br

D) $C_6H_5CH_2CH_2CH_2Br$

E) CH_3
3
$C_6H_5CCH_2Br$
3
CH_3

Chapter 16: Aldehydes and Ketones I

1.

```
                                    CH3   O
                                    3     :
A correct name for the compound, CH3CHCH2CCH2CH3, is?
```

A) 2-Methyl-4-hexanone
B) 2-Methyl-3-hexanone
* C) 5-Methyl-3-hexanone
D) Ethyl isopropyl ketone
E) Isobutylpropanone

2. Which of the following compounds is an acetal?

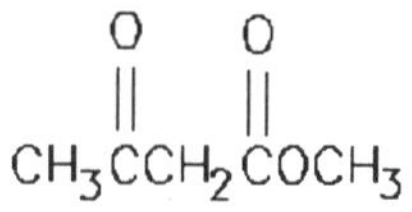

I: $CH_3\overset{O}{\overset{\|}{C}}CH_2\overset{O}{\overset{\|}{C}}OCH_3$

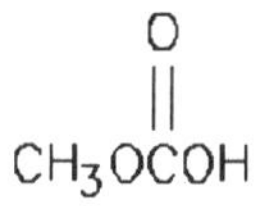

II: $CH_3O\overset{O}{\overset{\|}{C}}OH$

III: (benzene ring fused to O–CHCH$_3$–O ring)

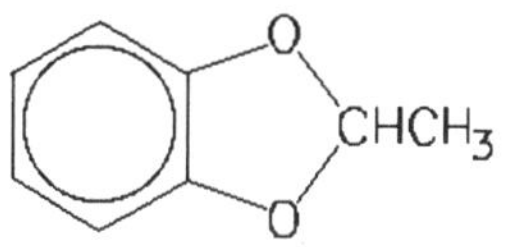

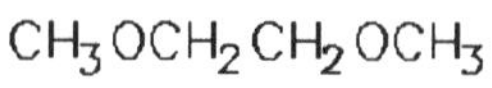

IV: $CH_3OCH_2CH_2OCH_3$

A) I
B) II
* C) III
D) IV
E) None of these

3. The product, C, of the following reaction sequence,

```
      O
      :                      H+                   H2O
  CH3CCH3   +   CN-   DDDDDD4   C4H7NO   DDDDDD4   C
                                                  H2SO4
                                                  heat
```

would be:

* A)

```
     CH3
     3
  CH2MCCOOH
```

B) $CH_3CH_2COOCH_3$

C)

```
  CH3CHCH3
     3
     CN
```

D) CH$_3$CHMCHCOOH

E) None of these

4. What would be a correct name for the compound shown below?

```
     CH3 O
      3  :
   CH3CCH2CCH3
      3
     CH3
```

A) tert-Butyl methyl ketone
B) 2,2-Dimethyl-4-pentanone
* C) 4,4-Dimethyl-2-pentanone
D) Isohexanone
E) Isopentyl methyl ketone

5. Dissolving benzaldehyde in methyl alcohol establishes an equilibrium with what compound?

* A)
```
      OH
      3
   C6H5CHOCH3
```

B)
```
      O
      :
   C6H5COCH3
```

C) $C_6H_5C(OCH_3)_3$
D) Answers A) and B)
E) Answers B) and C)

6. The product, H, of the following reaction sequence,

```
                                                      O
                                                      :
                                          C6H5Li   C6H5CCH3
C6H5CH2Br  +  (C6H5)3P  DDDDDD4  F  DDDDDD4  G  DDDDDDDD4  H
```

would be:

A)
```
         CH3
         3
   C6H5CH2CC6H5
         3
         OH
```

B)
```
         O
         :
   C6H5CH2CC6H5
```

C)
```
   C6H5CHMCHCC6H5
               :
               O
```

D) $C_6H_5CH_2CHMCHC_6H_5$

* E)
```
   C6H5CHMCCH3
           3
          C6H5
```

7. Which of the following reactions would yield benzaldehyde?

A) $C_6H_5CH_2Cl$ DDDDDD4 (OH^- / H_2O / heat)

* B) $C_6H_5CH(OCH_3)_2$ DDDDDD4 (H^+ / H_2O)

C) C_6H_5COOH DDDDDDDDDD4 ((1) $LiAlH_4$ / (2) H_2O)

D) Answers A) and B)

E) Answers A), B), and C)

8. The product, E, of the following reaction sequence,

CH_3CH_2COOH DDDDDD4 (PCl_3) C DDDDDD4 (C_6H_6 / $AlCl_3$) D DDDDDDDDD4 (NH_2NH_2 / base,heat) E

would be?

I: $-CH_2CH_2CH_3$

II: $-\overset{O}{\overset{\|}{C}}OCH_2CH_3$

III: $-\underset{OH}{\underset{|}{C}}HCH_2CH_3$

IV: $-\overset{O}{\overset{\|}{C}}CH_2CH_3$

V: $-CH(CH_3)_2$

* A) I B) II C) III D) IV E) V

9. What new compound will be formed when gaseous HCl is added to a solution of benzaldehyde in methanol?

* A) $C_6H_5CH(OCH_3)_2$

B) C_6H_5CHCl with OCH_3 (3)

C) $C_6H_5COCH_3$ with O (:)

D) $C_6H_5CH(OH)_2$

E) C_6H_5CCl with O (:)

10. The product, F, of the following reaction sequence,

```
       O
       :         SOCl2                  C6H6                  NH2NH2
CH3CHCOH   DDDDDD4   organic   DDDDDD4   organic   DDDDDD4   F,
    3                product    AlCl3    product    NaOH
   CH3                                              heat
```

would be?

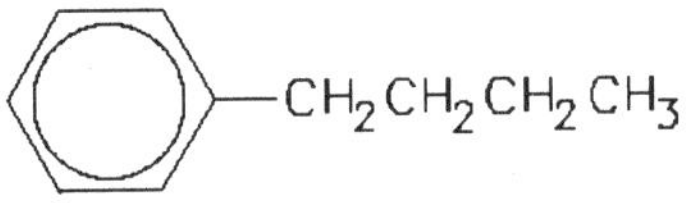

I: $C_6H_5-CH_2CH_2CH_2CH_3$

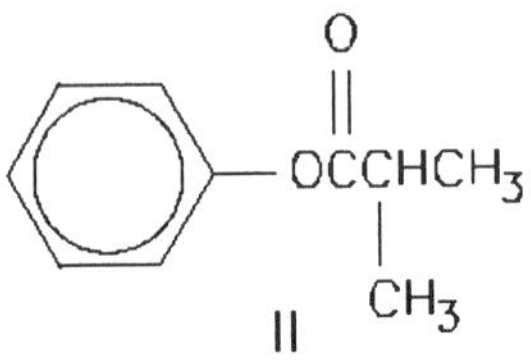

II: $C_6H_5-OC(=O)CH(CH_3)CH_3$

III: $C_6H_5-CH(OH)CH(CH_3)CH_3$

IV: $C_6H_5-CH(CH_3)CH_2CH_3$

V: $C_6H_5-CH_2CH(CH_3)CH_3$

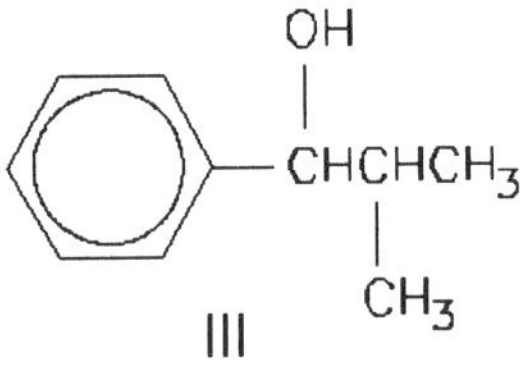

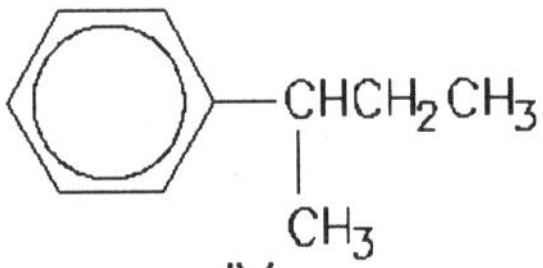

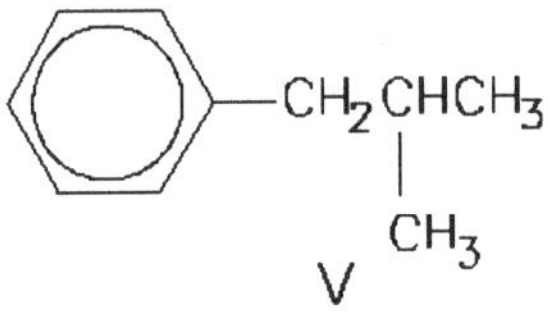

A) I   B) II   C) III   D) IV   * E) V

11. The product, K, of the following sequence of reactions

```
              CH3
              3
            CH3CHCOCl        CH3CH2MgBr          H+
benzene     DDDDDDDD4    I   DDDDDDDDD4    J   DDDDDD4   K,
             AlCl3
```

would be?

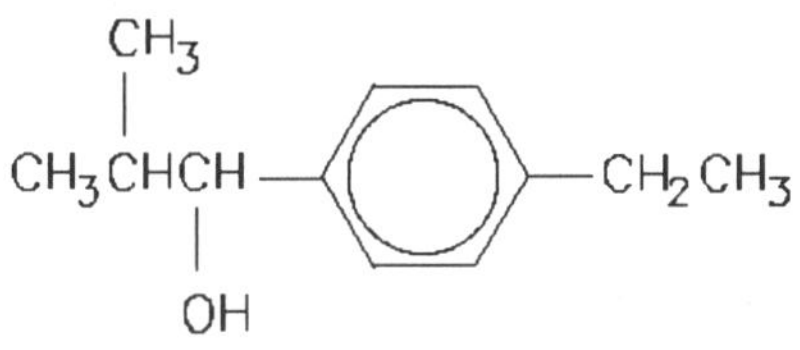

I: CH_3 / CH_3CHCH—(benzene ring)—CH_2CH_3 / OH

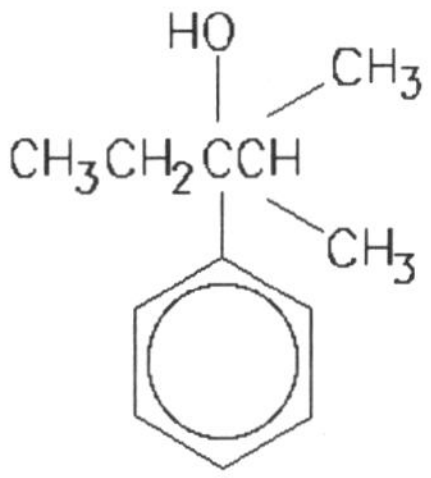

II: HO, CH_3 / CH_3CH_2CCH / CH_3 / (benzene ring)

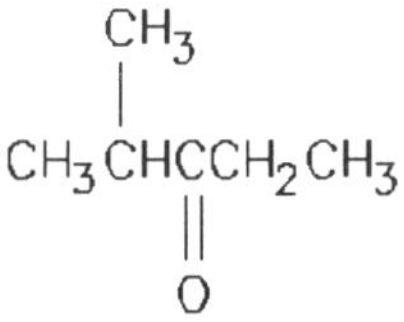

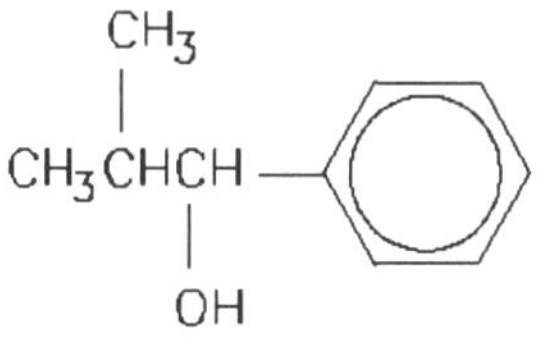

III: CH_3 / $CH_3CHCCH_2CH_3$ / O

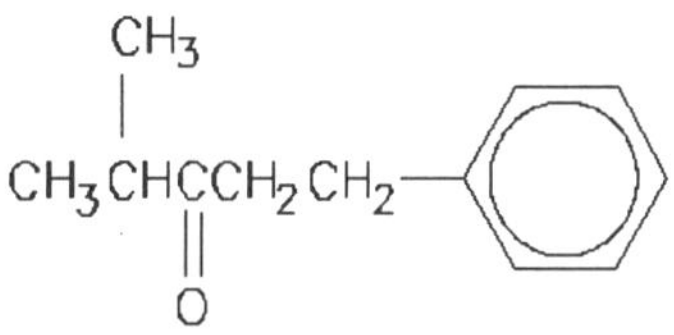

IV: CH_3 / CH_3CHCH—(benzene ring) / OH

V: CH_3 / $CH_3CHCCH_2CH_2$—(benzene ring) / O

A) I   * B) II   C) III   D) IV   E) V

12. A correct name for the compound shown below is:

```
     CH3      O
     3        :
CH3CHCH2CH2CCH3
```

A) Isobutyl methyl ketone
* B) Isopentyl methyl ketone
C) 2-Methyl-5-hexanone
D) Isoheptanone
E) 5-Methyl-2-hexanal

13. Which of the reactions listed below would serve as a synthesis of acetophenone,

$$C_6H_5\overset{O}{\overset{:}{C}}CH_3 \quad ?$$

A) $C_6H_5\overset{O}{\overset{:}{C}}Cl + (CH_3)_2CuLi$ DDDDDD4

B) $C_6H_6 + CH_3COCl \xrightarrow{AlCl_3}$ DDDDDD4

C) $C_6H_5CpN + CH_3Li$ DDDDDD4 $\xrightarrow{H_3O^+}$ DDDDDD4

D) Answers A) and B) only
* E) Answers A), B), and C)

14.

A correct name for $C_6H_5CH_2CH_2\overset{O}{\overset{:}{C}}H$ is?

A) 3-Benzylpropanone
* B) 3-Phenylpropanal
C) 3-Benzylpropanal
D) Nonanone
E) Nonanal

15. Predict the major organic product of the following reaction:

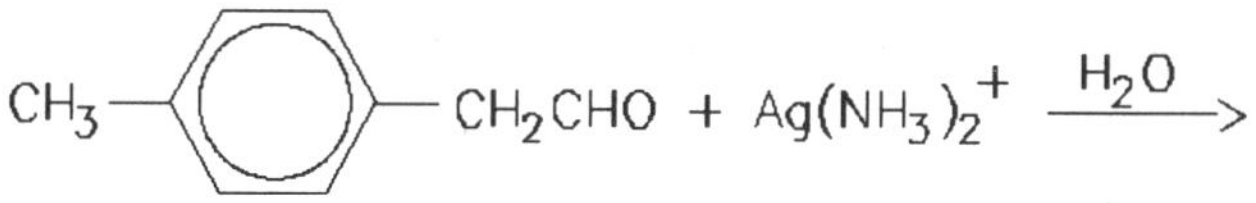

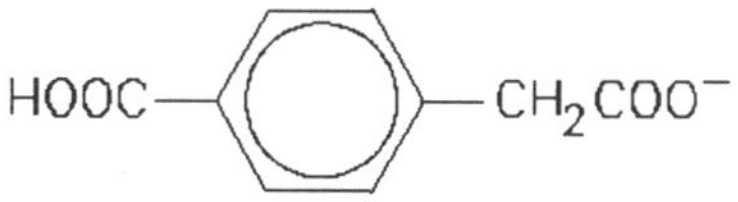

$$CH_3-C_6H_4-CH_2CHO + Ag(NH_3)_2^+ \xrightarrow{H_2O}$$

$HOOC-C_6H_4-CH_2COO^-$

I

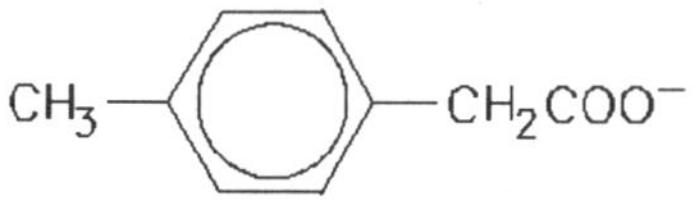

$CH_3-C_6H_4-CH_2COO^-$

II

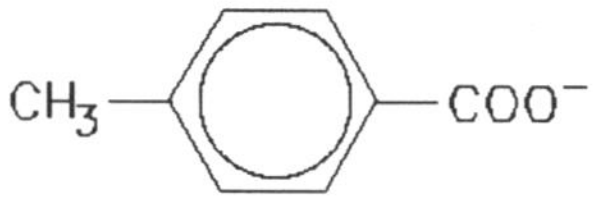

$CH_3-C_6H_4-COO^-$

III

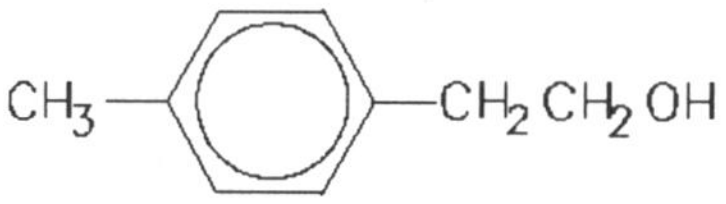

$CH_3-C_6H_4-CH_2CH_2OH$

IV

$HC(=O)-C_6H_4-CH_2CHO$

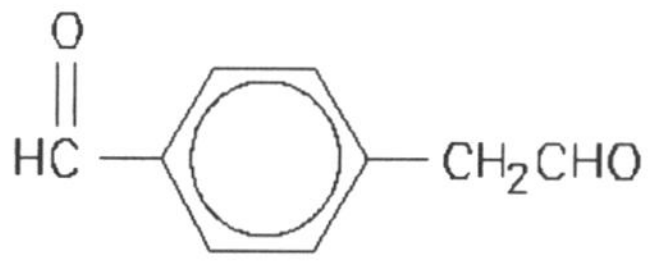

V

A) I * B) II C) III D) IV E) V

16. The product, E, of the following reaction sequence,

$$CH_3CH_2COOH \xrightarrow{PCl_5} C \xrightarrow[AlCl_3]{C_6H_6} D \xrightarrow{NaBH_4} E$$

would be?

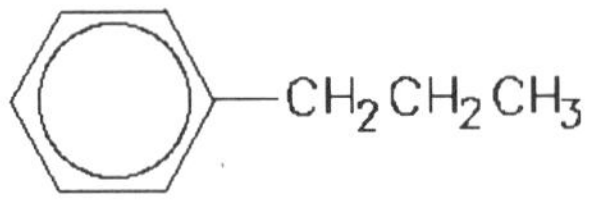

I: $C_6H_5-CH_2CH_2CH_3$

II: $C_6H_5-\overset{O}{\overset{||}{C}}OCH_2CH_3$

III: $C_6H_5-\overset{OH}{\overset{|}{C}}HCH_2CH_3$

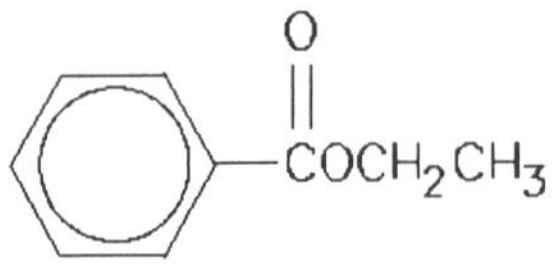

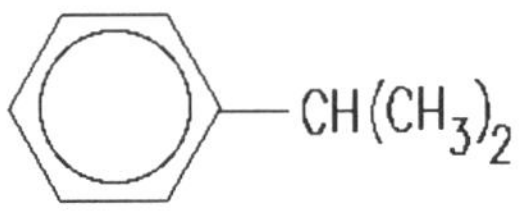

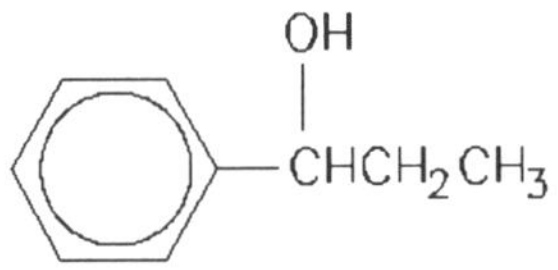

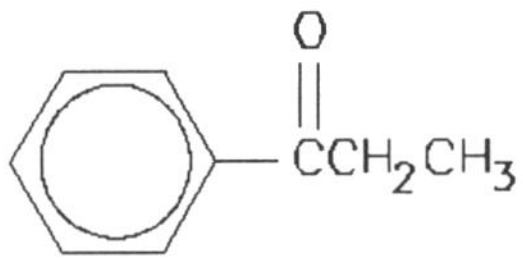

IV: $C_6H_5-\overset{O}{\overset{||}{C}}CH_2CH_3$

V: $C_6H_5-CH(CH_3)_2$

A) I B) II * C) III D) IV E) V

17. Which is the general formula for a thioacetal?

A) $R\overset{H}{\overset{|}{C}}=S$ B) $R\underset{OH}{\underset{|}{C}}HSR'$ C) $R\underset{SH}{\underset{|}{C}}HSR'$ * D) $R\underset{SR'}{\underset{|}{C}}HSR'$ E) $R\underset{SR'}{\underset{|}{C}}HOR'$

18. Which of the following would yield 3-pentanone as the major product?

A) $CH_3CH_2C\equiv N + CH_3CH_2Li \longrightarrow \text{Product} \xrightarrow{H_3O^+}$

B) $CH_3CH_2CO_2H + 2\ CH_3CH_2Li \longrightarrow \text{Product} \xrightarrow{H_3O^+}$

C) $CH_3CH_2\overset{O}{\overset{||}{C}}Cl + (CH_3CH_2)_2CuLi \longrightarrow$

D) Two of these
* E) All of these

19. Which of the following procedures would not yield $(CH_3)_2CHCHO$ (i.e., $(CH_3)_2CHC(=O)H$) as a product?

I $(CH_3)_2CHCH_2OH$ + PCC $\xrightarrow{CH_2Cl_2}$

II $(CH_3)_2CHCO_2H \xrightarrow{SOCl_2} \xrightarrow[\text{ether, } -78^\circ C]{LiAlH[OC(CH_3)_3]_3}$

III $(CH_3)_2CHCH{=}CH_2 \xrightarrow{O_3} \xrightarrow[H_2O]{Zn}$

IV $(CH_3)_2CHCH_2OH \xrightarrow[OH^-]{KMnO_4} \xrightarrow{H_3O^+}$

V $CH_3CH(CH_3)CH$ (cyclic 1,3-dioxolane acetal, $-C(H)(O-CH_2CH_2-O)$) $\xrightarrow{H_3O^+}$

A) I B) II C) III * D) IV E) V

20. Which of the following procedures would not yield 3-pentanone as a major product?

A) $CH_3CH_2C{\equiv}N \xrightarrow{CH_3CH_2MgBr} \xrightarrow{H_3O^+}$

B) $CH_3CH_2C(=O)OH \xrightarrow{2CH_3CH_2Li} \xrightarrow{H_2O}$

C) $CH_3CH_2C{\equiv}N \xrightarrow{CH_3CH_2Li} \xrightarrow{H_3O^+}$

D) $CH_3CH_2C(=O)Cl \xrightarrow{(CH_3CH_2)_2CuLi}$

* E) $CH_3CH_2CO_2H \xrightarrow{CH_3CH_2MgBr} \xrightarrow{H_3O^+}$

21. How could the following synthetic conversion be accomplished?

```
                                                O
                                                :
CH3CH2CH2CHO   DDDDDD4   CH3CH2CCH3
```

A) $HgSO_4/H_2SO_4$; then PCl_5/0xC; then $NaNH_2$
B) PCl_5/0xC; then $NaNH_2$; then Sia_2BH; then H_2O_2
* C) PCl_5/0xC; then $NaNH_2$; then $HgSO_4$, H_2SO_4/H_2O
D) $NaNH_2$; then PCl_5/0xC; then $HgSO_4$, H_2SO_4/H_2O
E) H_2O_2; then PCl_5/0xC; then $NaNH_2$; then Sia_2BH

22. Which sequence of reactions would be utilized to convert

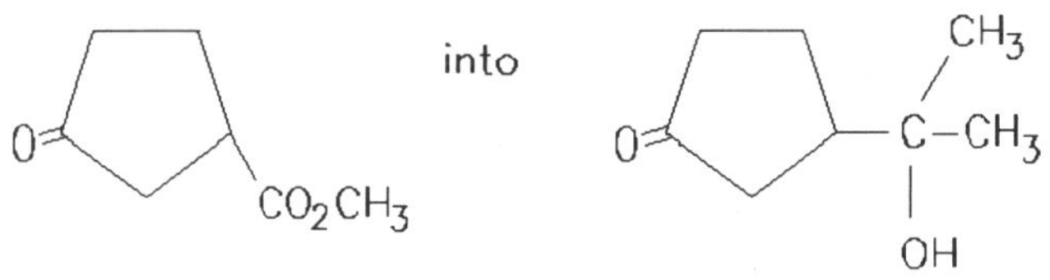

O= cyclopentanone–CO_2CH_3 into O= cyclopentanone–C(CH_3)(CH_3)(OH) ?

A) $2CH_3MgBr$, then H_3O^+
B) $HOCH_2CH_2OH$, H^+; $LiAlH_4$, ether; $2CH_3MgBr$, then H_3O^+
* C) $HOCH_2CH_2OH$, H^+; $2CH_3MgBr$, then H_3O^+
D) $HOCH_2CH_2OH$, H^+; H_2, Pt; CH_3OH, H^+
E) None of the above

23. Which of the compounds listed below would you expect to have the highest boiling point? (They all have approximately the same molecular weight.)

A) $CH_3CH_2CH_2CH_2CH_3$
* B) $CH_3CH_2CH_2CH_2OH$
C)
```
           O
           :
CH3CH2CH2CH
```
D) $CH_3CH_2CH_2Cl$
E) $CH_3CH_2OCH_2CH_3$

24. Select the structure of the major product in the following reaction.

```
                      (1) Sia2BH
C6H5CpCH  DDDDDDDDDDDDDDDDDDD4
                      (2) H2O2, OH-, HOH
```

A) $C_6H_5CH_2CH_3$

B)
```
C6H5CHCH3
       3
       OH
```

C)
```
C6H5CCH3
       :
       O
```

* D) $C_6H_5CH_2CHO$

E) $C_6H_5CHMCH_2$

25. The product, J, of the following reaction sequence,

```
                 SOCl2          (CH3)2CuLi
CH3CHCOOH   DDDDDD4   I   DDDDDDDDDD4   J
       3
       CH3
```

would be:

A)
```
         O
         :
CH3CHCOCH3
     3
     CH3
```

B)
```
   HO O
   3  :
CH3C-CCH3
   3
   CH3
```

C)
```
CH3CHCOCu
     3
     CH3
```

* D)
```
        O
        :
CH3CHCCH3
     3
     CH3
```

E)
```
        O
        :
CH3CHCCHCH3
     3   3
     CH3CH3
```

26. Which reaction sequence would be used to prepare

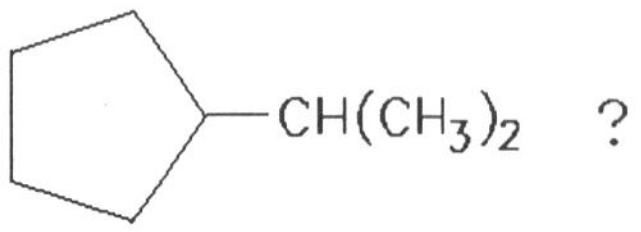

(cyclopentyl)$-CH(CH_3)_2$?

I (cyclopentanone)$=O + (CH_3)_2CHMgBr \xrightarrow[2)\ H_2O,H^+]{1)\ ether} \xrightarrow[Ni]{H_2}$

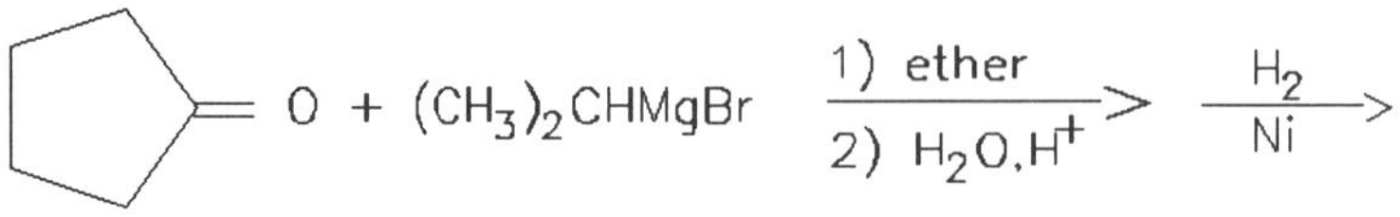

II (cyclopentyl)$-\overset{O}{\overset{||}{C}}H + 2CH_3MgBr \xrightarrow[2)\ H_2O,H^+]{1)\ ether} \xrightarrow[Ni]{H_2}$

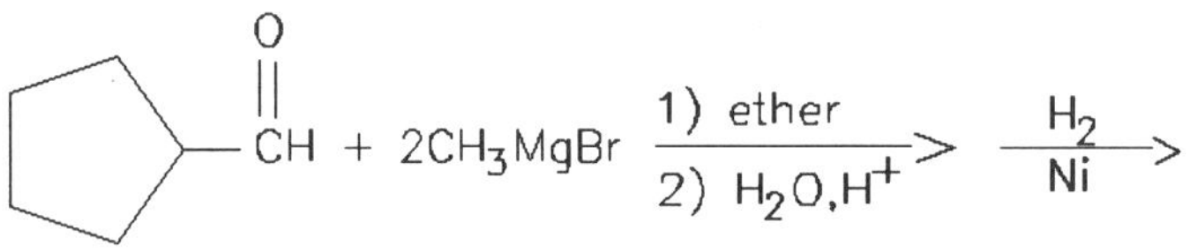

III $(CH_3)_2CHCl \xrightarrow{(C_6H_5)_3P} \xrightarrow{C_6H_5Li} \xrightarrow{cyclopentanone} \xrightarrow[Ni]{H_2}$

IV (cyclopentyl)$-\overset{O}{\overset{||}{C}}CH_3 \xrightarrow[ether]{CH_3MgBr} \xrightarrow{H_3O^+}$

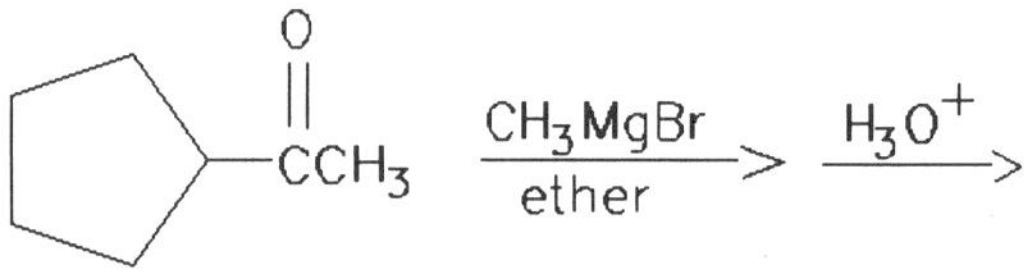

A) I
B) II
* C) III
D) IV
E) All of the above would yield the product.

27. Which Wittig reagent would be used to synthesize

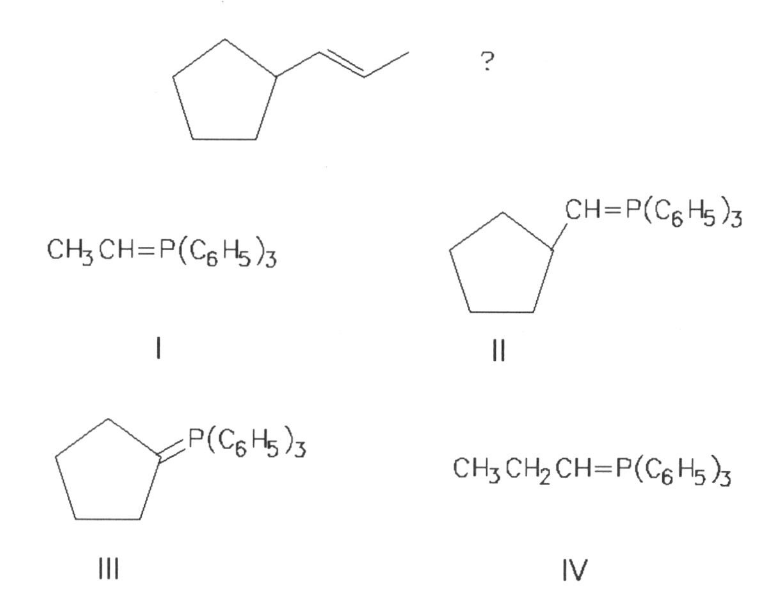

A) I
B) II
C) III
D) IV
* E) Either I or II could be used.

28. What is the major product of the following reaction sequence?

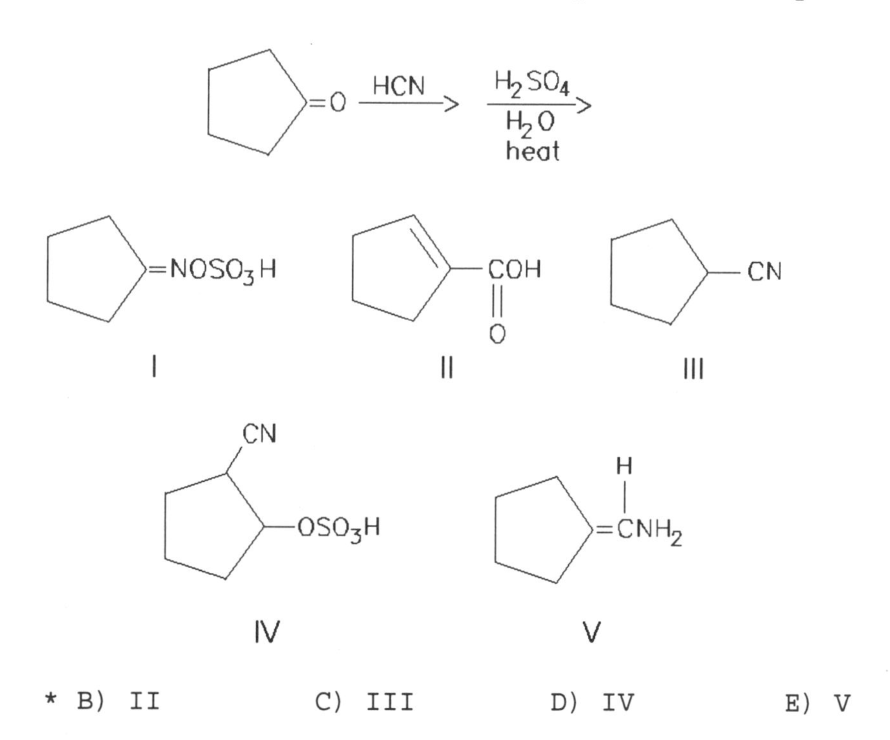

A) I
* B) II
C) III
D) IV
E) V

29. A good synthesis of

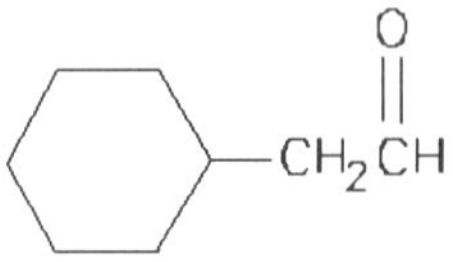

$$\text{C}_6\text{H}_{11}-CH_2\overset{O}{\overset{\|}{C}}H$$

would be:

I $\quad \text{C}_6\text{H}_{11}-C\equiv CH + H_2O \xrightarrow[HgSO_4]{H_2SO_4}$

II $\quad \text{C}_6\text{H}_{11}-C\equiv CH \xrightarrow{Sia_2BH} \xrightarrow[OH^-]{H_2O_2}$

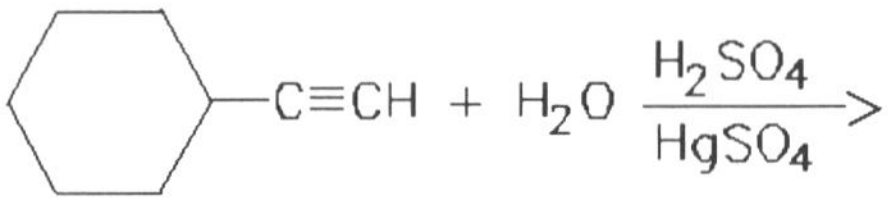

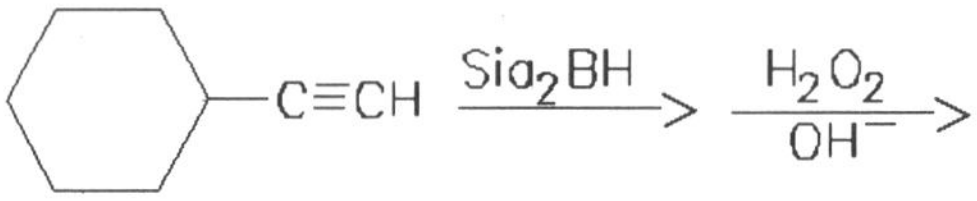

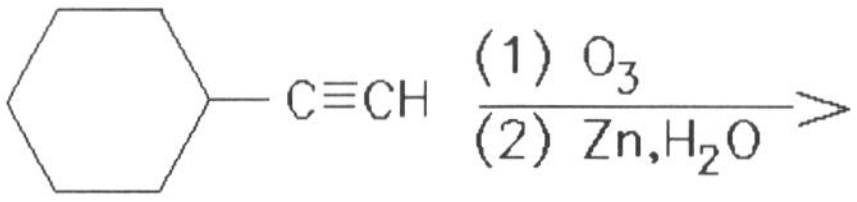

III $\quad \text{C}_6\text{H}_{11}-C\equiv CH \xrightarrow[(2)\ Zn, H_2O]{(1)\ O_3}$

IV $\quad \text{C}_6\text{H}_{11}-CH_2CH_2OH \xrightarrow[(2)\ H_3O^+]{(1)\ KMnO_4,\ OH^-}$

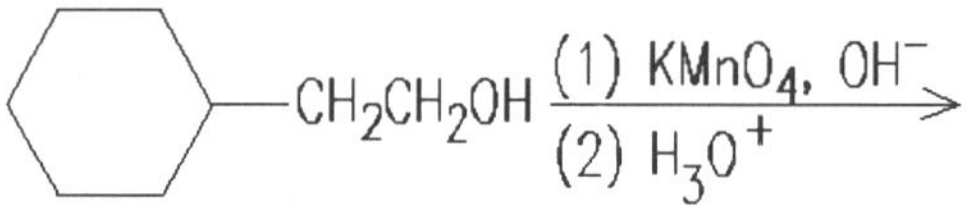

A) I
* B) II
C) III
D) IV
E) All of these are equally useful.

30. Which of the following is <u>not</u> a synthesis of benzophenone,

O
:
$C_6H_5CC_6H_5$?

A)
O
: $AlCl_3$
C_6H_6 + C_6H_5CCl DDDDDD4

B)
H_2CrO_4
$(C_6H_5)_2CHOH$ DDDDDD4

C)
(1) O_3
$(C_6H_5)_2CMCH_2$ DDDDDDDDDD4
(2) Zn, H_2O

D)
(1) ether
$C_6H_5CO_2H$ + 2 C_6H_5Li DDDDDDDDD4
(2) H_2O

* E) All of the above will give benzophenone.

31. Which of the following is a synthesis of 3-heptanone?

A)
PCC (1) CH_3CH_2MgBr H_2CrO_4
$CH_3CH_2CH_2CH_2CH_2OH$ DDDDDD4 DDDDDDDDDDDDDDD4 DDDDDD4
CH_2Cl_2 (2) H_3O^+

B)
$CH_3(CH_2)_2CH_2Li$ H_3O^+
CH_3CH_2CpN DDDDDDDDDDDDDDD4 DDDDDD4

C)
PBr_3 NaCN CH_3CH_2MgBr H_3O^+
$CH_3CH_2CH_2CH_2OH$ DDDDDD4 DDDDDD4 DDDDDDDDDD4 DDDDDD4

D)
$KMnO_4$ H_3O^+ $2CH_3CH_2Li$ H_2O
$CH_3CH_2CH_2CH_2CH_2OH$ DDDDD4 DDDDDD4 DDDDDDDDD4 DDDDD4
OH^-

* E) All of the above are syntheses of 3-heptanone.

32. Identify the reagent(s) that would bring about the following reaction:

$CH_3CH_2CH_2COCl$ DDDDDD4 $CH_3CH_2CH_2CHO$

A) H_2/Ni
B) Li/liq.NH_3
* C) $LiAlH[OC(CH_3)_3]_3$
D) $NaBH_4$
E) $LiAlH_4$

33. The product, D, of the following sequence of reactions

```
           O
           :          SOCl2         toluene        Zn(Hg)
CH3CH2COH  DDDDDD4  A  DDDDDD4  B  DDDDDD4  D
                                 AlCl3     +       HCl

                                         C (discard)
```

would be:

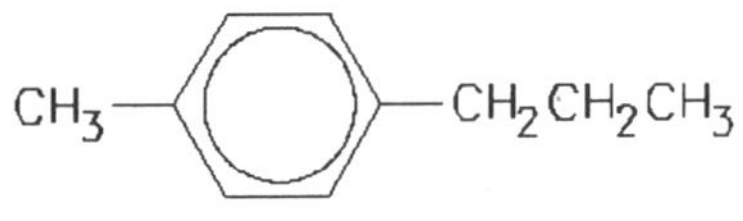

I: CH_3–C_6H_4–$CH_2CH_2CH_3$

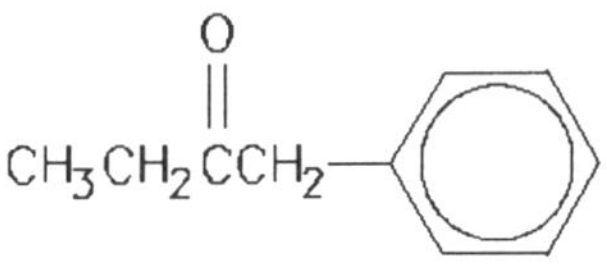

II: $CH_3CH_2C(=O)CH_2$–C_6H_5

III: CH_3–C_6H_4–$CH_2CH_2C(=O)Cl$

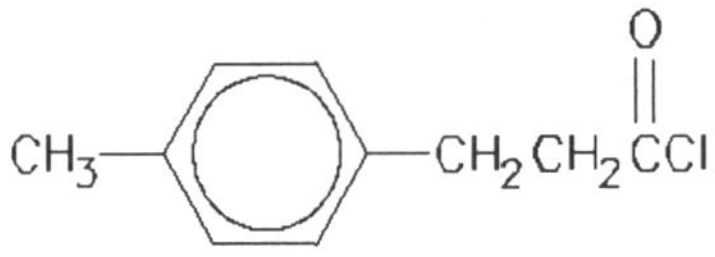

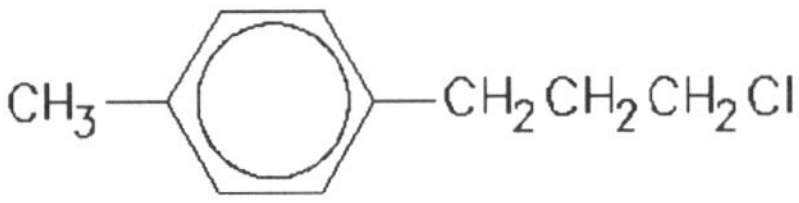

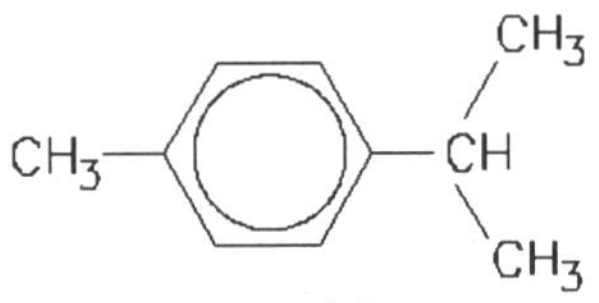

IV: CH_3–C_6H_4–$CH_2CH_2CH_2Cl$

V: CH_3–C_6H_4–$CH(CH_3)_2$

* A) I   B) II   C) III   D) IV   E) V

34. What new compound will be formed when gaseous HCl is added to a solution of propanal in methanol?

A) $CH_3CH_2C(=O)OCH_3$

B) $CH_3CH_2C(=O)OH$

* C) $CH_3CH_2CH(OCH_3)_2$

D) $CH_3CH_2CHOCH_3$ (with OH beneath, printed "3 / OH")

E) None of the above

35. What is the product of the following reaction sequence?

```
                PCC      BrCH2CO2Et,Zn     H3O+
C6H5CH2OH   DDDDDD4  DDDDDDDDDDDDD4  DDDDD4
                CH2Cl2
```

A) $C_6H_5CH_2OCH_2CO_2Et$

B)
```
        O
        :
C6H5COCH2CO2Et
```

* C)
```
        OH
        3
C6H5CHCH2CO2Et
```

D)
```
        OH
        3
C6H5CHOCH2CO2Et
```

E) $C_6H_5CH_2CH_2CO_2Et$

36. Which compound is an acetal?

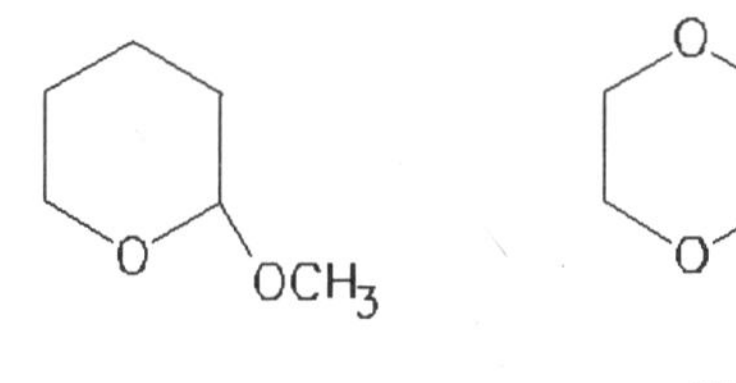

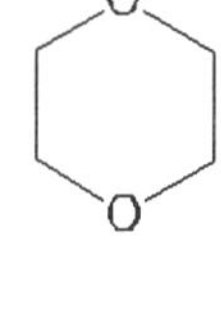

I   II   III

IV

A) I
* B) II
C) III
D) IV
E) All of the above

37. What would be the final product?

$$CH_3I \xrightarrow{(C_6H_5)_3P} \xrightarrow{C_6H_5Li} \xrightarrow{C_6H_5\overset{\overset{\Large O}{:}}{C}CH_3} ?$$

* A) $C_6H_5CMCH_2$
   3
   CH_3

B) C_6H_5
   3
   $C_6H_5CCH_3$
   3
   CH_3

C) $C_6H_5CHMCHCH_3$

D) OH
   3
   $C_6H_5CCH_3$
   3
   CH_3

E) $C_6H_5CH{=}CC_6H_5$
   3
   CH_3

38. Which synthesis or syntheses would yield propanal?

A) $CH_3CH_2CH_2OH \xrightarrow{PCC/CH_2Cl_2}$

B) $CH_3CH_2\overset{\overset{\Large O}{:}}{C}Cl \xrightarrow[\text{ether, -78xC}]{LiAlH[OC(CH_3)_3]_3}$

C) $CH_3CpCH \xrightarrow[(2)\ H_2O_2/OH^-]{(1)\ Sia_2BH}$

* D) All of these

E) None of these

39. Dissolving propanal in methyl alcohol establishes an equilibrium with what other compound?

A) $CH_3CH_2\overset{\overset{\Large O}{:}}{C}OCH_3$

* B) OH
   3
   $CH_3CH_2CHOCH_3$

C) $CH_3CH_2CH(OCH_3)_2$

D) OCH_3
   3
   CH_3CCH_3
   3
   OCH_3

E) OH
   3
   CH_3CCH_3
   3
   OCH_3

40. A correct name for the following compound would be which of those below?

A) 2,5-Dimethyl-6-hexanal
* B) 2,5-Dimethylhexanal
C) 2-Aldehydoisohexane
D) 3,5-Dimethylheptanone
E) 1-Hydro-2,5-dimethyl-1-hexanone

41. An aldehyde results from the reaction of which of these compounds with aqueous base?

A) $CH_3CH_2CH_2Cl$
B) $CH_3CHClCH_2Cl$
C) $CH_3CHMCCl_2$
* D) $CH_3CH_2CHCl_2$
E) $CH_3CCl_2CH_3$

42. The Baeyer-Villiger oxidation of propiophenone (ethyl phenyl ketone) produces chiefly:

A)
O
:
$CH_3CH_2OCC_6H_5$

B)
O
:
$CH_3CH_2OCOC_6H_5$

C)
O
:
p-$HOC_6H_4CCH_2CH_3$

* D)
O
:
$C_6H_5OCCH_2CH_3$

E)
O
:
$C_6H_5CCH_2CH_2OH$

43.

What is the IUPAC name for

O
:
$CH_3DCHDCH_2DCDCHDCH_3$?
3 3
OH C_6H_5

A) 4-Oxo-5-phenyl-2-hexanol
* B) 5-Hydroxy-2-phenyl-3-hexanone
C) 2-Hydroxy-5-phenyl-4-hexanone
D) 2-Hydroxypropyl 1-phenylethyl ketone
E) 5-Hydroxy-3-keto-2-phenylhexane

44. $LiAlH_4$ (LAH) cannot be used to convert carboxylic acids to the corresponding aldehydes because:

A) LAH is not sufficiently reactive.
B) RCOOH is converted into RCOOLi.
* C) RCOOH is reduced to RCH_2OH.
D) RCOOH is reduced to RCH_3.
E) RCOOH is converted into RCR.

45. The following reduction is best carried out with which reagent(s)?

$$\text{(1,3-dioxolane)}-CH_2CH_2\overset{O}{\overset{\|}{C}}CH_3 \longrightarrow \text{(1,3-dioxolane)}-CH_2CH_2CH_2CH_3$$

A) Zn(Hg), HCl
B) $LiAlH_4$
C) $NaBH_4$
D) H_2, Ni
* E) H_2NNH_2, DMSO, $KOC(CH_3)_3$

46. Which reagent will <u>not</u> differentiate between the two compounds

H
3
CH_2MCHCH_2CMO and $CH_3CH_2CCH_3$?
(O above the C of the second compound)

A) Br_2/CCl_4
B) $Ag(NH_3)_2^+$
* C) $H_2NNHC_6H_5$
D) $KMnO_4, OH^-$
E) None of these

47. Stereoisomers can exist in the case of which of the following?

* A) The hydrazone of 2-butanone
B) The oxime of acetone
C) The phenylhydrazone of cyclohexanone
D) The cyclic acetal formed from propanal and ethane-1,2-diol
E) The imine of cyclopentanone

48. Which is the proper name for the structure shown?

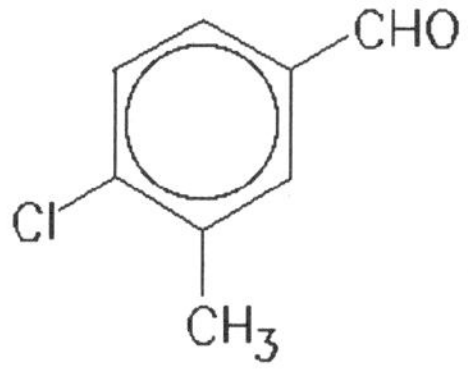

A) 2-Chloro-5-aldehydotoluene
B) 6-Chloro-3-aldehydotoluene
C) 2-Methyl-4-aldehydochlorobenzene
* D) 4-Chloro-3-methylbenzaldehyde
E) 4-Methyl-5-chloro-2-benzaldehyde

49. What is the order of decreasing reactivity of aldehydes, esters and ketones towards the organozinc compounds used in the Reformatsky reaction?
A) Aldehydes > esters > ketones
B) Esters > aldehydes > ketones
C) Ketones > esters > aldehydes
D) Ketones > aldehydes > esters
* E) Aldehydes > ketones > esters

50. The Reformatsky reaction involves the reaction of an aldehyde or ketone with this type of organometallic compound:
A) RLi
B) R_2CuLi
* C) R
3
$XZnCHCO_2R'$
D) RCpCNa
E) R_3P

51. Acetals are unstable in the presence of an aqueous solution of which of these?
* A) HCl B) NaOH C) $KHCO_3$ D) Na_2CO_3 E) NaCl

52. H 3 H 3
The compound C_6H_5CMNDNMCC_6H_5 is produced by the reaction of an excess of benzaldehyde with which compound?
A) Ammonia
* B) Hydrazine
C) Nitrogen
D) Phenylhydrazine
E) Hydroxylamine

53. Which of these will <u>not</u> catalyze the reaction of a weak nucleophile with the carbonyl group of an aldehyde or ketone?
* A) I^- B) H^+ C) $AlCl_3$ D) BF_3 E) $ZnCl_2$

54. Which of these gem-diols is expected to be the most stable?

* A)
```
    OH
    3
 CF3CCF3
    3
    OH
```

B) $CH_3CH(OH)_2$

C)
```
    OH
    3
 CH3CCH3
    3
    OH
```

D) $C_6H_5CH(OH)_2$

E) $ClCH_2CH(OH)_2$

55. The relationship of propanone and propen-2-ol is designated by the term:

* A) Tautomers.
B) Conformational isomers.
C) Diastereomers.
D) Resonance structures.
E) Stereoisomers.

56. Only one of these reactions will produce a product which, at least in principle, can be resolved into the separate enantiomers. Which reaction is it?

A)
```
                      H
                      3
 C6H5MgBr   +   C6H5CMO
```

B)
```
              O
              :
 HCN   +   CH3CCH3
```

* C)
```
      H                   O
      3                   :
 CH3CH2CMO   +   BrZnCH2COC2H5
```

D)
```
    H
    3
 CH3CMO   +   xs CH3OH   +   HCl
```

E)
```
         O
         :
 CH3CHCH2COCH3   +   (i-Bu)2AlH
    3
    CH3
```

57. What is compound V in the following synthesis?

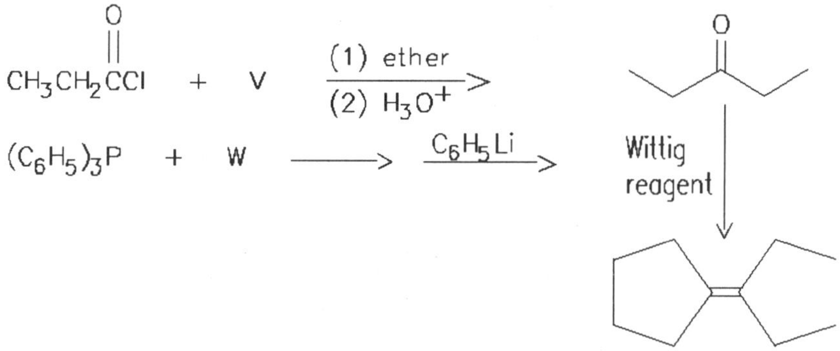

$$CH_3CH_2\overset{O}{\overset{\|}{C}}Cl + V \xrightarrow[(2)\ H_3O^+]{(1)\ ether}$$

$$(C_6H_5)_3P + W \longrightarrow \xrightarrow{C_6H_5Li}$$

Wittig reagent

A) CH_3CH_2COCl
B) CH_3CH_2MgBr
C) CH_3COCl
D) $(CH_3)_2CuLi$
* E) $(CH_3CH_2)_2CuLi$

58. What is the reactant W in the synthesis given above?
A) Cyclopentanone
B) Cyclopentene
C) Cyclopentanol
* D) Bromocyclopentane
E) Triphenylphosphine oxide

59. Which reagent(s) could be used to carry out the following transformation?

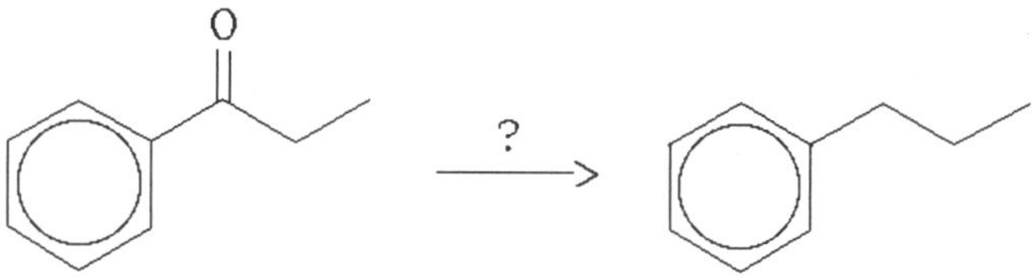

?

A) Zn(Hg), HCl, reflux
B) H_2NNH_2, NaOH, trimethylene glycol, 200xC
C) $HSCH_2CH_2SH$, BF_3; then Raney Ni (H_2)
* D) All of the above
E) None of the above

60. What is the final product, Z, of the following synthesis?

$$C_6H_5CH_3 \xrightarrow[\text{OH}^-,\ \text{heat};\ (2)\ H_3O^+]{(1)\ KMnO_4} X \xrightarrow{SOCl_2} Y \xrightarrow[\text{ether},\ -78^\circ C;\ (2)\ H_2O]{(1)\ LiAlH(O\text{-}\underline{t}\text{-}Bu)_3} Z$$

I: $C_6H_5COAl(\underline{t}\text{-}Bu)_2$ (with C=O)

II: $C_6H_5COCH_3$ (with C=O)

III: C_6H_5CH (with C=O)

IV: C_6H_5COLi (with C=O)

V: C_6H_5COH (with C=O)

A) I B) II * C) III D) IV E) V

61. What is the final product of this synthetic sequence?

$$\text{benzene} \xrightarrow[FeCl_3]{Br_2} \xrightarrow[\text{ether}]{Mg} \xrightarrow[(2)\ H_3O^+]{(1)\ C_6H_5CHO} \xrightarrow[\text{acetone}]{H_2CrO_4} ?$$

* A) $C_6H_5CC_6H_5$ (with C=O)

B) $\underline{p}\text{-}BrC_6H_4CH_2C_6H_5$

C) $C_6H_5CH_2COOH$

D) $C_6H_5CH_2CC_6H_5$ (with C=O)

E) $C_6H_5CH_2C_6H_5$

62. The reaction of *tert*-butyl methyl ketone with a peroxy acid produces which of these as the principal product(s)?

* A)
```
            O
            :
  (CH3)3COCCH3
```
B)
```
            O
            :
  (CH3)3CCOCH3
```
C)
```
              CH3
              3
  (CH3)2CDDDCOCH3
            \ /
             O
```
D) $(CH_3)_3COH$ + HO_2CCH_3

E)
```
          O
          :
  (CH3)3COCOCH3
```

63. A compound with formula $C_5H_{10}O$ gives two signals only, both singlets, in the ^1H NMR spectrum. Which of these structures is a possible one for this compound?

A)
```
         O
         :
  CH3CH2CCH2CH3
```
B)
```
            O
            :
  (CH3)2CHCCH3
```
* C)
```
          H
          3
  (CH3)3CC=O
```
D)
```
              H
              3
  CH3CH2CHC=O
           3
           CH3
```
E)
```
                 H
                 3
  (CH3)2CHCH2C=O
```

64. What, in general, is the order of decreasing reactivity of these carbonyl compounds towards nucleophilic reagents?

```
     O            H              O          H           O
     :            3              :          3           :
  CH3CCH3     CH3C=O     (CH3)3CCCH3      HC=O    (CH3)3CCC(CH3)3

     I            II            III         IV          V
```

A) I > III > V > II > IV
* B) IV > II > I > III > V
C) V > III > I > II > IV
D) II > I > V > III > IV
E) III > V > IV > II > I

1. Which compound would be most acidic?

A) $CH_3CH_2CH_3$

B) CH_3CHMCH_2

C) Cyclohexane

* D)
```
    O
    :
CH3CCH3
```

E) Benzene

2. What would be the major product of the following reaction?

```
     O           O
     :           :      OH-
C6H5CH   +   CH3CH   DDDDDD4
                       heat
```

A)
```
        O
        :
C6H5CH2CCH3
```

B)
```
    O   O
    :   :
C6H5CCH2CH
```

C)
```
    OH    OH
    3     3
C6H5CHCH2CH2
```

D)
```
            O
            :
C6H5CH2CH2CH
```

* E)
```
           O
           :
C6H5CHMCHCH
```

3. Which reagent would best serve as the basis for a simple chemical test to distinguish between

```
     O                            O
     :                            :
CH3CH2CCH2CH3    and    CH3CH2CH2CCH3   ?
```

* A) NaOI (I_2 in NaOH)
B) Br_2/CCl_4
C) CrO_3/H_2SO_4
D) $NaHCO_3/H_2O$
E) $Ag(NH_3)_2^+$

4. Which reagent would best serve as the basis for a simple chemical test to distinguish between

CH_3CHO and CH_3COCH_3 ?

A) NaOI (I_2 in NaOH)
B) Br_2/CCl_4
C) $C_6H_5NHNH_2$
D) $NaHCO_3/H_2O$
* E) $Ag(NH_3)_2^+$

5. Which reagent would best serve as the basis for a simple chemical test to distinguish between

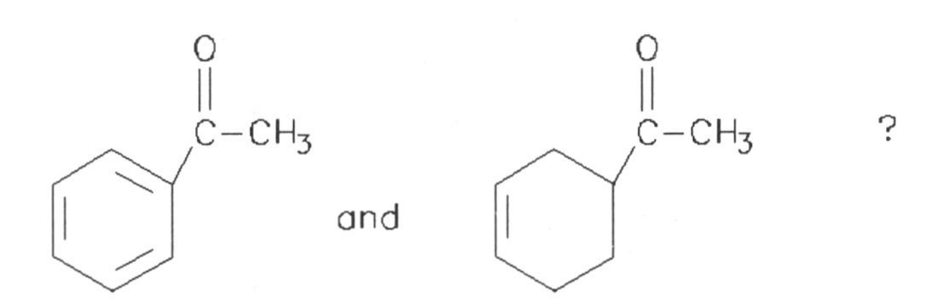

A) NaOI (I_2 in NaOH)
* B) Br_2/CCl_4
C) CrO_3/H_2SO_4
D) $NaHCO_3/H_2O$
E) $Ag(NH_3)_2^+$

6. The product, C, of the following sequence of reactions,

CH_3
3
CH_3CCHO + CH_3CH_2CHO $\overset{OH^-}{DDDDDD4}$ A $\overset{H^+}{\underset{(-H_2O)}{DDDDDD4}}$ B $\overset{NaBH_4}{DDDDDD4}$ C,
3
CH_3

would be:

A)
CH_3
3
$CH_3CCHMCHCH_2CH_2OH$
3
CH_3

* B)
CH_3
3
$CH_3CCHMCCH_2OH$
3 3
CH_3 CH_3

C)
CH_3
3
$CH_3CCH_2CH_2CH_2OH$
3
CH_3

D)
CH_3
3
$CH_3CCH_2CH_2CH_2CH_2OH$
3
CH_3

E)
CH_3
3
CH_3CCH_2CHCHO
3 3
CH_3 CH_3

7. Which of the following represent tautomers?

A)

CH_3CHMCH-Ȯ:⁻ and $CH_3\dot{C}HDCHMO$:

B) CH_3DCMCH_2 and CH_3DCDCH_2:⁻

3 :

:O: :O

* C) CH_3DCMCH_2 and CH_3DCDCH_3

3 :

:O: :O

3

H

D) All of these

E) None of these

8. Which of these compounds would exist in an enol form to the greatest extent?

A) O
:
$CH_3COC_2H_5$

* B) O O
: :
$CH_3CCH_2CCH_3$

C) O
:
CH_3CCH_3

D) CH_3CH
:
O

E) $CH_3CCH_2CH_2CCH_3$
: :
O O

9. What would be the major product of the following reaction?

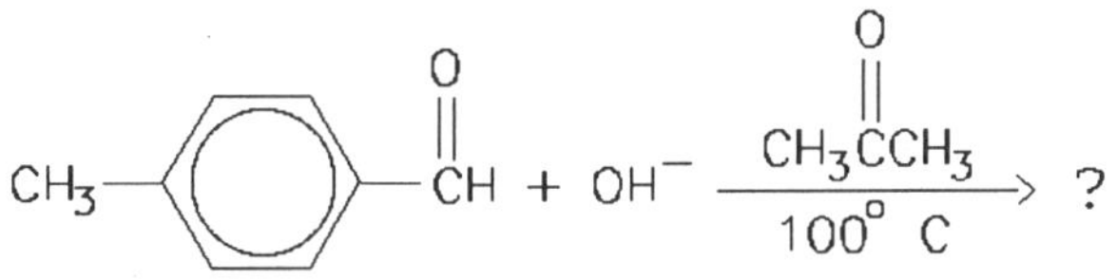

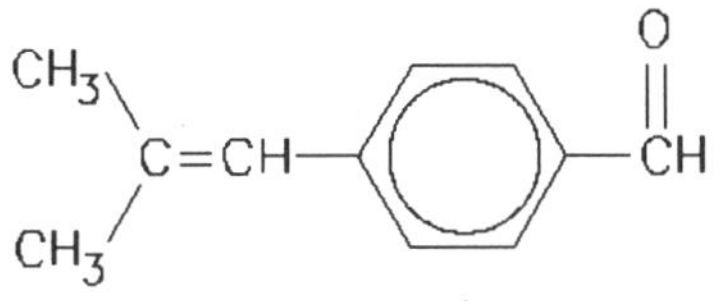

$$CH_3-C_6H_4-\overset{O}{\overset{\|}{C}}H + OH^- \xrightarrow[100^\circ\ C]{CH_3\overset{O}{\overset{\|}{C}}CH_3} ?$$

I: $(CH_3)_2C=CH-C_6H_4-\overset{O}{\overset{\|}{C}}H$

II: $CH_3-C_6H_4-CH=C(CH_3)_2$

III: $CH_3-C_6H_4-\underset{OH}{\underset{|}{C}}HCH_2\overset{O}{\overset{\|}{C}}CH_3$

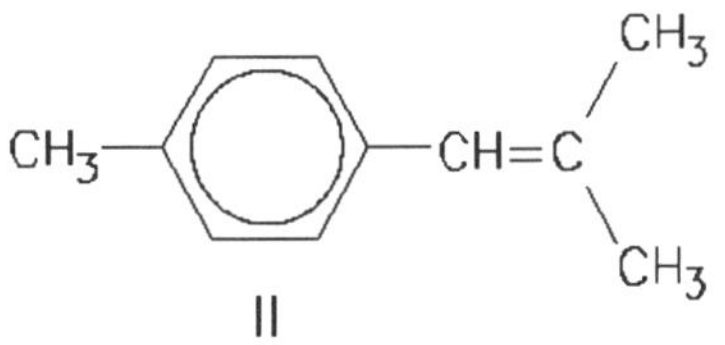

IV: $CH_3-C_6H_4-CH=CH\overset{O}{\overset{\|}{C}}CH_3$

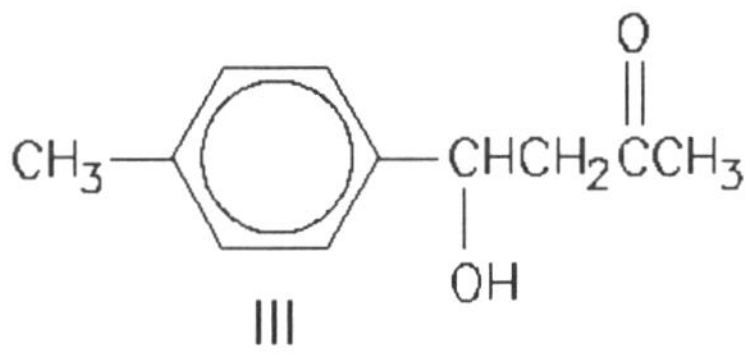

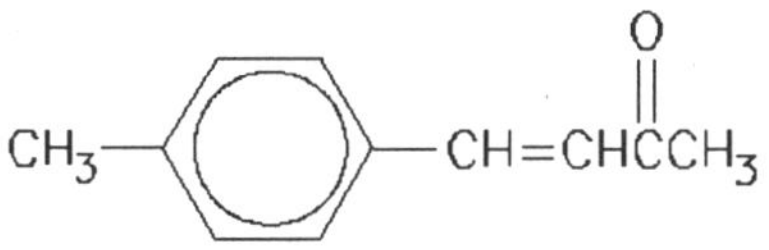

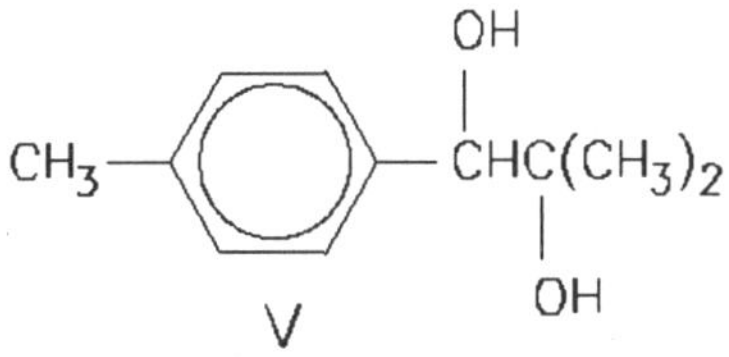

V: $CH_3-C_6H_4-\overset{OH}{\overset{|}{C}}H\underset{OH}{\underset{|}{C}}(CH_3)_2$

A) I B) II C) III * D) IV E) V

10. Which of the following is a keto-enol tautomeric pair?

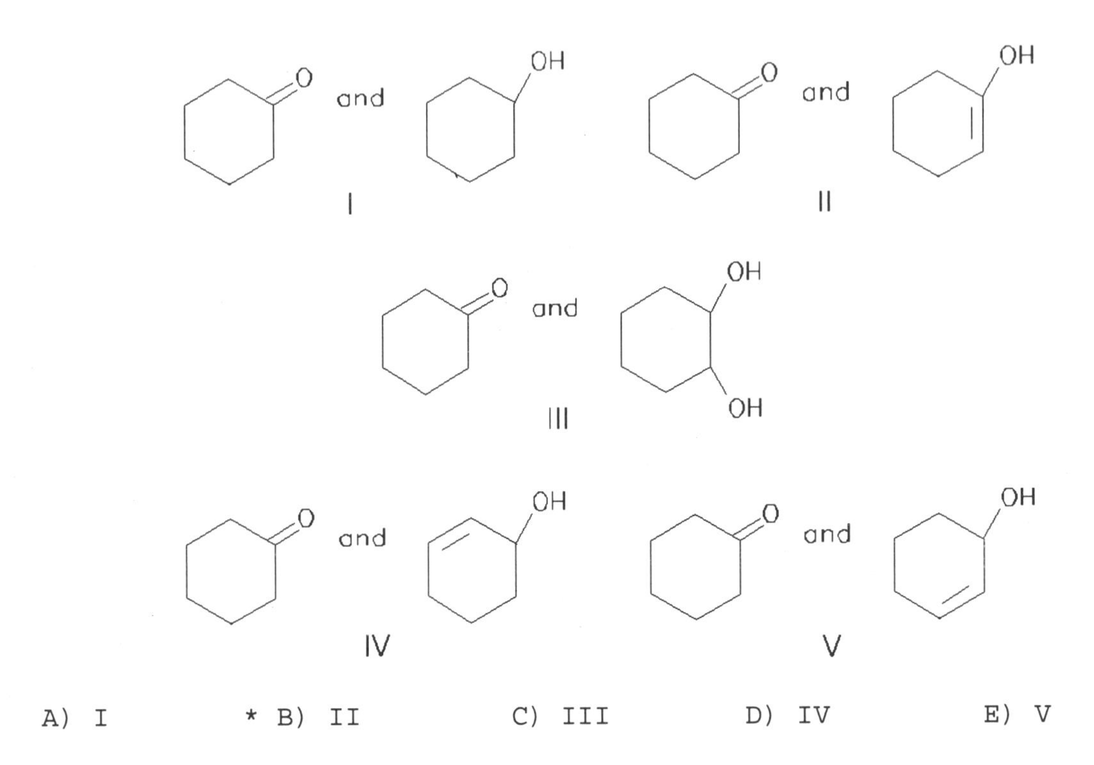

A) I * B) II C) III D) IV E) V

11. What would be the major product of the following reaction,

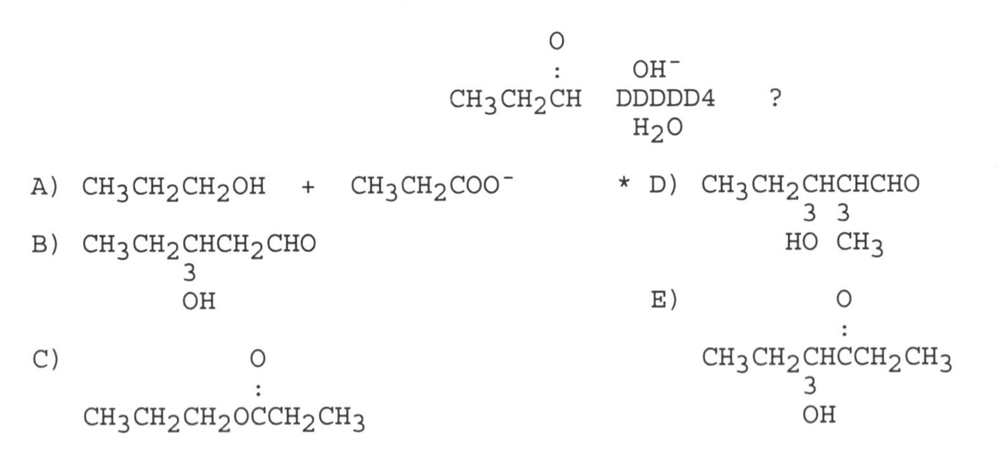

$$CH_3CH_2\overset{O}{\overset{:}{C}}H \xrightarrow[H_2O]{OH^-} ?$$

A) $CH_3CH_2CH_2OH$ + $CH_3CH_2COO^-$

B) $CH_3CH_2CHCH_2CHO$ (with OH on the CH)

C) $CH_3CH_2CH_2O\overset{O}{\overset{:}{C}}CH_2CH_3$

* D) $CH_3CH_2CHCHCHO$ (with HO and CH_3 substituents)

E) $CH_3CH_2CH\overset{O}{\overset{:}{C}}CH_2CH_3$ (with OH on the CH)

12. What would be the major product of the following reaction?

O CH_3
: 3
C_6H_5C-$CHCH_3$ + D_2O $\xrightarrow[\text{room temp.}]{OD^-}$?

A) O CD_3
: 3
$C_6H_5CDCHCH_3$

* B) O CH_3
: 3
$C_6H_5CDCDCH_3$

C) O CH_3
: 3
$C_6D_5CDCDCH_3$

D) O CD_3
: 3
$C_6H_5CDCDCD_3$

E) OD
3
$C_6H_5CDCHCH_3$
3
CH_3

13. What would be the product of the following reaction?

$CH_3CHMCHCO_2C_2H_5$ $\xrightarrow[C_2H_5OH,\ CH_3COOH]{NaCN}$?

A) $CH_3CH_2CHCO_2C_2H_5$
3
CN

B) $CH_3CHMCHCO_2CN$

C) $CH_3CHMCHCN$

* D) $CH_3CHCH_2CO_2C_2H_5$
3
CN

E) CN
3
$CH_3CHMCHCOC_2H_5$
3
OH

14. Select the structure of the major product formed in the following reaction.

```
                              (1) CH3Cu
(CH3)2CMCHCCH3   DDDDDDDDD4
                  :       (2) H2O
                  O
```

* A) $(CH_3)_3CCH_2COCH_3$

B) $(CH_3)_2CMCHC(CH_3)_2$
```
           3
           OH
```

C) $(CH_3)_2CHCHCOCH_3$
```
          3
          CH3
```

D) 2 CH_3CCH_3
```
        :
        O
```

E)
```
           OH
           3
(CH3)2CMCHCCH3
           3
           H
```

15. Which of the reagents listed below would serve as the basis for a simple chemical test to distinguish between $(CH_3)_2CHCH_2OH$ and $CH_3CHCH_2CH_3$?
```
    3
    OH
```

* A) NaOI (I_2 in NaOH)

B) $KMnO_4$ in H_2O

C) Br_2 in CCl_4

D) Cold concd H_2SO_4

E) CrO_3 in H_2SO_4

16. What would be the product of the following reaction?

```
    O
    :                      EtO⁻, EtOH
C6H5CH  +  CH3CH2CN   DDDDDDDDDD4   ?
```

A) $C_6H_5CHMCHCH_2CN$

B)
```
    O
    :
C6H5CCHCN
      3
      CH3
```

* C) $C_6H_5CHMCDCN$
```
          3
          CH3
```

D)
```
      OH
      3
C6H5CHCH2CH2CN
```

E)
```
      OH
      3
C6H5CCH2CH2CN
      3
      OC2H5
```

17. What would be the final product of the following reaction sequence?

$$\overset{\displaystyle O}{\overset{:}{C_6H_5CH}} \quad + \quad \overset{\displaystyle O}{\overset{:}{CH_3CCH_3}} \quad \xrightarrow{OH^-,\ heat} \quad A \quad \xrightarrow{LiAlH_4} \quad \text{Final product}$$

A) $C_6H_5CH_2CH_2CH_2CH_3$

B) OH O
3 :
$C_6H_5CHCH_2CCH_3$

C) O
:
$C_6H_5CH_2CH_2CCH_3$

* D) OH
3
$C_6H_5CHMCHCHCH_3$

E) None of these

18. What would be the major product of the following reaction?

O
:
$C_6H_5CCHCH_2CH_3$ + Br_2 $\xrightarrow{H_3O^+}$?
3
CH_3

* A) O
:
$C_6H_5CCBrCH_2CH_3$
3
CH_3

B) O Br
: 3
$C_6H_5CCHCHCH_3$
3
CH_3

C) O OH
: 3
$C_6H_5CCHCHCH_3$
3
CH_3

D) $C_6H_5CBr_2CHCH_2CH_3$
3
CH_3

E) O
:
m-$BrC_6H_4CCHCH_2CH_3$
3
CH_3

19. What would be the major product of the following reaction?

O O
: : OH^-
C_6H_5CH + CH_3CCH_3 DDDDDD4 ?
heat

A) O OH
: 3
$C_6H_5CCH_2CHCH_3$

B) O O
: :
$C_6H_5CCH_2CCH_3$

C) OH OH
3 3
$C_6H_5CHCH_2CHCH_3$

D) O O
: :
$C_6H_5CCH_2CCH_3$

* E) O
:
$C_6H_5CHMCHCCH_3$

20. What would be the product, C, of the following reaction sequence?

O O
: : OH^- H^+, heat H_2, Ni
$(CH_3)_3CCH$ + CH_3CH_2CH DDDDDD4 A DDDDDDDD4 B DDDDDD4 C

A) $(CH_3)_3CCH_2CH_2CH_2OH$

B) OH O
3 :
$(CH_3)_3CCHCH_2CH_2CH$

* C) $(CH_3)_3CCH_2CHCH_2OH$
3
CH_3

D) $(CH_3)_3CCHMCCH_2OH$
3
CH_3

E) O
:
$(CH_3)_3CCHMCCH$
3
CH_3

21. What would be the product, C, of the following reaction sequence?

O O
: : OH^- H^+, heat $^-:CH_2-^+P(C_6H_5)_3$
$(CH_3)_3CCH$ + CH_3CH_2CH DDDDDD4 A DDDDDDDD4 B DDDDDDDDDDDDDD4 C

A) $(CH_3)_3CCH_2CH_2CH_2OH$

B) OH
3
$(CH_3)_3CCHCH_2CHMCH_2$

C) $(CH_3)_3CCH_2CHMCH_2$
3
CH_3

* D) $(CH_3)_3CCHMCCHMCH_2$
3
CH_3

E) OH
3
$(CH_3)_3CCHCHCH=CH_2$
3
CH_3

22. Which reagents would you use to synthesize this compound by an aldol condensation?

O
:
$C_6H_5CHMCHCC_6H_5$

A)
O　　　　　　O
:　　　　　　:
C_6H_5CH and $C_6H_5CH_2CH$

B)
O　　　　　　O
:　　　　　　:
$C_6H_5CH_2CH$ and $C_6H_5CCH_3$

C)
O
:
$C_6H_5CHMCHCH$ and C_6H_5OH

* D)
O　　　　　　O
:　　　　　　:
$C_6H_5CCH_3$ and C_6H_5CH

E)
O
:
$(C_6H_5)_2CuLi$ and $CH_2{=}CHCC_6H_5$

23. What would be the product, B, of the following reaction sequence?

O
:
C_6H_5CH + CH_3CN $\overset{EtO^-,\ EtOH}{DDDDDDDDDD4}$ A (C_9H_7N) $\underset{heat}{\overset{H_3O^+,\ H_2O}{DDDDDDDDD4}}$ B

* A) $C_6H_5CHMCHCO_2H$

B)
O
:
$C_6H_5CCH_2CO_2H$

C) $C_6H_5CH_2CH_2CO_2H$

D)
OH
3
$C_6H_5CHCH_2CO_2H$

E)
OH
3
$C_6H_5CHCH_2CN$

24. What starting compound(s) would you use in an aldol reaction to prepare

1-cyclopentenecarbaldehyde (cyclopentene ring with CH=O on the double-bond carbon) ?

I: $CH_3CH(OH)CH_2CH_2CCH_3$ (with C=O at the second-to-last carbon)

II: $CH_3\overset{O}{\overset{\|}{C}}CH_2CH_2\overset{O}{\overset{\|}{C}}CH_3$

III: $CH_3\overset{O}{\overset{\|}{C}}CH_2CH_2CH_2\overset{O}{\overset{\|}{C}}H$

IV: $H\overset{O}{\overset{\|}{C}}CH_2CH_2CH_2CH_2\overset{O}{\overset{\|}{C}}H$

V: cyclopentanone + $H_2C{=}O$

A) I B) II C) III * D) IV E) V

25. Which reagents would you use to synthesize this compound by an aldol condensation?

$C_6H_5CH{=}C(C_6H_5)CH$ (with O on the terminal CH)

* A) $C_6H_5\overset{O}{\overset{:}{C}}H$ and $C_6H_5CH_2\overset{O}{\overset{:}{C}}H$

B) $C_6H_5CH_2\overset{O}{\overset{:}{C}}H$ and $C_6H_5\overset{O}{\overset{:}{C}}CH_3$

C) $C_6H_5CH{=}CH\overset{O}{\overset{:}{C}}H$ and C_6H_5OH

D) $C_6H_5\overset{O}{\overset{:}{C}}CH_3$ and $C_6H_5\overset{O}{\overset{:}{C}}H$

E) $C_6H_5CH_2Cl$ and $C_6H_5CH{=}\overset{O}{\overset{:}{C}}H$, ONa

26. What would be the product, C, of the following reaction sequence?

```
             O             O
             :             :     OH-          H+, heat        NaBH4
        C6H5CH  +  CH3CH2CH  DDDDDD4  A  DDDDDDDD4  B  DDDDD4  C
                                 25xC         (-H2O)
```

```
A)      O                              D)          O
        :                                          :
    C6H5CCHCH2OH                           C6H5CHMCCH
          3                                          3
          CH3                                        CH3

B)  C6H5CHCHCH2OH                    * E)  C6H5CHMCCH2OH
        3  3                                         3
       HO  CH3                                       CH3

C)             O
               :
    C6H5CH2CHCH
             3
             CH3
```

27. What would be the product, C, of the following reaction sequence?

```
             O             O
             :             :     OH-          H+, heat        H2, Ni
        C6H5CH  +  CH3CH2CH  DDDDDD4  A  DDDDDDDD4  B  DDDDDD4  C
                                 25xC         (-H2O)          high
                                                            pressure
```

```
  A)      O                              D)          O
          :                                          :
      C6H5CCHCH2OH                           C6H5CHMCCH
            3                                          3
            CH3                                        CH3

* B)  C6H5CH2CHCH2OH                     E)  C6H5CHMCCH2OH
               3                                       3
               CH3                                     CH3

  C)             O
                 :
      C6H5CH2CHCH
               3
               CH3
```

28. Which of the following would undergo racemization in base?

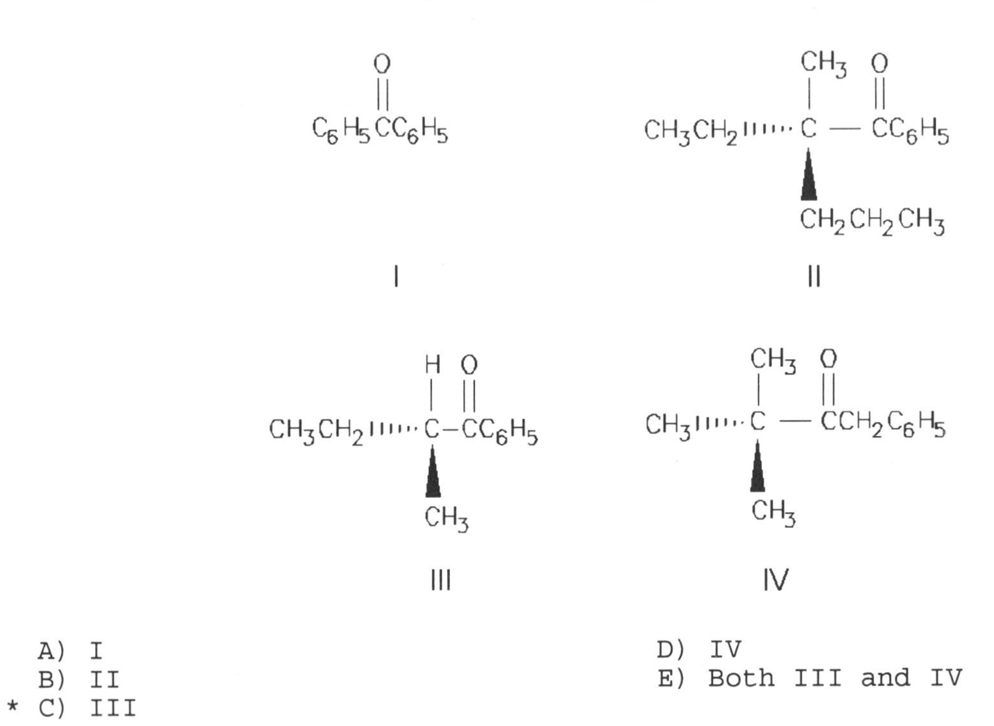

A) I
B) II
* C) III
D) IV
E) Both III and IV

29. What would be the major product of the following reaction?

```
                O
                :                      CH3CH2OH
C6H5CHMCHCCH3        +     CN-      DDDDDDDD4     ?
                                       CH3CO2H
```

A)
```
            O
            :
C6H5CH2CHCCH3
         3
         CN
```

* B)
```
            O
            :
C6H5CHCH2CCH3
     3
     CN
```

C)
```
              OH
              3
C6H5CH2CH2CCH3
              3
              CN
```

D)
```
              O
              :
C6H5CH2CH2CCN
```

E)
```
              OH
              3
C6H5CHCH=CCH3
       3
       CN
```

30. Which of the following represent keto-enol tautomers?

A)
O
:
$CH_3CCH_2CH_3$ and OH / 3 / $CH_3CMCHCH_3$

B)
OH
3
$CH_2MCCH_2CH_3$ and O / : / $CH_3CCH_2CH_3$

C)
O
:
$HOCH_2CCHMCH_2$ and O / : / $CH_3CCH_2CH_3$

* D) More than one of these

E) None of these

31. A compound, X, $C_9H_{10}O$, gives a strong IR absorption peak at 1690 cm^{-1} and gives the following ^1H NMR spectrum.

Triplet, k 1.2
Quartet, k 3.0
Multiplet, k 7.7

Which is a possible structure for X?

A)
H
3
p-$CH_3C_6H_4CH_2C$=O

B)
O
:
$C_6H_5CH_2CCH_3$

* C)
O
:
$C_6H_5CCH_2CH_3$

D)
H
3
$C_6H_5CH_2CH_2CMO$

E)
H
3
C_6H_5CHCMO
3
CH_3

32. What is the missing reagent?

? —OH^- / heat→ [cyclohexenone: O]

A)
O O
: :
$CH_3CCH_2CH_2CCH_3$

* B)
O O
: :
$CH_3CCH_2CH_2CH_2CH$

C)
O O
: :
$HCCH_2CH_2CH_2CH_2CH$

D)
O O
: :
$CH_3CCH_2CH_2CH_2COEt$

E)
H O
3 :
$OMCCH_2CH_2CH_2CH_2COEt$

33. Which would be formed when 2-methylpropanal is dissolved in D_2O containing NaOD?

A)
```
       O
       :
CH3CHCD
   3
   CH3
```

B)
```
       O
       :
CH3CHCH
   3
   CH2D
```

* C)
```
       O
       :
CH3CDCH
   3
   CH3
```

D)
```
       O
       :
CH3CDCD
   3
   CH3
```

E)
```
       D
       3
CH3CHCHOD
   3
   CH3
```

34. What would be the major product of the following reaction?

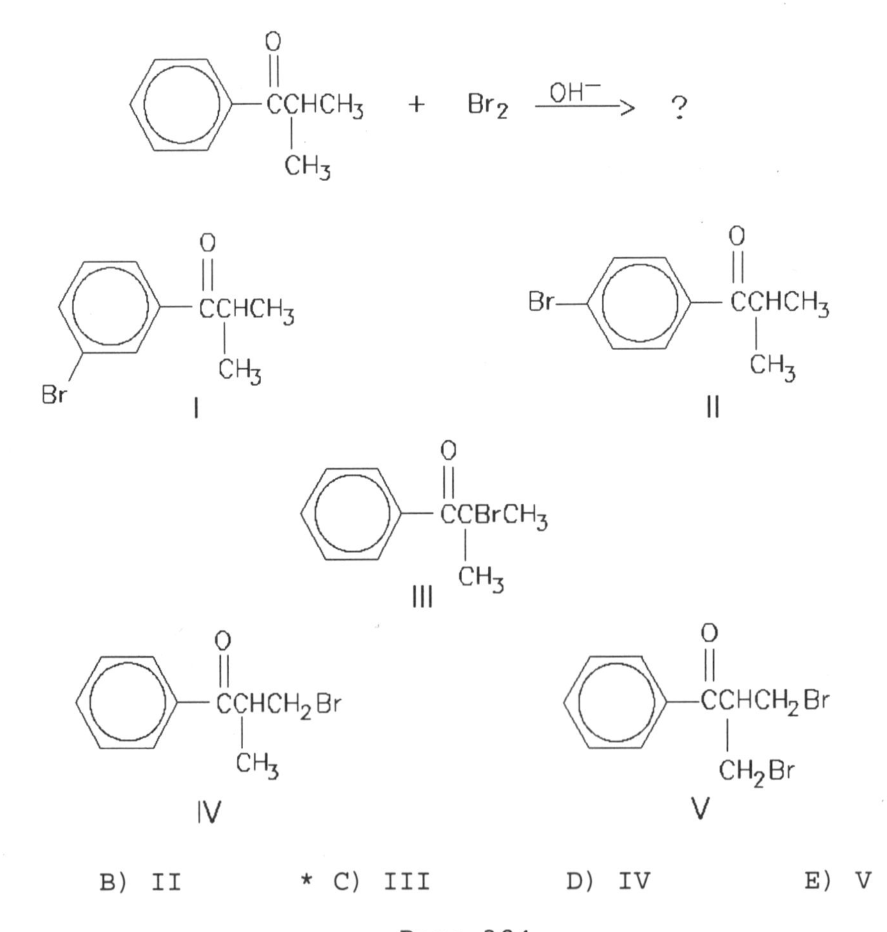

A) I B) II * C) III D) IV E) V

35. Which compound would be formed when 2-methylbutanal is treated with a solution of NaOD in D_2O?

* A) CH_3CH_2CDCHO
3
CH_3

B) CH_3CH_2CHCDO
3
CH_3

C) $CH_3CHDCHCHO$
3
CH_3

D) CH_2DCH_2CHCHO
3
CH_3

E) CH_3CH_2CDCDO
3
CH_3

36. What is the final product of this reaction sequence?

$CH_3CH_2NO_2$ H_2, Ni
$H_2C{=}O$ DDDDDDDDD4 DDDDDD4 ?
OH^-

A) NH_2
3
CH_3CMCH_2

B) $CH_3CH_2CH_2NH_2$

* C) NH_2
3
CH_3CHCH_3

D) $CH_3CH_2CH_2NO_2$

E) NO_2
3
CH_3CHCH_3

37. Which is the only one of these compounds which cannot self-condense in the presence of dilute aqueous alkali?

A) Phenylethanal
B) Propanal
C) 2-Methylpropanal
D) 3-Methylpentanal
* E) 2,2-Dimethylpropanal

38. The aldol condensation product formed from 3-pentanone in the presence of base has the IUPAC name:

* A) 5-Ethyl-4-methyl-4-hepten-3-one
B) 5-Ethyl-4-methyl-5-hepten-3-one
C) 4-Methyl-4-nonen-3,7-dione
D) 3-Ethyl-4-methyl-3-hepten-5-one
E) 3-Ethyl-4-methyl-2-hepten-5-one

39. Which of these is <u>not</u> among the reaction products when a crossed aldol addition occurs between ethanal and butanal?

A)
```
   OH  H
   3   3
CH3CHCH2CMO
```

* B)
```
          OH     H
          3      3
CH3CH2CH2CHCHCH2CMO
            3
            CH3
```

C)
```
   OH  H
   3   3
CH3CHCHCMO
     3
     CH2CH3
```

D)
```
          OH   H
          3    3
CH3CH2CH2CHCH2CMO
```

E)
```
          OH  H
          3   3
CH3CH2CH2CHCHCMO
            3
            CH2CH3
```

40. The retro-aldol reaction of

```
          H
          3
C6H5CHMCCMO
        3
        CH2CH3
```

forms:

A)
```
           H            H
           3            3
C6H5CH2CH2CMO   +   CH3CMO
```

B)
```
        H               H
        3               3
C6H5CH2CMO   +   CH3CH2CMO
```

C)
```
     H               O
     3               :
C6H5CMO   +   CH3CH2CCMO
                     3
                     H
```

D)
```
           O
           :
C6H5CH2CCH2CH3   +   H2CMO
```

* E)
```
     H                    H
     3                    3
C6H5CMO   +   CH3CH2CH2CMO
```

41. If butanal is added slowly to an aqueous solution of sodium hydroxide and 2,2-dimethylpropanal at 25xC, the principal product is which of these?

```
* A)      OH  H
          3   3
    (CH3)3CCHCHCMO
             3
             CH2CH3

  B)         OH  H
             3   3
    CH3CH2CH2CHCHCMO
                3
                CH2CH3

  C)      OH
          3
    (CH3)3CCHCC(CH3)3
             :
             O

  D)            OH
                3
    CH3CH2CH2CHC(CH3)3

  E)            OH  H
                3   3
    CH3CH2CH2CHCHC=O
                 3
                 C(CH3)3
```

42. The conversion of $C_6H_5\overset{O}{\overset{:}{C}}CH_3$ to $C_6H_5\overset{O}{\overset{:}{C}}OH$ is accomplished by the use of which oxidizing agent?

A) $Ag(NH_3)_2^+$

B) O_3

* C) NaOI (I_2 in NaOH)

D) $R\overset{O}{\overset{:}{C}}OOH$

E) Cu^{++}

43. Which of these is a product of the reaction of C_6H_5MgBr with $C_6H_5CH{=}CH\overset{O}{\overset{:}{C}}CH_3$?

```
  A)          O
              :
    C6H5CHMCCCH3
           3
           C6H5

* B)          OH
              3
    C6H5CHMCHCCH3
              3
              C6H5

  C)            O
                :
    C6H5CHMCHCCH2C6H5

  D)                      O
                          :
    pDC6H5DC6H4DCHMCHCCH3

  E)            OC6H5
                3
    C6H5CHMCHCHCH3
```

44. What compound results from the cyclic aldolization of

$$CH_3\overset{O}{\overset{\|}{C}}CH_2CH_2\underset{CH_3}{\underset{|}{CH}}CH_2\overset{O}{\overset{\|}{C}}CH_3 \quad ?$$

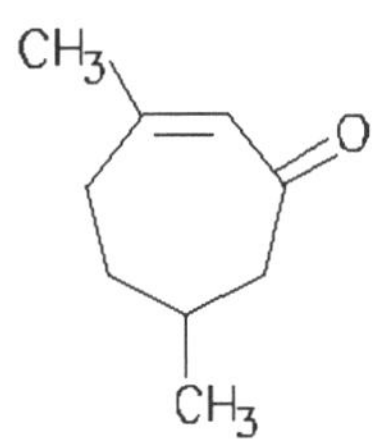

I

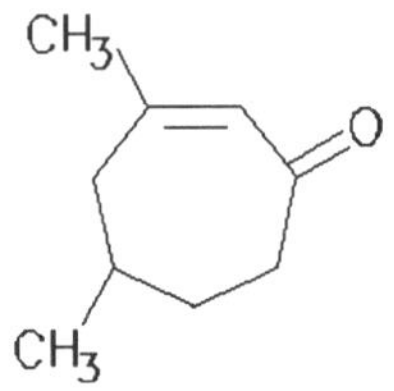

II

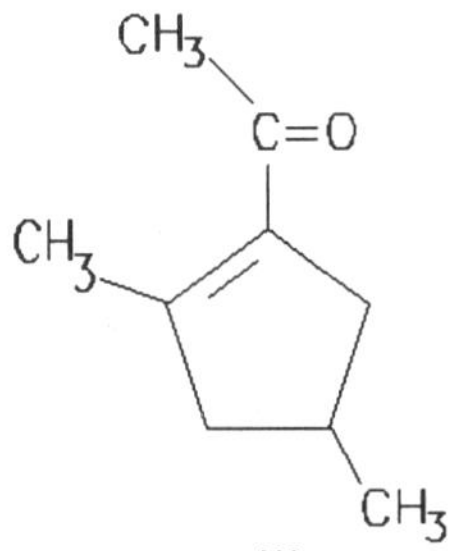

III

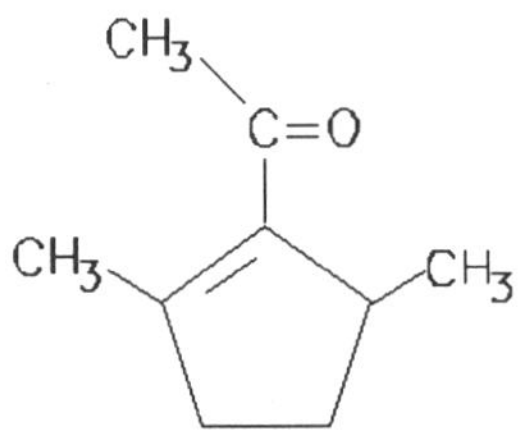

IV

A) I
B) II
C) III
D) IV
* E) Both III and IV

45. The cyclic aldolization of

```
                  O          H
                  :          3
CH3CCH2CH2CH2CMO
```

The cyclic aldolization of $CH_3CCH_2CH_2CH_2CMO$ produces which of these?

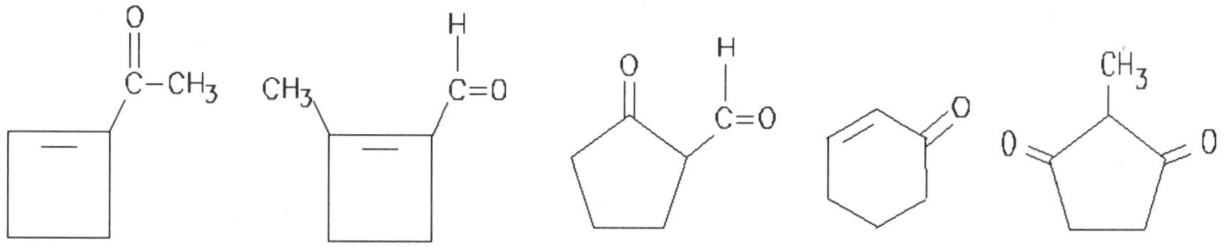

I II III IV V

A) I B) II C) III * D) IV E) V

46.

```
       O                 O
       :                 :
CH3CCH2CH2CH2CH2CH2CCH3
```

When $CH_3CCH_2CH_2CH_2CH_2CH_2CCH_3$ cyclizes in basic solution, which of these compounds will be formed?

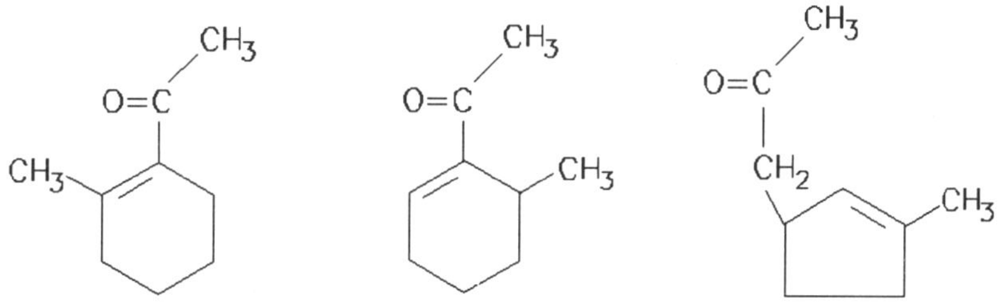

I II III

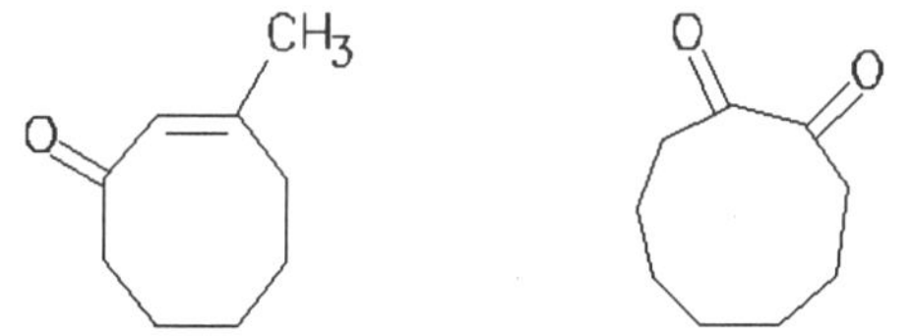

IV V

* A) I B) II C) III D) IV E) V

47. Simple enols are less stable than the tautomeric keto forms because:
A) severe angle strain exists in the enol forms.
B) fewer atoms are coplanar in the keto form.
C) the enol cannot be chiral.
* D) the C-C c bond is weaker than the C-O c bond.
E) Actually, simple enols are the <u>more</u> stable.

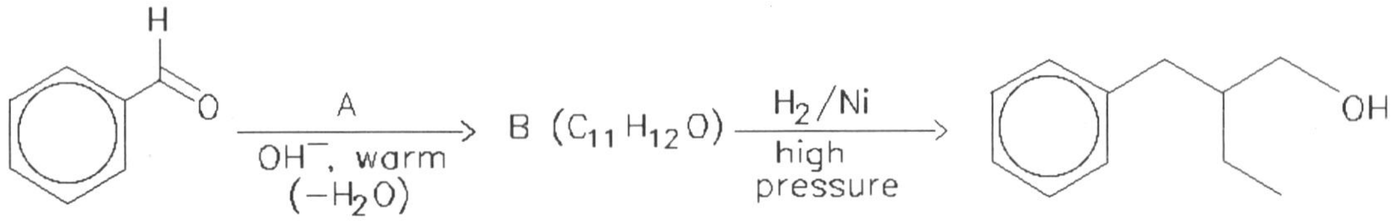

$\xrightarrow[\text{OH}^-,\ \text{warm}\ (-H_2O)]{A}$ B ($C_{11}H_{12}O$) $\xrightarrow[\text{high pressure}]{H_2/Ni}$

48. Consider the synthesis above in answering this question. What is compound A?
   A) 2-Butanone
 * B) Butanal
   C) Propanal
   D) 1-Butanol
   E) 2-Methylpropanal

49. What is the intermediate B in the synthesis shown above?

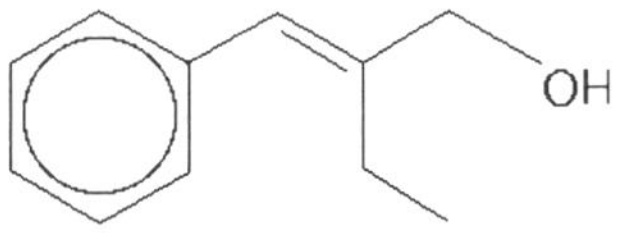

I

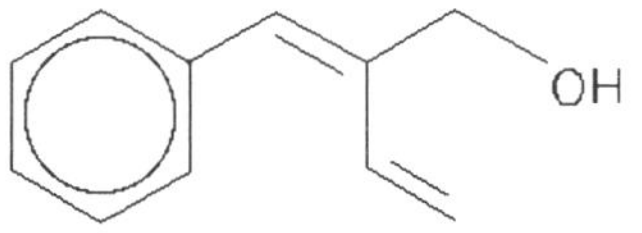

II

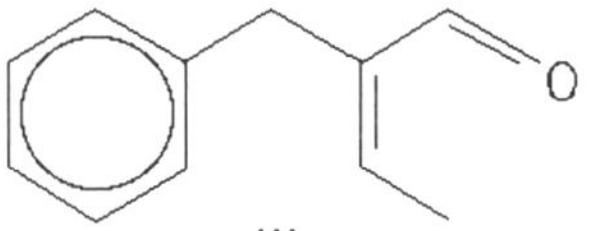

III

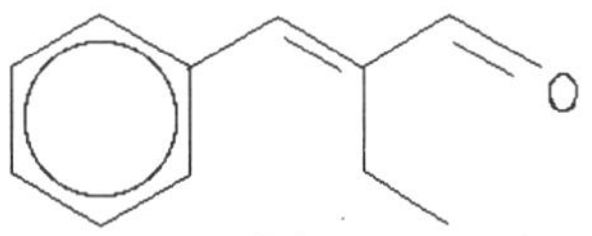

IV

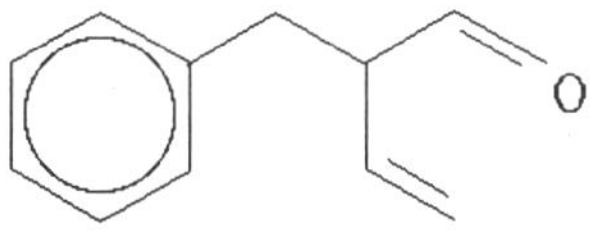

V

A) I   B) II   C) III   * D) IV   E) V

50. A negative iodoform test will be observed in the case of which of these?
   A) Acetone
   B) Ethanal
   C) Ethanol
   D) 2-Butanol
 * E) All of these will give a positive test.

51. Which of these would not be an intermediate or final product when $C_6H_5CH_2CN$ and HCHO react in basic solution and the product is treated with acid?

A)
CH_2OH
3
C_6H_5CHCN

B)
CH_2
:
C_6H_5CCN

C)
C_6H_5CHCN
3
$C_6H_5CH_2CMNH$

D)
C_6H_5CHCN
3
$C_6H_5CH_2CMO$

* E) Each of these would be either an intermediate or a final product.

52. The Robinson annulation reaction which produces

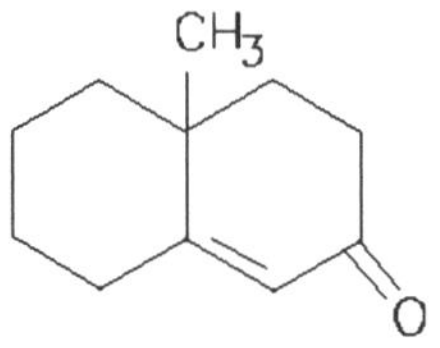

CH_3

O

uses which of the following as starting materials?

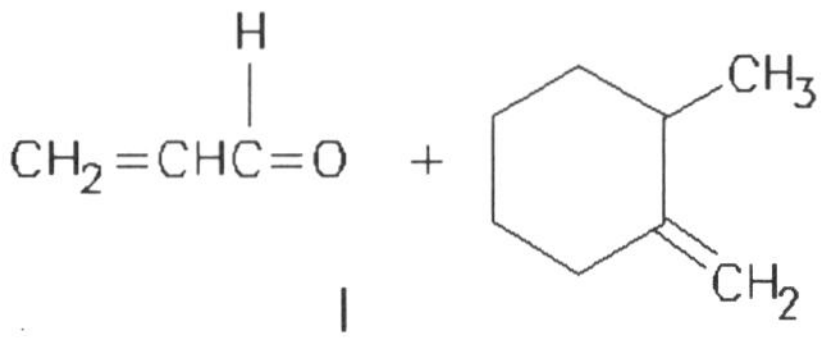

H | $CH_2=CHC=O$ + (ring with CH_3 and $=CH_2$)

I

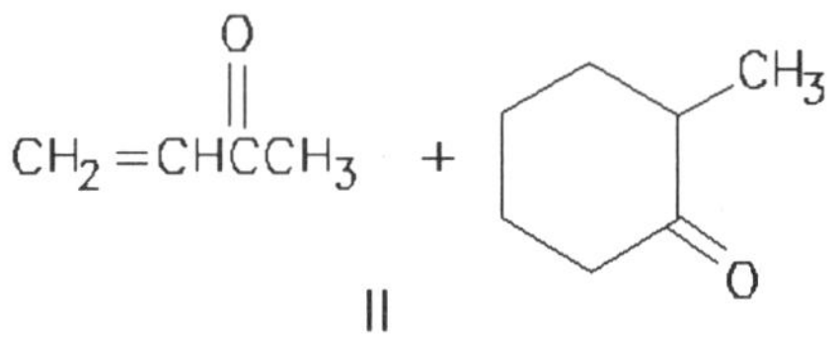

O || $CH_2=CHCCH_3$ + (ring with CH_3 and $=O$)

II

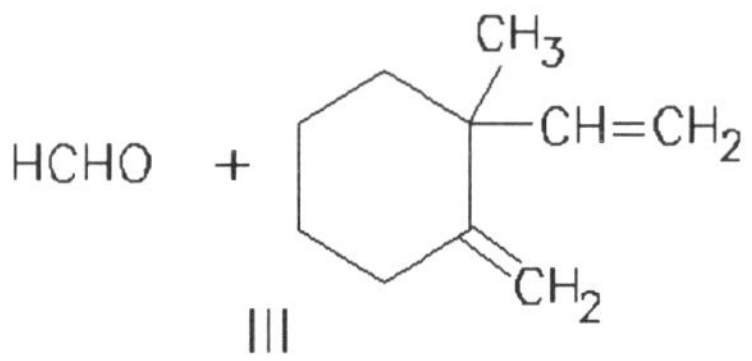

HCHO + (ring with CH_3, $-CH=CH_2$ and $=CH_2$)

III

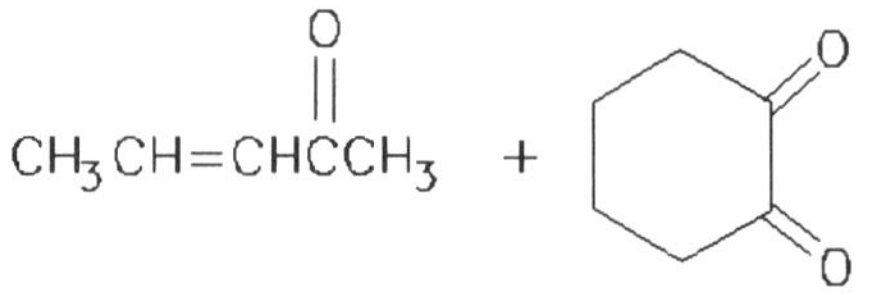

O || $CH_3CH=CHCCH_3$ + (ring with two $=O$)

IV

O || $CH_3CH=CHCCH_3$ + (ring with $=O$)

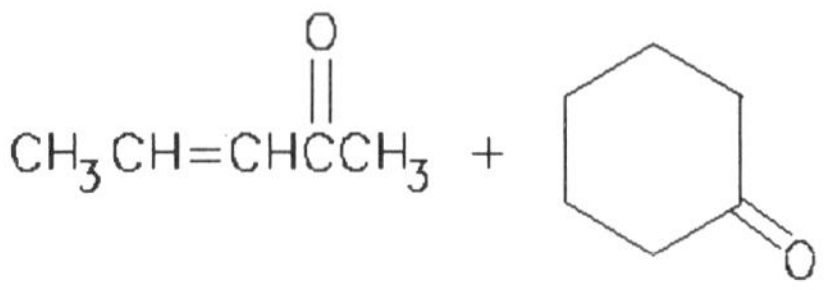

V

A) I * B) II C) III D) IV E) V

53. A compound, $C_5H_{10}O$, reacts with phenylhydrazine and gives a positive iodoform test. The compound could be which of these?

A)
```
              H
              3
CH3CH2CH2CH2C=O
```

B)
```
          H
          3
CH3CHCH2C=O
   3
   CH3
```

C) $CH_2=CHCH_2CHOHCH_3$

* D)
```
       O
       :
CH3CHCCH3
   3
   CH3
```

E)
```
         O
         :
CH3CH2CCH2CH3
```

54. What product results from the intramolecular aldol reaction of 2,5-hexanedione?

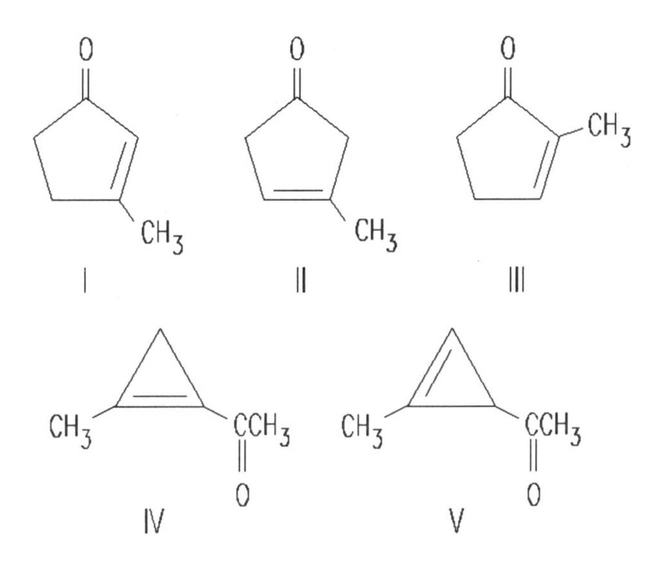

* A) I B) II C) III D) IV E) V

55. The reaction of $CH_3\overset{O}{C}CH_2CH_2CH_2\overset{O}{C}CH_3$ with base affords which of these products?

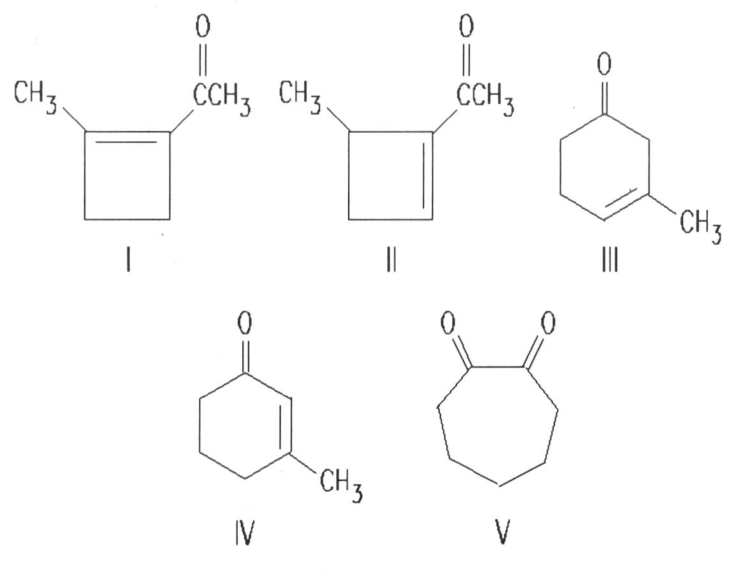

A) I B) II C) III * D) IV E) V

56. The aldol reaction of cyclohexanone produces which of these self-condensation products?

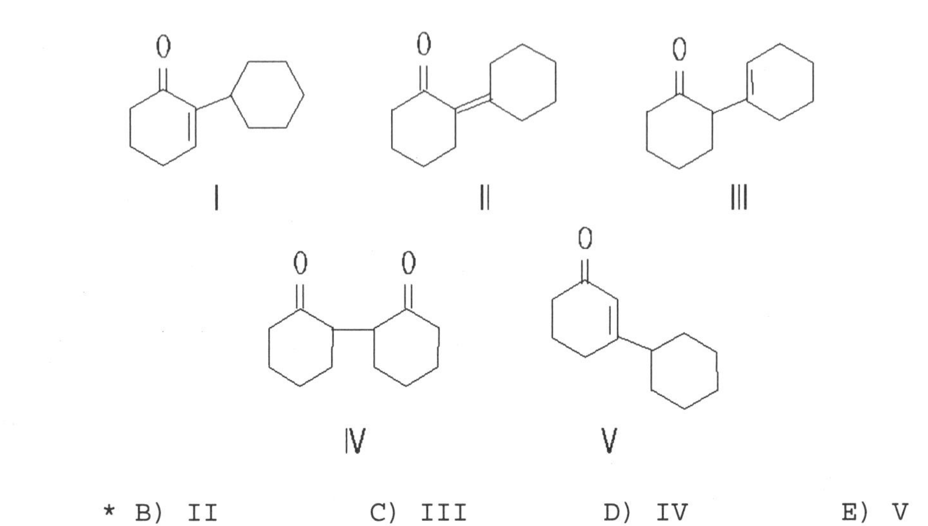

A) I * B) II C) III D) IV E) V

57. Which compound could be subjected to a haloform reaction to produce m-chlorobenzoic acid?

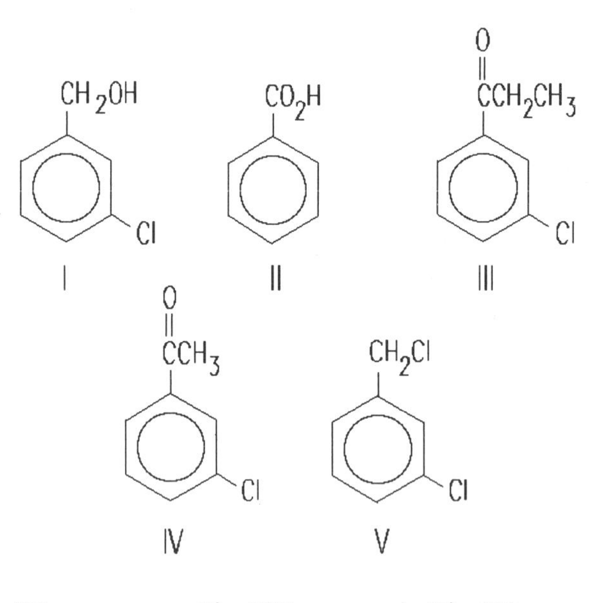

A) I B) II C) III * D) IV E) V

58. What is the product of the reaction below?

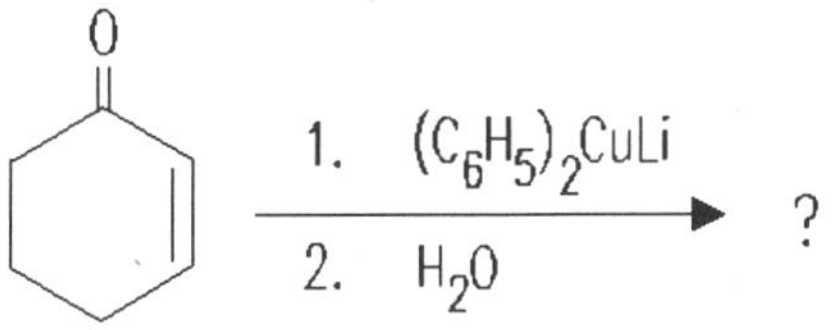

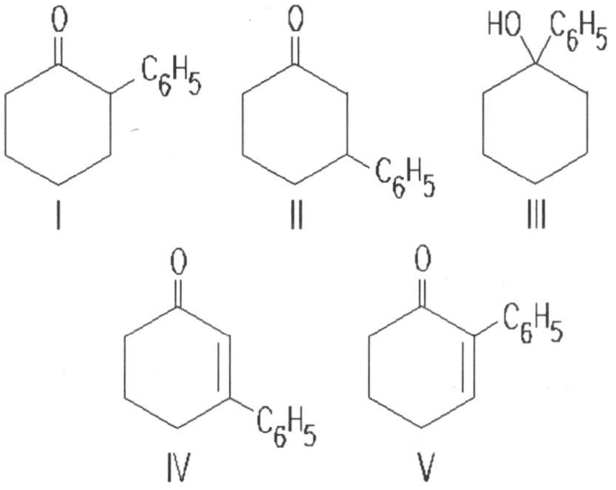

$$\text{2-cyclohexen-1-one} \xrightarrow[\text{2. } H_2O]{\text{1. } (C_6H_5)_2CuLi} ?$$

I: 2-C_6H_5-cyclohexanone; II: 3-C_6H_5-cyclohexanone; III: 1-C_6H_5-cyclohexanol (HO, C_6H_5); IV: 3-C_6H_5-2-cyclohexen-1-one; V: 2-C_6H_5-2-cyclohexen-1-one

A) I * B) II C) III D) IV E) V

1. The product, A, of the following reaction sequence,

$$C_6H_5CH_2Cl \xrightarrow{NaCN} \text{organic product} \xrightarrow[\text{reflux}]{70\%\ H_2SO_4} A + NH_4^+$$

would be:

A) $C_6H_5CH_2OCH$ (with O double-bonded to the terminal C)

* B) $C_6H_5CH_2COOH$

C) $C_6H_5CH_2OSO_3H$

D) $C_6H_5CHClCOOH$

E) $C_6H_5CH_2CCH_2C_6H_5$ (with O double-bonded to the central C)

2. Predict the major organic product of the reaction sequence below

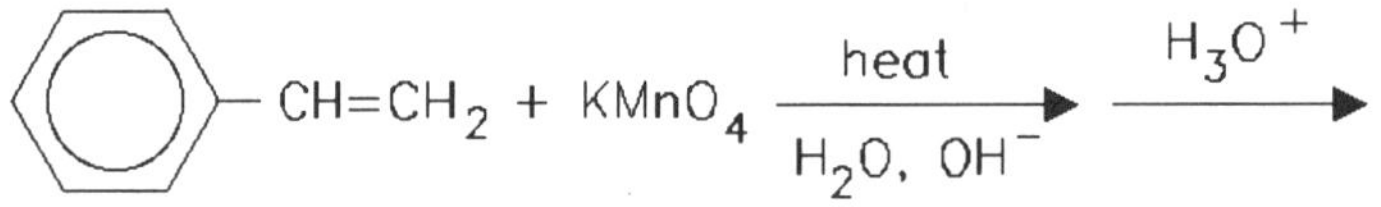

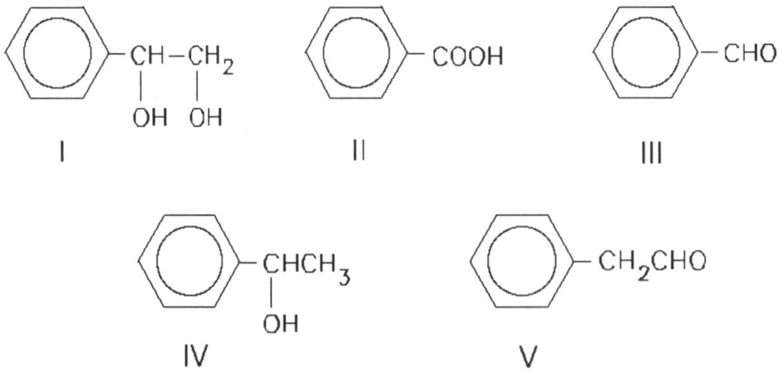

$$C_6H_5CH{=}CH_2 + KMnO_4 \xrightarrow[H_2O,\ OH^-]{heat} \xrightarrow{H_3O^+}$$

I: $C_6H_5CH(OH)CH_2OH$

II: C_6H_5COOH

III: C_6H_5CHO

IV: $C_6H_5CH(OH)CH_3$

V: $C_6H_5CH_2CHO$

A) I * B) II C) III D) IV E) V

3. Which reagent would serve as the basis for a simple chemical test to distinguish between benzoic acid and benzamide?

A) Cold dilute NaOH
B) Cold dilute $NaHCO_3$
C) Cold concd H_2SO_4
* D) More than one of these
E) None of these

4. In which of the following sequences are the compounds listed in order of decreasing acidity?

* A) $CH_3COOH > H_2O > CH_3CH_2OH > HCpCH > NH_3$
B) $CH_3CH_2OH > CH_3COOH > H_2O > HCpCH > NH_3$
C) $CH_3COOH > CH_3CH_2OH > H_2O > NH_3 > HCpCH$
D) $H_2O > CH_3COOH > CH_3CH_2OH > HCpCH > NH_3$
E) $CH_3CH_2OH > H_2O > CH_3COOH > HCpCH > NH_3$

5. The reason for the difference in behavior of p-cresol and benzoic acid toward sodium bicarbonate solutions is:

A) The pk_a of HCO_3^- lies between the pk_a of benzoic acid and the pk_a of p-cresol.
B) The pk_a of HCO_3^- is greater then the pk_a of p-cresol.
* C) The pk_a of H_2CO_3 lies between the pk_a of benzoic acid and the pk_a of p-cresol.
D) The pk_a of H_2CO_3 is greater than the pk_a of p-cresol.
E) CO_2 is more soluble in phenols than it is in carboxylic acids.

6. Which reagent would best serve as a basis for a simple chemical test to distinguish between CH_3CH_2COOH and CH_3COOCH_3?

A) Concd H_2SO_4
B) Br_2/CCl_4
C) CrO_3/H_2SO_4
* D) $NaHCO_3/H_2O$
E) $KMnO_4/H_2O$

7. Which compound would be the strongest acid?

A) $CHCl_2CH_2CH_2CO_2H$
B) $ClCH_2CHClCH_2CO_2H$
C) $CH_3CCl_2CH_2CO_2H$
D) $CH_3CHClCHClCO_2H$
* E) $CH_3CH_2CCl_2CO_2H$

8. The product of the following reaction

$$CH_3CH_2\overset{\overset{O}{:}}{C}OH + CH_3{}^{18}OH \xrightarrow{H^+}$$

would be:

* A) $CH_3CH_2\overset{\overset{O}{:}}{C}{}^{18}OCH_3$

B) $CH_3CH_2\overset{\overset{^{18}O}{:}}{C}OCH_3$

C) $CH_3CH_2\overset{\overset{^{18}O}{:}}{C}{}^{18}OCH_3$

D) $CH_3CH_2\overset{\overset{O}{:}}{C}OCH_3$

E) $CH_3CH_2O\overset{\overset{O}{:}}{C}{}^{18}OCH_3$

9. Which of the following would be the strongest acid?

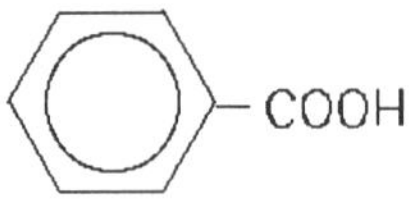

I: benzoic acid (C_6H_5–COOH)

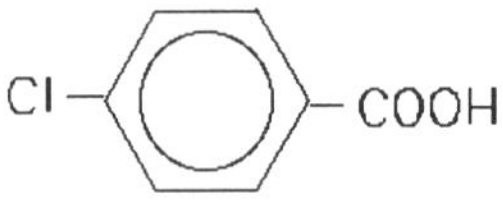

II: Cl–C_6H_4–COOH (para-Cl)

III: 2,4,6-trichloro: Cl at both ortho positions and para position relative to COOH

IV: Cl at ortho position relative to COOH

V: Cl at ortho and para positions relative to COOH

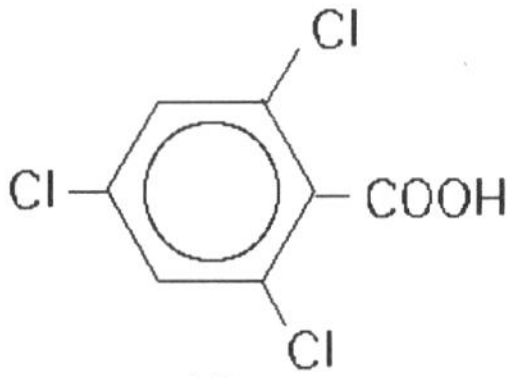

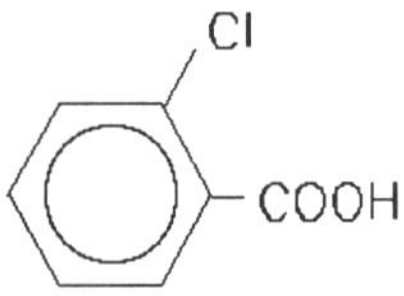

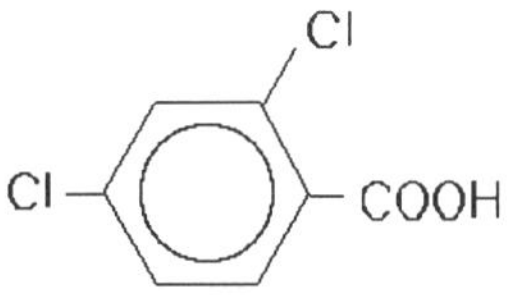

A) I B) II * C) III D) IV E) V

10. The product, F, of the following sequence of reactions,

CH_3
3 $LiAlH_4$ H_2O PBr_3 Mg CO_2
$CH_3CHCOOH$ DDDDDD4 A DDDDDD4 B DDDDDD4 C DDDDDD4 D DDDDDD4 E
ether

H^+
DDDDDD4 F, would be:
H_2O

A) CH_3
3
CH_3CHCH_2Br

B) CH_3
3
CH_3CCH_2COOH
3
Br

C) $CH_3CH_2CH_2COOH$

* D) CH_3
3
CH_3CHCH_2COOH

E) CH_3
3
$CH_3CHCOOCH_3$

11. The product, Z, of the following sequence of reactions is which compound?

	$KMnO_4$		H^+		$SOCl_2$		CH_3CH_2OH	
p-chlorotoluene	DDDDDDDD4	W	DDDDDD4	X	DDDDDDD4	Y	DDDDDDDD4	Z
	OH^-, heat						base	

I: CH_3CH_2–(benzene ring)–Cl

II: $CH_3CH_2OC(=O)$–(benzene ring)–Cl

III: (benzene ring)–$C(=O)OCH_2CH_3$

IV: Cl–(benzene ring)–$C(=O)CH_2CH_3$

V: Cl–(benzene ring)–$C(=O)CH_2CH_2Cl$

A) I * B) II C) III D) IV E) V

12. A compound has the molecular formula, $C_6H_{12}O_2$. Its IR spectrum shows a strong absorption band near 1740 cm^{-1}; its ^1H NMR spectrum consists of two singlets, at k 1.4 and k 2.0. The most likely structure for this compound is:

A)
```
        O
        :
CH3CHOCCH2CH3
   3
   CH3
```

* D)
```
  CH3 O
  3   :
CH3CDO-CCH3
  3
  CH3
```

B)
```
              O
              :
CH3CH2CH2OCCH2CH3
```

E)
```
        CH3 O
        3   :
CH3CH2CDDOCH
        3
        CH3
```

C)
```
          O CH3
          : 3
CH3CH2OCDCHCH3
```

13. A compound has the molecular formula $C_8H_{14}O_4$. Its IR spectrum shows a strong absorption band near 1740 cm^{-1}. Its ^1H NMR spectrum consists of:

triplet, k 1.3
singlet, k 2.6
quartet, k 4.2

The most likely structure for the compound is:

* A)
```
         O     O
         :     :
CH3CH2OCCH2CH2COCH2CH3
```

B)
```
     O         O
     :         :
CH3OCCH2CH2CH2COCH2CH3
```

C)
```
     O     O
     :     :
CH3OCCHDCHCOCH3
       3   3
       CH3CH3
```

D)
```
   O                       O
   :                       :
HOCCH2CH2CH2CH2CH2CH2COH
```

E)
```
     O                 O
     :                 :
CH3OCCH2CH2CH2CH2COCH3
```

14. A compound with the molecular formula $C_5H_{10}O_2$ gave the following ^1H NMR spectrum:

triplet, k 0.90
multiplet, k 1.60
singlet, k 1.95
triplet, k 3.95

The IR spectrum showed a strong absorption band near 1740 cm^{-1}. The most likely structure for the compound is:

* A)
```
     O
     :
CH3COCH2CH2CH3
```

B)
```
       O
       :
CH3CH2COCH2CH3
```

C)
```
     O   CH3
     :  /
CH3COCH
        \
         CH3
```

D)
```
 O CH3
 : 3
HCOCCH3
    3
    CH3
```

E)
```
    CH3 O
    3   :
CH3CDDDCOH
    3
    CH3
```

15. A compound with the molecular formula $C_{18}H_{18}O_4$ has a ^1H NMR spectrum that consists of:

singlet, k 2.7
singlet, k 3.1
multiplet, k 7.3

The IR spectrum shows a strong absorption band near 1750 cm^{-1}. The most likely structure for the compound is:

A)
O O
: :
$C_6H_5CH_2CH_2CH_2CH_2OCDCOC_6H_5$

B)
O O
: :
$C_6H_5CH_2CH_2CH_2OCCH_2COC_6H_5$

* C)
O O
: :
$C_6H_5CH_2OCCH_2CH_2COCH_2C_6H_5$

D)
O O
: :
$C_6H_5OCCH_2CH_2CH_2CH_2COC_6H_5$

E)
O O
: :
$C_6H_5COCH_2CH_2CH_2CH_2OCC_6H_5$

16. Which of the following reactions would constitute a reasonable synthesis of propyl acetate?

A)
O
: H^+
$CH_3CH_2CH_2OH$ + CH_3COH DDDDDD4

B)
O
: pyridine
$CH_3CH_2CH_2OH$ + CH_3CCl DDDDDDDD4

C)
O
:
$CH_3CH_2CH_2OH$ + $(CH_3C)_2O$ DDDDDD4

* D) All of these
E) None of these

17. Which of the following would serve as a synthesis of 2,2-dimethylpropanoic acid,

$$
\begin{array}{c}
CH_3 \\
3 \\
CH_3CCOOH \quad ? \\
3 \\
CH_3
\end{array}
$$

A)

$$
\begin{array}{cl}
CH_3 & \\
3 & (1)\ Mg,\ Et_2O \\
CH_3CBr & DDDDDDDDDDDDD4 \\
3 & (2)\ CO_2 \\
CH_3 & (3)\ H^+
\end{array}
$$

B)

$$
\begin{array}{cl}
CH_3 & \\
3 & (1)\ KMnO_4,\ OH^-,heat \\
CH_3CCH_2OH & DDDDDDDDDDDDDDDDDDDDD4 \\
3 & (2)\ H^+ \\
CH_3 &
\end{array}
$$

C)

$$
\begin{array}{cl}
CH_3 & \\
3 & (1)\ CN^- \\
CH_3CBr & DDDDDDDDDDDDDDDDDDD4 \\
3 & (2)\ OH^-,\ H_2O,\ heat \\
CH_3 & (3)\ H_3O^+
\end{array}
$$

D) All of these

* E) Answers A) and B) only

18. Predict the major organic product, A, of the following sequence of reactions:

$$
\begin{array}{ccccc}
CH_3 & & & & \\
3 & 1\ mole\ Cl_2 & SOCl_2 & CH_3OH & \\
CH_3CHCH_2CH_2COOH & DDDDDDDDDDD4 & DDDDD4 & DDDDDD4 & A(C_7H_{13}ClO_2) \\
 & P & & &
\end{array}
$$

* A)

$$
\begin{array}{c}
Cl \\
3 \\
(CH_3)_2CHCH_2CHCOOCH_3
\end{array}
$$

B)

$$
\begin{array}{c}
OCH_3 \\
3 \\
(CH_3)_2CHCH_2CHCOCl
\end{array}
$$

C)

$$
\begin{array}{c}
Cl \\
3 \\
(CH_3)_3CCHCOOCH_3
\end{array}
$$

D)

$$
\begin{array}{c}
OCH_3 \\
3 \\
(CH_3)_2CHCHCH_2COCl
\end{array}
$$

E)

$$
\begin{array}{c}
Cl \\
3 \\
(CH_3)_2CCH_2CH_2COOCH_3
\end{array}
$$

19. Predict the major organic product of the reaction sequence,

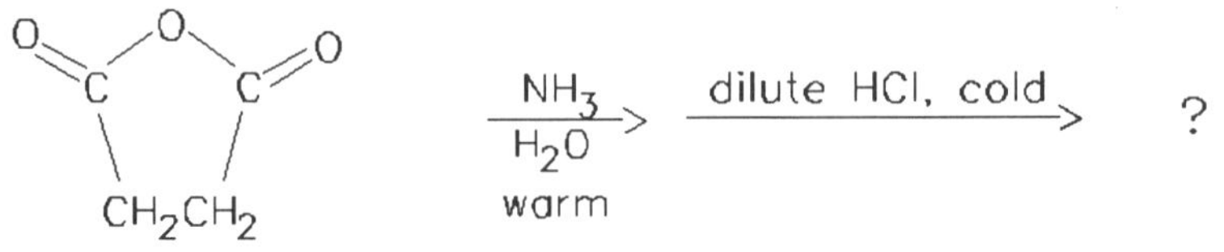

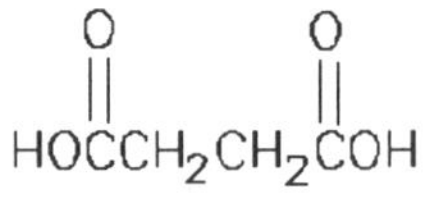

Succinic anhydride (cyclic: O=C–O–C=O with CH_2CH_2) $\xrightarrow[\text{warm}]{NH_3,\ H_2O}$ $\xrightarrow{\text{dilute HCl, cold}}$?

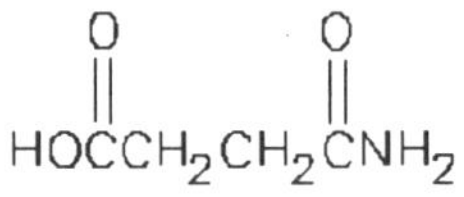

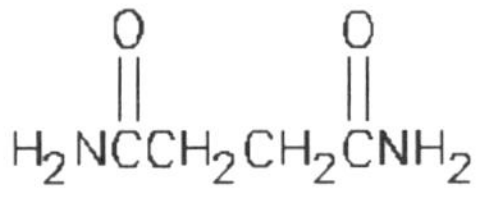

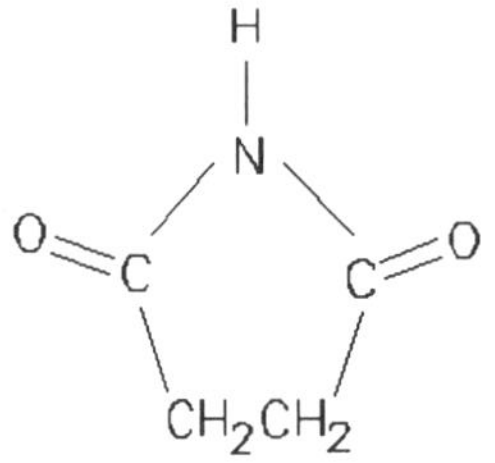

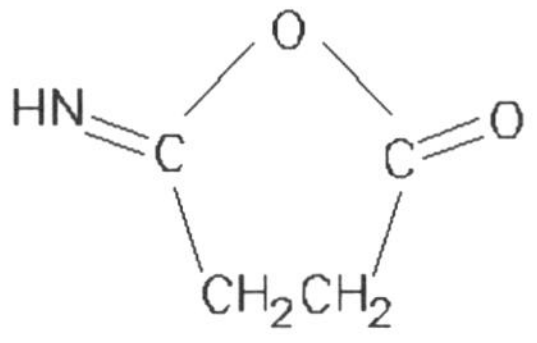

I: $HO\overset{O}{\overset{\|}{C}}CH_2CH_2\overset{O}{\overset{\|}{C}}OH$

II: $HO\overset{O}{\overset{\|}{C}}CH_2CH_2\overset{O}{\overset{\|}{C}}NH_2$

III: $H_2N\overset{O}{\overset{\|}{C}}CH_2CH_2\overset{O}{\overset{\|}{C}}NH_2$

IV: cyclic imide (H–N with two C=O, CH_2CH_2)

V: cyclic (O ring, HN=C and C=O, CH_2CH_2)

A) I * B) II C) III D) IV E) V

20. Which of the reactions listed below would serve as a synthesis of acetophenone,

$C_6H_5\overset{O}{\overset{:}{C}}CH_3$?

A) $C_6H_5\overset{O}{\overset{:}{C}}Cl$ + $(CH_3)_2CuLi$ DDDDDD4

B) C_6H_6 + $CH_3\overset{O}{\overset{:}{C}}Cl$ $\xrightarrow{AlCl_3}$ DDDDDD4

C) $C_6H_5\overset{O}{\overset{:}{C}}OCH_3$ + CH_3MgI DDDDDD4

* D) Two of these

E) All of these

21. What product, B, would you expect from the following sequence of reactions?

$$\text{(R)-2-Butanol} + C_6H_5\overset{O}{\overset{\|}{C}}Cl \xrightarrow{\text{Pyridine}} A \xrightarrow[\text{heat}]{OH^-,\ H_2O} B + C_6H_5\overset{O}{\overset{\|}{C}}O^-$$

* A) (R)-2-butanol
  B) (S)-2-butanol
  C) Equal amounts of A) and B)
  D) 2-Butene (cis and trans)
  E) 1-Butene

22. Which of the following would serve as syntheses of $(CH_3)_3CCO_2H$?

A) $(CH_3)_3\overset{O}{\overset{\|}{C}}CH_3 \xrightarrow[(2)\ H^+]{(1)\ Cl_2/OH^-\ \text{(excess)}}$

B) $(CH_3)_3CBr + CN^- \longrightarrow \text{product} \xrightarrow[\text{heat}]{H_3O^+,\ H_2O}$

C) $(CH_3)_3CBr + Mg \xrightarrow{\text{ether}} \text{product} \xrightarrow[(2)\ H^+]{(1)\ CO_2}$

D) Answers A) and B) only
* E) Answers A) and C) only

23. Which of the following acids would have the smallest value for pK_a?
A) CH_3COOH B) $ClCH_2COOH$ C) $Cl_2CHCOOH$* D) Cl_3CCOOH E) ICH_2COOH

24. What would be the final product?

$$C_6H_5CH_2CONH_2 \xrightarrow[\text{heat}]{P_4O_{10}} \text{product} \xrightarrow[(2)\ H_3O^+]{(1)\ CH_3MgI} ?$$

A) $C_6H_5CH_2CO_2CH_3$
B) $C_6H_5CH_2CH_2NHCH_3$
* C) $C_6H_5CH_2\overset{O}{\overset{\|}{C}}CH_3$
D) $C_6H_5CH_2\underset{CH_3}{\underset{|}{C}H}CN$
E) $C_6H_5CH_2CH{=}NCH_3$

25. Which compound would be most acidic?

```
* A)    O         B) CH3CH2OH   C) C6H5OH     D)     O        E) H2O
        :                                            :
     CH3COH                                       CH3CCH3
```

26. Which compound would be most reactive toward nucleophilic acyl substitution?

```
  A)    O                                D)    O
        :                                      :
     CH3CO⁻                                 CH3CNH2

* B)    O                                E)    O
        :                                      :
     CH3CCl                                 CH3COCH3

  C)    O O
        : :
     CH3COCCH3
```

27. Which of the reactions listed below would serve as a synthesis of benzyl acetate,

```
                          O
                          :
                   C6H5CH2OCCH3 ?
```

```
  A) C6H5CH2OH  +  (CH3CO)2O  DDDDDD4

  B)                         H+
     C6H5CH2OH  +  CH3COOH  DDDDDD4

  C)                  O
                      :
     C6H5CH2OH  +  CH3CCl  DDDDDD4
```

  D) Answers A) and C) only
* E) Answers A), B), and C)

28. Which reaction is a Hunsdiecker reaction?

A) heat
$CH_2(CO_2H)_2$ DDDDDD4 CH_3CO_2H

B) O
heat :
CH_3CCO_2H DDDDDDDDDDD4 CH_3CH
: glass beads
O

C) heat
$CH_3CCH_2CO_2H$ DDDDDD4 CH_3CCH_3
: :
O O

* D) (1) $AgNO_3$
$CH_3CH_2CO_2H$ DDDDDDDDDDDDD4 CH_3CH_2Br
(2) Br_2, CCl_4

E) KOH
CH_3CCH_3 DDDDDD4 $CH_3CO_2{}^-K^+$ + $CHBr_3$
: Br_2
O

29. The relative reactivity of acyl compounds toward nucleophilic substitution is:

A) Amide > ester > acid anhydride > acyl chloride
B) Acyl chloride > ester > acid anhydride > amide
C) Ester > acyl chloride > acid anhydride > amide
* D) Acyl chloride > acid anhydride > ester > amide
E) Acid anhydride > acyl chloride > ester > amide

30. N,N-Dimethylbenzamide can be made from which of the following?

A) O
:
$C_6H_5COC_2H_5$ + $(CH_3)_2NH$ DDDDDD4

B) O
:
C_6H_5CCl + $(CH_3)_2NH$ DDDDDD4

C) O O
: :
$C_6H_5COCC_6H_5$ + $(CH_3)_2NH$ DDDDDD4

D) O
:
$C_6H_5CNH_2$ + CH_3MgCl DDDDDD4

* E) A), B), and C) only

31. The product, G, of the following reaction sequence is:

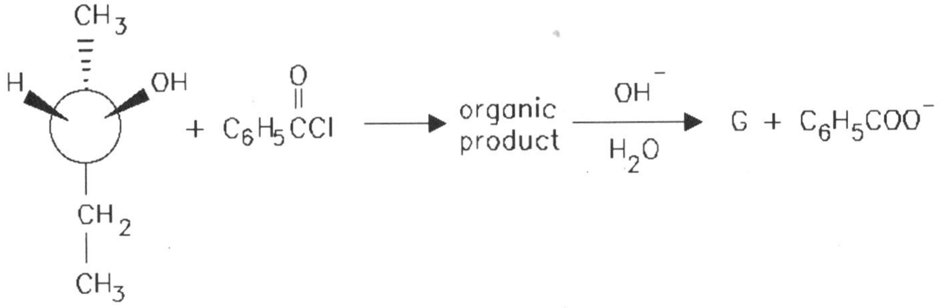

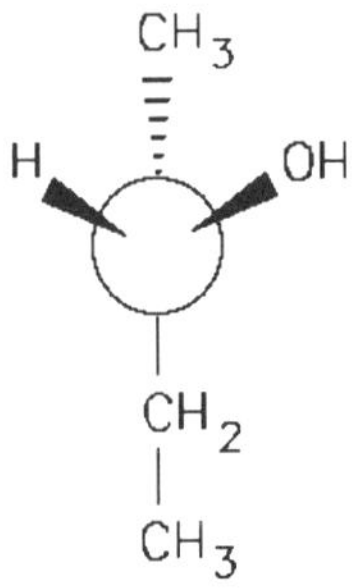

$$\text{(Newman projection: front carbon with H, OH, and CH}_3\text{; rear: CH}_2\text{CH}_3) + C_6H_5\overset{O}{\overset{\|}{C}}Cl \longrightarrow \text{organic product} \xrightarrow[H_2O]{OH^-} G + C_6H_5COO^-$$

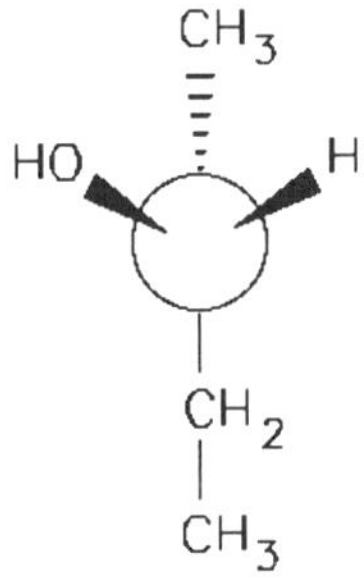

I: Newman projection with CH$_3$ (top), H (left), OH (right); CH$_2$CH$_3$ (bottom)

II: Newman projection with CH$_3$ (top), HO (left), H (right); CH$_2$CH$_3$ (bottom)

III: $CH_2{=}CHCH_2CH_3$

IV: $CH_3CH{=}CHCH_3$

* A) I
B) II
C) Equal amounts of I and II
D) III
E) IV

32. Which of the following compounds would be the strongest acid?
A) $CHF_2CH_2CH_2COOH$
B) $CH_2FCHFCH_2COOH$
C) $CH_3CF_2CH_2COOH$
* D) $CH_3CH_2CF_2COOH$
E) $CH_3CH_2CH_2COOH$

33. The product of the following reaction is:

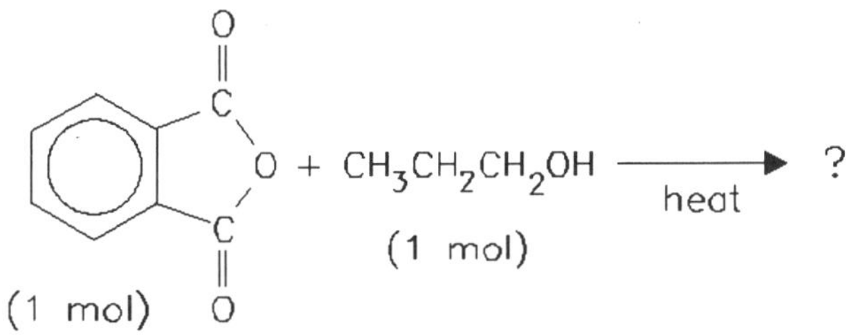

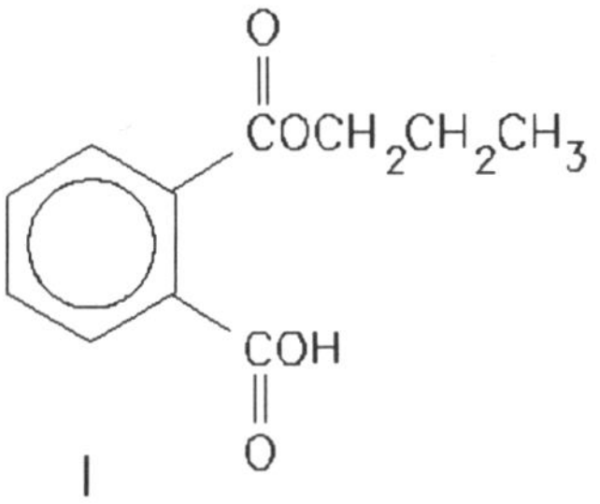

(1 mol) phthalic anhydride + $CH_3CH_2CH_2OH$ (1 mol) $\xrightarrow{\text{heat}}$?

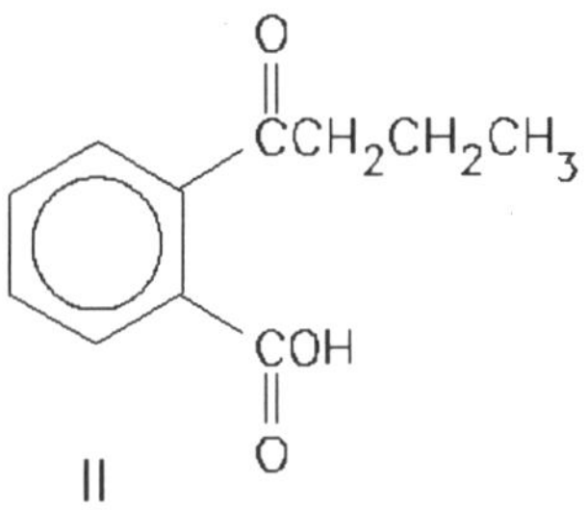

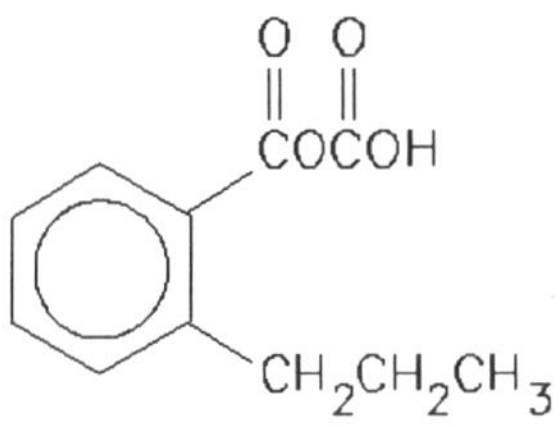

I: $C_6H_4(COOCH_2CH_2CH_3)(COOH)$ (ortho)

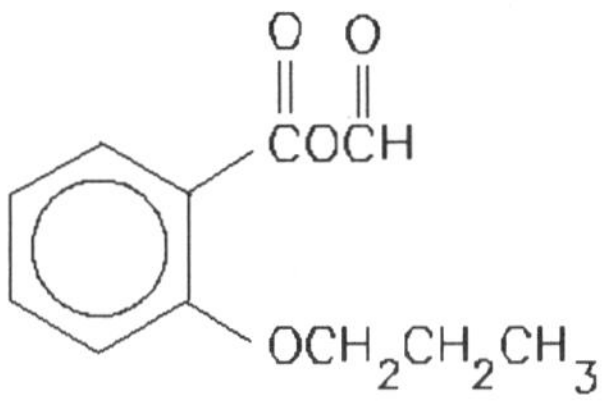

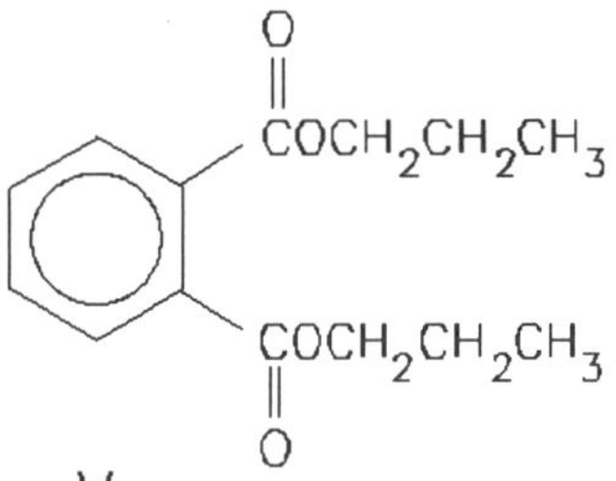

II: $C_6H_4(COCH_2CH_2CH_3)(COOH)$ (ortho)

III: $C_6H_4(COCOOH)(CH_2CH_2CH_3)$ (ortho)

IV: $C_6H_4(COCHO)(OCH_2CH_2CH_3)$ (ortho)

V: $C_6H_4(COOCH_2CH_2CH_3)_2$ (ortho)

* A) I B) II C) III D) IV E) V

34. Choose the reagent(s) that would bring about the following reaction:

$CH_3CH_2CH_2COOH$ DDDDDD4 $CH_3CH_2CH_2CH_2OH$

A) H_2/Ni
B) Li/liq NH_3
C) $LiAlH[OC(CH_3)_3]_3$
D) $NaBH_4$
* E) $LiAlH_4$

35. Which of the following compounds is capable of forming a k-lactone?

* A) $HOCH_2CH_2CH_2CH_2COOH$

B) $HOOCCH_2CH_2CH_2COOH$

C) $CH_3CHCH_2CH_2COOH$
3
OH

D) $CH_3CH_2CHCH_2COOH$
3
OH

E) $CH_3CH_2CH_2CHCOOH$
3
OH

36. Which of the following reactions could be used to synthesize

O
:
$CH_3COCH_2CH_3$?

A) CH_3CH_2OH + CH_3COCl pyridine DDDDDDDD4

B) CH_3CH_2OH + $(CH_3CO)_2O$ DDDDDD4

C) CH_3CH_2OH + CH_3CO_2H H^+ DDDDDD4

D) Answers A) and C) only

* E) Answers A), B), and C)

37. What would be the final product?

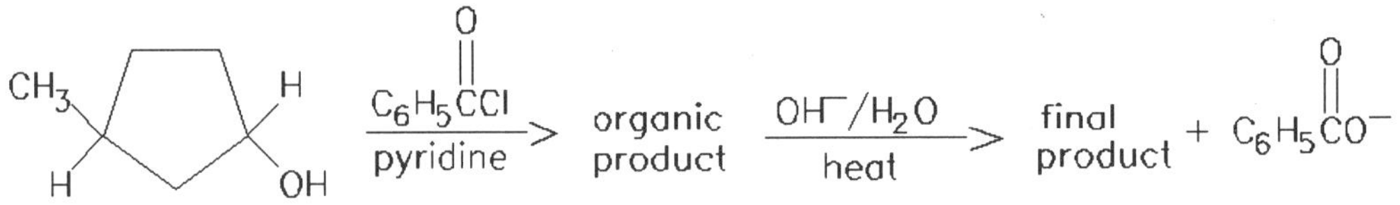

CH_3, H, H, OH (cyclopentane) $\xrightarrow[\text{pyridine}]{C_6H_5\overset{O}{\overset{\|}{C}}Cl}$ organic product $\xrightarrow[\text{heat}]{OH^-/H_2O}$ final product + $C_6H_5\overset{O}{\overset{\|}{C}}O^-$

A) <u>cis</u>-3-Methylcyclopentanol

* B) <u>trans</u>-3-Methylcyclopentanol

C) Equal amounts of A) and B)

D) 3-Methylcyclopentanone

E) None of these

38. Which carboxylic acid would decarboxylate when heated to 100-150xC?

A) O O
: :
$HOCCH_2CH_2COH$

B) O O
: :
CH_3CCH_2COH

C) O O
: :
$HOCCH_2COH$

* D) More than one of these

E) None of these

39. What would be the final organic product of the following reaction?

$$C_6H_5CH_2Cl \xrightarrow{NaCN} \text{organic product} \xrightarrow[(2)\ H_2O]{(1)\ \text{excess } LiAlH_4} ?$$

(arrows printed as DDDDDD4 and DDDDDDDDDDDDDDDDDD4)

A) $C_6H_5CH_2CH_2CO_2H$

* B) $C_6H_5CH_2CH_2NH_2$

C) $C_6H_5CH_2CHCN$
          3
          CH_3

D) $C_6H_5CH_2CHMNH$

E) $C_6H_5CH_2NH_2$

40. What would be the final organic product of the following reaction?

$$(CH_3)_3CBr \xrightarrow[(2)\ CO_2\quad (3)\ H_3O^+]{(1)\ Mg,\ ether} \text{organic product} \xrightarrow[(2)\ H_2O]{(1)\ LiAlH_4} ?$$

(arrows printed as DDDDDDDDDDDDDDDDDD4 and DDDDDDDDDD4)

A) $(CH_3)_3CCO_2H$

B)   O
      :
$(CH_3)_3CCCH_3$

* C) $(CH_3)_3CCH_2OH$

D) $(CH_3)_3COCH_3$

E) $(CH_3)_3CCO_2CH_3$

41. Which of the following is an anhydride?

O O
|| ||
$CH_3CCH_2CCH_3$

I

O O O

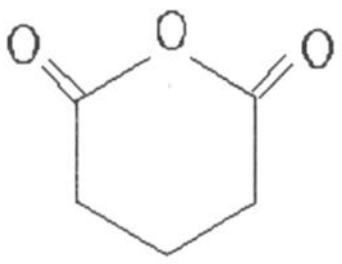

II

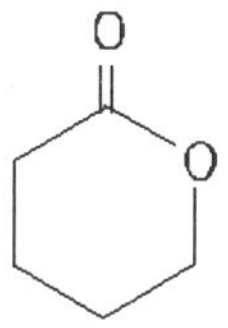

O
O

III

O O
|| ||
COOC

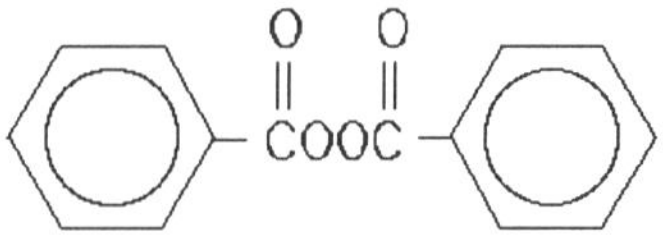

IV

A) I
* B) II
C) III
D) IV
E) More than one of these

42. Identify the product(s) of the following reaction.

```
         O  O
         :  :        heat
CH3CH2CCHCOH    DDDDDD4
          3
         CH3
```

A)
```
     CH3     O
     3      N
     CHDDDC
     3    3
     CDDDDO
    / \
CH3CH2  OH
```

B)
```
     CH3    O
     3     N
     CDDDC    +   H2O
     :   3
     CDDDO
    /
CH3CH2
```

* C)
```
       O
       :
CH3CH2CCH2CH3   +   CO2
```

D)
```
        O
        :
CH3CH2CHCOH   +   CO
        3
       CH3
```

E)
```
           O
           :
2 CH3CH2COH
```

43. In which of the following are the compounds listed in order of decreasing acidity?

A) CH_3CO_2H > CH_3CH_2OH > C_6H_5OH > H_2O
B) C_6H_5OH > CH_3CO_2H > H_2O > CH_3CH_2OH
C) CH_3CO_2H > H_2O > C_6H_5OH > CH_3CH_2OH
D) H_2O > CH_3CO_2H > C_6H_5OH > CH_3CH_2OH
* E) None of the above

44. What would be the final product?

```
          O
          :      PCl5              NH3(xs)              P4O10
CH3CH2COH    DDDDD4   product   DDDDDDD4   product   DDDDDD4   ?
                                                      heat
```

A) $CH_3CH_2CH_2NH_2$

B)
```
       O
       :
CH3CH2CCN
```

C)
```
       O
       :
CH3CH2CNH2
```

* D) CH_3CH_2CN

E) $CH_3CH_2COO^-NH_4^+$

45. Which reagent would best serve as the basis for a simple chemical test to distinguish between $C_6H_5CH{=}CHCOOH$ and $C_6H_5CH{=}CHCH_3$?

A) Concd. H_2SO_4
B) Br_2/CCl_4
C) CrO_3/H_2SO_4
* D) $NaHCO_3/H_2O$
E) $KMnO_4/H_2O$

46. Which of the following would yield (S)-2-butanol?

A) (R)-2-Bromobutane + $CH_3CO_2^-Na^+$ DDDDDD4 product DDDDDDDD4 (OH^-, H_2O, heat)

B) (R)-2-Bromobutane DDDDDDDD4 (OH^-, H_2O, heat)

C) (S)-sec-Butyl acetate DDDDDDDD4 (OH^-, H_2O, heat)

* D) All of these
E) None of these

47. Which of the following would be the strongest acid?

A) Benzoic acid
* B) 4-Nitrobenzoic acid
C) 4-Methylbenzoic acid
D) 4-Methoxybenzoic acid
E) 4-Iodobenzoic acid

48. Which of the following would be the strongest acid?

A) CH_3CO_2H B) ICH_2CO_2H C) $BrCH_2CO_2H$ D) $ClCH_2CO_2H$* E) FCH_2CO_2H

49. Which is the best name for the following compound?

```
       O
       :
CH3CH2COCHCH3
          3
          CH3
```

A) Propyl isopropanoate
B) Propanoyl isopropoxide
* C) Isopropyl propanoate
D) Ethyl isopropyl ketone
E) 1-Methylethyl propionate

50. What is the final product of this sequence of reactions?

$$(CH_3)_3CBr \xrightarrow[(2)\ CO_2\ (3)\ H_3O^+]{(1)\ Mg,\ ether} A \xrightarrow[(2)\ xs\ NH_3]{(1)\ SOCl_2} \text{Final product}$$

A) $(CH_3)_3CSONH_2$ (with C=O)

B) $(CH_3)_3CCN$

* C) $(CH_3)_3CCNH_2$ (with C=O)

D) $(CH_3)_3CCH_2NH_2$

E) $(CH_3)_3CCNHCl$ (with C=O)

51. Which of the following would serve as a reasonable synthesis of ethyl benzoate?

A) $C_6H_5COH + xs\ CH_3CH_2OH \xrightarrow[reflux]{H^+}$ (with C=O)

B) $C_6H_5CCl + CH_3CH_2OH \xrightarrow{base}$ (with C=O)

C) $(C_6H_5C)_2O + CH_3CH_2OH \longrightarrow$ (with C=O)

* D) All of the above
E) None of the above

52. Which of the following will not undergo hydrolysis, whether acid or base is present?

A) CH_3CCl (with C=O)

B) $CH_3CH_2CNH_2$ (with C=O)

C) CH_3COCCH_3 (with two C=O)

* D) $CH_3CCH_2CH_3$ (with C=O)

E) $CH_3COCH_2CH_2CH_3$ (with C=O)

53. On theoretical grounds, one would predict that the (1:1) reaction of acetic propionic anhydride with methyl alcohol would form chiefly:

* A) Methyl acetate + propionic acid
  B) Acetone + propionic acid
  C) Methyl propionate + acetic acid
  D) Methyl acetate + acetic acid
  E) Methyl ethyl ketone + acetic acid

54. Which of the following combinations of reagents would <u>not</u> produce an ester?

A) $CH_3C(=O)OAg$ + CH_3CH_2Br

* B) $CH_3C(=O)ONa$ + CH_3CH_2OH

C) $CH_3C(=O)OH$ + CH_3OH + H^+

D) $(CH_3C(=O))_2O$ + $CH_3CHOHCH_3$

E) $CH_3C(=O)Cl$ + $CH_3CH_2CH_2OH$

55. Which of these compounds could <u>not</u> be formed by nucleophilic attack by an appropriate reagent on acetyl chloride?

A) $CH_3C(=O)NH_2$

B) $CH_3C(=O)OC_2H_5$

* C) $ClCH_2C(=O)Cl$

D) $CH_3C(=O)OH$

E) $CH_3C(=O)OC(=O)CH_3$

56. The IR spectrum of a compound exhibits a broad absorption band at 2500-3000 cm^{-1} and a sharp band at 1710 cm^{-1}. Which of these compounds could it be?

  A) 1-Butanol
  B) Propyl acetate
* C) Butanoic acid
  D) Acetyl chloride
  E) Acetic anhydride

57. An acid chloride is prepared from the related carboxylic acid by reaction with which of these?

A) HCl B) Cl_2 * C) $SOCl_2$ D) HOCl E) $AlCl_3$

58. Which of the following is <u>not</u> a method for preparing butanoic acid?

A) $CH_3CH_2CH_2Br$ + NaCN; then H_3O^+, reflux

B) $CH_3CH_2CH_2MgBr$ + CO_2; then H_3O^+

* C) $CH_3CH_2CH_2OH$ + CO

D) $CH_3CH_2CH_2C(=O)OCH_2CH_3$ + OH^-/H_2O; then H_3O^+

E) $CH_3CH_2CH_2CH_2OH$ + $KMnO_4/OH^-/H_2O$/heat; then H_3O^+

59. Alkaline hydrolysis of an ester involves initial attack by hydroxide ion on the carbonyl carbon. In what order should the five substituents below be arranged to represent the <u>decreasing</u> order of the rates of hydrolysis of ethyl <u>p</u>-substituted benzoates?

A) $-NO_2$ > -H > -Cl > $-CH_3$ > $-OCH_3$
* B) $-NO_2$ > -Cl > -H > $-CH_3$ > $-OCH_3$
C) $-OCH_3$ > $-CH_3$ > -Cl > -H > $-NO_2$
D) -Cl > $-NO_2$ > -H > $-OCH_3$ > $-CH_3$
E) -H > -Cl > $-CH_3$ > $-OCH_3$ > $-NO_2$

60. Intramolecular dehydration to form a cyclic monoester is most likely to occur when which of the following is heated with acid?

A) $CH_3CH_2CH_2CHOHCO_2H$
B) $CH_3CH_2CHOHCH_2CO_2H$
C) $CH_3CH_2CH_2CH_2CO_2H$
* D) $CH_3CHOHCH_2CH_2CO_2H$
E) $HO_2CCH_2CH_2CO_2H$

61. Which of these combinations will <u>not</u> produce benzoic acid?

A) $C_6H_5CH_2OH$ + $KMnO_4/OH^-/H_2O$, heat; then H_3O^+

B) $C_6H_5CH_3$ + $KMnO_4/OH^-/H_2O$, heat; then H_3O^+

* C) C_6H_6 + CO_2, high pressure

D) $C_6H_5C(=O)CH_3$ + Cl_2/OH^-; then H_3O^+

E) $C_6H_5C(=O)Cl$ + OH^-/H_2O; then H_3O^+

62. When silver isobutyrate reacts with bromine in the Hunsdiecker reaction, this alkyl halide is produced.

A) Isobutyl bromide
* B) Isopropyl bromide
C) Propyl bromide
D) <u>sec</u>-Butyl bromide
E) <u>tert</u>-Butyl bromide

63. Reasoning by analogy, one would predict that the reaction of carbon disulfide with sec-butylmagnesium bromide should yield which of the following (after acidification)?

* A)
```
          S
          :
CH3CH2CHCSH
        3
        CH3
```
B)
```
        SH
        3
CH3CH2CHCH3
```
C)
```
          S
          :
CH3CHCH2CSH
    3
    CH3
```
D)
```
            H
            3
CH3CH2CHCMS
        3
        CH3
```
E)
```
  H3C S
    3 :
CH3C-CSH
    3
  H3C
```

64. Fischer esterification of a carboxylic acid (1.00 mol acid and 2.00 mol alcohol) produces 0.100 mol ester at equilibrium. What is the equilibrium constant for this reaction?

* A) 0.00585 B) 0.0585 C) 0.0500 D) 0.00500 E) 0.500

65. "- And k-hydroxy acids can be esterified intramolecularly to form compounds known as which of these?

A) Anhydrides
B) Cycloalkenes
* C) Lactones
D) Lactams
E) Cyclic ketones

66.
```
                         O
                         :
A correct name for (CH3)2CHCH2COCH2CHCH2CH3 is:
                                  3
                                  CH3
```

A) 2-Methylbutyl 2-methylbutanoate
* B) 2-Methylbutyl 3-methylbutanoate
C) 3-Methylbutyl isovalerate
D) Isopentyl isovalerate
E) Isopentyl isobutyrate

67. What is the product of the reaction of 1-propanol with phenyl isocyanate, C_6H_5NMCMO?

* A)
```
        O
        :
C6H5NHCOCH2CH2CH3
```

  B)
```
     CO2H
     3
C6H5NCH2CH2CH3
```

  C)
```
     CHO
     3
C6H5NOCH2CH2CH3
```

  D)
```
        O
        :
C6H5NHCOCH(CH3)2
```

  E)
```
        O
        :
C6H5NHCCH2CH2CH3
```

68. Which of the following statements concerning nitriles is incorrect?
  A) Nitriles can be hydrolyzed to carboxylic acids.
  B) Nitriles can be formed from (many) alkyl halides by nucleophilic substitution by cyanide ion.
  C) Nitriles can be reduced with excess lithium aluminum hydride to primary amines, RNH_2.
* D) Nitriles react with Grignard reagents to form tertiary alcohols.
  E) Nitriles can be made by the dehydration of amides.

69.

What is the IUPAC name for
```
              Cl
              3
CH3CH2CH2CHCCl   ?
               :
               O
```

  A) ! -Chlorovaleryl chloride
* B) 2-Chloropentanoyl chloride
  C) 1-Chloropentanoyl chloride
  D) 1,2-Dichloropentanal
  E) 1-Chloro-1-butanecarbonyl chloride

70. What would be the final product of this reaction sequence?

C_6H_5COH (with C=O) $\xrightarrow{PCl_3}$ organic product $\xrightarrow{NH_3}$ organic product $\xrightarrow{P_4O_{10}}$ final product

I: C_6H_5–C≡N

II: C_6H_5–$\overset{O}{\overset{\|}{C}}NH_2$

III: C_6H_5–NH_2

IV: C_6H_5–$\overset{O}{\overset{\|}{C}}O^- \ NH_4^+$

V: C_6H_5–$NH\overset{O}{\overset{\|}{C}}NH$–$C_6H_5$

* A) I B) II C) III D) IV E) V

71. What is the ultimate product of this sequence of reactions?

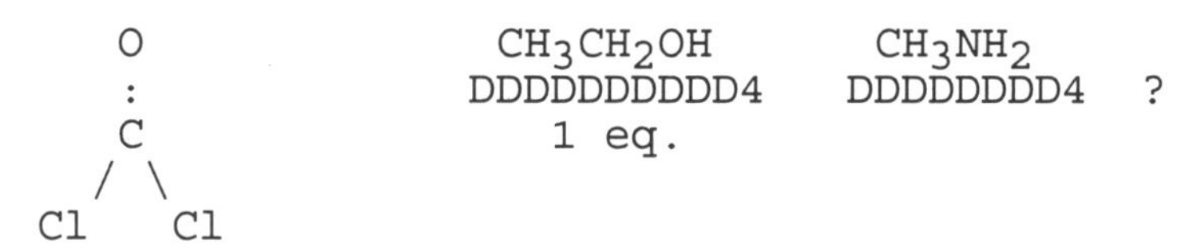

* A) $CH_3CH_2O\overset{O}{C}NHCH_3$

B) $CH_3NH\overset{O}{C}NHCH_3$

C) $Cl\overset{O}{C}NHCH_3$

D) $CH_3CH_2O\overset{O}{C}OCH_2CH_3$

E) $CH_3CH_2O\overset{O}{C}Cl$

72. What is the product of this reaction?

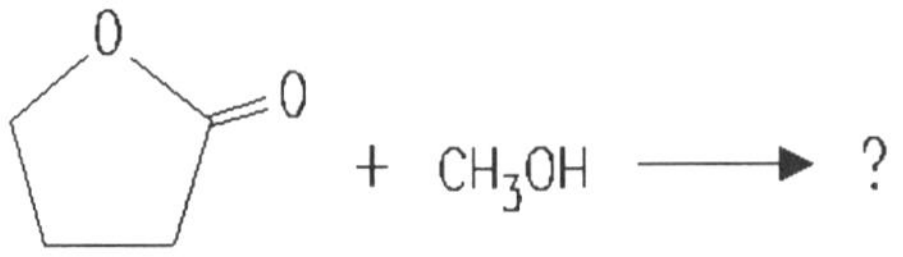

[γ-butyrolactone] + CH_3OH ⟶ ?

$CH_3OCH_2CH_2CH_2CO_2H$ (I) $HOCH_2CH_2CH_2CO_2CH_3$ (II) $CH_3OCH_2CH_2CH_2CO_2CH_3$ (III)

$CH_3O_2CCH_2CH_2CH_2CO_2CH_3$ (IV) [tetrahydrofuran ring with OH and OCH_3 on C-2] (V)

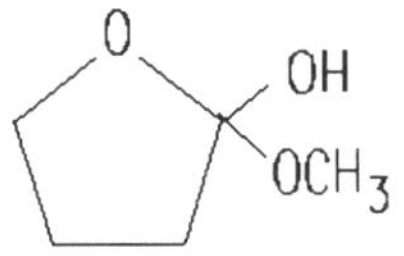

A) I * B) II C) III D) IV E) V

73. What is the product of the reaction of propanamide with methylmagnesium bromide (1 eq.)?

A) $CH_3CH_2C(=O)NHCH_3$

B) $CH_3CH_2C(=O)N(CH_3)_2$

C) $CH_3CH_2C(=O)CH_3$

* D) CH_4 + $CH_3CH_2C(=O)NHMgBr$

E) $CH_3CH_2C(OMgBr)(CH_3)NH_2$

74. In which of these species are all the carbon-oxygen bonds of equal length?

A) Diethyl carbonate
B) Methyl butanoate
* C) Lithium acetate
D) Propionic anhydride
E) Pentanoic acid

75. Neopentyl isobutyrate has the IUPAC name:
  A) 1,1-Dimethylpropyl 1-methylpropanoate
  B) 1,1-Dimethylpropyl 2-methylpropanoate
  C) 2,2-Dimethylpropyl 1-methylpropanoate
* D) 2,2-Dimethylpropyl 2-methylpropanoate
  E) 2-Methylpropyl 3,3-dimethylbutanoate

76. What is the final product of this sequence of reactions?

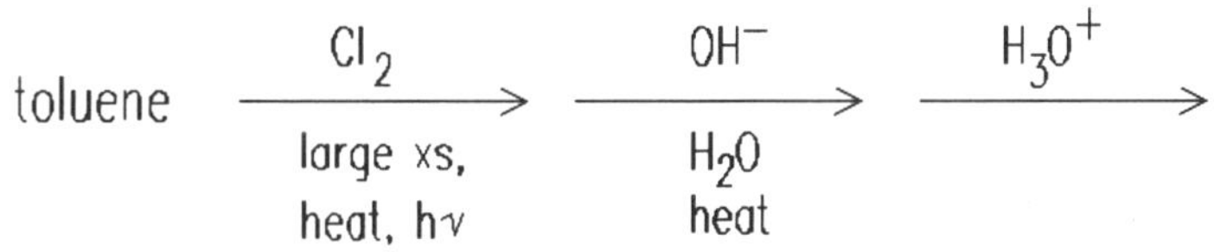

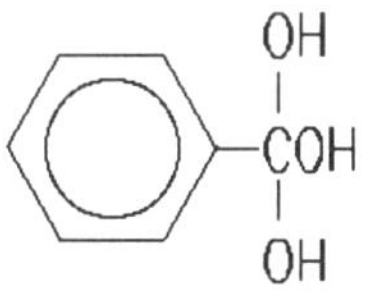

$$\text{toluene} \xrightarrow[\text{large xs, heat, } h\nu]{Cl_2} \xrightarrow[H_2O \text{ heat}]{OH^-} \xrightarrow{H_3O^+}$$

I: OH, COH, OH

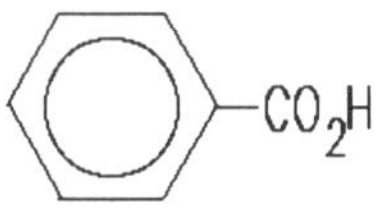

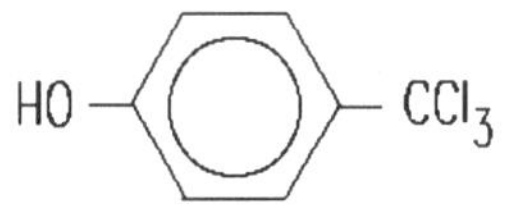

II: CO_2H

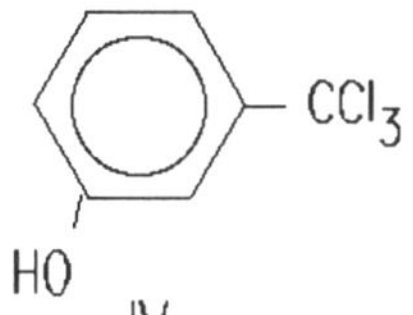

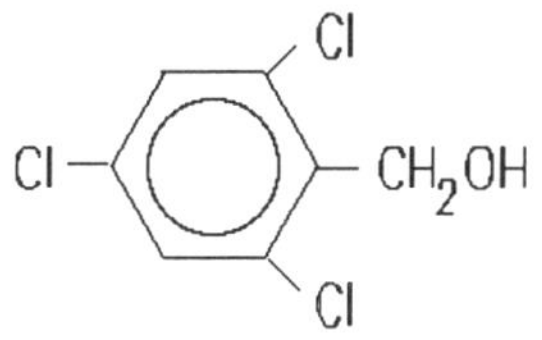

III: HO, CCl_3

IV: CCl_3, HO

V: Cl, Cl, CH_2OH, Cl

A) I * B) II C) III D) IV E) V

77. A 0.2505 g sample of an organic acid is titrated to the stoichiometric endpoint with 20.10 mL of 0.0750 M NaOH. Which of these is a possible structure for the acid?
  A) $C_6H_5CO_2H$
  B) p-$CH_3C_6H_4CO_2H$
  C) m-$ClC_6H_4CO_2H$
* D) o-$NO_2C_6H_4CO_2H$
  E) p-$BrC_6H_4CO_2H$

Chapter 19: Amines

1. Which reagent would serve as the basis for a simple chemical test that would distinguish between the pair of compounds listed below?

CH_3 — NH_2 and — $NHCH_3$

A) $AgNO_3$ in H_2O
B) Dilute $NaHCO_3$
C) Dilute NaOH
* D) $C_6H_5SO_2Cl/OH^-$, then H_3O^+
E) Dilute HCl

2. Which reagent could be used to separate a mixture of the pair of compounds listed below?

A) $AgNO_3$ in H_2O
B) Dilute NaOH
C) Dilute $NaHCO_3$
D) $Ag(NH_3)_2OH$
* E) Dilute HCl

3. What is the product, A, of the following reaction sequence?

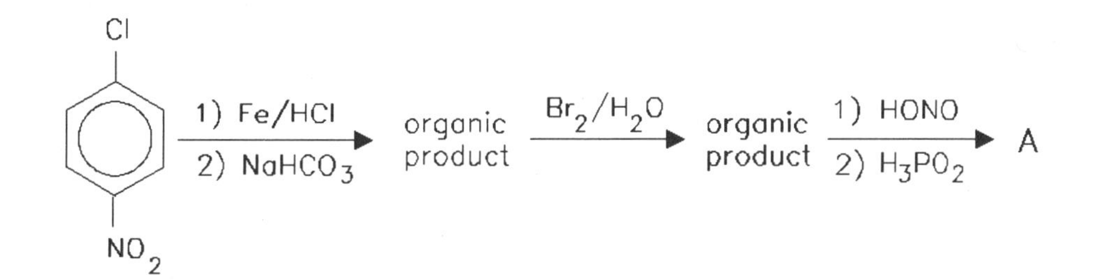

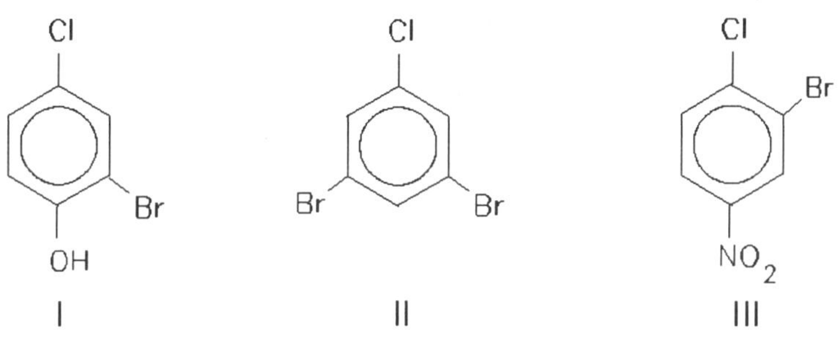

Cl, Br, Br, PO_2H — IV

Cl, PO_2H, Br, NO — V

A) I * B) II C) III D) IV E) V

4. Which of the following reactions would yield $C_6H_5CH_2NH_2$?

A) $C_6H_5CH_2Br$ + CN^- DDDDDD4 DDDDDD4 ($LiAlH_4$)

B) C_6H_5CHO + NH_3 DDDDDDD4 (H_2, Ni)

C) $C_6H_5CH_2Br$ DDDDDDDDDD4 ((1) NaN_3 (2) $LiAlH_4$)

D) All of these

* E) Answers B) and C) only

5. Which of the following reaction sequences would yield aniline?

A) benzene $\xrightarrow[H_2SO_4]{HNO_3}$ organic product $\xrightarrow[(2)\ NaOH/H_2O]{(1)\ Fe/HCl}$

B) C_6H_5CCl (with O above the C) $\xrightarrow{NH_3}$ organic product $\xrightarrow{Br_2,\ OH^-}$

C) $C_6H_5CH_2Br$ $\xrightarrow{(xs)NH_3}$ organic product $\xrightarrow[(2)\ H_3PO_2]{(1)\ HONO}$

D) All of the above
* E) Answers A) and B) only

6. The product, C, of the following reaction sequence is what compound?

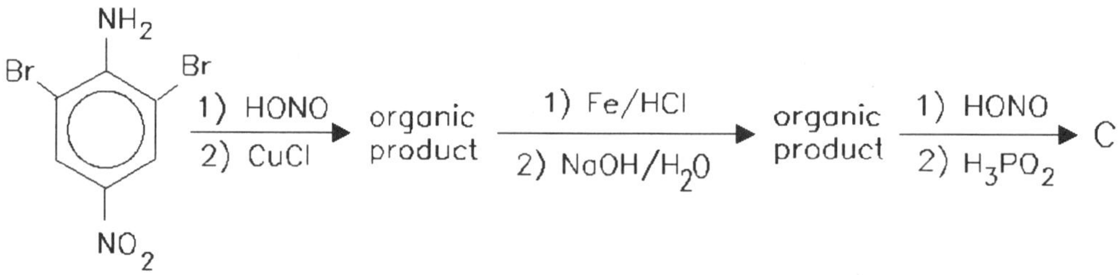

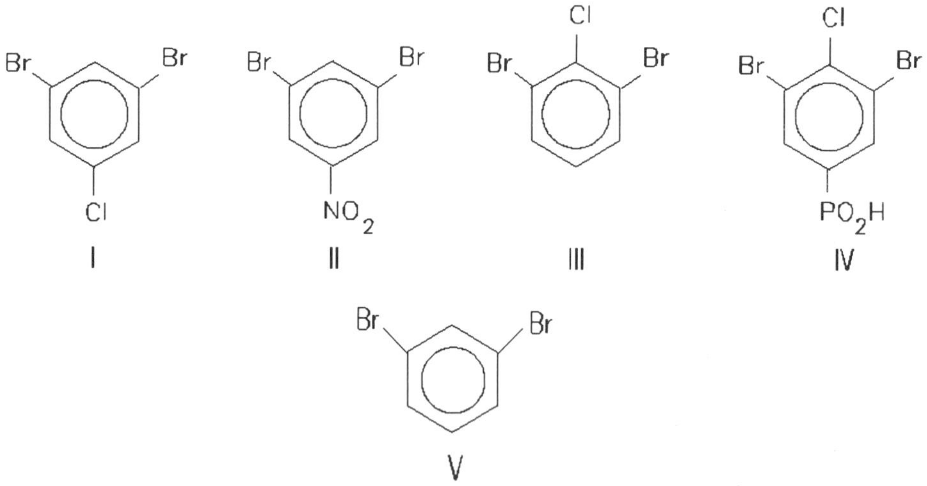

NH$_2$ Br Br NO$_2$ $\xrightarrow[2)\ CuCl]{1)\ HONO}$ organic product $\xrightarrow[2)\ NaOH/H_2O]{1)\ Fe/HCl}$ organic product $\xrightarrow[2)\ H_3PO_2]{1)\ HONO}$ C

I: Br, Br, Cl

II: Br, Br, NO$_2$

III: Cl, Br, Br

IV: Cl, Br, Br, PO$_2$H

V: Br, Br

A) I B) II * C) III D) IV E) V

7. Which of the following compounds would be the strongest base?

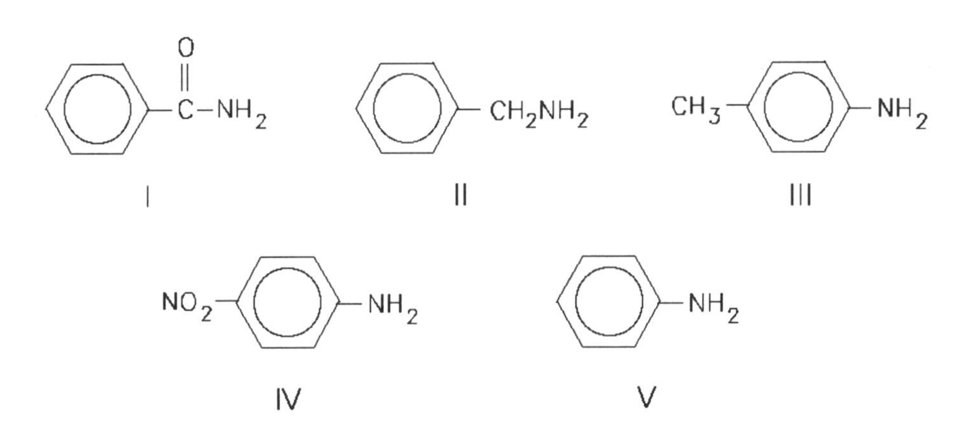

A) I * B) II C) III D) IV E) V

8. Which of the following might be used to synthesize m-bromoaniline?

A)
HNO3 (1) Fe/HCl Br2
C_6H_6 DDDDDD4 DDDDDDDDDDDDDDD4 DDDDDD4
H_2SO_4 (2) $NaOH/H_2O$ H_2O

B)
Br_2 HNO_3 H_2
C_6H_6 DDDDDD4 DDDDDD4 DDDDD4
$FeBr_3$ H_2SO_4 Pt

* C)
Br_2 $SOCl_2$ NH_3 Br_2
C_6H_5COOH DDDDDD4 DDDDDD4 DDDDDD4 DDDDDD4
$FeBr_3$ NaOH

D) Answers A) and B)
E) Answers B) and C)

9. Which reagent would serve as the basis for a simple chemical test that would distinguish between compounds listed below?

CH_3–C_6H_4–NH_2 and C_6H_5–CH_2NH_2

A) $AgNO_3$ in H_2O
* B) HONO, then a-naphthol
C) Dilute NaOH
D) $C_6H_5SO_2Cl$ and OH^- in H_2O
E) Dilute HCl

10. The best synthesis of 3,5-dibromotoluene would be?

A) Toluene, Br_2, Fe and heat
* B) p-$CH_3C_6H_4NH_2$, Br_2, H_2O; then HONO; then H_3PO_2
C) Toluene, fuming HNO_3, H_2SO_4; then NH_3, H_2S; then HONO; then CuBr
D) m-Dibromobenzene, CH_3Cl, $AlCl_3$, heat
E) m-Bromotoluene, HNO_3, H_2SO_4; then NH_3, H_2S; then HONO; then CuBr

11. What would be the product, L, of the following reaction?

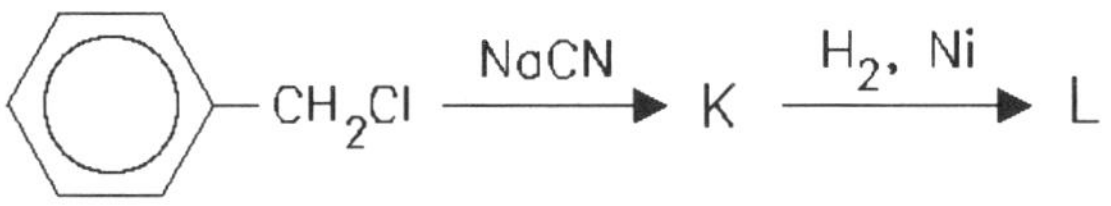

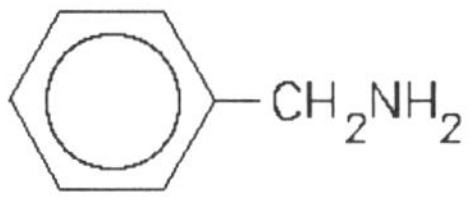

$$C_6H_5CH_2Cl \xrightarrow{NaCN} K \xrightarrow{H_2,\ Ni} L$$

I: $C_6H_5CH_2NH_2$

II: $C_6H_5CH_2NHCH_2C_6H_5$

III: $C_6H_5CH_2CH_2CN$

IV: $C_6H_5CH_2CH_2C_6H_5$

V: $C_6H_5CH_2CH_2NH_2$

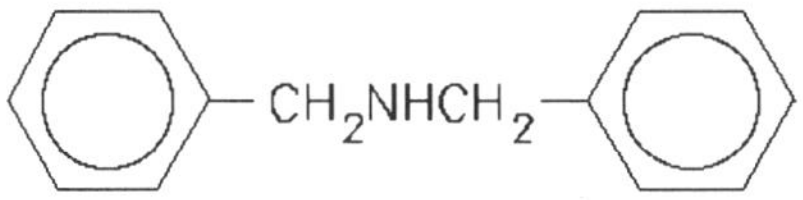

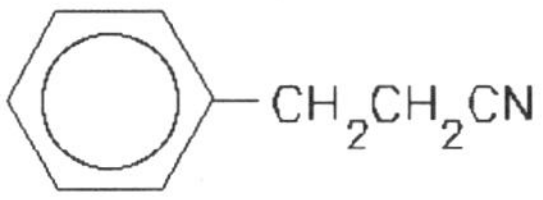

A) I B) II C) III D) IV * E) V

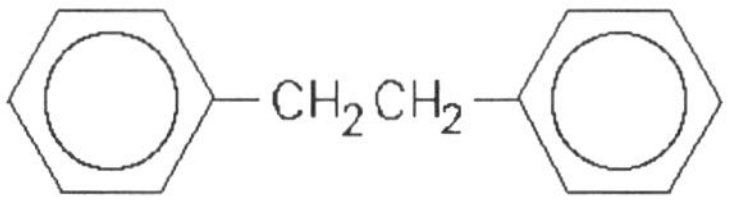

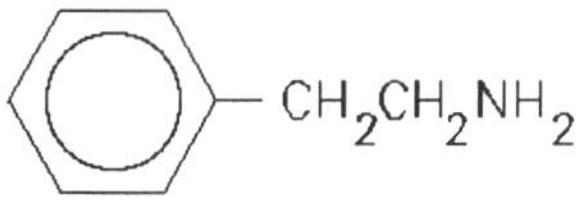

12. What would be the product, G, of the following reaction sequence?

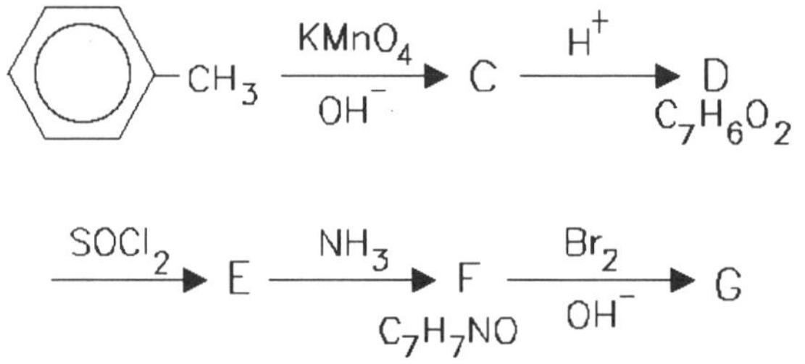

$$C_6H_5CH_3 \xrightarrow[OH^-]{KMnO_4} C \xrightarrow{H^+} \underset{C_7H_6O_2}{D} \xrightarrow{SOCl_2} E \xrightarrow{NH_3} \underset{C_7H_7NO}{F} \xrightarrow[OH^-]{Br_2} G$$

A) $C_6H_5CONH_2$
B) $C_6H_5CH_2NH_2$
C) p-$CH_3C_6H_4SO_2NH_2$
D) p-$CH_3C_6H_4NH_2$
* E) $C_6H_5NH_2$

13. What would be the product, B, of the following reaction sequence?

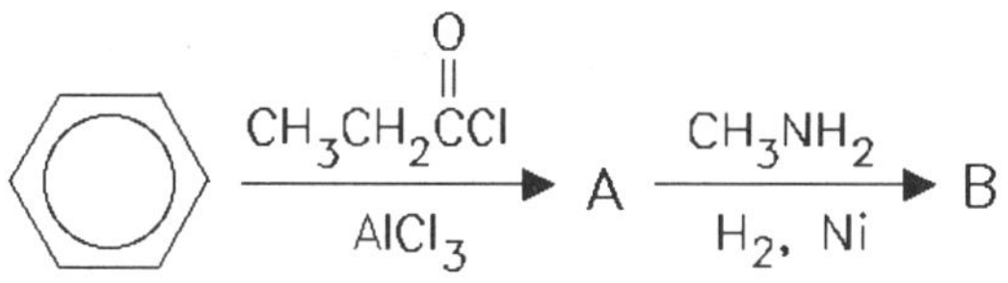

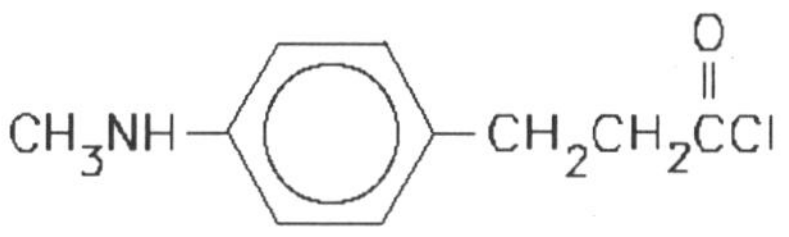

$$C_6H_6 \xrightarrow[AlCl_3]{CH_3CH_2\overset{O}{\overset{\|}{C}}Cl} A \xrightarrow[H_2,\ Ni]{CH_3NH_2} B$$

I: $CH_3NH-C_6H_4-CH_2CH_2\overset{O}{\overset{\|}{C}}Cl$

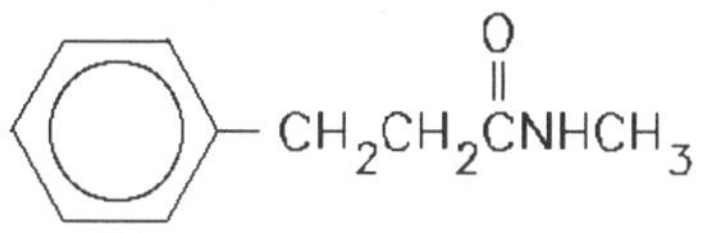

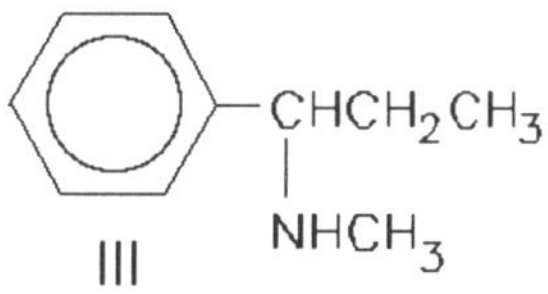

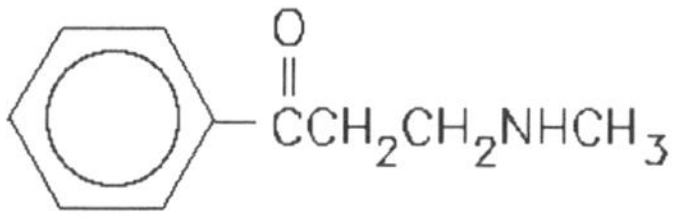

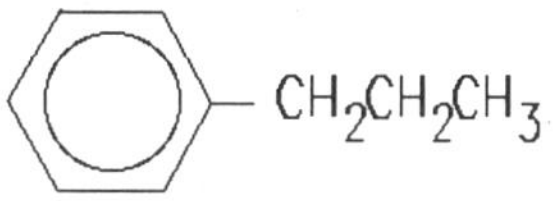

II: $C_6H_5-CH_2CH_2\overset{O}{\overset{\|}{C}}NHCH_3$

III: $C_6H_5-CH(NHCH_3)CH_2CH_3$

IV: $C_6H_5-\overset{O}{\overset{\|}{C}}CH_2CH_2NHCH_3$

V: $C_6H_5-CH_2CH_2CH_3$

A) I B) II * C) III D) IV E) V

14. Which of the following compounds would be the strongest base?

A)
```
    CH3
    3
CH3CNH2
    3
    CH3
```

B)
```
    CH3
    3
CH3CHNHCH3
```

C)
```
CH3NCH3
    3
    CH3
```

* D)
```
    CH3
    3
CH3N+CH3 OH-
    3
    CH3
```

E)
```
    O   CH3
    :  /
CH3CN
       \
        CH3
```

15. Which of the following is a tertiary amine?

A) $CH_3CH_2NH_2$

B) $CH_3CH_2NHCH_2CH_3$

* C) $CH_3CH_2NCH_2CH_3$
3
CH_2CH_3

D) $(CH_3CH_2)_4N^+ \ OH^-$

E) CH_3
3
CH_3CNH_2
3
CH_3

16. Compound W has the molecular formula $C_{11}H_{17}N$. Treatment of W with benzenesulfonyl chloride in base gives no reaction. Acidification of the resulting mixture gives a clear solution. The ^1H NMR spectrum of W consists of:

triplet, k 1.0
quartet, k 2.5
singlet, k 3.6 (2H)
multiplet, k 7.3 (5H)

The most likely structure for W is:

A) $C_6H_5CH_2CH_2CH_2NHCH_2CH_3$

* B) $C_6H_5CH_2NCH_2CH_3$
3
CH_2CH_3

C) $C_6H_5CH_2CH_2NCH_2CH_3$
3
CH_3

D) $C_6H_5CH_2CH_2CH_2N(CH_3)_2$

E) p-$CH_3C_6H_4N(CH_2CH_3)_2$

17. The best synthesis of m-dibromobenzene would be?

A) Benzene, Br_2, $FeBr_3$, heat
B) Aniline, Br_2, H_2O; then HONO; then CuBr
* C) Nitrobenzene, HNO_3, H_2SO_4, heat; then Fe/HCl (excess); then 2 HONO, then 2 CuBr
D) Bromobenzene, HNO_3, H_2SO_4; then Fe, HCl, C_2H_5OH, reflux; then HONO; then CuBr
E) Answers C) and D)

18. The best synthesis of m-fluoronitrobenzene would be?

* A) Nitrobenzene, fuming HNO_3, H_2SO_4, heat; then NH_3, H_2S; then HONO; then HBF_4; then heat
B) Aniline, F_2, heat
C) Fluorobenzene, HNO_3, H_2SO_4, heat
D) o-Nitroacetanilide, NH_3, H_2S; then Br_2, OH^-; then HF
E) m-$C_6H_4(NH_2)_2$, HONO; then $CuNO_2$; then HBF_4

19. The product, B, of the following reaction sequence,

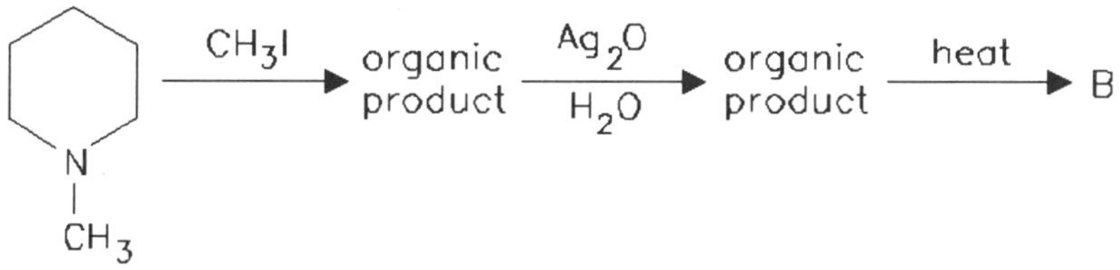

$$\text{N-methylpiperidine} \xrightarrow{CH_3I} \text{organic product} \xrightarrow[H_2O]{Ag_2O} \text{organic product} \xrightarrow{heat} B$$

$CH_2{=}CHCH_2CH_2CH_2N(CH_3)_2$ (I); N-methyl-3-hydroxypiperidine (II); N-methyl-1,2,3,4-tetrahydropyridine (III); N-methyl-1,2,3,6-tetrahydropyridine (IV)

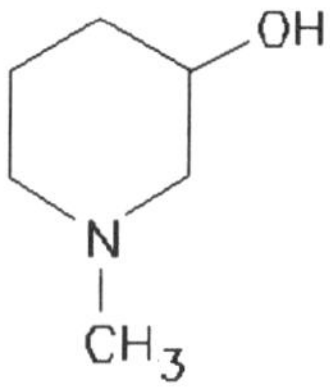

I II 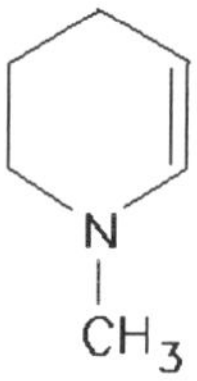 III 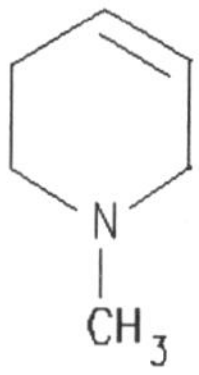 IV

would be:

* A) I
B) II
C) III
D) IV
E) Answers A) and C) only

20. Which is not an intermediate in the Hofmann degradation reaction?

$$RCONH_2 \xrightarrow[Br_2]{KOH} RNH_2$$

A) RNMCMO
B) RCNHBr (with C=O)
C) $RC\overset{-}{N}Br$ (with C=O)
D) RNCOH (N–H, with C=O)
* E) RCNBr (with N–Br and C=O)

21. Which product could not be formed during the following reaction?

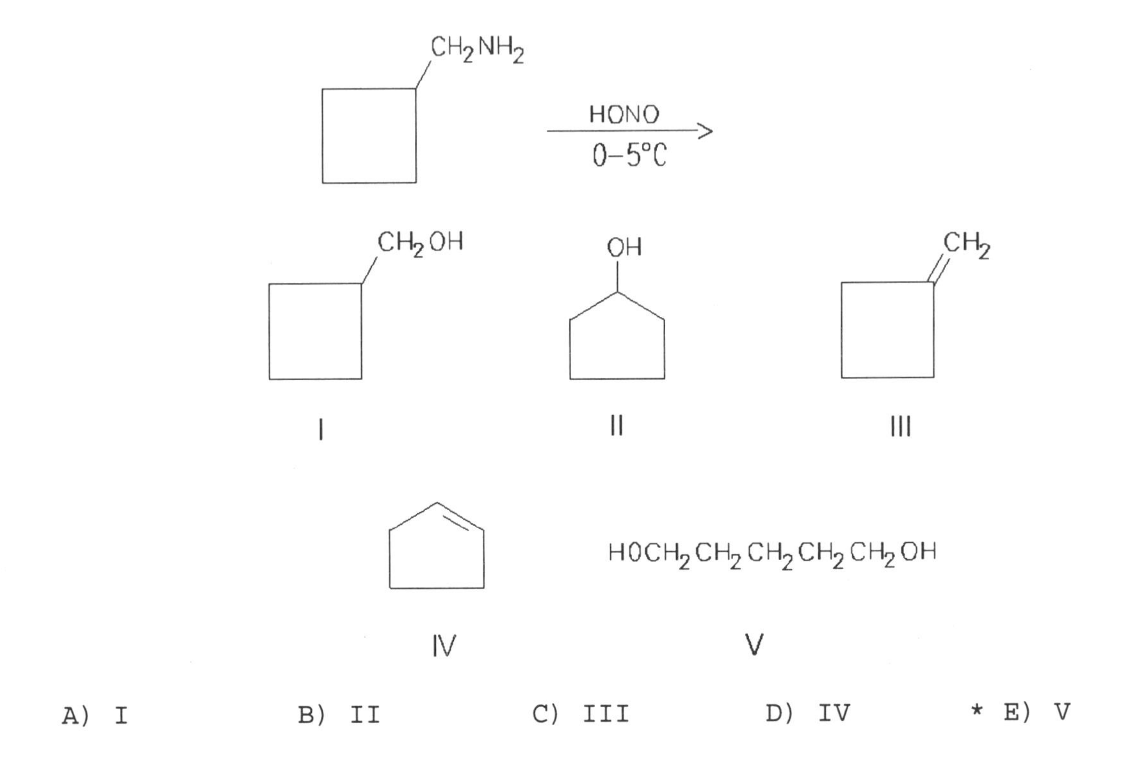

A) I B) II C) III D) IV * E) V

22. What type of amine is <u>tert</u>-butylamine?

* A) Primary
B) Secondary
C) Tertiary
D) Quaternary
E) None of these

23. The product, A, of the following reaction sequence,

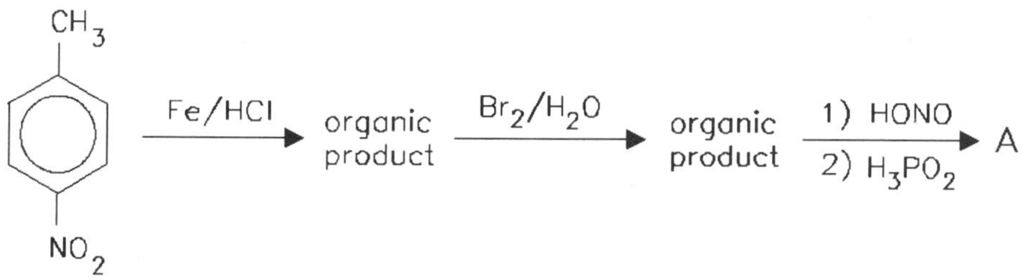

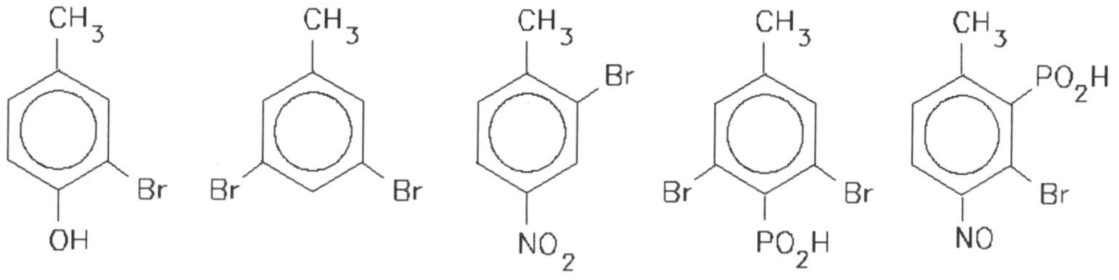

CH_3 / NO_2 (benzene) $\xrightarrow{Fe/HCl}$ organic product $\xrightarrow{Br_2/H_2O}$ organic product $\xrightarrow[2)\ H_3PO_2]{1)\ HONO}$ A

I: CH_3, Br, OH

II: CH_3, Br, Br

III: CH_3, Br, NO_2

IV: CH_3, Br, Br, PO_2H

V: CH_3, PO_2H, Br, NO

would be:

A) I * B) II C) III D) IV E) V

24. Which is the best preparation of benzonitrile, C_6H_5CN, from benzene?

* A) HNO_3; then Fe/HCl; then HONO; then CuCN
 B) CH_3I, $AlCl_3$; then hot $KMnO_4$; then H^+; then $SOCl_2$; then NH_3; then NaOH + Cl_2
 C) CH_3I, $AlCl_3$; then Br_2, h0; then KCN
 D) Br_2, Fe; then KCN
 E) CH_3I, $AlCl_3$; then Br_2 (2 eq.), h0; then hot NaOH; then HCN

25. Identify the compound (Z) formed by the following reaction.

$C_6H_5NH_2$ + $NaNO_2$, dil. H_2SO_4 $\xrightarrow[H_2O]{0-5xC}$ DDDDDDD4 $\xrightarrow[Cu(NO_3)_2,\ H_2O]{Cu_2O}$ DDDDDDDDDD4 Z

 A) $C_6H_5NH_2$
 B) C_6H_5Cl
* C) C_6H_5OH
 D) C_6H_6
 E) None of these

26. Which of the following can be used to prepare propylamine (pure)?

I (1 mol) 1-bromopropane + (1 mol) ammonia ⟶

II (1 mol) 1-bromopropane + (1 mol) NaN_3 ⟶ $\xrightarrow{LiAlH_4}$

III [phthalimide] $\xrightarrow{KOH}$ $\xrightarrow{CH_3CH_2CH_2Br}$ $\xrightarrow{NH_2NH_2}$

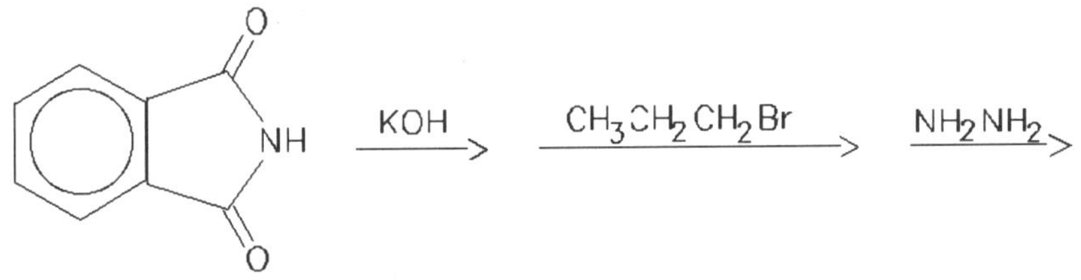

A) I
B) II
C) III
D) I and II
* E) II and III

27. Which is the predominant product (K) of the following reaction?

m-dinitrobenzene $\xrightarrow{H_2S/NH_3}$ K

A) m-$HSC_6H_4NH_2$
B) m-$C_6H_4(NH_2)_2$
C) m-$HOC_6H_4NH_2$
* D) m-$H_2NC_6H_4NO_2$
E) $C_6H_5NH_2$

28. Which is the best method to prepare $C_6H_5CH(NH_2)CH_3$?

* A) $C_6H_5C(=O)CH_3$ + xs NH_3 $\xrightarrow{H_2/Ni}$

B) $C_6H_5CHBrCH_3$ + NH_3 ⟶

C) $C_6H_5CH(=O)$ + CH_3NH_2 $\xrightarrow{H_2/Ni}$

D) $C_6H_5NO_2$ + $CH_3CH_2NH_2$ + HONO ⟶

E) $C_6H_5CH_2CH_3$ + HNO_3 $\xrightarrow{h\nu}$ $\xrightarrow{H_2/Ni}$

29.

H
3
Identify the best method(s) to prepare $C_6H_5CH_2CH_2NCH_3$.

A) $C_6H_5CH_2CH_2NH_2$(1 mol) + CH_3I(1 mol)

B)
O
:
C_6H_5CH + $(CH_3)_2NH$ $\xrightarrow{H_2/Ni}$

* C)
O
:
$C_6H_5CH_2CH$ + CH_3NH_2 $\xrightarrow{H_2/Ni}$

D) A) and B)
E) B) and C)

30. Which is the major product of the following reaction?

$C_6H_5NH_2$ + $NaNO_2$ + HCl $\xrightarrow{0\text{-}5xC}$ $\xrightarrow[NaOH]{\underline{p}\text{-}CH_3C_6H_4OH}$

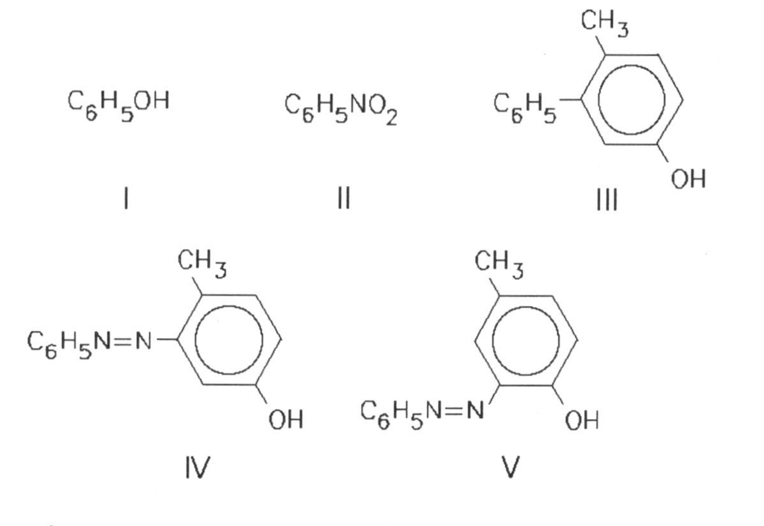

A) I B) II C) III D) IV * E) V

31. What is the final product?

$$\text{p-}CH_3C_6H_4NO_2 \xrightarrow[\text{2) } OH^-]{\text{1) Fe/HCl}} C_7H_9N \xrightarrow[H_2O]{Br_2\text{(excess)}} \text{product}$$

$$\xrightarrow[\text{2) } H_2O,\ Cu_2O,\ Cu(NO_3)_2]{\text{1) } NaNO_2,\ H_2SO_4,\ H_2O,\ 0-5°C} \text{final product}$$

A) 2-Bromo-4-methylaniline
B) 2,6-Dibromo-4-methylaniline
* C) 2,6-Dibromo-4-methylphenol
D) 2,4-Dibromophenol
E) 3,5-Dibromotoluene

32. Which reagent will distinguish between $C_6H_5NH_2$ and $(C_6H_5)_2NH$?
A) HCl (aq)
B) NaOH
* C) $C_6H_5SO_2Cl/OH^-$, then H_3O^+
D) Br_2/CCl_4
E) $KMnO_4$

33. Which reagent will distinguish between propylamine and allylamine?
A) HONO
B) $C_6H_5SO_2Cl/OH^-$, then H_3O^+
C) NaOH
D) HCl
* E) Br_2/CCl_4

34. Which would be the weakest base?
A) p-Methylaniline
B) p-Methoxyaniline
C) Hexylamine
* D) p-Nitroaniline
E) Dipropylamine

35. What is the final product, C?

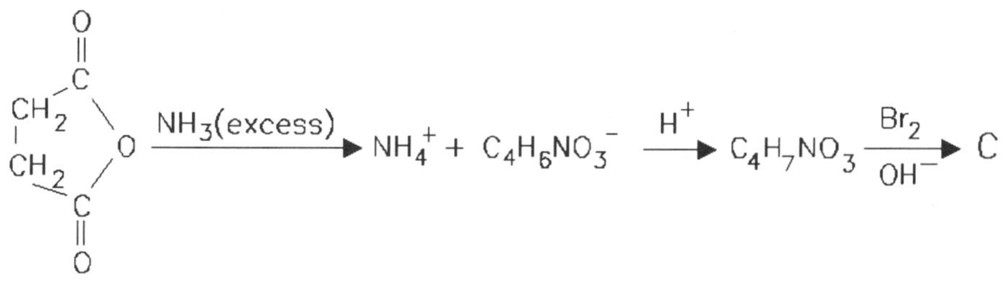

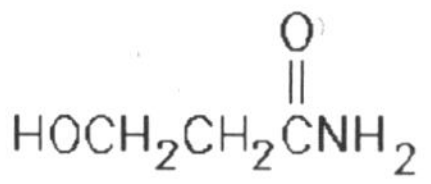

$$\text{succinic anhydride} \xrightarrow{NH_3(\text{excess})} NH_4^+ + C_4H_6NO_3^- \xrightarrow{H^+} C_4H_7NO_3 \xrightarrow[OH^-]{Br_2} C$$

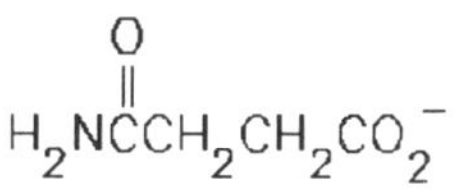

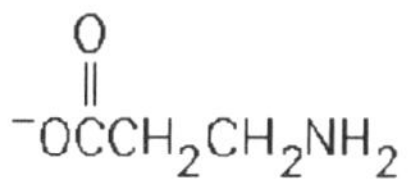

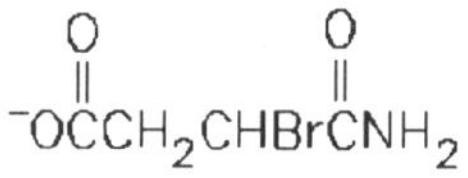

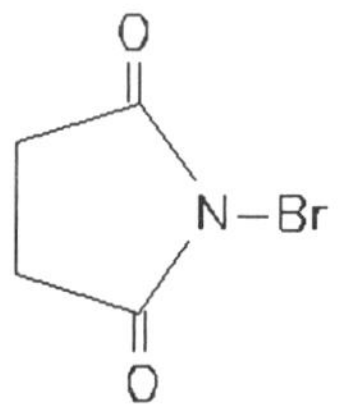

I: $HOCH_2CH_2CNH_2$ (C=O)

II: $H_2NCCH_2CH_2CO_2^-$ (C=O)

III: $^-OCCH_2CH_2NH_2$ (C=O)

IV: $^-OCCH_2CHBrCNH_2$ (both C=O)

V: N-bromosuccinimide (N–Br)

A) I B) II * C) III D) IV E) V

36. What would be the product, F, of the following reaction sequence?

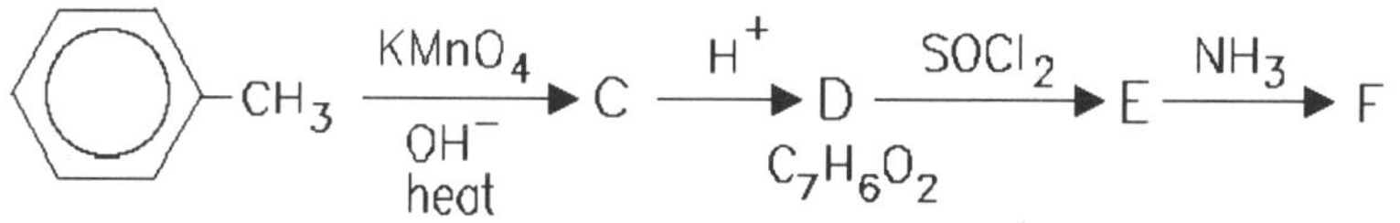

$$C_6H_5CH_3 \xrightarrow[OH^-,\ heat]{KMnO_4} C \xrightarrow{H^+} \underset{C_7H_6O_2}{D} \xrightarrow{SOCl_2} E \xrightarrow{NH_3} F$$

* A) $C_6H_5CONH_2$
B) $C_6H_5CHMNOH$
C) p-$CH_3C_6H_4SO_2NH_2$
D) p-$CH_3C_6H_4NH_2$
E) $C_6H_5NH_2$

37. Your task is to convert toluene into aniline. Which sequence of reagents constitutes the best method?
A) $NaNH_2$ and heat
B) NBS/CCl_4; then NH_3; then Br_2/OH^-
* C) $KMnO_4$, OH^-, heat; then H_3O^+; then PCl_5; then NH_3, then Br_2/OH^-
D) $KMnO_4$, OH^-, heat; then H_3O^+; then PCl_5; then NH_3; then $LiAlH_4$; then Br_2/OH^-
E) $KMnO_4$, OH^-, heat; then H_3O^+; then NH_3 with H_2/Ni

38. How would one carry out the following transformation?

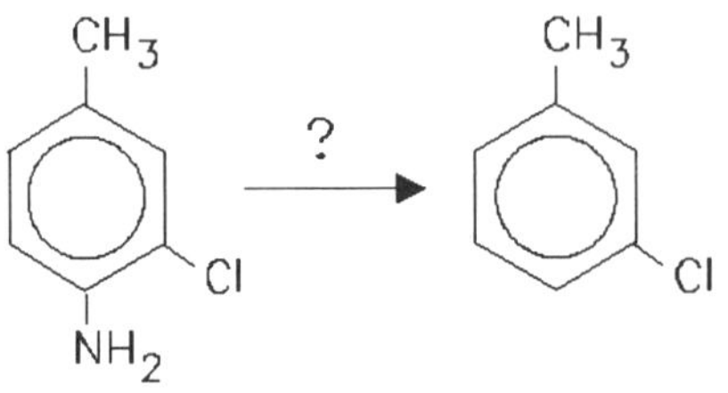

A) $NaNO_2$, HCl, 0-5xC; then HNO_3
* B) $NaNO_2$, HCl, 0-5xC; then H_3PO_2
C) $NaNO_2$, HCl, 0-5xC; then H_2, Ni
D) $C_6H_5SO_2Cl$, OH^-; then HCl
E) NaH, DMSO

39. How could one carry out this synthesis?

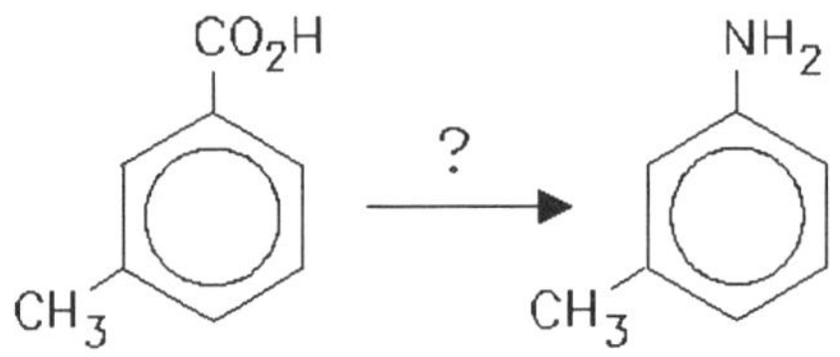

A) $SOCl_2$; then NH_3; then H_3PO_2
B) CH_3Li; then NH_3, H_2, Ni
* C) $SOCl_2$; then NH_3; then Br_2, NaOH
D) PCl_5; then NH_3; then HCl, $NaNO_2$
E) PCl_5; then CH_3NH_2; then $KMnO_4$, OH^-, heat

40. What is the final product of the following sequence of reactions?

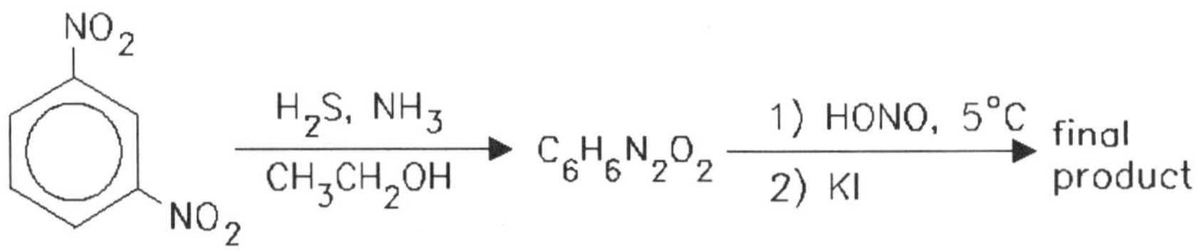

H_2S, NH_3 / CH_3CH_2OH → $C_6H_6N_2O_2$ → 1) HONO, 5°C 2) KI → final product

* A) 1-Iodo-3-nitrobenzene
B) 3-Nitroaniline
C) 3-Nitrophenol
D) 3-Nitrosoaniline
E) 3-Nitrosophenol

NO_2 / CH_3 —A→ B —C→ (Br, NH_2, Br / CH_3) —1) HCl, $NaNO_2$; 2) D→

(Br, CN, Br / CH_3) —E→ (Br, CO_2H, Br / CH_3) —F→ (Br, CO_2H, Br / CO_2H)

41. Consider the synthesis above. What is reagent A?
   A) Br_2, $FeBr_3$
 * B) Fe, HCl; then OH^-
   C) NH_2Cl, $AlCl_3$
   D) H_3PO_2
   E) $LiNH_2$

42. Consider the synthesis above. What is compound B?

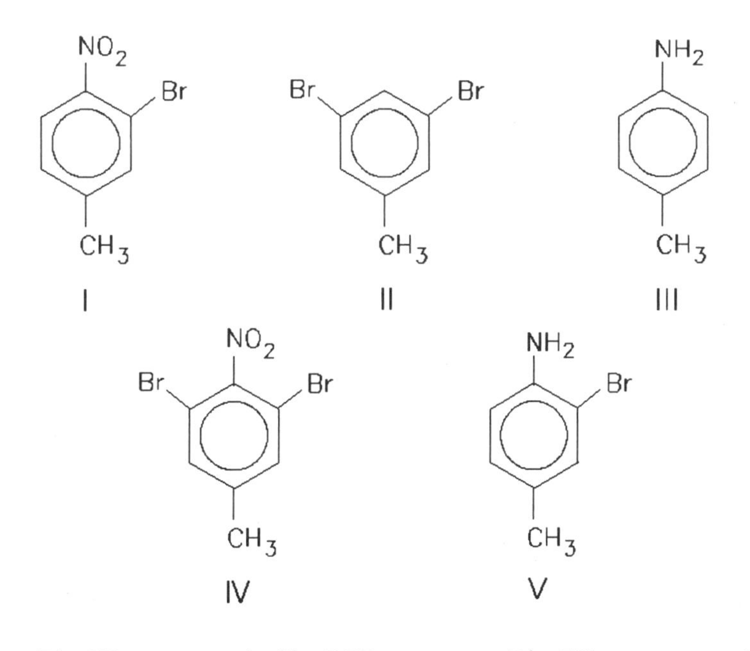

A) I   B) II   * C) III   D) IV   E) V

43. Consider the synthesis above. What is reagent C?
* A) xs Br_2, H_2O
  B) Fe, HCl; then OH^-
  C) NH_2Cl, $AlCl_3$
  D) CuBr
  E) HNO_3, H_2SO_4, Fe

44. Consider the synthesis above. What is reagent D?
  A) H_3PO_2
  B) HCN
  C) P_4O_{10}
* D) CuCN
  E) $CuCl_2$

45. Consider the synthesis above. What is reagent E?
  A) CO_2; then H_3O^+
* B) H_3O^+, H_2O, heat
  C) Mg; then CO_2; then H^+
  D) $LiAlH_4$
  E) $KMnO_4$, OH^-, heat; then H_3O^+

46. Consider the synthesis above. What is reagent F?
  A) CO_2; then H_3O^+
  B) H_3O^+, H_2O, heat
  C) Mg; then CO_2; then H^+
  D) H_2O_2, heat
* E) $KMnO_4$, OH^-, heat; then H_3O^+

47. Consider the synthesis above. What is reagent P?
  A) Fe/HCl; then OH^-
* B) H_2S, NH_3, C_2H_5OH
  C) H_2, Ni, high pressure
  D) Zn(Hg)/HCl
  E) Raney Ni

48. Consider the synthesis above. What is reagent Q?
* A) CuCl
  B) $CuCl_2$
  C) NaCl
  D) KCl
  E) HCl/heat

49. Consider the synthesis above. What is reagent R?

* A) Fe/HCl; then OH^-
  B) NH_2Cl
  C) H_3PO_2
  D) CuCN
  E) HONO, then NH_3

50. Consider the synthesis above. What is compound S?

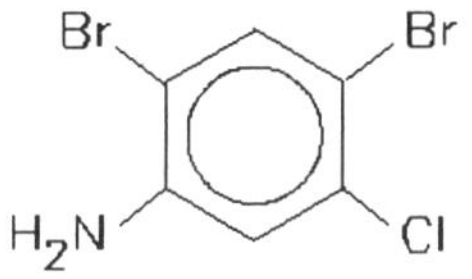

I 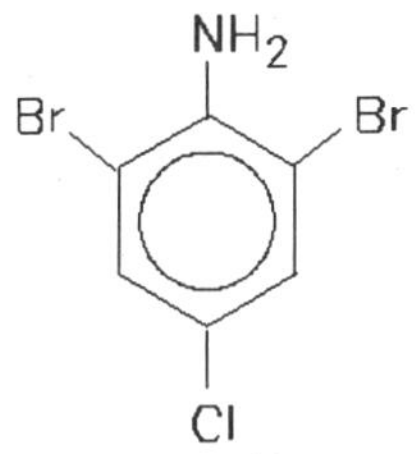 II 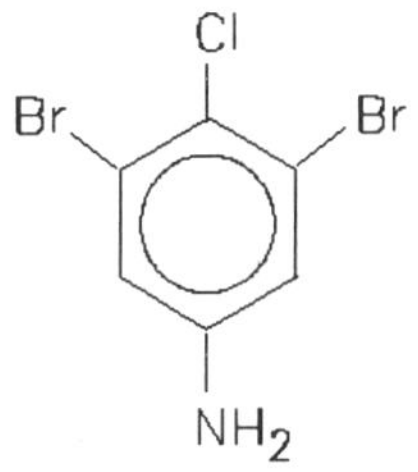 III

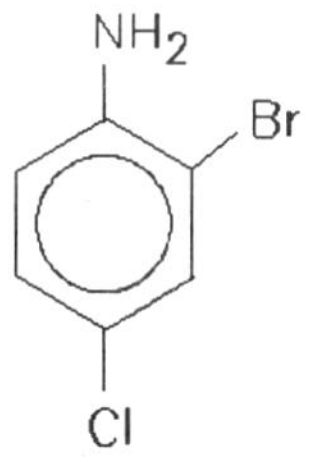

IV 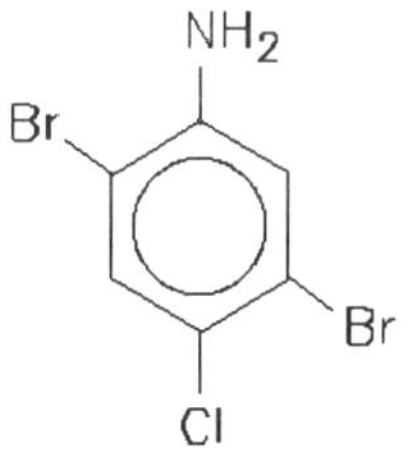 V

A) I * B) II C) III D) IV E) V

51. Consider the synthesis above. What is reagent T?

A) $Br_2/FeBr_3$ B) CuBr C) $CuBr_2$ * D) H_3PO_2 E) H_3PO_4

52. When an equimolar mixture of ammonia and butyl bromide reacts, which of these products will form?

  A) Butylamine
  B) Dibutylamine
  C) Tributylamine
  D) Tetrabutylammonium bromide
* E) All of these

53. Which of these compounds is expected to possess the <u>lowest</u> boiling point?

  A) $CH_3CH_2CH_2NH_2$
  B) $CH_3CH_2NHCH_3$
* C) $(CH_3)_3N$
  D) $CH_3CH_2CH_2OH$
  E) $(CH_3)_3NH^+ Cl^-$

54. Which of these is an acceptable alternative name to "(1-methylbutyl)amine"?

A) 2-Aminopentane
B) 2-Pentanamine
C) Isopentylamine
D) sec-Pentylamine
* E) Both A) and B)

55. What is the chief product of the Hofmann elimination reaction applied to the compound shown?

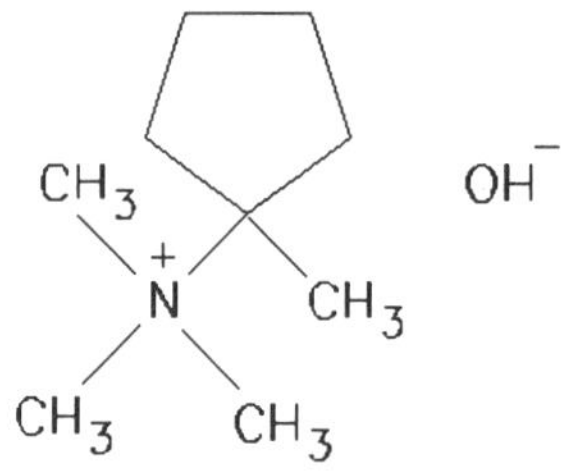

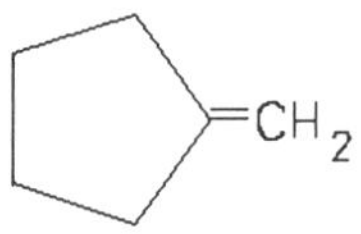

CH_3 OH^- $\overset{+}{N}$ CH_3 CH_3 CH_3

$=CH_2$ I

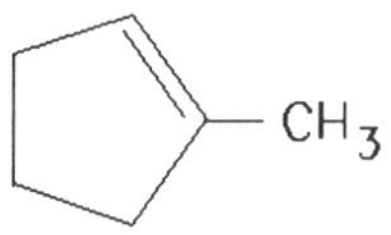

$-CH_3$ II

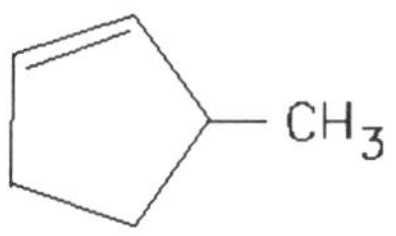

$-CH_3$ III

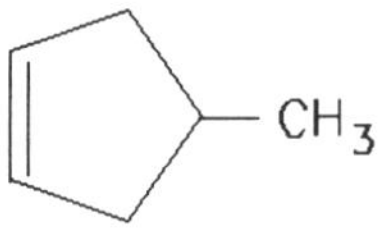

$-CH_3$ IV

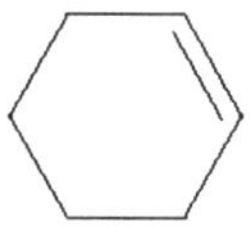

V

* A) I
B) II
C) III
D) IV
E) V

56. What is the principal product when aniline is treated with sodium nitrite and hydrochloric acid at 0-5xC and this mixture is added to p-ethylphenol?

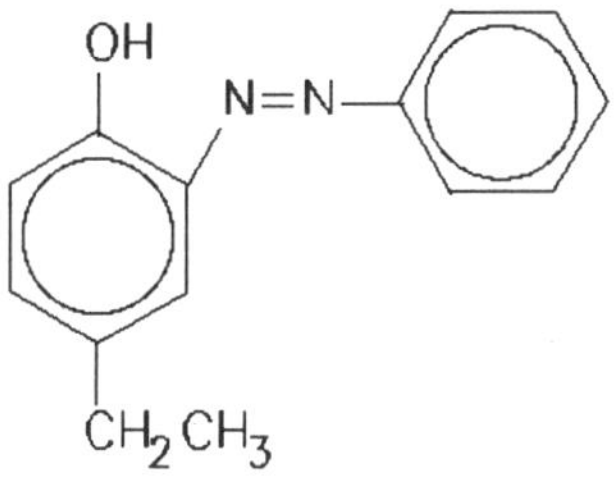

I

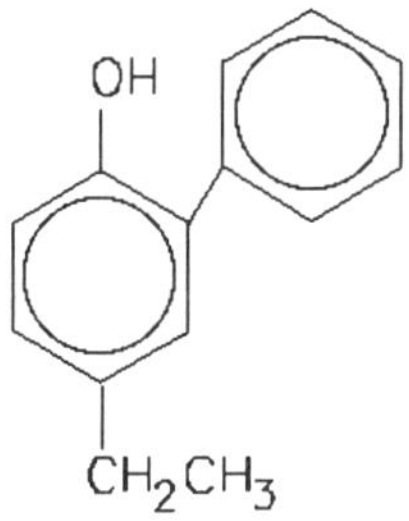

II

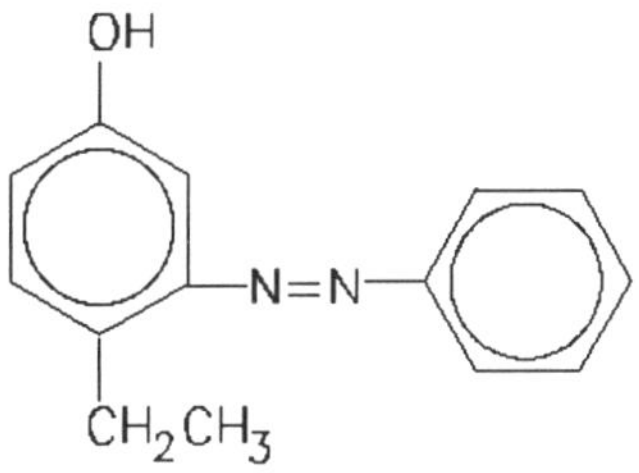

III

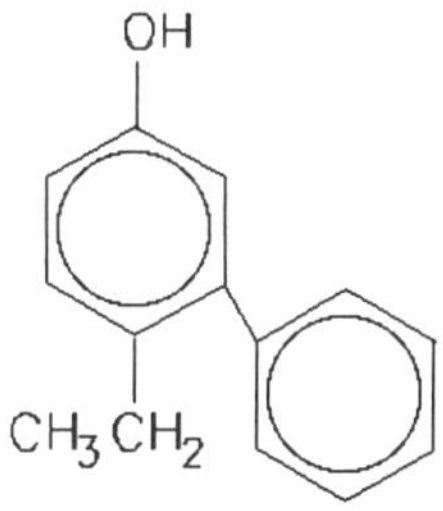

IV

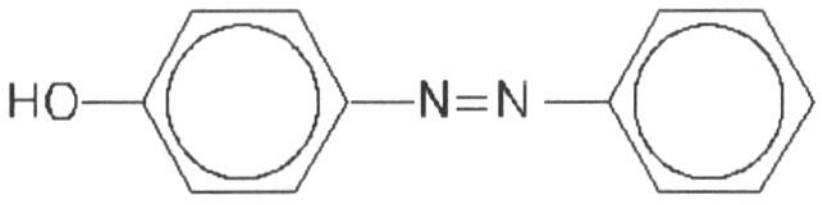

V

* A) I B) II C) III D) IV E) V

57. Which of these could be resolved into the separate enantiomers?

A) $CH_3CHCH_2NH_2$
3
CH_3

* B) $CH_3CH_2CHNH_2$
3
CH_3

C) H
3
$CH_3CH_2CH_2NCH_3$

D) CH_3
3
$CH_3CH_2NCH_2CH_3$

E) $CH_2CH_2CH_3$
+3
$CH_3CH_2NCH_2CH_3$ Br^-
3
CH_3

58. Which of these compounds is soluble in dilute sodium hydroxide solution?

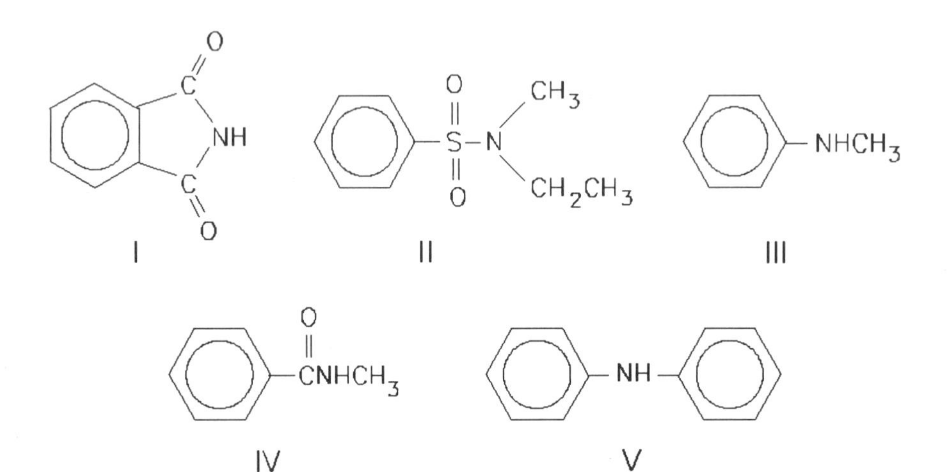

* A) I B) II C) III D) IV E) V

59. What is the final product in the Curtius rearrangement of the acyl azide formed from $CH_3CH_2CH_2CCl$ (with O on the C)?

* A) $CH_3CH_2CH_2NH_2$
B) $CH_3CH_2CH_2CH_2NH_2$
C) $CH_3CH_2CH_2NMCMO$
D) $CH_3CH_2CH_2CNHCH_2CH_2CH_3$ (with O on the C)
E) $CH_3CH_2CH_2OCNHCH_2CH_2CH_3$ (with O on the C)

60. This type of compound is the only one of these which can be converted by reduction into a 1x, 2x or 3x amine, according to its particular structure:
A) Nitrile
B) Oxime
C) Azide
* D) Amide
E) Nitroalkane

61. When the aromatic $-NH_2$ group is to be converted to an -OH group via a diazonium salt, the nitrous acid used is generated in situ by the reaction of a nitrite salt with which acid?
A) HCl B) HBr C) HNO_3 * D) H_2SO_4 E) CH_3CO_2H

62. The reaction of which of these compounds with nitrous acid results in a stable N-nitroso compound?

A) $C_6H_5NH_2$

B) $C_6H_5N(CH_3)_2$

C) $CH_3CH_2CH_2CH_2CH_2NH_2$

* D) $C_6H_5NHCH_3$

E)
```
        O
        :
CH3CH2CNH2
```

63. When the process $ArNH_2$ DDDDDD4 ArY is carried out via an intermediate diazonium salt, this salt is isolated only in the case in which Y is which of these groups?

* A) -F B) -Cl C) -Br D) -I E) -CN

64.

Which combination of reactants will not produce
```
      H
      3
CH3CH2NCH2CH2CH3  ?
```

A)
```
   O
   :
CH3CNHCH2CH2CH3   +   LiAlH4
```

B)
```
     O
     :
CH3CH2CNHCH2CH3   +   LiAlH4
```

C)
```
   H
   3
CH3CMO   +   CH3CH2CH2NH2   +   NaBH3CN
```

D)
```
                      H
                      3
CH3CH2NH2   +   CH3CH2CMO   +   NaBH3CN
```

* E)
```
     O
     :
CH3CH2CNHCH2CH2CH3   +   Br2, NaOH
```

65. In aqueous solution, which of the following possesses the smallest value for pK_b?

A) $C_6H_5NH_2$

B) NH_3

C) $(CH_3CH_2)_3N$

* D) $(CH_3CH_2)_2NH$

E) $CH_3CH_2CH_2NH_2$

66. Which is a correct common name for

```
              CH2CH3
              3
CH3CH2NCHCH2CH3   ?
                3
                CH3
```

A) Ethylethylisobutylamine
B) Diethylisobutylamine
* C) sec-Butyldiethylamine
D) Ethylethyl-sec-butylamine
E) 2-Diethylaminobutane

67. The overall conversion RBr DDDDDD4 RCH_2NH_2 can be accomplished by successive application of which of these sets of reagents?
A) Mg, ether; then NH_3
B) NaN_3; then $LiAlH_4$
* C) NaCN; then $LiAlH_4$
D) H_2CMO; then NH_3
E) H_2NOH; then $LiAlH_4$

68. What is the basis for the successful resolution of racemic $C_6H_5CHCO_2H$ through use of the chiral amine,

```
C6H5CHCO2H          H
    3               3
    OH           CH3CNH2   ?
                    3
                    C6H5
```

A) One enantiomer is more soluble than the other.
B) The racemic mixture is converted into a single isomer in the basic solvent.
* C) The diastereomeric salts formed have different solubilities.
D) The diastereomeric salts have different boiling points.
E) The diastereomeric salts have different melting points.

69. Which would be a good method to synthesize m-nitroaniline?
* A) 1,3-Dinitrobenzene + H_2S and NH_3 in EtOH
B) Aniline + HNO_3/H_2SO_4
C) Aniline + CH_3COCl; then HNO_3/H_2SO_4
D) Nitrobenzene + NH_3 and $AlCl_3$
E) More than one of the above

70. Which of these is the strongest acid?

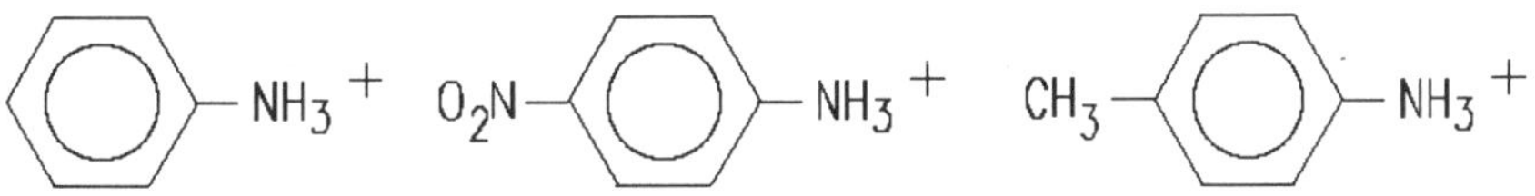

I: $C_6H_5NH_3^+$ II: $O_2N-C_6H_4-NH_3^+$ III: $CH_3-C_6H_4-NH_3^+$

IV: $I-C_6H_4-NH_3^+$ V: $CH_3(CH_2)_4CH_2NH_3^+$

A) I * B) II C) III D) IV E) V

71. Which of these is the strongest acid?

A) $CH_3C(=O)NH_2$

B) $CH_3C(=O)NHCH_3$

C) $CH_3C(=O)N(CH_3)_2$

* D) $CH_3C(=O)NHC(=O)CH_3$

E) NH_3

72. What is the chief alkene product when butylethylmethylpropylammonium hydroxide is heated to 150xC?

* A) $CH_2=CH_2$
B) $CH_3CH=CH_2$
C) $CH_3CH_2CH=CH_2$
D) H 3 $CH_3C=CCH_3$ 3 H
E) H H 3 3 $CH_3C=CCH_3$

73. Arrange the following amines in order of decreasing basicity in aqueous solution?

NH_3	$CH_3CH_2CH_2NH_2$	$(CH_3CH_2CH_2)_2NH$	$(CH_3CH_2CH_2)_3N$
I	II	III	IV

A) IV > III > II > I
* B) III > II > IV > I
C) I > IV > II > III
D) II > III > I > IV
E) II > I > III > IV

74. Which of these alkyl halides cannot be used effectively in a Gabriel amine synthesis?

A) CH_3Br
B) CH_3CH_2Br
C) $(CH_3)_2CHCH_2Br$
D) CH_3CHCH_3
   3
   Br
* E) $(CH_3)_3CBr$

75. Which of these is properly termed a "quaternary ammonium salt"?

A) $CH_3NH_3^+\ Cl^-$
B) $(CH_3)_2NH_2^+\ Cl^-$
C) $(CH_3)_3NH^+\ Cl^-$
* D) $(CH_3)_4N^+\ Cl^-$
E) None of these

Chapter 20: Synthesis and Reactions of a-Dicarbonyl Compounds

1. Which of the compounds listed below is more acidic than ethanol?

* A) $CH_3C(=O)CH_2C(=O)OCH_2CH_3$ ($CH_3CCH_2COCH_2CH_3$, with O above each carbonyl C)

B) $CH_3C(=O)CH_3$ (CH_3CCH_3)

C) $CH_3C(=O)OCH_2CH_3$ ($CH_3COCH_2CH_3$)

D) All of these answers

E) Answers A) and B) only

2. What would be the major product of the following reaction?

$$CH_3CH_2C(=O)OC_2H_5 \xrightarrow[(2)\ H^+]{(1)\ NaOC_2H_5} ?$$

A) $CH_3CH_2COCH_2CH_2CO_2C_2H_5$

B) $CH_3CH_2CO_2CH_2CH_2CO_2C_2H_5$

* C) $CH_3CH_2COCH(CH_3)CO_2C_2H_5$

D) $CH_3CH_2CH(OH)CH(CH_3)CO_2C_2H_5$

E) $CH_3CH_2CH(OH)CH_2CH_2CO_2C_2H_5$

3. What would be the product, E, of the following sequence?

$$C_6H_5CO_2C_2H_5 + CH_3CO_2C_2H_5 \xrightarrow{NaOC_2H_5} [D] \xrightarrow{H^+} E$$

* A) $C_6H_5COCH_2CO_2C_2H_5$

B) $C_6H_5CH(OH)CH_2CO_2C_2H_5$

C) $C_6H_5CH_2CO_2C_2H_5$

D) m-$CH_3COC_6H_4CO_2C_2H_5$

E) $C_6H_5CO_2CH_2CO_2C_2H_5$

4. Which of the following might be used to synthesize

$$\begin{array}{l} \quad\quad CH_3 \\ \quad\quad 3 \\ C_6H_5CHCHCH_2OH \quad ? \\ \quad\quad 3 \\ \quad\quad OH \end{array}$$

* A) $C_6H_5CO_2C_2H_5 \xrightarrow[NaOC_2H_5]{CH_3CH_2CO_2C_2H_5} \xrightarrow{H^+} \xrightarrow{LiAlH_4} \xrightarrow{H_2O}$

(printed as: $CH_3CH_2CO_2C_2H_5$ DDDDDDDDDDDDDD4 $NaOC_2H_5$, H^+ DDDDD4, $LiAlH_4$ DDDDDD4, H_2O DDDDD4)

B) C_6H_5CHO + $CH_3CHCH_2CO_2C_2H_5$ (3 / Br on the second carbon) $\xrightarrow[Ether]{Mg} \xrightarrow[H^+]{H_2O} \xrightarrow{LiAlH_4} \xrightarrow{H_2O}$

(printed as: Mg DDDDDD4 Ether, H_2O DDDDDD4 H^+, $LiAlH_4$ DDDDDD4, H_2O DDDDDD4)

C) CH_3CHCH_2OH (3 / Br on the second carbon) $\xrightarrow[Ether]{Mg}$ $\xrightarrow{C_6H_5CH}$ (O : above CH) $\xrightarrow[H_2O]{}$

(printed as: Mg DDDDDD4 Ether, C_6H_5CH DDDDDD4, DDDDDD4 H_2O)

D) Answers A) and B)

E) Answers A), B), and C)

5. The product, J, of the following reaction sequence,

$C_2H_5OCOC_2H_5$ (O : above central C) + $C_6H_5CH_2COC_2H_5$ (O : above C) $\xrightarrow{NaOC_2H_5}$ [I] $\xrightarrow{H^+}$ J would be:

(arrows printed as: $NaOC_2H_5$ DDDDDD4 [I] H^+ DDDDDD4 J)

* A) $C_2H_5OCCHCOC_2H_5$ (O : above each carbonyl C; 3 / C_6H_5 on the central CH)

B) $C_6H_5CH_2CDCOC_2H_5$ (O O : : above the two carbonyl carbons)

C) p-$C_2H_5O_2CC_6H_4CH_2CO_2C_6H_5$

D) $C_6H_5CCH_2COC_2H_5$ (O : above each carbonyl C)

E) $C_6H_5CCH_2CC_6H_5$ (O : above each carbonyl C)

6. Which of the following compounds is the strongest acid?

A) C_2H_5OH

B) C_2H_6

* C) $C_2H_5O_2CCH_2CO_2C_2H_5$

D) $C_2H_5O_2CCH_2CH_2CO_2C_2H_5$

E) $CH_3CO_2C_2H_5$

7. Which of the following could be used to synthesize

CH_2COOH (2-oxocyclohexyl)

in good yield?

A) Cyclohexanone, $ClCH_2COOH$, $AlCl_3$, heat
* B) Cyclohexanone, $(CH_3CH_2)_2NH$, H^+, $(-H_2O)$; then $BrCH_2COOC_2H_5$; then OH^-, H_2O, heat; then H^+
C) Cyclohexylacetic acid, $KMnO_4$, OH^-, heat; then H^+
D) Answers A) and B)
E) Answers A), B), and C)

8. What is the product, C, of the following reaction sequence?

Cyclopentanone + pyrrolidine (N–H) $\xrightarrow[-H_2O]{H^+}$ A $\xrightarrow{C_6H_5\overset{O}{\overset{\|}{C}}Cl}$ B $\xrightarrow[H^+]{H_2O}$ C

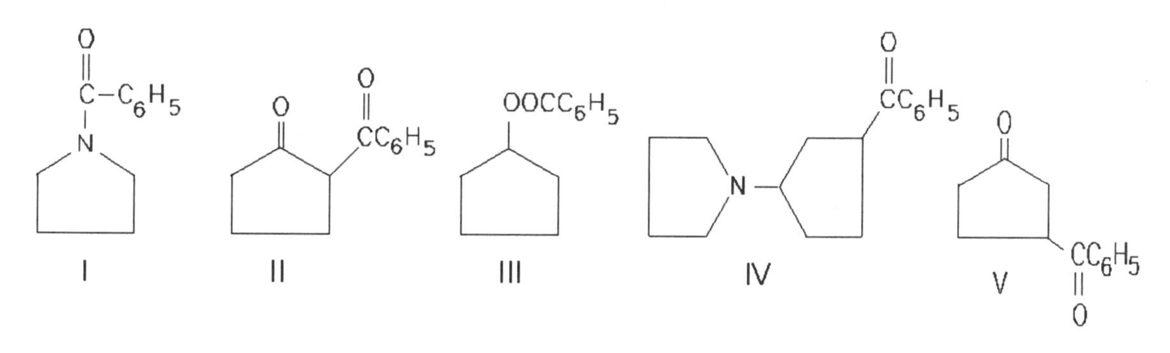

A) I * B) II C) III D) IV E) V

9. The product, X, of the following reaction sequence,

$$CH_3\overset{O}{\overset{:}{C}}CH_2\overset{O}{\overset{:}{C}}OC_2H_5 \xrightarrow[(2)\ C_6H_5CH_2Br]{(1)\ NaOC_2H_5} \xrightarrow[(2)\ H_3O^+]{(1)\ NaOH} \xrightarrow[-CO_2]{heat} X$$

would be:

A) $CH_3\overset{O}{\overset{:}{C}}CH_2C_6H_5$

B) $C_6H_5CH_2\overset{O}{\overset{:}{C}}OC_2H_5$

* C) $C_6H_5CH_2CH_2\overset{O}{\overset{:}{C}}CH_3$

D) $C_6H_5CH_2CH_2\overset{O}{\overset{:}{C}}OC_2H_5$

E) $C_6H_5CH_2\overset{O}{\overset{:}{C}}CH_3$

10. Which of the following would provide the best synthesis of 3,5-dimethyl-2-hexanone?

A) Ethyl acetoacetate + $NaOC_2H_5$ + CH_3I; then KO-t-Bu + $(CH_3)_3CCH_2Br$; then NaOH; then H^+; then heat

B) Ethyl acetoacetate + $NaOC_2H_5$ + $(CH_3)_3CBr$; then KO-t-Bu + CH_3I; then NaOH; then H^+; then heat

* C) Ethyl acetoacetate + $NaOC_2H_5$ + $(CH_3)_2CHCH_2Br$; then KO-t-Bu + CH_3I; then NaOH; then H^+; then heat

D) Ethyl acetoacetate + $NaOC_2H_5$ + $(CH_3)_2CHCH_2CHBrCH_3$; then NaOH; then H^+; then heat

E) Ethyl acetoacetate + $NaOC_2H_5$ + CH_3I; then KO-t-Bu + $(CH_3)_2CHBr$; then NaOH; then H^+; then heat

11. The product, I, of the following reaction sequence,

$$BrCH_2CH_2Br + 2\ Na^+\ {}^-{:}CH(CO_2C_2H_5)_2 \longrightarrow E \xrightarrow[H_2O]{OH^-} F \xrightarrow{H^+} G \xrightarrow{heat} I$$

would be:

$HOOC-C_6H_{10}-COOH$ (1,4-cyclohexanedicarboxylic acid)	$HOOCCH_2CH_2CH_2CH_2COOH$	$C_2H_5O_2C-C_6H_{10}-CO_2C_2H_5$ (diethyl 1,4-cyclohexanedicarboxylate)	$CH_3\overset{O}{\overset{\|}{C}}CH_2CH_2\overset{O}{\overset{\|}{C}}CH_3$	▷– COOH
I	II	III	IV	V

A) I * B) II C) III D) IV E) V

12. Which of the following would afford the best synthesis of diethyl phenylmalonate, $C_6H_5CH(CO_2C_2H_5)_2$?

A)
$$C_6H_5Br + Na^+\ {}^-{:}CH(CO_2C_2H_5)_2 \longrightarrow$$

* B)
$$C_6H_5CH_2CO_2C_2H_5 + C_2H_5O\overset{O}{\overset{\|}{C}}OC_2H_5 \xrightarrow{^-OC_2H_5} \xrightarrow{H^+}$$

C)
$$C_6H_5CH_2CO_2C_2H_5 + C_2H_5O\overset{O}{\overset{\|}{C}}-\overset{O}{\overset{\|}{C}}OC_2H_5 \xrightarrow{^-OC_2H_5} \xrightarrow{H^+}$$

D)
$$C_6H_5CH_2CO_2C_2H_5 + CH_3\overset{O}{\overset{\|}{C}}OC_2H_5 \xrightarrow{^-OC_2H_5} \xrightarrow{H^+}$$

E)
$$C_6H_5\overset{H}{\overset{|}{C}}=O + 2\ H\overset{O}{\overset{\|}{C}}OC_2H_5 \xrightarrow{^-OC_2H_5} \xrightarrow{H^+}$$

13. Predict the product, P, of the following reaction sequence.

$$CH_3CH{=}CHCO_2C_2H_5 \quad + \quad \begin{array}{c} CO_2C_2H_5 \\ 3 \\ CH_2 \\ 3 \\ CO_2C_2H_5 \end{array} \quad \overset{^{-}OC_2H_5}{DDDDDD4} \quad M \quad \overset{OH^{-}}{DDDDDD4} \quad N \quad \overset{H^{+}}{DDDDDD4} \quad O$$

$$\underset{(-CO_2,\, -H_2O)}{\overset{heat}{DDDDDDDDDDD4}} \quad P$$

I: $CH_3CH{=}CH\overset{O}{\overset{\|}{C}}CH_2\overset{O}{\overset{\|}{C}}OH$

II: CH_3; O; O; O

III: CH_2CH_3; O; O; O

IV: $CH_3CH_2\underset{CH_2COOH}{\underset{|}{C}}HCOOH$

V: $CH_3\underset{CH_2COOH}{\underset{|}{C}}HCH_2COOH$

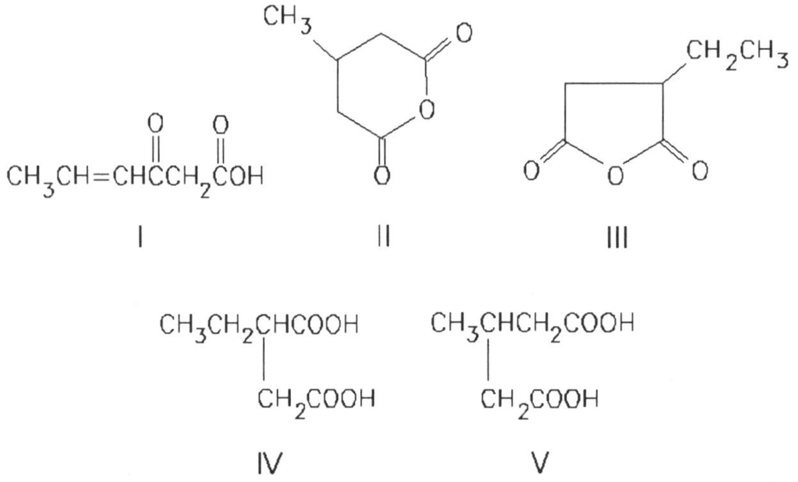

A) I * B) II C) III D) IV E) V

14. The product, L, of the following reaction sequence,

$CO_2C_2H_5$
3
CH_2 + $^-OC_2H_5$ $\xrightarrow{CH_3CH_2Br}$ J $\xrightarrow[CH_3I]{^-OC(CH_3)_3}$ K $\xrightarrow[^-OC_2H_5]{H_2NCONH_2}$ L
3
$CO_2C_2H_5$

I: $CH_3O_2CCHCO_2C_2H_5$ with $NHCNH_2$ (C=O) substituent

II: $CH_3C(CO_2C_2H_5)_2-NHCONH-C(CO_2C_2H_5)_2C_2H_5$

III: 5-ethyl-5-methyl pyrimidinetrione ring (CH_3, C_2H_5, O, NH, O, N–H, O)

IV: $(NHCONHCOCH_2CONHCONH)_n$

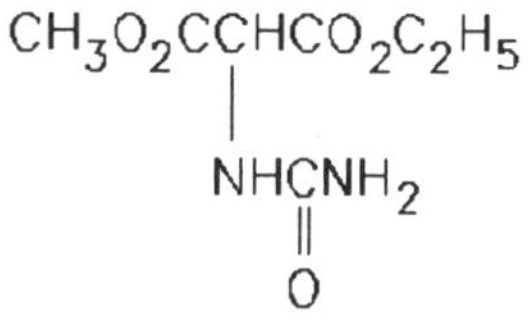

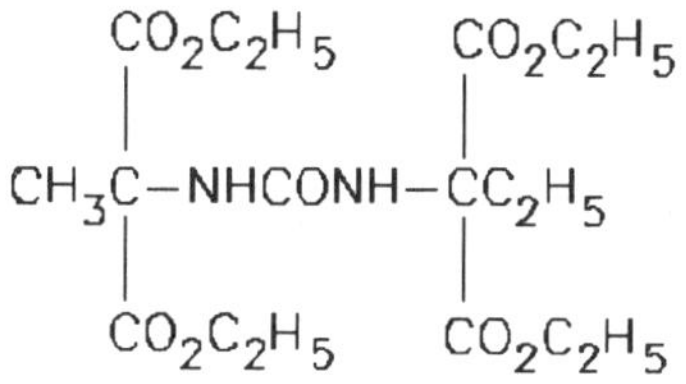

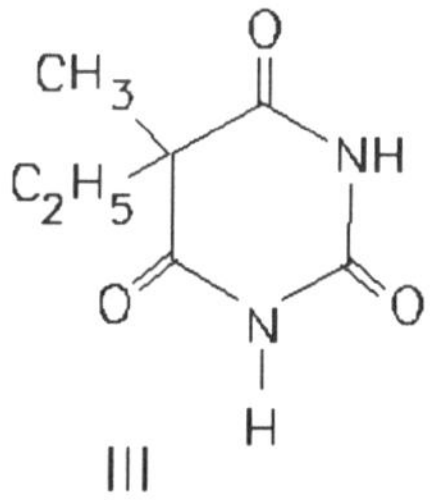

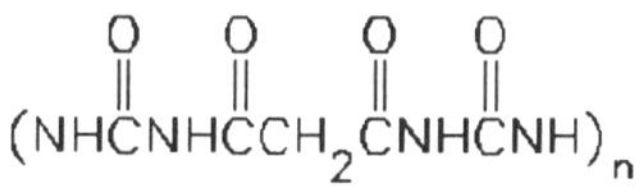

V: $CH_3(C_2H_5)C(CONHCONH_2)_2$

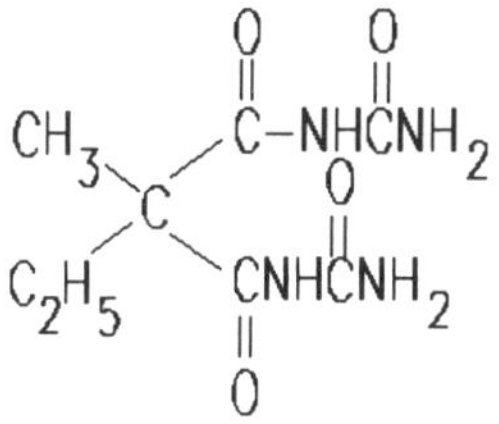

would be:

A) I B) II * C) III D) IV E) V

15. Which compound may be prepared using a Mannich reaction?

$C_6H_5CH{=}CHCO_2H$

I

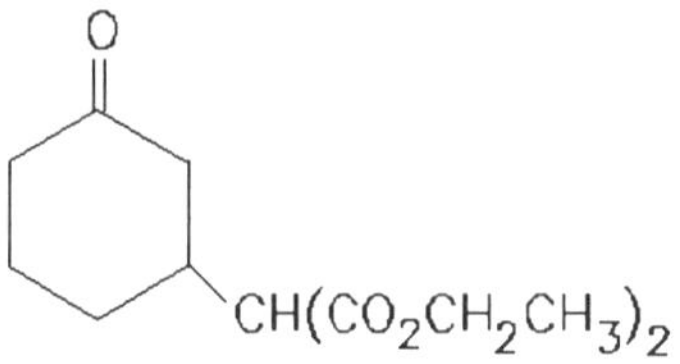

(3-oxocyclohexyl) $CH(CO_2CH_2CH_3)_2$

II

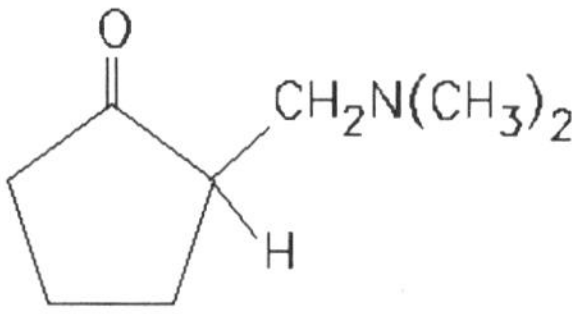

(2-oxocyclopentyl, with H) $CH_2N(CH_3)_2$

III

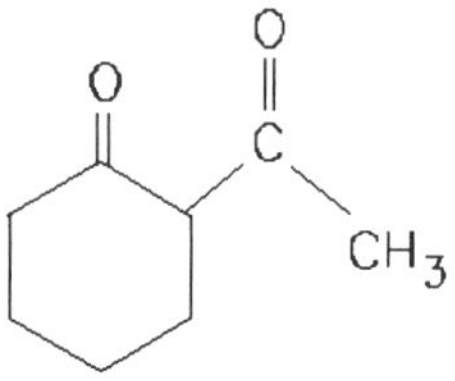

(2-oxocyclohexyl) $C(=O)CH_3$

IV

$CH_3C(=O)CH_2CH_2N$ (pyrrolidine ring)

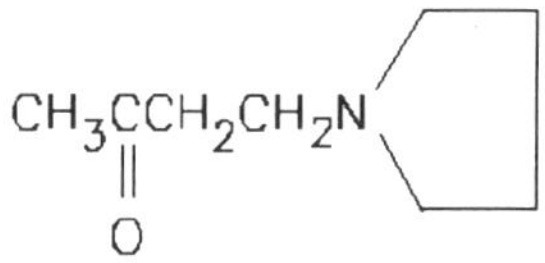

V

A) I B) II * C) III D) IV E) V

16. What is the structure for J?

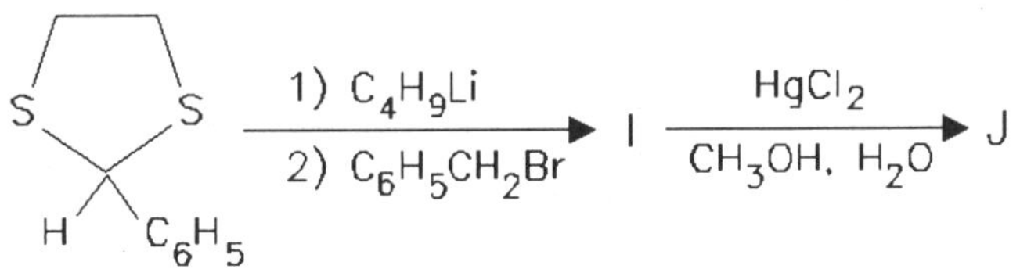

S
S
H
C_6H_5
1) C_4H_9Li
2) $C_6H_5CH_2Br$
I
$HgCl_2$
CH_3OH, H_2O
J

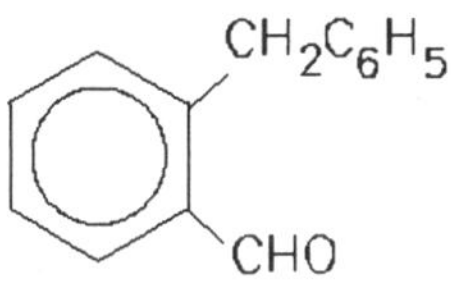

$CH_2C_6H_5$
CHO

I

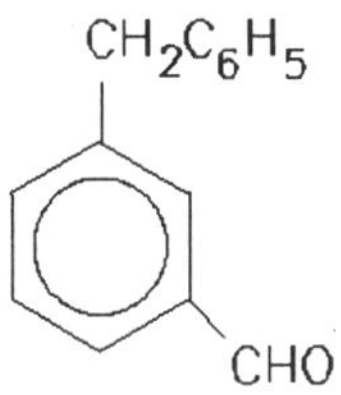

$CH_2C_6H_5$
CHO

II

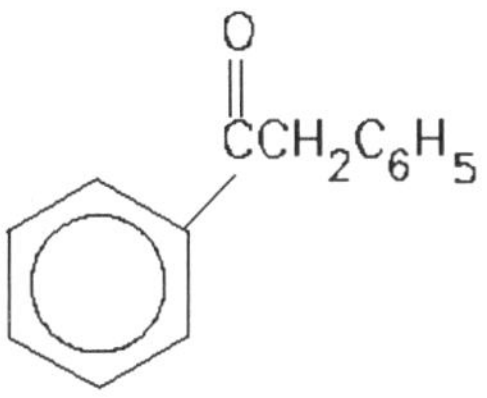

O
$CCH_2C_6H_5$

III

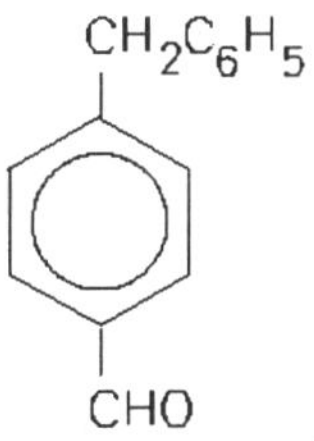

$CH_2C_6H_5$
CHO

IV

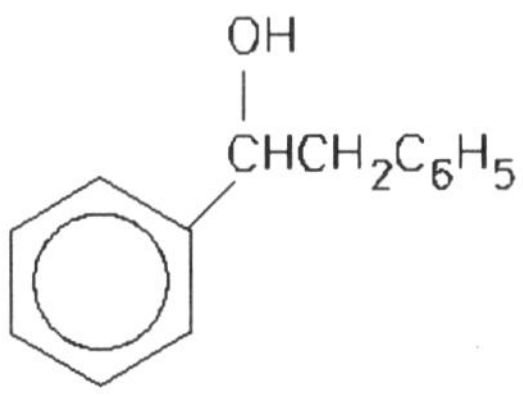

OH
$CHCH_2C_6H_5$

V

A) I B) II * C) III D) IV E) V

17. Which compound could be prepared using a Michael reaction?

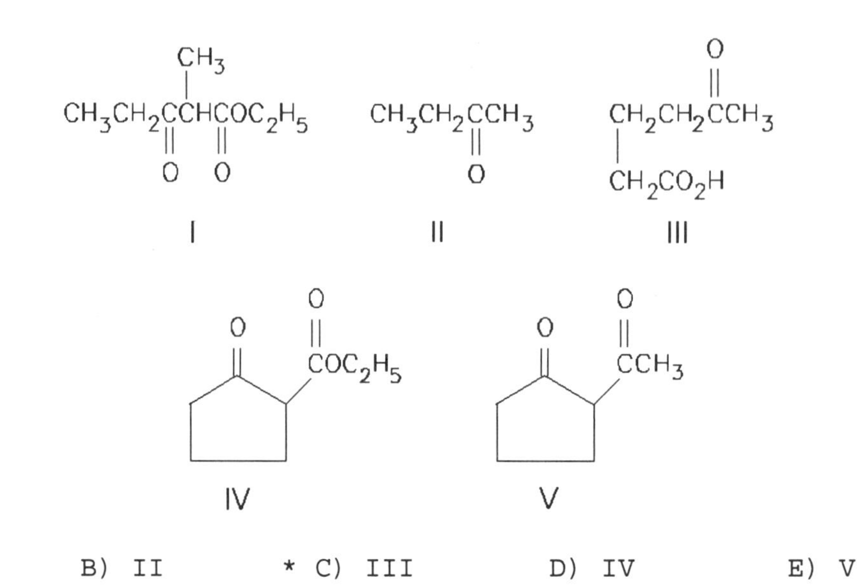

A) I   B) II   * C) III   D) IV   E) V

18. The product(s) of the reaction of 2 mol of ethyl butanoate with sodium ethoxide is(are):

A) $CH_3CH_2CH_2CO_2H$ + $NaOH$

B)
```
          O O
          : :
CH3CH2CH2COCCH2CH2CH3
```

* C)
```
          O  O
          :  :
CH3CH2CH2CCHCOC2H5
          3
          CH2CH3
```

D)
```
          O    O
          :    :
CH3CH2CH2CCHCH2COC2H5
          3
          CH3
```

E)
```
          O            O
          :            :
CH3CH2CH2CCH2CH2CH2COC2H5
```

19. The reaction of diethyl heptanedioate with sodium ethoxide would give as the product:

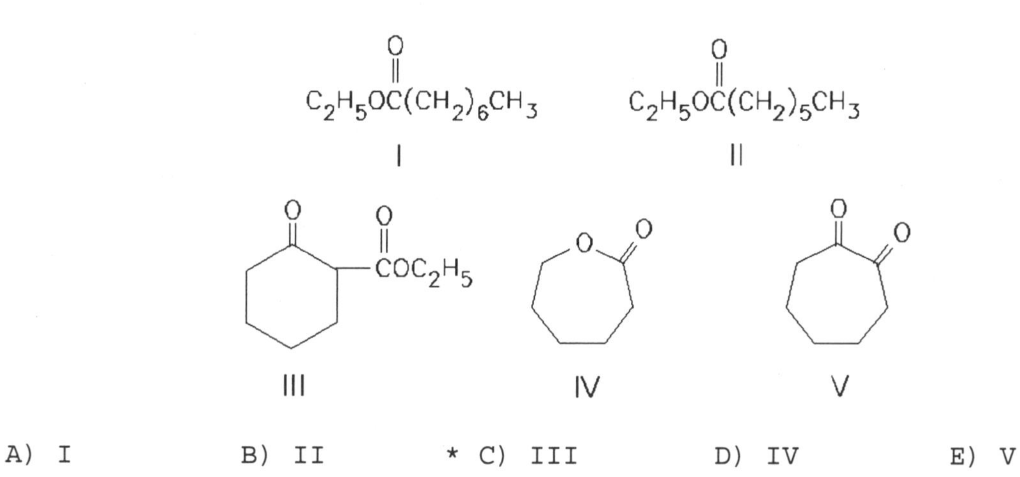

A) I B) II * C) III D) IV E) V

20. Which of the following hydrogens is the most acidic?

```
            O                      O
            :                      :
CH3CH2OCCH2CH2CH2CH2CH2CH2COCH2CH3
 t  t      t  t  t
 I  II    III IV V
```

A) I B) II * C) III D) IV E) V

21. Which reagents would you use to prepare $C_6H_5C(=O)CH_2C(=O)CH_3$ from ethyl acetoacetate?

* A) $\xrightarrow[DMSO]{NaH}$ $\xrightarrow{C_6H_5C(=O)Cl}$ $\xrightarrow[(2)\ H_3O^+]{(1)\ OH^-,\ H_2O,\ heat}$ $\xrightarrow{heat}$

B) $\xrightarrow[C_2H_5OH]{NaOC_2H_5}$ $\xrightarrow{C_6H_5C(=O)CH_2Br}$ $\xrightarrow[(2)\ H_3O^+]{(1)\ OH^-,\ H_2O,\ heat}$ $\xrightarrow{heat}$

C) $\xrightarrow{heat}$ $\xrightarrow{NaOC_2H_5}$ $\xrightarrow{C_6H_5C(=O)CH_2Br}$

D) $\xrightarrow[C_2H_5OH]{NaOC_2H_5}$ $\xrightarrow{C_6H_5Cl}$

E) $\xrightarrow[C_2H_5OH]{NaOC_2H_5}$ $\xrightarrow{C_6H_5C(=O)Cl}$ $\xrightarrow[(2)\ H_3O^+]{(1)\ OH^-,\ H_2O,\ heat}$ $\xrightarrow{heat}$

22. The product of the following reaction is?

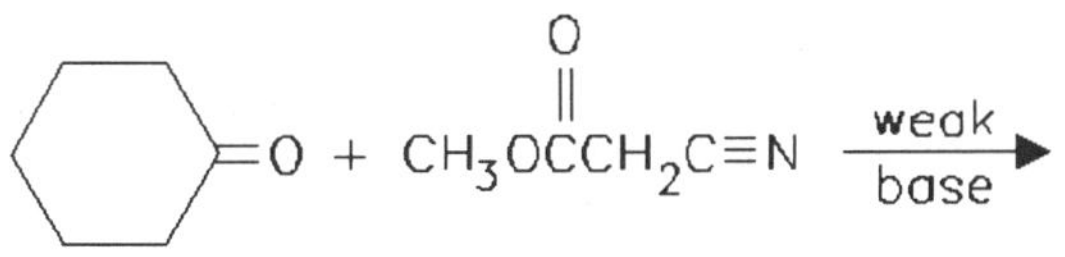

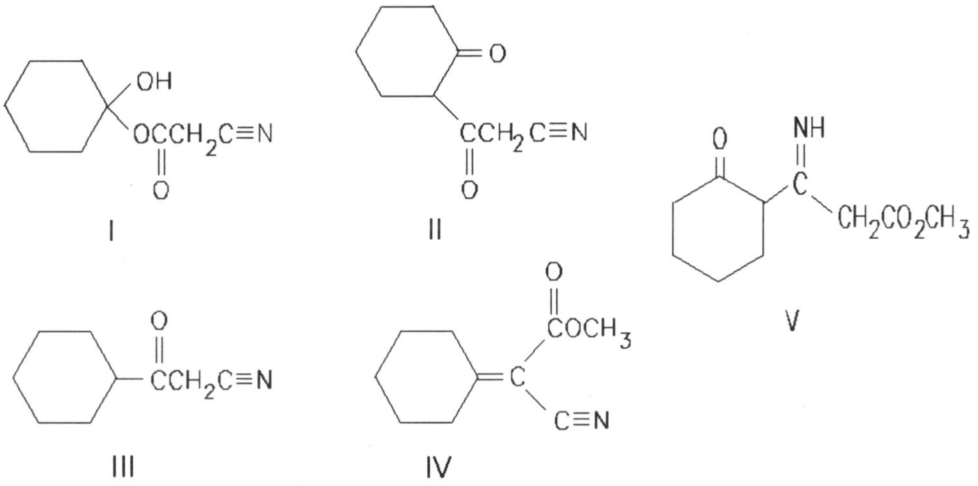

$$\text{cyclohexanone} + CH_3O\overset{O}{\overset{\|}{C}}CH_2C\equiv N \xrightarrow[\text{base}]{\text{weak}}$$

I: 1-hydroxycyclohexyl with $OCCH_2C\equiv N$ (C=O)

II: 2-($CCH_2C\equiv N$, C=O) cyclohexanone

III: cyclohexyl–$\overset{O}{\overset{\|}{C}}CH_2C\equiv N$

IV: cyclohexylidene=C($\overset{O}{\overset{\|}{C}}OCH_3$)($C\equiv N$)

V: 2-(C(=NH)$CH_2CO_2CH_3$) cyclohexanone

A) I B) II C) III * D) IV E) V

23. Which of the indicated hydrogens can be replaced by alkylation?

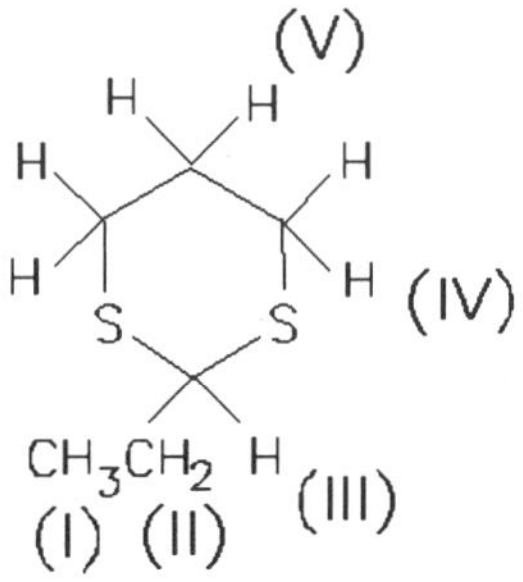

1,3-dithiane ring with H labels (V), (IV); CH_3CH_2 (I) (II), H (III) at C-2

A) I B) II * C) III D) IV E) V

24. The following generalized reaction,

```
                      O
                      :              NaOC2H5
RI  +  C2H5OCCH2CpN   DDDDDDD4
                                C2H5OH
```

would give which product?

```
* A)    O  R
        : /
   C2H5OCCH
           \
            CpN
```

```
  B)    O
        :
   C2H5OCCH2R
```

```
  C) O
     :
   RCCH2CpN
```

```
  D)      O
          :
   C2H5OCCHCHMNR
           3
           I
```

```
  E)  O
      :
   ROCCH2CpN
```

25. Which of the following statements is true about the anion formed from the reaction of diethyl malonate with sodium ethoxide?

A) It can react with an !, a-unsaturated ester by conjugate addition.
B) It can condense with aldehydes and ketones.
C) It can be alkylated with an alkyl halide.
D) It is resonance stabilized.
* E) All of the above statements are true.

26. Which reagents would be used in a Mannich reaction to synthesize

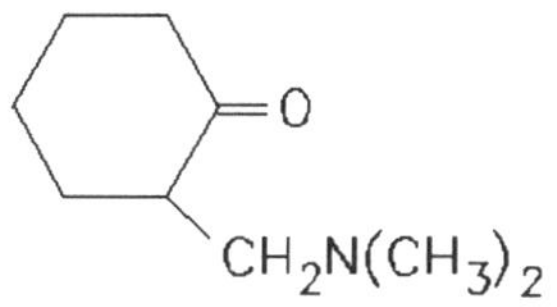

[structure: 2-(dimethylaminomethyl)cyclohexanone, $CH_2N(CH_3)_2$] ?

I cyclohexanone + CH_3CHO + $(CH_3)_2NH$

II cyclohexanone + CH_3CHO + $(CH_3)_3N$

III cyclohexanone + $(CH_3)_3N$ + base

IV cyclohexanone + CH_2O + $(CH_3)_2NH$

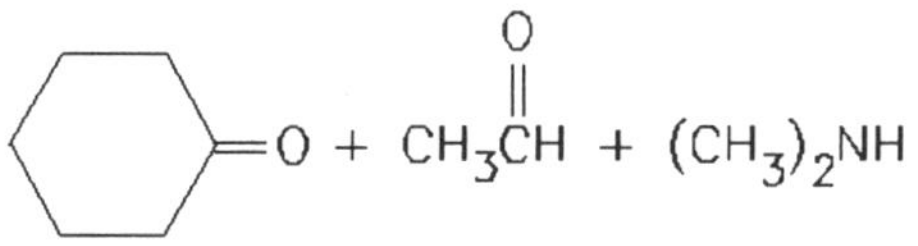

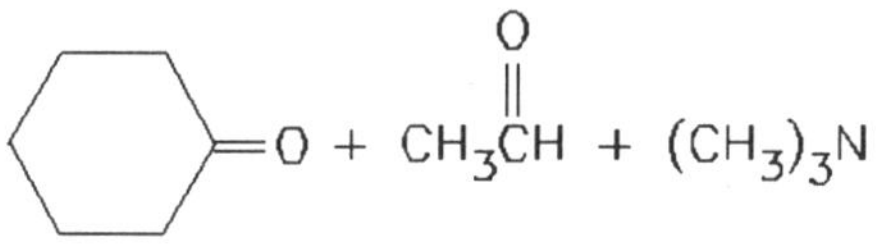

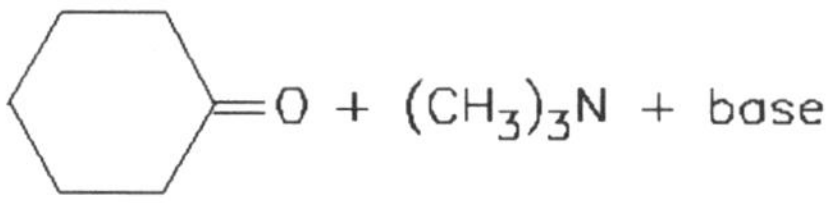

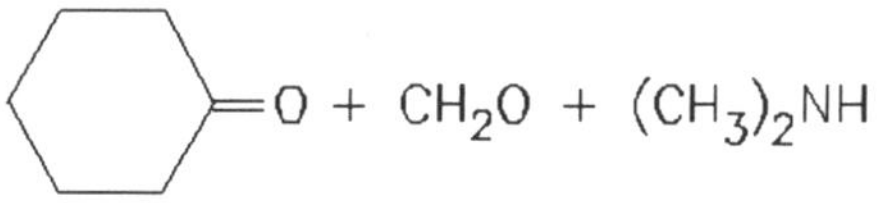

V 2-methylenecyclohexanone ($=CH_2$) + $(CH_3)_2NH$

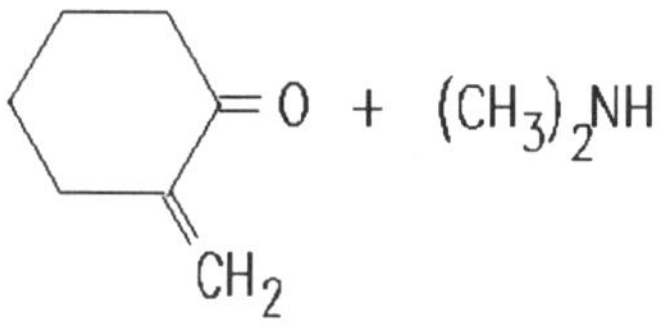

A) I B) II C) III * D) IV E) V

27. Which of the following statements is true of the enamine,

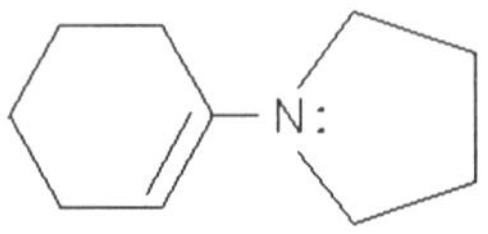

?

A) The enamine can be made from cyclohexanone + pyrrolidine.
B) It has another resonance structure

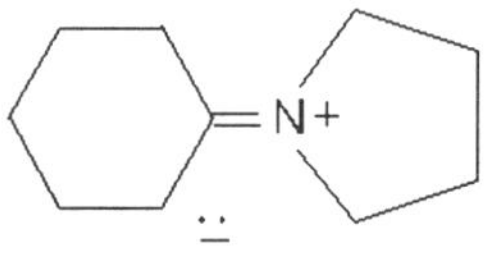

.

C) It can be acylated at the !-carbon of the original carbonyl compound.
D) It can be alkylated.
* E) All of the above are true.

28. Utilizing an enamine,

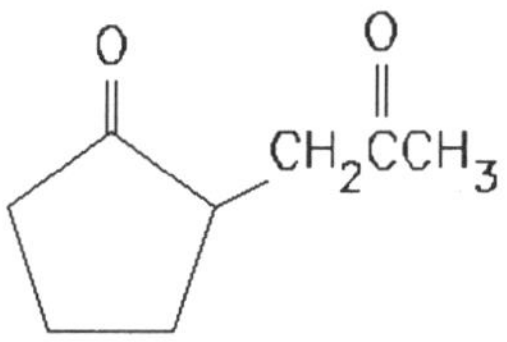

can be made from:

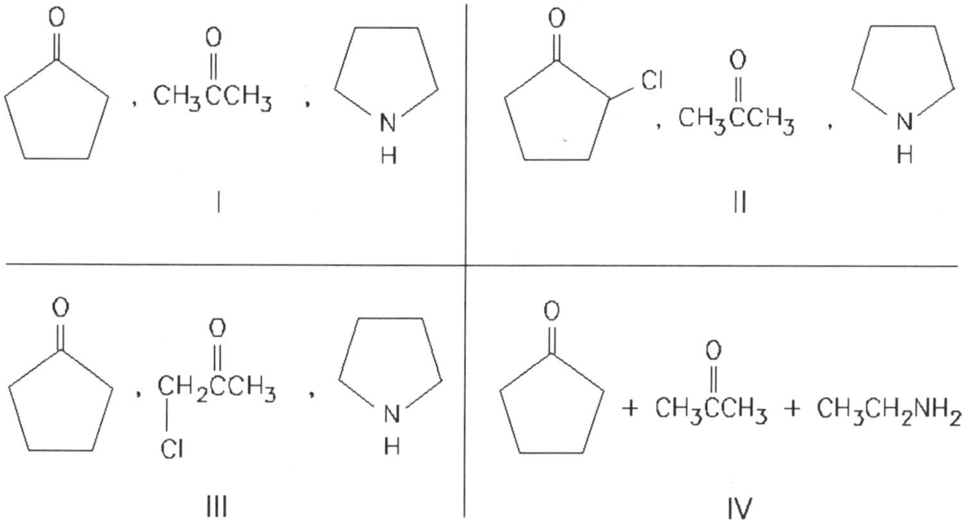

I: cyclopentanone, CH_3CCH_3 (C=O), pyrrolidine (N–H)

II: 2-chlorocyclopentanone (Cl), CH_3CCH_3 (C=O), pyrrolidine (N–H)

III: cyclopentanone, CH_2CCH_3 (C=O) with Cl on CH_2, pyrrolidine (N–H)

IV: cyclopentanone + CH_3CCH_3 (C=O) + $CH_3CH_2NH_2$

A) I
B) II
* C) III
D) IV
E) Both reactions II and III

29. In the synthesis of barbiturates, the R groups would originate in which step?

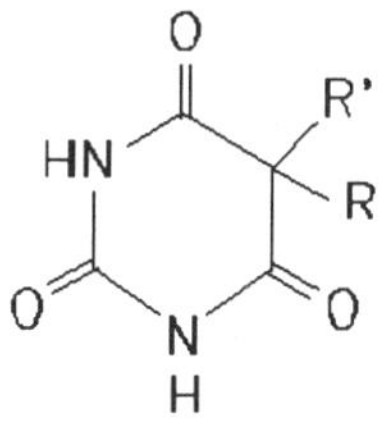

A) Alkylation of the unsubstituted barbituric acid
B) Alkylation of the urea
* C) Alkylation of the starting diethyl malonate
D) Alkylation of an enamine
E) Alkylation via a Grignard reagent

30. The reaction

$$\overset{O}{\underset{}{C_6H_5\overset{:}{C}OC_2H_5}} + CH_3CH_2\overset{O}{\overset{:}{C}}OC_2H_5 \xrightarrow[(2)\ H^+]{(1)\ NaOC_2H_5}$$

produces which of these?

A) $C_6H_5\overset{O}{\overset{:}{C}}CH_2CH_2\overset{O}{\overset{:}{C}}OC_2H_5$

* B) $C_6H_5\overset{O}{\overset{:}{C}}\underset{\underset{CH_3}{3}}{CH}\overset{O}{\overset{:}{C}}OC_2H_5$

C) $C_6H_5\overset{O}{\overset{:}{C}}\text{-}\overset{O}{\overset{:}{C}}CH_2CH_3$

D) $C_6H_5\overset{O}{\overset{:}{C}}O\overset{O}{\overset{:}{C}}CH_2CH_3$

E) $C_6H_5\overset{OH}{\overset{3}{C}H}\underset{\underset{CH_3}{3}}{CH}\overset{O}{\overset{:}{C}}OC_2H_5$

31. Which organic reagents would you need to make 2-ethylpentanoic acid from diethyl malonate?

* A) Bromoethane, 1-bromopropane, sodium ethoxide, and potassium <u>tert</u>-butoxide
B) 3-Bromohexane and sodium ethoxide
C) 3-Bromopentane and sodium ethoxide
D) Bromoethane, 2-bromopropane, sodium ethoxide, and potassium <u>tert</u>-butoxide
E) 2-Bromopentane and sodium ethoxide

32. 2-Methylcyclohexane-1,3-dione can be synthesized from:

I (cyclohexane-1,3-dione) $\xrightarrow{CH_3MgBr}$ $\xrightarrow{H_3O^+}$

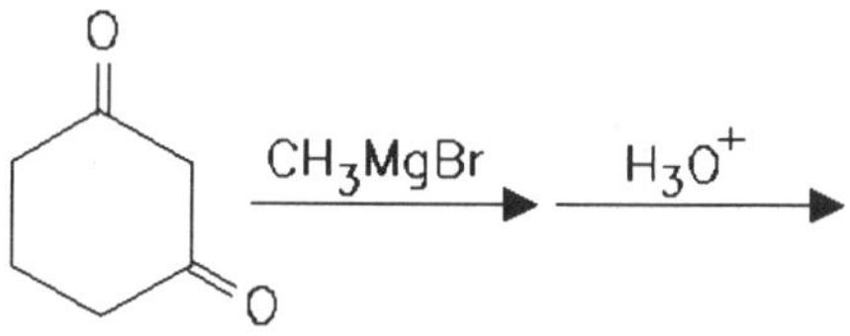

II (2-methylcyclohexanone) $\xrightarrow{KMnO_4,\ OH^-}$

III $CH_3CH_2\overset{O}{\overset{\|}{C}}CH_2CH_2CH_2\overset{O}{\overset{\|}{C}}OC_2H_5$ $\xrightarrow{NaOC_2H_5}$ $\xrightarrow{H_3O^+}$

IV $CH_3CH_2\overset{O}{\overset{\|}{C}}CH_2\underset{CH_3}{\underset{|}{C}}HCH_2\overset{O}{\overset{\|}{C}}OC_2H_5$ $\xrightarrow{NaOC_2H_5}$ $\xrightarrow{H_3O^+}$

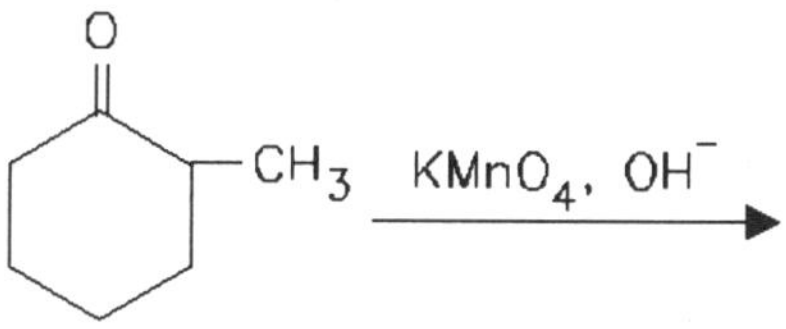

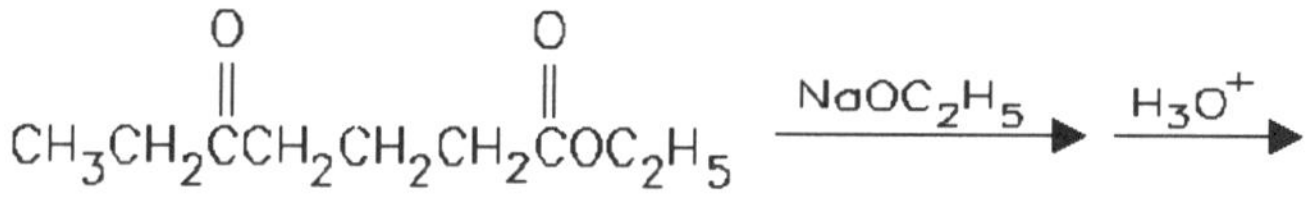

V $CH_3CH_2\overset{O}{\overset{\|}{C}}CH_2CH_2\underset{CH_3}{\underset{|}{C}}H\overset{O}{\overset{\|}{C}}OC_2H_5$ $\xrightarrow{NaOC_2H_5}$ $\xrightarrow{H_3O^+}$

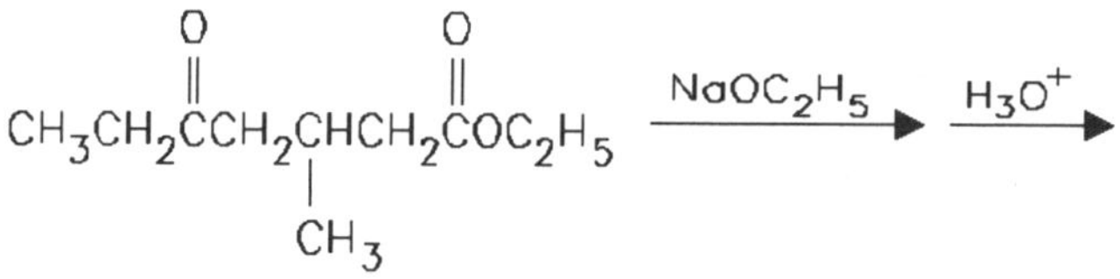

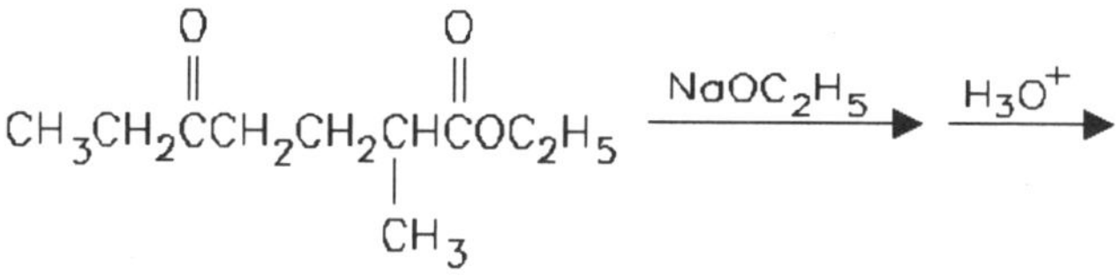

A) I　　B) II　　* C) III　　D) IV　　E) V

33. 2-Heptanone can be synthesized by which reaction sequence?

* A)
Ethyl acetoacetate DDDDDDD4 ($NaOC_2H_5$) DDDDDDDDDDD4 ($CH_3(CH_2)_3Br$) DDDDDDDDDDD4 ((1) dil. OH^- (2) H_3O^+) DDDDDD4 (heat)

B)
Ethyl acetoacetate DDDDDD4 (heat) DDDDDDD4 ($NaOC_2H_5$) DDDDDDDDDDD4 ($CH_3(CH_2)_3Br$)

C)
Ethyl acetoacetate DDDDDDDDDDDD4 ((1) dil. OH^- (2) H_3O^+) DDDDDD4 (heat) DDDDDDD4 ($NaOC_2H_5$) DDDDDDDDDDD4 ($CH_3(CH_2)_3Br$)

D)
Ethyl acetate DDDDDDDDDDD4 ($CH_3(CH_2)_3Br$) DDDDDDDDDDDD4 ((1) dil. OH^- (2) H_3O^+) DDDDDD4 (heat)

E)
Ethyl hexanoate DDDDDDDDDDD4 ((1) $NaOC_2H_5$ (2) CH_3COCl) DDDDDDDDDDDD4 ((1) dil. OH^- (2) H_3O^+) DDDDDD4 (heat)

$$C_6H_5CH_2\overset{O}{\overset{\|}{C}}OEt + X \xrightarrow[(2)\ H^+]{(1)\ NaOEt} C_6H_5CH(CO_2Et)_2 \xrightarrow[(2)\ Y]{(1)\ KOC(CH_3)_3} Z$$

$$Z \xrightarrow{(1)\ OH^-/H_2O,\ heat;\ (2)\ H_3O^+;\ (3)\ Heat\ (-CO_2)} C_6H_5CH(CH_2C_6H_5)CO_2H$$

34. Consider the synthesis above. What is compound X?

* A) EtOCOEt (C=O)

B) HCOEt (C=O)

C) EtOCDCOEt (both C=O)

D) CH_3COEt (C=O)

E) $BrCH_2COEt$ (C=O)

35. Consider the synthesis above. What is compound Y?

A)
```
      O
      :
BrCH2COEt
```

B)
```
  O O
  : :
EtOCDCOEt
```

C) CH_3CH_2Br

* D) $C_6H_5CH_2Br$

E) C_6H_5Br

36. Consider the synthesis above. What is compound Z?

A)
```
C6H5CHCH2CO2Et
    3
    CH2C6H5
```

* B)
```
    CO2Et
    3
C6H5CCH2C6H5
    3
    CO2Et
```

C)
```
    CO2Et
    3
C6H5CCH2CH2C6H5
    3
    CO2Et
```

D)
```
      CO2Et
      3
C6H5CHCHCO2Et
    3
    CH2C6H5
```

E)
```
      O
      :
C6H5CHCCO2Et
    3
    CH2C6H5
```

37. Which of these is <u>not</u> a reversible process?

* A) Base-promoted ester hydrolysis
B) Acid-catalyzed ester hydrolysis
C) Aldol addition
D) Claisen condensation
E) Acetal formation

38. Malonic ester (diethyl malonate) is treated successively with sodium ethoxide (1 eq.), ethyl bromide, potassium *tert*-butoxide, isobutyl chloride, hot aqueous NaOH, HCl, and heat. What is the final product?

A) 4-Ethyl-2-methylpentanoic acid
B) 6-Methylheptanoic acid
C) 2-Ethyl-3-methylpentanoic acid
* D) 2-Ethyl-4-methylpentanoic acid
E) Ethylisobutylmalonic acid

39. The Claisen condensation of which of these esters demands the use of $(C_6H_5)_3CNa$ as the base (as opposed to sodium ethoxide)?

A) $CH_3CH_2CO_2C_2H_5$

B) $CH_3CO_2C_2H_5$

C) $C_6H_5CH_2CO_2C_2H_5$

* D)
```
C6H5CHCO2C2H5
    3
    CH3
```

E) None of these

40. What is the product of the Dieckmann condensation of this diester,

$$C_2H_5O\overset{\overset{O}{:}}{C}CH_2\underset{\underset{CH_3}{3}}{C}HCH_2CH_2\overset{\overset{O}{:}}{C}OC_2H_5 \quad ?$$

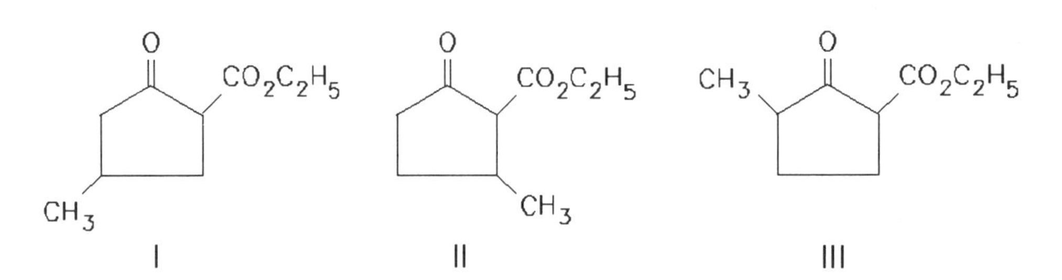

A) I
B) II
C) III
* D) I and II
E) I, II, and III

41. Which of these halides is predicted to alkylate malonic ester (as the anion) in highest yield?
* A) CH_3I
B) C_6H_5Br
C) $(CH_3)_3CCH_2Cl$
D) $CH_3CHClCH_3$
E) All of these should be equally effective.

42. The Claisen condensation produces which of these?
A) An !-keto ester
* B) A a-keto ester
C) A a-hydroxy ester
D) A a-hydroxyaldehyde
E) A a-diketone

43. Which is the most acidic hydrogen in the compound shown?

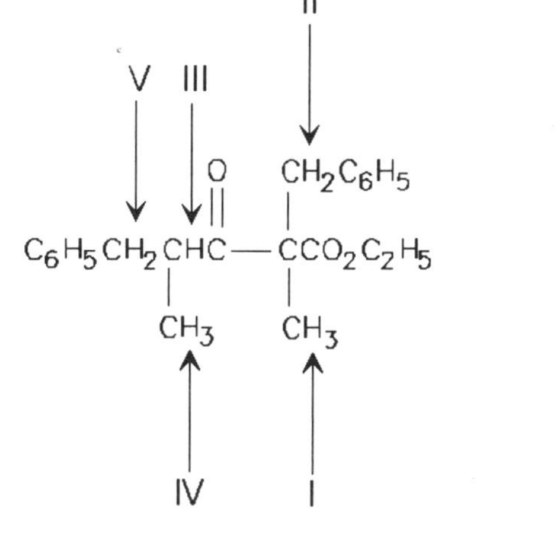

A) I B) II * C) III D) IV E) V

44. The Thorpe reaction of dinitriles is analogous to the Dieckmann condensation of diesters. What is the product predicted to result from the use of $NpCCH_2CH_2CH_2CH_2CpN$ and sodium ethoxide?

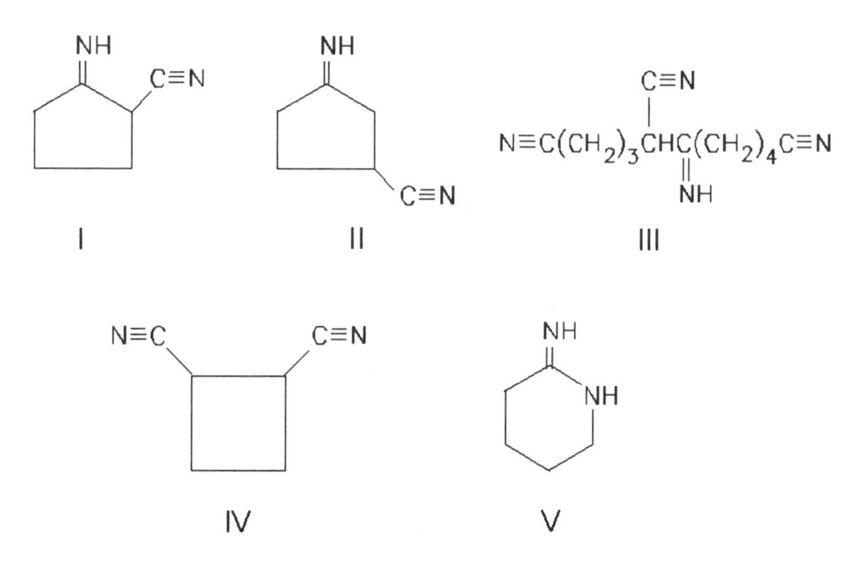

* A) I B) II C) III D) IV E) V

45. What product(s) result from the Claisen condensation carried out with an equimolar mixture of ethyl acetate and ethyl propanoate?

A)
```
    O  O
    :  :
CH3CCH2COC2H5
```
$CH_3CCH_2COC_2H_5$

B)
```
    O  O
    :  :
CH3CCHCOC2H5
      3
      CH3
```
$CH_3CCH(CH_3)COC_2H_5$

C)
```
       O  O
       :  :
CH3CH2CCHCOC2H5
         3
         CH3
```
$CH_3CH_2CCH(CH_3)COC_2H_5$

D)
```
       O   O
       :   :
CH3CH2CCH2COC2H5
```
$CH_3CH_2CCH_2COC_2H_5$

* E) All of these

46. Cyclization reactions, such as the Dieckmann condensation, are best carried out using fairly dilute solutions of the compound to be cyclized. Why is this so?

A) It then is possible to use less base.
B) The reagents generally are expensive.
C) A smaller amount of the compound to be cyclized can be used.
* D) Intermolecular condensation is minimized at low concentration.
E) The concentration factor is unimportant.

47. Why is CH_3ONa not used in the Claisen condensation of ethyl acetate?

A) CH_3O^- is a weaker base than the $CH_3CH_2O^-$ which <u>is</u> used.
B) $CH_3O^-Na^+$ is more difficult to prepare than $CH_3CH_2O^-Na^+$.
C) CH_3O^- would abstract a proton from the ethyl group of the ester.
* D) Use of $CH_3O^-Na^+$ would result in transesterification.
E) $CH_3O^-Na^+$ can be used as well as $CH_3CH_2O^-Na^+$.

48. Which of these combinations is <u>not</u> one which would result in the formation of essentially one Claisen condensation product when one compound is added slowly to the mixture of the other and the base employed?

A)
```
O                O
:                :
HCOC2H5   +   CH3CH2COC2H5
```
$HCOC_2H_5 + CH_3CH_2COC_2H_5$

B)
```
    O              O
    :              :
C6H5COC2H5   +   CH3COC2H5
```
$C_6H_5COC_2H_5 + CH_3COC_2H_5$

C)
```
     O O              O
     : :              :
C2H5OC-COC2H5   +   C6H5CH2COC2H5
```
$C_2H_5OC\text{-}COC_2H_5 + C_6H_5CH_2COC_2H_5$

D)
```
          O             O
          :             :
(CH3)3CCOC2H5   +   CH3COC2H5
```
$(CH_3)_3CCOC_2H_5 + CH_3COC_2H_5$

* E)
```
          O             O
          :             :
C6H5CH2COC2H5   +   CH3COC2H5
```
$C_6H_5CH_2COC_2H_5 + CH_3COC_2H_5$

49. Which structure represents an ester enolate?

A)
```
        O⁻
        3
CH2MCHOCCH2CH3
```

B)
```
        O
        :_
CH3CH2OCCHCH3
```

C)
```
        O⁻
        3
CH3CH2OCMCHCH3
```

D)
```
           O
        _  :
CH3CHOCCH2CH3
```

* E) Both B) and C)

50. The Knoevenagel condensation of p-methoxybenzaldehyde with ethyl acetoacetate in the presence of diethylamine produces which of these?

I $CH_3O-C_6H_4-CH=C(COCH_3)(CO_2C_2H_5)$

II $CH_3O-C_6H_4-CHOH-CH(COCH_3)(CO_2C_2H_5)$

III $CH_3O-C_6H_3(CHO)-CH(COCH_3)(CO_2C_2H_5)$

IV $CH_3O-C_6H_4-CH-C(O)(CO_2C_2H_5)(COCH_3)$ (epoxide)

V $CH_3O-C_6H_4-CH=CH-C(=O)-CH_2CO_2C_2H_5$

* A) I B) II C) III D) IV E) V

51. What product is finally formed when the initial compound formed from cyclohexanone and pyrrolidine is mixed with allyl chloride and that product is heated and then hydrolyzed?

HO $CH_2CH=CH_2$ O $CH_2CH=CH_2$ $\overset{+}{N}$ Cl^- $CH_2CH=CH_2$

I   II   III

$OCH_2CH=CH_2$ O $CH_2CH=CH_2$

IV   V

A) I   * B) II   C) III   D) IV   E) V

52. Which of these amines is/are used with aldehydes and ketones to form enamines?

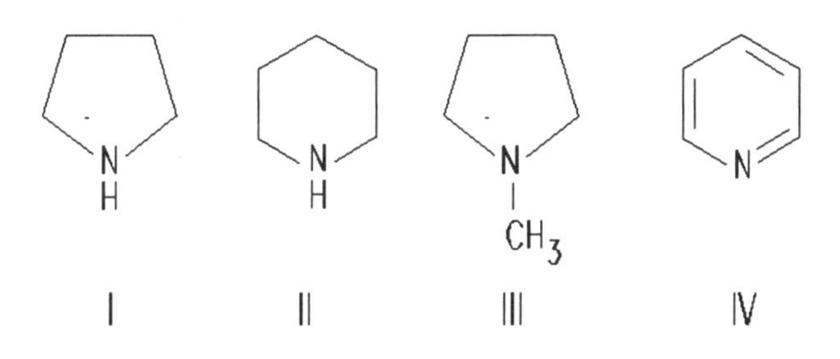

A) I
B) II
C) III
D) IV
* E) Both I and II

53. What alkylating agent would be used with phenylethanal in the Corey-Seebach method for the preparation of 4-methyl-1-phenyl-2-pentanone?
A) Isopropyl bromide
B) Butyl bromide
C) <u>sec</u>-Butyl bromide
* D) Isobutyl bromide
E) <u>tert</u>-Butyl bromide

54. Which base is employed in the alkylation of ethyl pentanoate with methyl iodide?

A) Sodium methoxide
B) Sodium ethoxide
C) Sodium hydride
D) Potassium <u>tert</u>-butoxide
* E) Lithium diisopropylamide

55. What is the major product of the following reaction?

```
              O
              :                   NaNH2           (1)  (DCO2C2H5)2
    CH3CHCH2CCH3                DDDDDDD4        DDDDDDDDDDDDDDDDDDD4
       3                          Et2O            (2)  H+
       CH3
```

A)
```
       COCH3
       3
    CH3CHCHC-COC2H5
       3    : :
       CH3  O O
```

* B)
```
              O    O O
              :    : :
    CH3CHCH2CCH2CDCOC2H5
       3
       CH3
```

C)
```
              O    O
              :    :
    CH3CHCH2CCH2COC2H5
       3
       CH3
```

D)
```
         O  H3C O O
         :    3 : :
    CH3CCH2DCDCDCOC2H5
              3
            H3C
```

E)
```
         COCH3
         3
    CH3CHCHCOC2H5
       3    :
       CH3  O
```

Chapter 21: Phenols and Aryl Halides

1. Indicate the correct product, if any, of the following reaction.

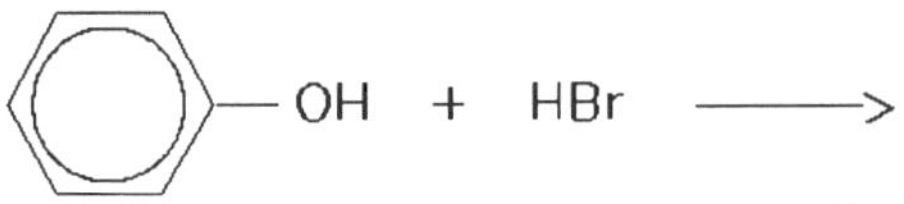

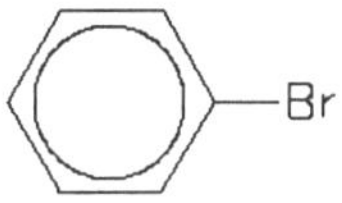

$$C_6H_5OH + HBr \longrightarrow$$

I: bromobenzene (ring–Br)

II: ring with –OH and ortho –Br

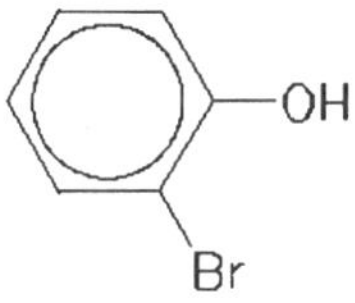

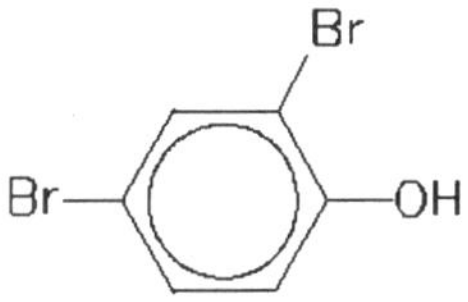

III: ring with –OH, –Br ortho to OH, and Br– para to OH

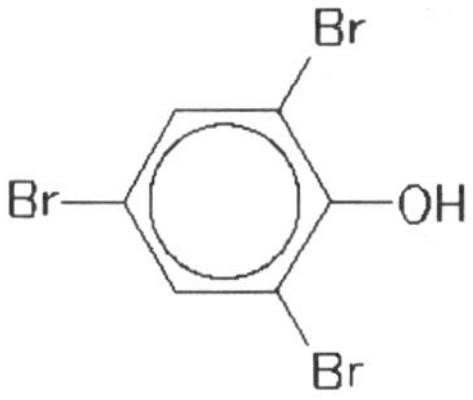

IV: ring with –OH, two –Br ortho to OH, and Br– para to OH

A) I
B) II
C) III
D) IV
* E) There is no net reaction.

2. The common name for 4-methylphenol is which of these?
A) Catechol
* B) p-Cresol
C) Resorcinol
D) Hydroquinone
E) p-Xylenol

3. Which of the following phenols would have the largest pk_a?

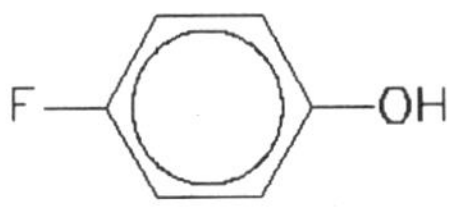

$F-C_6H_4-OH$ (I) $O_2N-C_6H_4-OH$ (II) $CH_3-C_6H_4-OH$ (III)

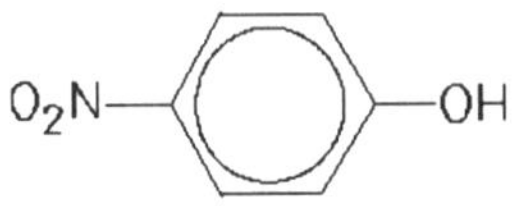

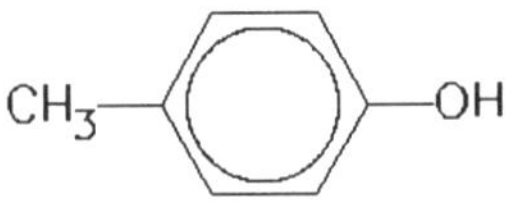

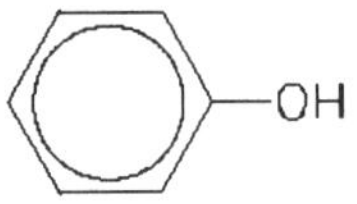

C_6H_5-OH (IV) $Cl-C_6H_4-OH$ (V)

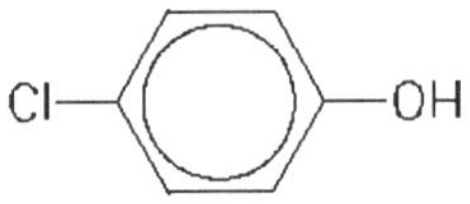

A) I B) II * C) III D) IV E) V

4. Which of the following compounds would you expect to be the strongest acid?

A) CH_3OH
B) $C_6H_5CH_2OH$
C) p-$CH_3C_6H_4OH$
D) C_6H_5OH
* E) p-$NO_2C_6H_4OH$

5. Refluxing anisole, $CH_3OC_6H_5$, with excess concentrated HBr would yield which of these product mixtures?

A) C_6H_5Br + CH_3OH
B) C_6H_5OH + CH_4
C) C_6H_5OH + CH_3OH
D) C_6H_5Br + CH_3Br
* E) C_6H_5OH + CH_3Br

6. What product(s) would you expect from the following reaction?

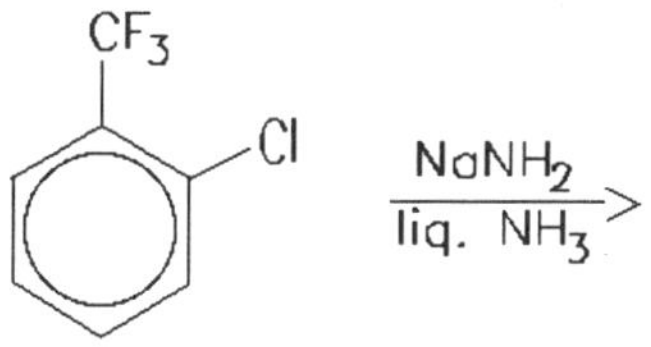

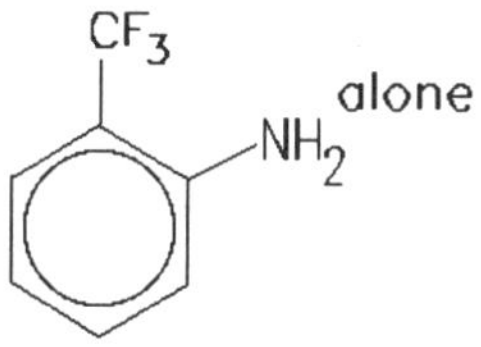

2-chloro(trifluoromethyl)benzene $\xrightarrow[\text{liq. } NH_3]{NaNH_2}$

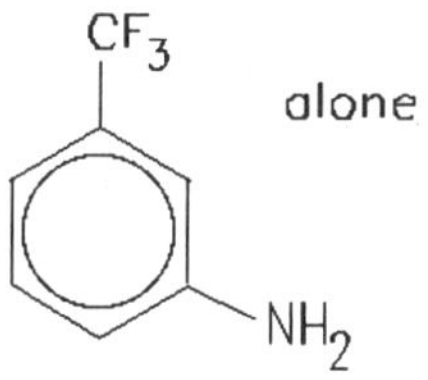

I: 2-(trifluoromethyl)aniline (CF_3, NH_2 ortho) alone

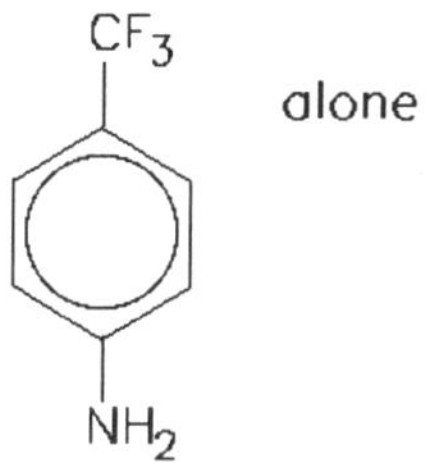

II: 3-(trifluoromethyl)aniline (CF_3, NH_2 meta) alone

III: 4-(trifluoromethyl)aniline (CF_3, NH_2 para) alone

A) I
* B) II
C) III
D) Substantial amounts of I and II
E) Substantial amounts of I, II, and III

7. Which compound would be most acidic?

A) Cyclohexanol
B) 1-Hexanol
C) Phenol
D) 4-Methylphenol
* E) 4-Chlorophenol

8. Which of the following would be most likely to undergo a nucleophilic substitution reaction with aqueous sodium hydroxide by an addition-elimination mechanism?

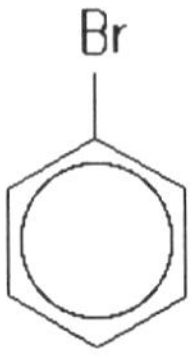

I

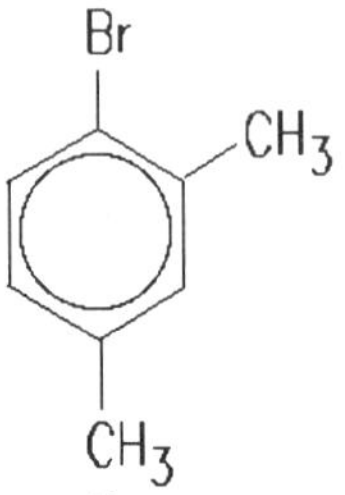

II

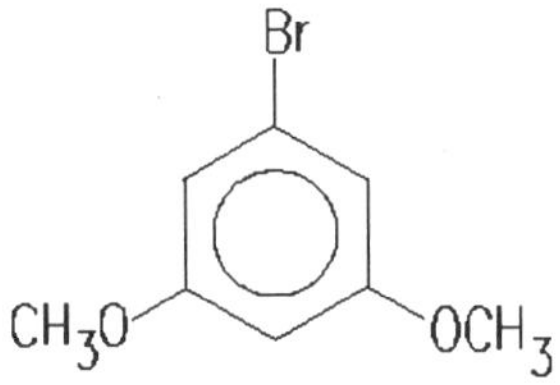

III

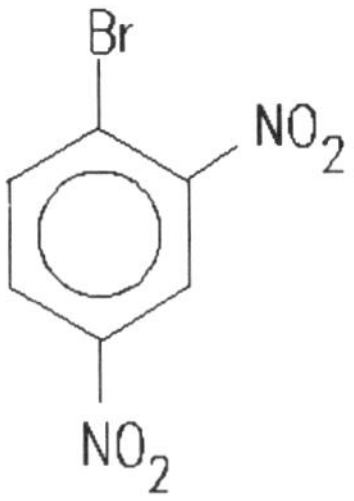

IV

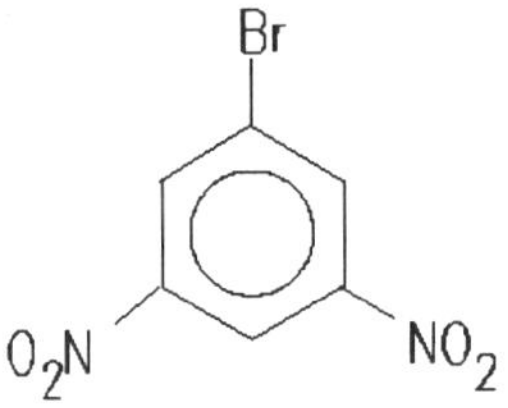

V

A) I   B) II   C) III   * D) IV   E) V

9. Which of the following would be the strongest acid?

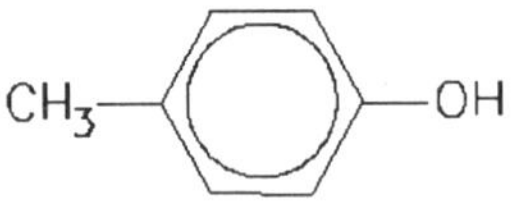

I

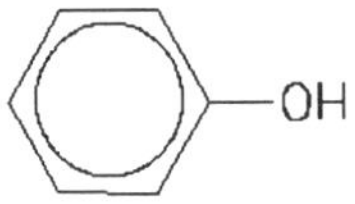

II

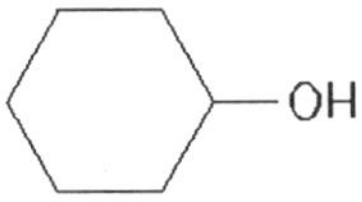

III

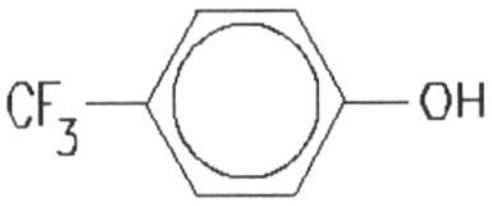

IV

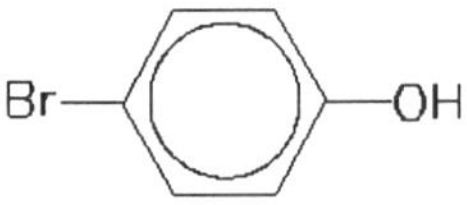

V

A) I   B) II   C) III   * D) IV   E) V

10. What would be the major product of the following reaction?

$$\text{phenol} + Br_2 \text{ (excess)} \xrightarrow[H_2O]{} ?$$

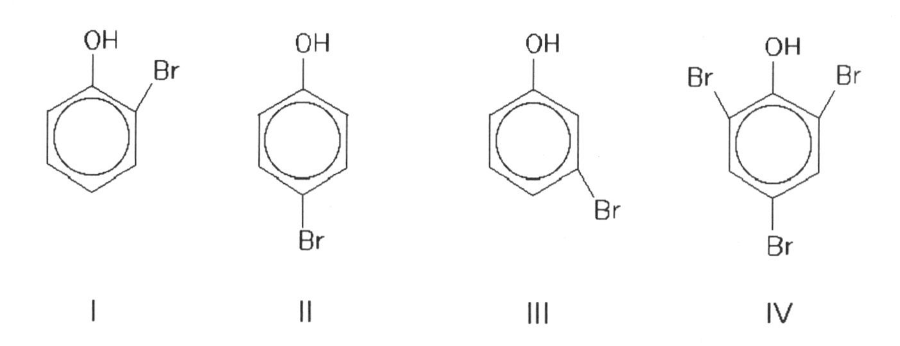

A) I
B) II
C) III
* D) IV
E) A mixture of I and II

11. What products would you expect from the following reaction?

$$\text{4-Chlorotoluene} \xrightarrow[(2)\ H_3O^+]{(1)\ OH^-/H_2O,\ 400°C} ?$$

A) 2-Methylphenol
B) 3-Methylphenol
C) 4-Methylphenol
D) A) and B)
* E) B) and C)

12. Which of the following reactions would yield p-tert-butylphenol?

A) $\text{Phenol} + CH_3C(CH_3){=}CH_2 \xrightarrow{H^+}$

B) $\text{Phenol} + (CH_3)_3COH \xrightarrow{H^+}$

C) $\text{Phenol} + (CH_3)_3CCl \xrightarrow{AlCl_3}$

* D) All of these
E) None of these

13. Compound L has the molecular formula $C_{11}H_{16}O$. L is insoluble in water but dissolves in aqueous NaOH. The infrared spectrum of L shows a broad absorption band in the 3200-3600 cm^{-1} region; its ^1H NMR spectrum consists of:

triplet,	δ 0.80	
singlet,	δ 1.2	(6H)
quartet,	δ 1.5	
singlet,	δ 4.5	(1H)
multiplet,	δ 7.0	(4H)

The most likely structure for compound L is:

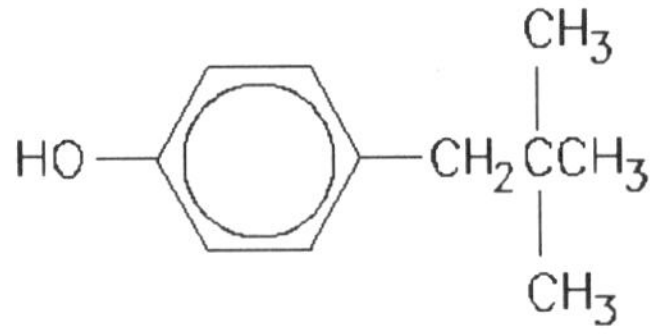

I: $HO-C_6H_4-CH_2C(CH_3)_2CH_3$

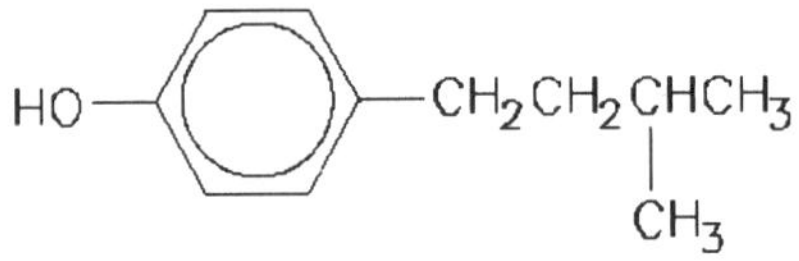

II: $HO-C_6H_4-CH_2CH_2CH(CH_3)CH_3$

III: $CH_3O-C_6H_4-CH_2CH_2CH_2CH_3$

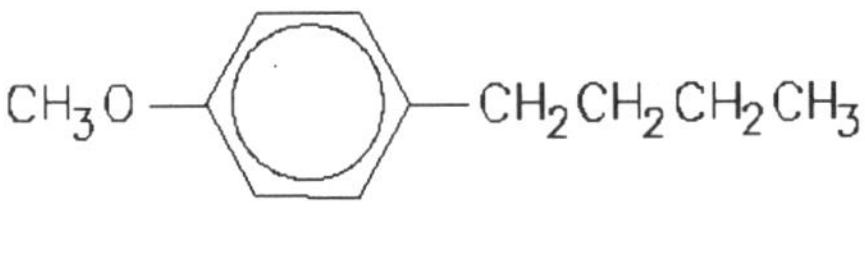

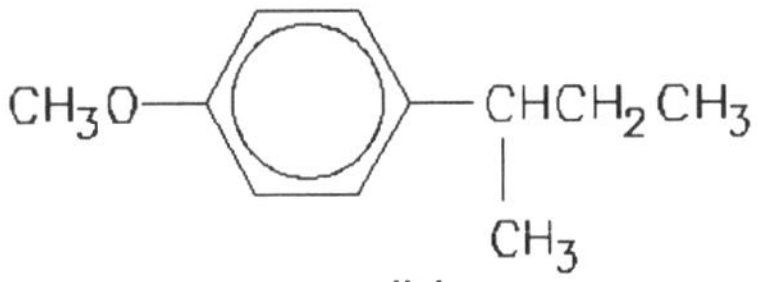

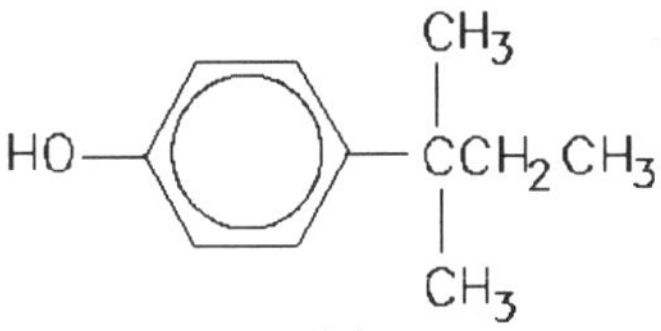

IV: $CH_3O-C_6H_4-CH(CH_3)CH_2CH_3$

V: $HO-C_6H_4-C(CH_3)_2CH_2CH_3$

A) I B) II C) III D) IV * E) V

14. What is the major product of the following reaction?

$$\text{phenol} \xrightarrow[100^\circ C]{\text{concd } H_2SO_4} \xrightarrow[\text{heat}]{HNO_3/H_2SO_4} \xrightarrow[\text{steam distill}]{H_3O^+/H_2O,\ \text{heat}} ?$$

A) p-Hydroxybenzenesulfonic acid
B) p-Nitrophenol
* C) o-Nitrophenol
D) 4-Hydroxy-3-nitrobenzenesulfonic acid
E) 2-Hydroxy-4-nitrobenzenesulfonic acid

15. Which products would be formed in the following reaction?

$$\text{2-}CH_3C_6H_4OCH_3 \xrightarrow[\text{heat}]{\text{concd HBr (excess)}} ?$$

A) Methoxybenzene + methyl bromide
* B) 2-Methylphenol + methyl bromide
C) 2-Bromotoluene + methanol
D) 2-Bromotoluene + methyl bromide
E) Bromobenzene + methyl bromide

16. Which would be soluble in aqueous sodium bicarbonate?
A) Benzoic acid
B) 4-Methylphenol
C) 2,4-Dinitrophenol
* D) More than one of these
E) All of these

17. What is the final product?

$$\text{phenol} \xrightarrow[AlCl_3]{\text{tert-butyl chloride}} C_{10}H_{14}O \text{ (para-isomer)} \xrightarrow{NaOH} \xrightarrow{CH_3CH_2I} ?$$

* A) 1-tert-Butyl-4-ethoxybenzene
B) 1-tert-Butyl-4-ethylbenzene
C) 1-tert-Butoxy-4-ethoxybenzene
D) tert-Butyl ethyl ether
E) 1-tert-Butoxy-3-ethylbenzene

18. Which reagent will distinguish between C_6H_5OH and $C_6H_5CH_2OH$?
A) $NaHCO_3$ (aq)
* B) NaOH (aq)
C) H_2SO_4
D) A) and B)
E) B) and C)

19. Which of the following would provide a synthesis of aspirin,

o-$CH_3COOC_6H_4COOH$?

A) C_6H_5COOH, CH_3COOH, $AlCl_3$, heat; then H_2O
B) $CH_3COOC_6H_5$, CO_2, heat; then H^+
C) $CH_3COOC_6H_5$, $HCOOC_2H_5$, $C_2H_5O^-$; then H_3O^+; then OH^-
D) C_6H_5OH, CO_2, H_3O^+; separate isomers; then CH_3COOH, $AlCl_3$
* E) C_6H_5OH, OH^-, CO_2, heat, pressure; then H_3O^+; then $(CH_3CO)_2O$

20. Identify the commercial preparation(s) of phenol.
A) Chlorobenzene + NaOH at 350°C; then H^+
B) Benzene + O_2
C) $C_6H_5CH(CH_3)_2$ + O_2; then H^+, heat
D) A) and B)
* E) A) and C)

21. Which reagent will distinguish between $C_6H_5OCH_3$ and p-$CH_3C_6H_4OH$?
* A) NaOH (aq)
B) $NaHCO_3$ (aq)
C) H_2CrO_4
D) A) and B)
E) None of these

22. What products(s) would you expect from the following reaction?

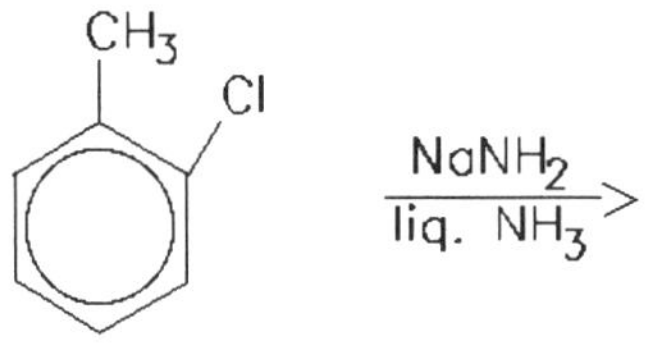

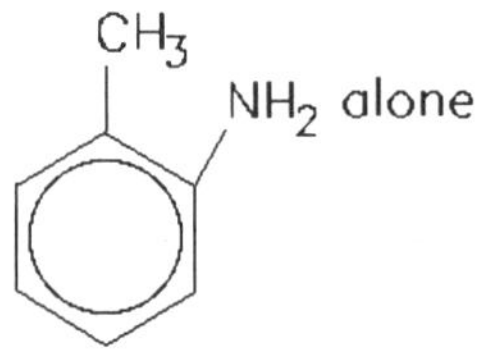

CH_3, Cl (ortho-chlorotoluene) $\xrightarrow[\text{liq. } NH_3]{NaNH_2}$

I: CH_3, NH_2 (ortho) alone
II: CH_3, NH_2 (meta) alone
III: CH_3, NH_2 (para) alone

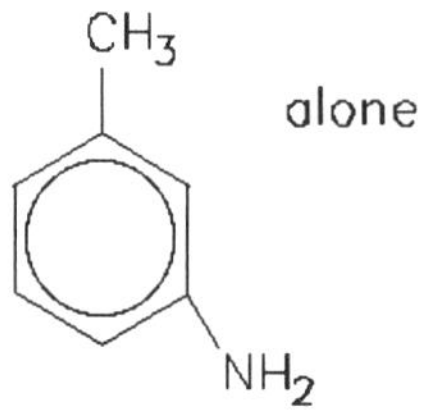

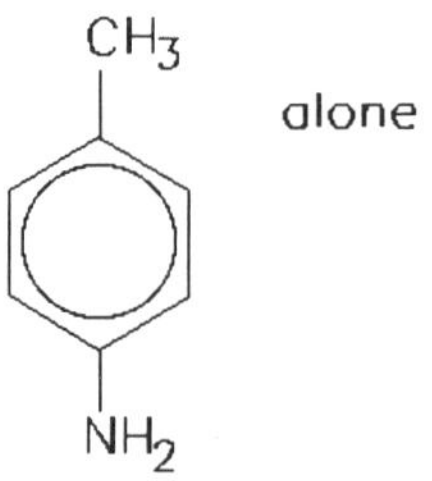

A) I
B) II
C) III
* D) Products I and II
E) All of the above

23. Which method could be used for the following synthesis?

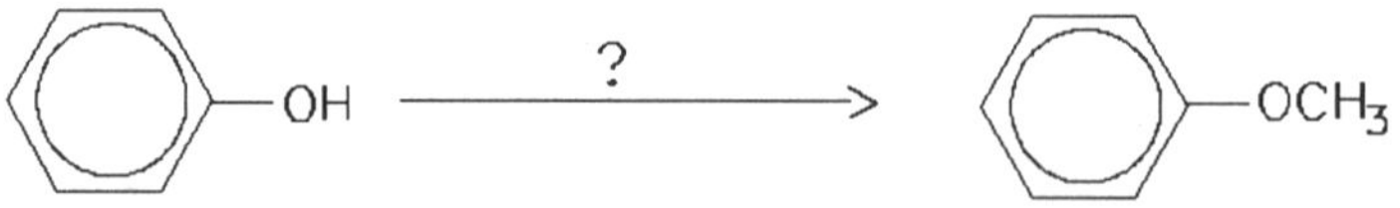

C_6H_5–OH $\xrightarrow{?}$ C_6H_5–OCH_3

A) NaOH, then CH_3I
B) NaOH, then $CH_3OSO_2CH_3$
C) NaOH, then CH_3OCH_3
* D) More than one of these
E) All of these

24. What is the product of the following synthesis?

$$\text{phenol} \xrightarrow{H_2SO_4,\ 100^\circ C} \xrightarrow{2Br_2/H_2O} \xrightarrow{\text{dil. } H_2SO_4,\ \text{heat}}$$

A) 2,3-Dibromophenol
B) 2,4-Dibromophenol
* C) 2,6-Dibromophenol
D) 2-Hydroxy-3,5-dibromobenzenesulfonic acid
E) 2,4,6-Tribromophenol

25. In terms of reactivity towards nucleophiles, bromobenzene is most similar to which of these?

A) Allyl bromide
* B) Vinyl bromide
C) tert-Butyl bromide
D) Propyl bromide
E) Methyl bromide

26. In nucleophilic aromatic substitution, the attacking species (the nucleophile) necessarily is:

* A) A Lewis base
B) Neutral
C) Positively charged
D) Negatively charged
E) Electron deficient

27. Which compound reacts most rapidly with CH_3ONa?

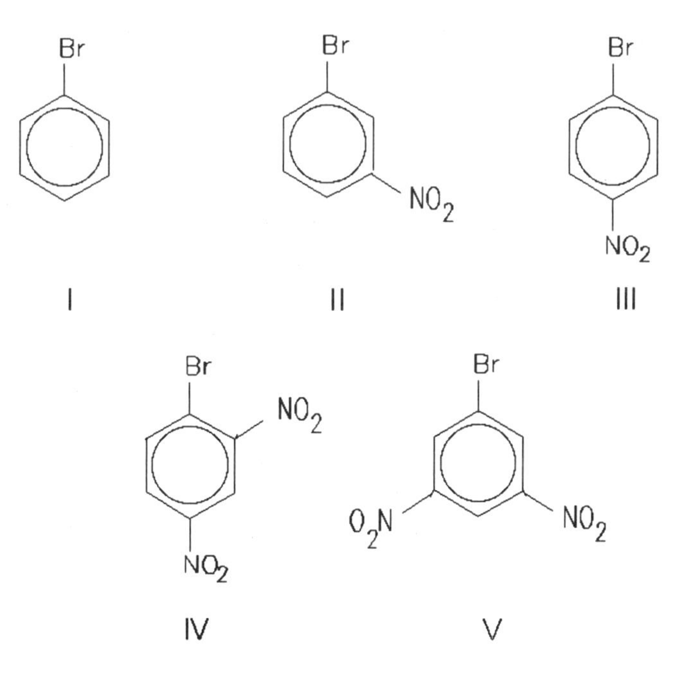

A) I
B) II
C) III
* D) IV
E) V

28. What product should be obtained if benzyne is generated in the presence of 1,3-butadiene?

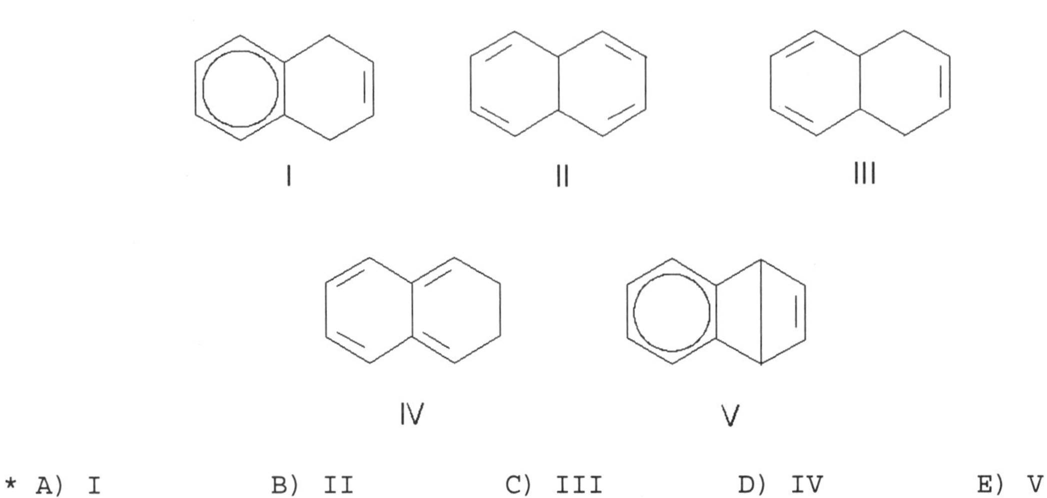

* A) I B) II C) III D) IV E) V

29. Which compound results from the heating of

C_6H_5—O—$^{14}CH_2CH{=}CH_2$

to 200°C?

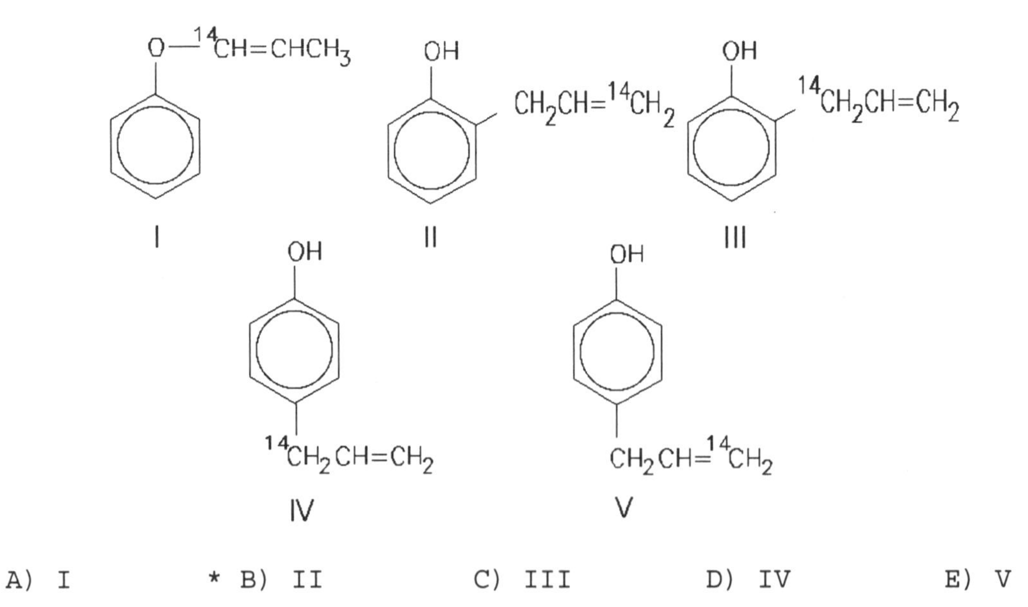

A) I * B) II C) III D) IV E) V

30. When sodium phenoxide is heated to 125°C with carbon dioxide under pressure and the product mixture acidified, there is produced which of these?

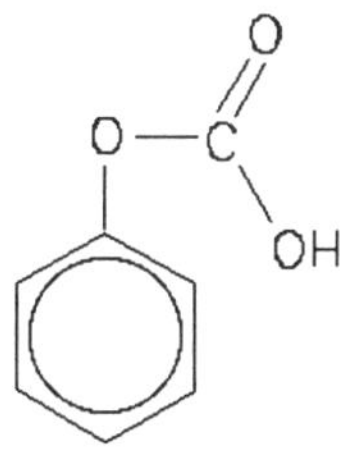

I: phenyl hydrogen carbonate ($C_6H_5O{-}C(=O){-}OH$)

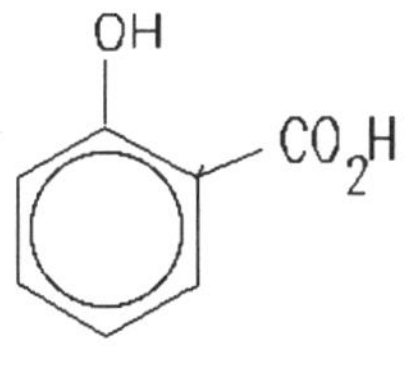

II: 2-hydroxybenzoic acid (OH and CO_2H ortho)

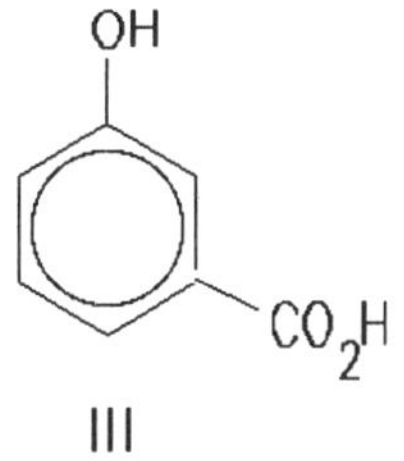

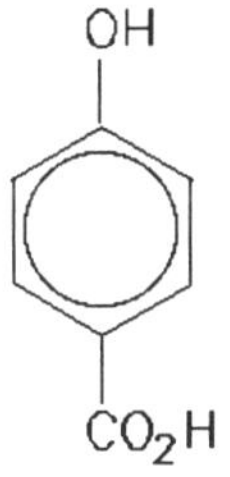

III: 3-hydroxybenzoic acid (OH and CO_2H meta)

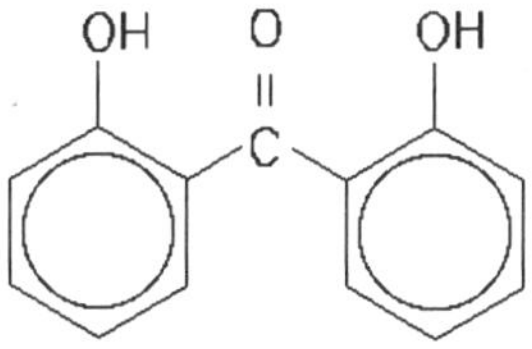

IV: 4-hydroxybenzoic acid (OH and CO_2H para)

V: bis(2-hydroxyphenyl) ketone (two 2-hydroxyphenyl rings joined by C=O)

A) I * B) II C) III D) IV E) V

31. Which of these is a satisfactory method for the preparation of m-nitrophenol from benzene?

A) $\xrightarrow[H_2SO_4]{HNO_3}$ $\xrightarrow[HCl]{Sn}$ $\xrightarrow[0\text{-}5°C]{NaNO_2,\ H_2SO_4,\ H_2O}$ $\xrightarrow{Cu_2O,\ Cu^{2+},\ H_2O}$ $\xrightarrow{HNO_3}$

B) $\xrightarrow[Fe]{Cl_2}$ $\xrightarrow[H_2SO_4]{HNO_3}$ $\xrightarrow[H_2O]{NaOH}$ $\xrightarrow{H_3O^+}$

C) $\xrightarrow[H_2SO_4]{HNO_3}$ $\xrightarrow[Fe]{Cl_2}$ $\xrightarrow[H_2O]{NaOH}$ $\xrightarrow{H_3O^+}$

* D) $\xrightarrow[2\ H_2SO_4]{2\ HNO_3}$ $\xrightarrow[H_2S]{NH_3}$ $\xrightarrow[0\text{-}5°C]{NaNO_2,\ H_2SO_4,\ H_2O}$ $\xrightarrow{Cu_2O,\ Cu^{2+},\ H_2O}$

E) $\xrightarrow[H_2SO_4]{HNO_3}$ $\xrightarrow[fusion]{NaOH}$ $\xrightarrow{H_3O^+}$

32. The action of Ag^{+} or Fe^{+3} on

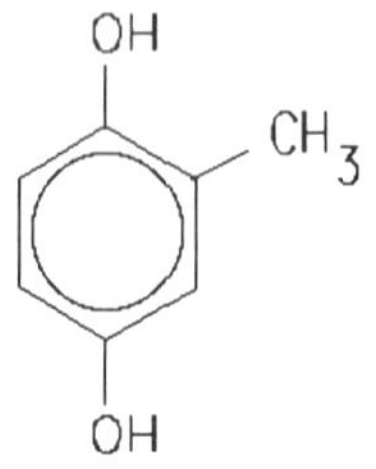

OH
CH3
OH

produces which of these products?

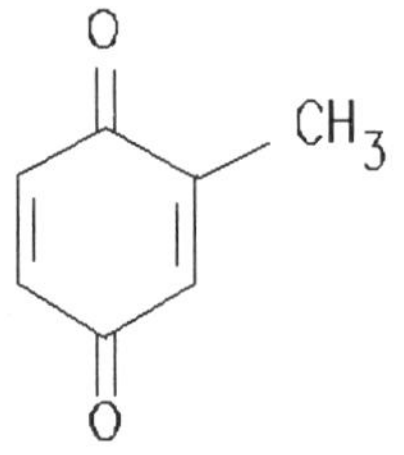

O
CH3
O

I

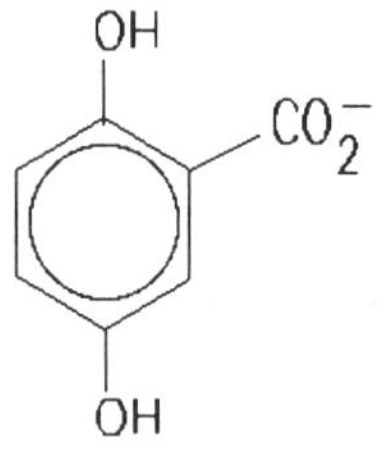

OH
CO2−
OH

II

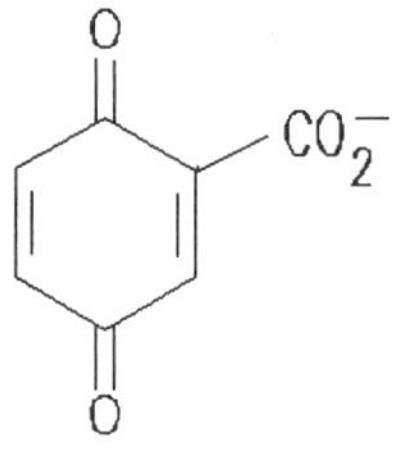

O
CO2−
O

III

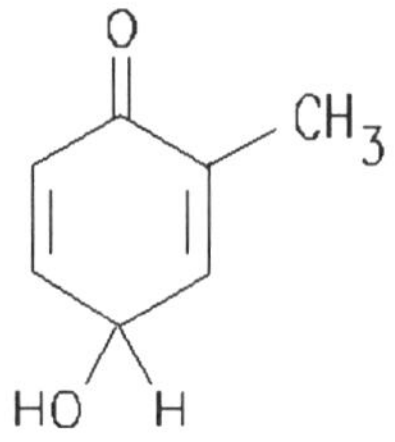

O
CH3
HO
H

IV

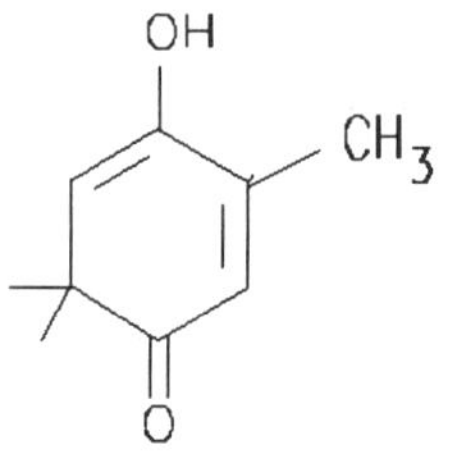

OH
CH3
O

V

* A) I B) II C) III D) IV E) V

33. Which is the leaving group when

Br I, Br II, H III, NO_2 IV, V H

reacts with sodium cyanide in DMSO solution?

* A) I B) II C) III D) IV E) V

34. What is the product of the reaction of 1 mol of p-benzoquinone with 1 mol of isoprene (2-methyl-1,3-butadiene)?

I II III

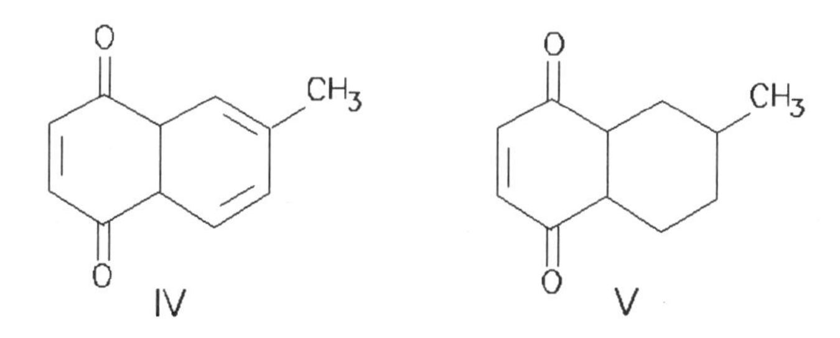

* A) I B) II C) III D) IV E) V

35. What are the products of the reaction of phenol with propanoic anhydride in the presence of base?

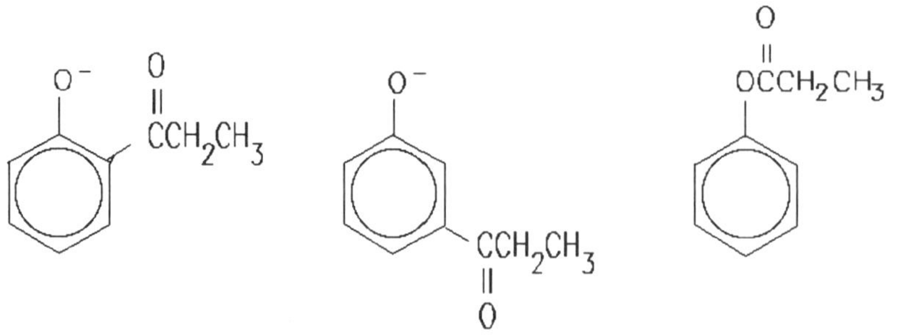

I: o-propanoylphenoxide (ring with O^- and ortho $\overset{O}{\overset{\|}{C}}CH_2CH_3$) + $CH_3CH_2CO_2^-$

II: m-propanoylphenoxide (ring with O^- and meta $\underset{O}{\underset{\|}{C}}CH_2CH_3$) + $CH_3CH_2CO_2^-$

III: phenyl propanoate ($C_6H_5O\overset{O}{\overset{\|}{C}}CH_2CH_3$) + $CH_3CH_2CO_2^-$

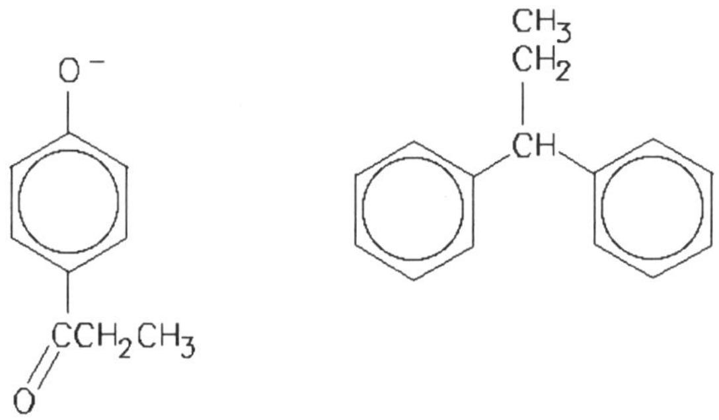

IV: p-propanoylphenoxide (ring with O^- and para $\overset{O}{\overset{\|}{C}}CH_2CH_3$) + $CH_3CH_2CO_2^-$

V: 1,1-diphenylpropane ($CH_3CH_2CH(C_6H_5)_2$) + $CH_3CH_2CH_2OH$

A) I B) II * C) III D) IV E) V

36. When

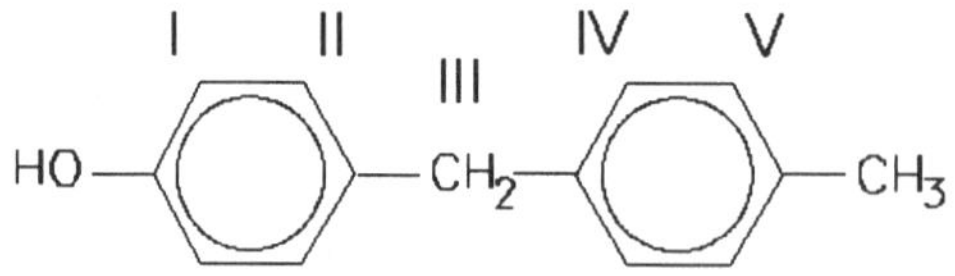

HO—C_6H_4—CH_2—C_6H_4—CH_3 (positions labeled I, II on the phenol ring; III at CH_2; IV, V on the toluene ring)

reacts with bromine in carbon disulfide at 10°C, which position is predicted to be the chief site of substitution?

* A) I B) II C) III D) IV E) V

37. What is the product of the reaction of phenol and chloroacetic acid in basic solution, followed by acidification?

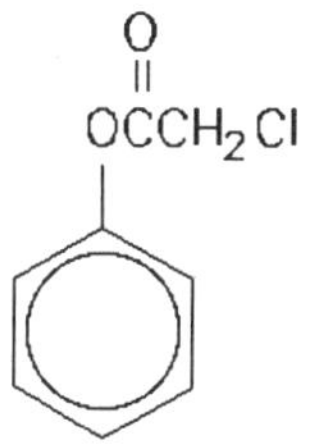

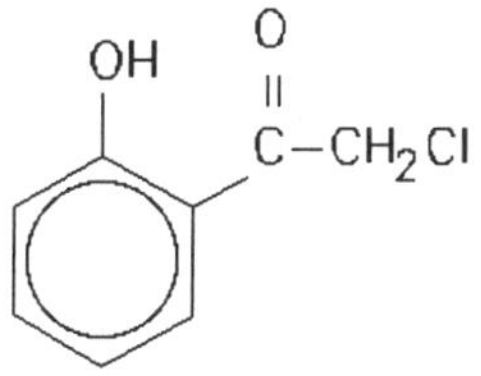

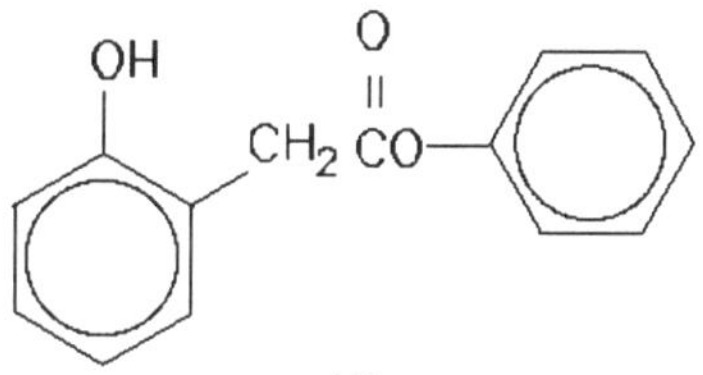

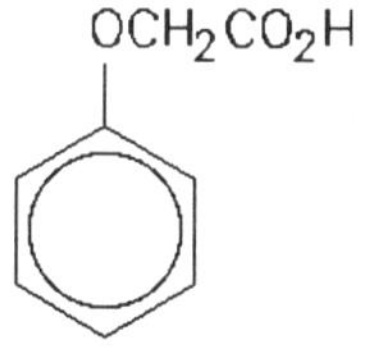

A) I B) II C) III * D) IV E) V

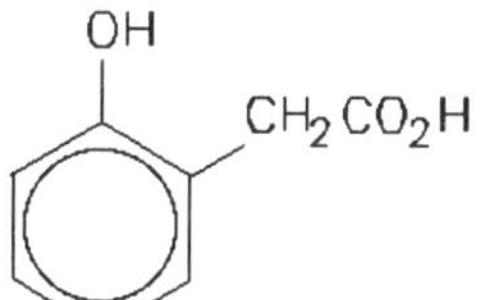

38. Which of these species is the strongest base?

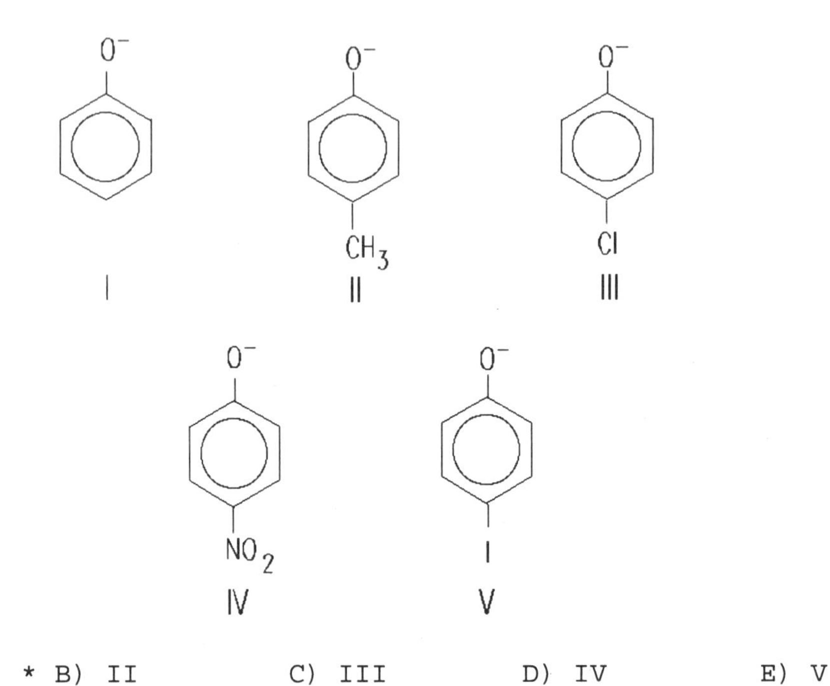

A) I   * B) II   C) III   D) IV   E) V

39. Which of these is an acceptable synthesis of phenetole (ethyl phenyl ether)?

A) $C_6H_5Cl + CH_3CH_2ONa \xrightarrow{100^\circ C}$

B) $C_6H_5ONa + CH_3CH_2I \xrightarrow{70^\circ C}$

C) $C_6H_5ONa + (CH_3CH_2O)_2SO_2 \xrightarrow{100^\circ C}$

D) $C_6H_5OH + CH_3CH_2OH \xrightarrow[70^\circ C]{H_2SO_4}$

* E) More than one of these

40. Which of these reactions does <u>not</u> produce phenol?

A) $C_6H_5OC(=O)CH_3 + NaOH/H_2O;$ then $H_3O^+ \longrightarrow$

B) $C_6H_5ONa + H_3O^+ \longrightarrow$

C) $C_6H_5OCH_3 + HI \longrightarrow$

* D) $C_6H_5Cl + H_2O \xrightarrow{100^\circ C}$

E) $C_6H_5NH_2 + NaNO_2$, dil H_2SO_4, 0-5°C; then Cu_2O, $Cu(NO_3)_2$, $H_2O \longrightarrow$

41. What is the systematic (IUPAC) name for

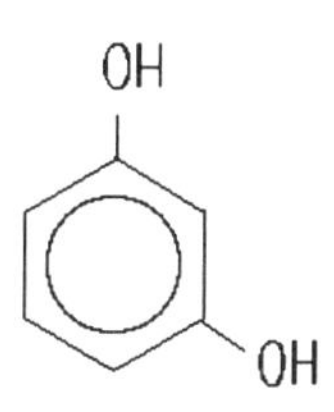

A) m-Hydroxyphenol
B) m-Dihydroxybenzene
C) Resorcinol
D) 1,3-dihydroxybenzene
* E) 1,3-Benzenediol

42. Predict the product of this Cope arrangement:

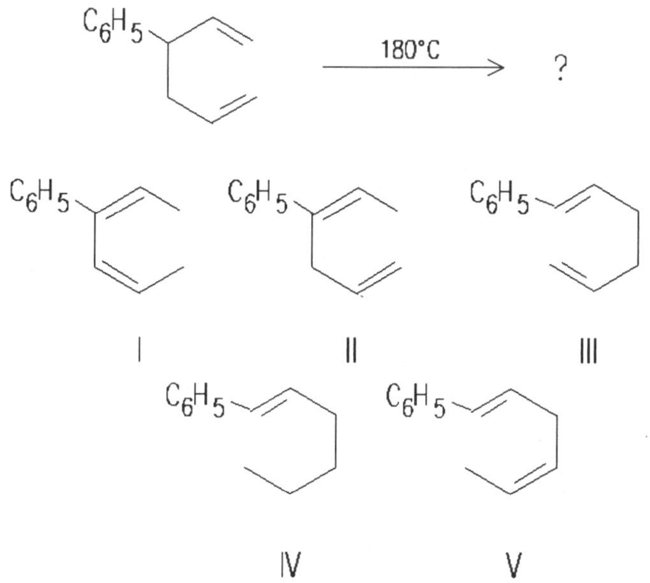

C_6H_5 —— 180°C → ?

C_6H_5 I

C_6H_5 II

C_6H_5 III

C_6H_5 IV

C_6H_5 V

A) I　　B) II　　* C) III　　D) IV　　E) V

43. Which of these resonance structures makes the greatest contribution to the hybrid for the intermediate in the S_NAr reaction of o-chloronitrobenzene with methoxide ion?

I II III

IV V

A) I B) II C) III D) IV * E) V

44. The formation of equal amounts of m-toluidine (m-aminophenol) and p-toluidine in the reaction of p-bromotoluene with sodium amide in liquid ammonia at -33°C suggests this species as the reaction intermediate:

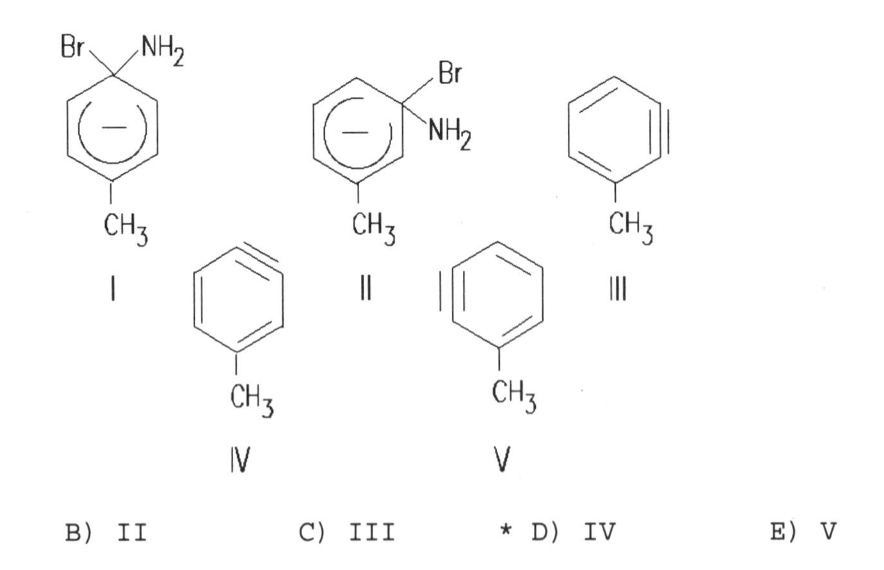

A) I B) II C) III * D) IV E) V

45. Phenol is less reactive than expected in Friedel-Crafts acylations for which of these reasons?

A) The acid chloride or anhydride hydrolyzes under the reaction conditions.

B) The electronegativity of the -OH group reduces the reactivity of phenol.

* C) Interaction of the -OH group with $AlCl_3$ forms a species less reactive than phenol.

D) The reaction gives O-acylation only, resulting in an ester.

E) The aromatic ring complexes with $AlCl_3$ with a resulting decrease in phenol reactivity.

Chapter 22: Carbohydrates

1. Sugars that would yield the same phenylosazone are:

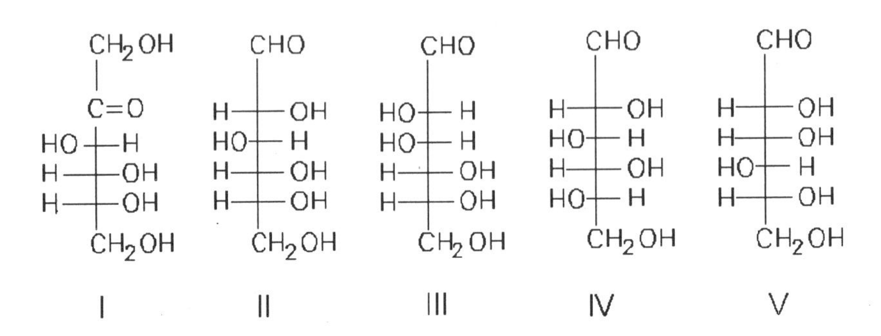

A) I and II
B) II and III
* C) I, II, and III
D) III and IV
E) I, II, and V

2. Epimers are represented by:

CH_2OH
C=O
HO—H
H—OH
H—OH
CH_2OH
I

CHO
H—OH
HO—H
H—OH
H—OH
CH_2OH
II

CHO
HO—H
HO—H
H—OH
H—OH
CH_2OH
III

CHO
H—OH
H—OH
HO—H
HO—H
CH_2OH
IV

CHO
HO—H
HO—H
HO—H
HO—H
CH_2OH
V

A) I and II
* B) II and III
C) I, II, and III
D) III and IV
E) I, II, and V

3. Reaction of I (below) with sodium borohydride ($NaBH_4$) would yield:

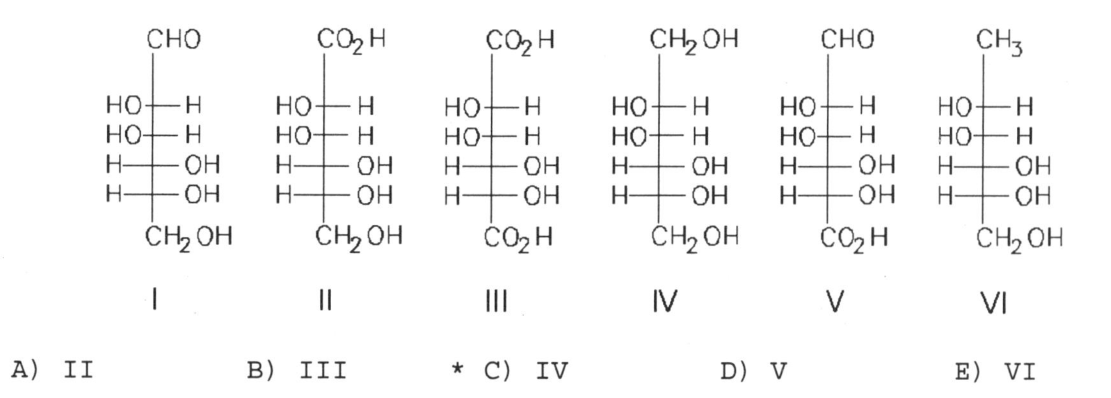

A) II B) III * C) IV D) V E) VI

4. Reaction of I (below) with bromine water would yield:

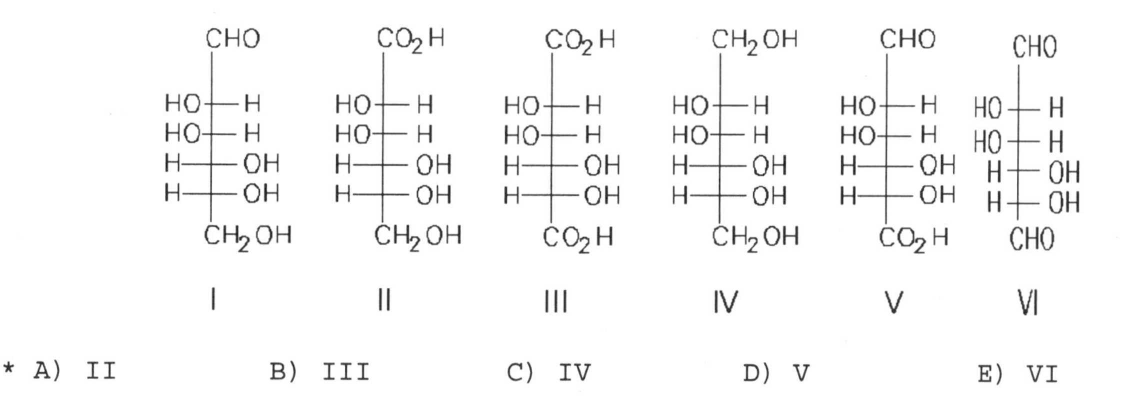

* A) II B) III C) IV D) V E) VI

5. An aldaric acid is represented by:

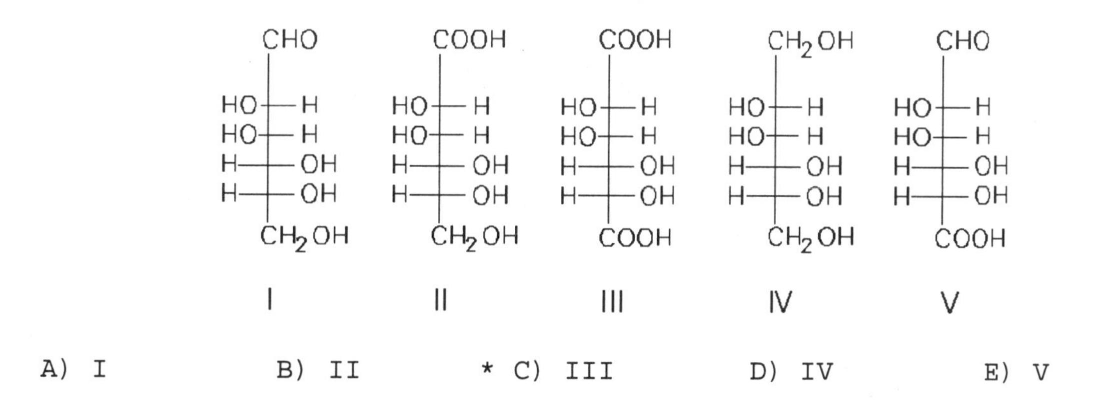

A) I B) II * C) III D) IV E) V

6. Sucrose reacts with which of these reagents?

A) $C_6H_5NHNH_2$
B) Cu^{2+}
C) Br_2/H_2O
* D) H_3O^+
E) $Ag(NH_3)_2OH$

7. Which of these compounds, I, II, III, IV, is a reducing disaccharide?

I

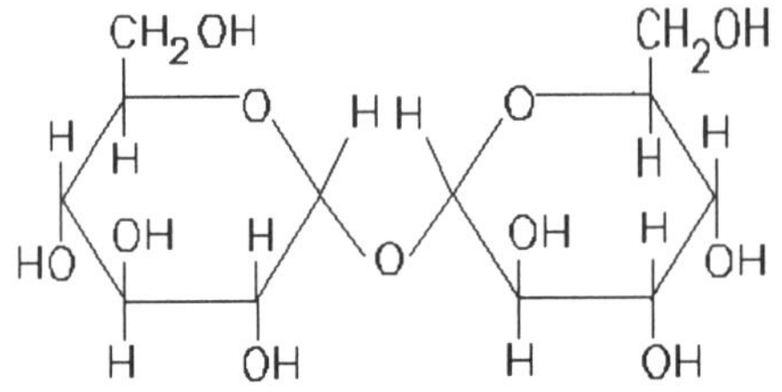

II

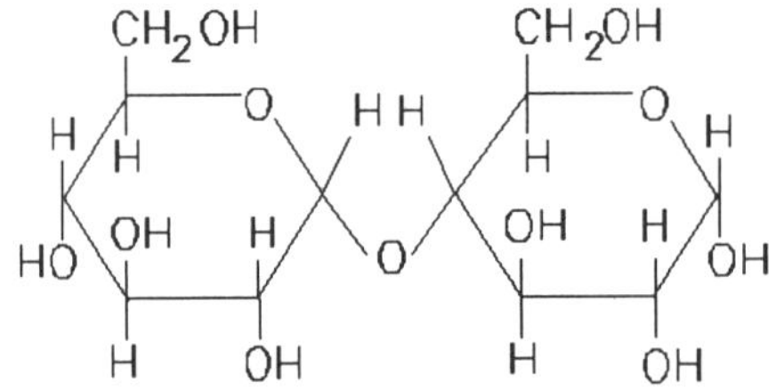

III

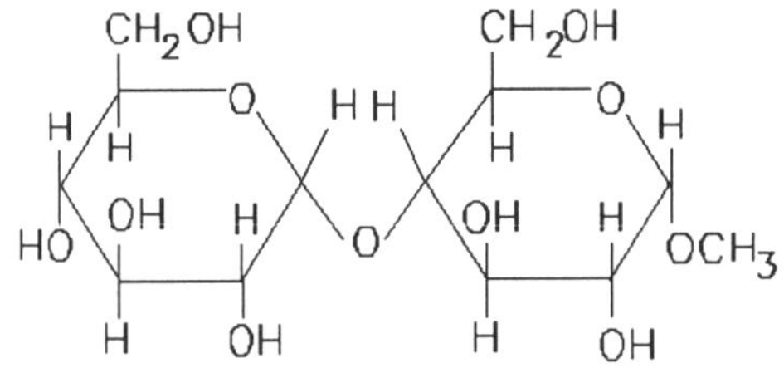

IV

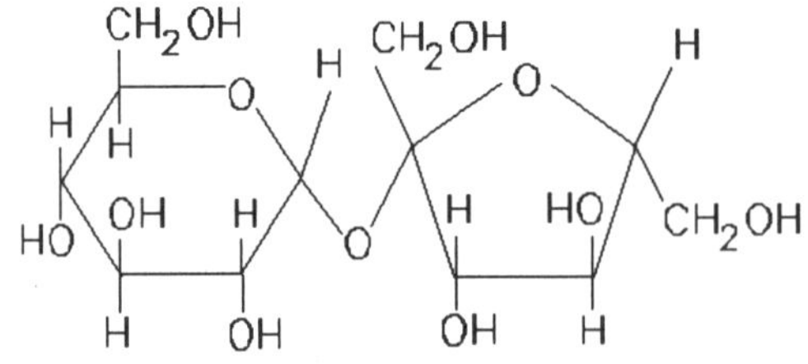

A) I alone
* B) II alone
C) III alone
D) IV alone
E) I, II, III, and IV

8. Sugars that undergo mutarotation in neutral aqueous solution are:

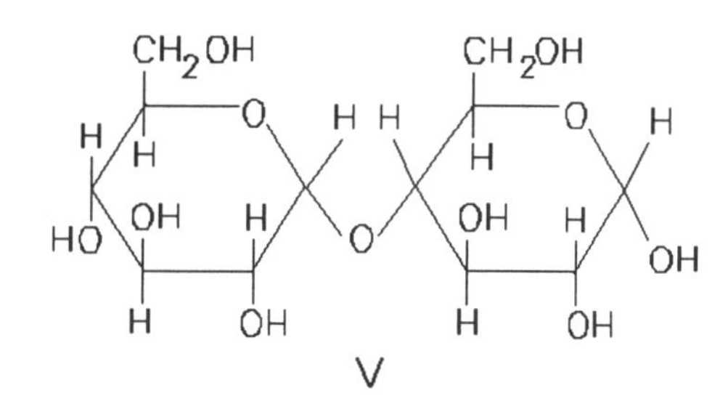

A) I and III
B) III and IV
C) II, III, and IV
D) I and IV
* E) I, II, and V

9. If J (below) were treated with dilute aqueous hydrochloric acid and the solution allowed to stand, what compounds (other than methanol) would be formed in the solution?

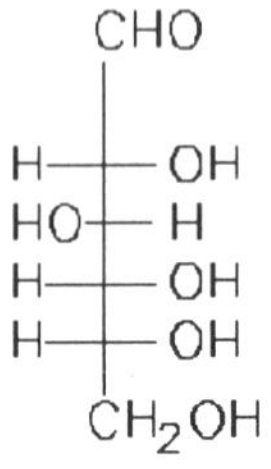

I	II
CHO	CHO
H—OH	HO—H
HO—H	HO—H
H—OH	H—OH
H—OH	H—OH
CH_2OH	CH_2OH

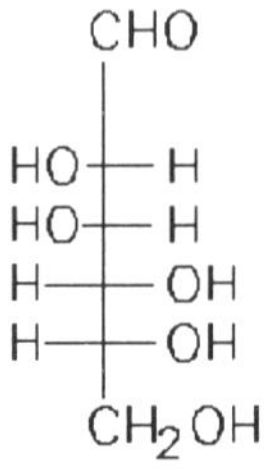

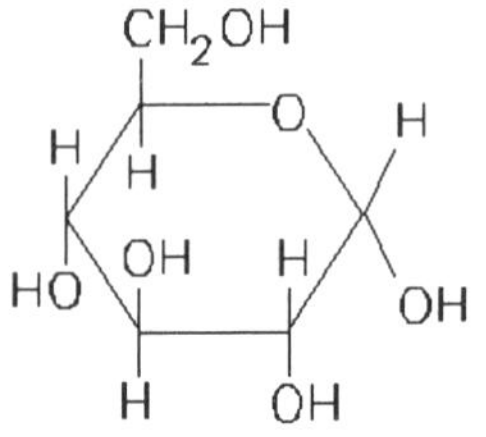

III, 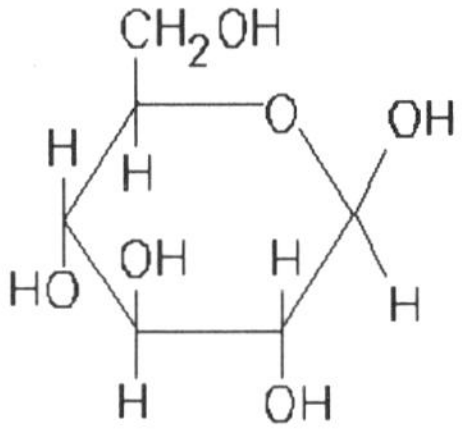 IV, 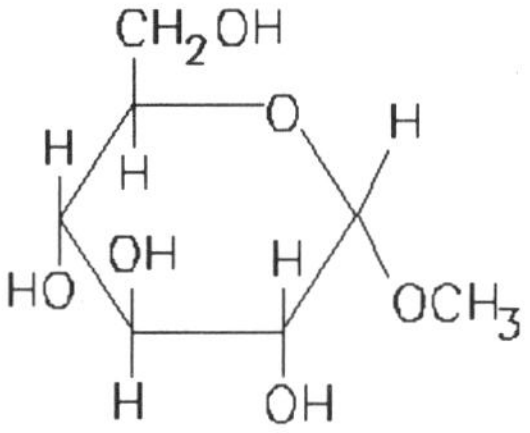 J (pyranose ring structures; J bears OCH_3 at the anomeric carbon)

A) I and II
B) I and III
C) II and III
* D) I, III, and IV
E) II, III, and IV

10. A compound X reacts with 3 mol of HIO_4 to yield 2 mol of HCO_2H and 2 mol of HCHO. What is the structure of X?

* A)	B)	C)	D)	E)
CH_2OH	O	O	CH_2OH	O
3	:	:	3	:
CHOH	CDH	CDH	CMO	CDH
3	3	3	3	3
CHOH	CHOH	CHOH	CMO	CMO
3	3	3	3	3
CH_2OH	CHOH	CHOH	CH_2OH	CMO
	3	3		3
	CH_2OH	CDH		CDH
		:		:
		O		O

11. A D-aldohexose, X is subjected to a Ruff degradation. The degradation product is treated with nitric acid to yield an optically inactive aldaric acid. A possible structure for X is:

I	II	III	IV	V
CHO	CHO	CHO	CHO	CHO
HO–H	H–OH	HO–H	H–OH	H–OH
HO–H	H–OH	H–OH	HO–H	HO–H
H–OH	H–OH	H–OH	HO–H	HO–H
H–OH	H–OH	HO–H	H–OH	HO–H
CH_2OH	CH_2OH	CH_2OH	CH_2OH	CH_2OH

A) I * B) II C) III D) IV E) V

12. Which of the following would yield D-glucose and D-mannose when subjected to a Kiliani-Fischer synthesis?

I	II	III	IV	V
CHO	CHO	CHO	CHO	CHO
HO–H	H–OH	H–OH	H–OH	HO–H
H–OH	H–OH	H–OH	HO–H	HO–H
H–OH	H–OH	HO–H	H–OH	H–OH
CH_2OH	CH_2OH	CH_2OH	CH_2OH	CH_2OH

* A) I B) II C) III D) IV E) V

13. Which of the following is an L-aldotetrose?

I	II	III	IV	V
CHO	CHO	CHO	CO_2H	CH_2OH
H–OH	HO–H	H–OH	H–OH	H–OH
H–OH	H–OH	HO–H	HO–H	HO–H
CH_2OH	CH_2OH	CH_2OH	CH_2OH	CH_2OH

A) I B) II * C) III D) IV E) V

14. Which aldohexose would yield an optically active aldaric acid when treated with nitric acid?

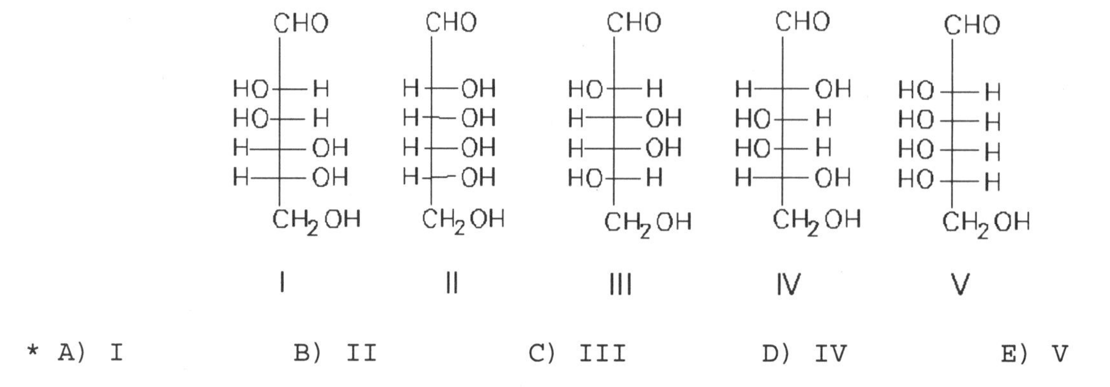

* A) I B) II C) III D) IV E) V

15. Which is not an intermediate monosaccharide in the Kiliani-Fischer synthesis of D-mannose from D-glyceraldehyde?

A) D-Ribose
B) D-Threose
C) D-Arabinose
D) D-Erythrose
* E) More than one of these

16. Which reagent would cause the following conversion to take place?

A) Excess CH_3OH and KOH
B) Excess CH_3OH and H^+
* C) Excess $(CH_3)_2SO_4$ and OH^-
D) Excess CH_3I and H_3O^+
E) Excess $(CH_3CO)_2O$

17. Which monosaccharide is recovered from the hydrolysis of glycogen?

A) D-Galactose
* B) D-Glucose
C) D-Gulose
D) Cellobiose
E) Maltose

18. What is the ratio of products formed by the reaction of periodic acid with the following compound?

```
         CHO
      H──┼──OH
      H──┼──OH
      H──┼──OH
      H──┼──OH
        CH2OH
```

	$\overset{O}{\underset{\cdot\cdot}{}}$ HCH (O:HCH)	HCO_2H	CO_2
A)	5	1	0
B)	3	3	0
* C)	1	5	0
D)	1	4	1
E)	0	4	2

19. Which compound will not reduce $Ag(NH_3)_2^+$?

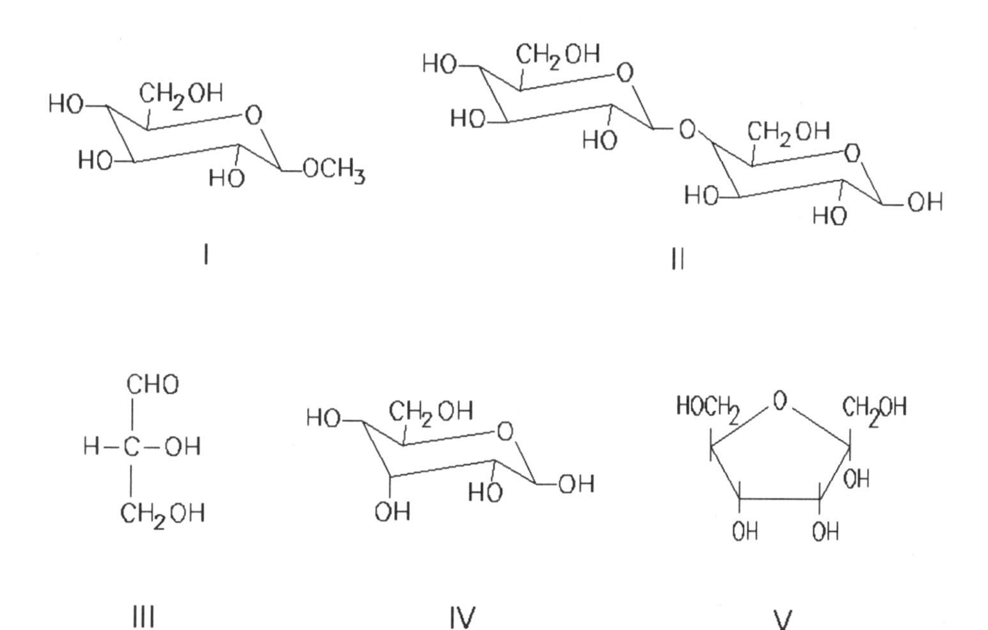

* A) I B) II C) III D) IV E) V

20. Which reagent will distinguish between the members of the following pair?

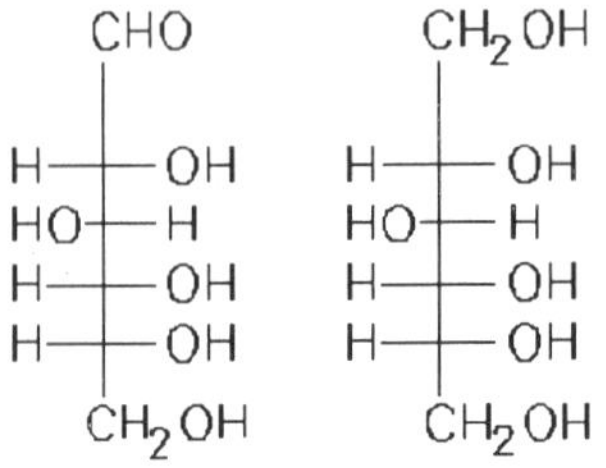

CHO
H—OH
HO—H
H—OH
H—OH
CH_2OH

CH_2OH
H—OH
HO—H
H—OH
H—OH
CH_2OH

* A) $Ag(NH_3)_2OH$
B) $AgNO_3/C_2H_5OH$
C) Br_2/CCl_4
D) $KMnO_4$
E) Hot $KMnO_4$

21. Which reagent will distinguish between the members of the following pair?

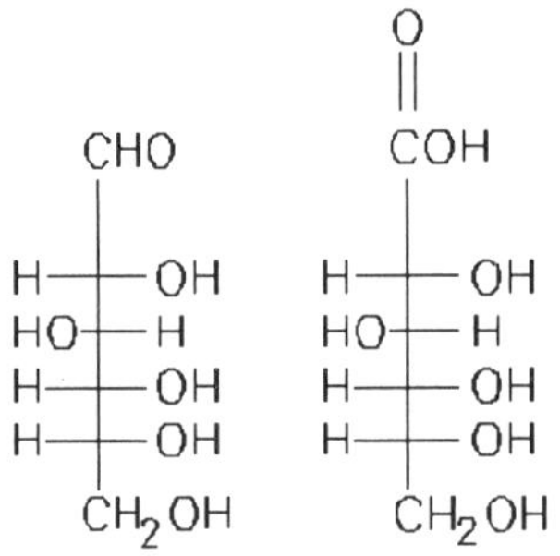

CHO
H—OH
HO—H
H—OH
H—OH
CH_2OH

O
||
COH
H—OH
HO—H
H—OH
H—OH
CH_2OH

* A) $Ag(NH_3)_2OH$
B) $AgNO_3/C_2H_5OH$
C) Br_2/CCl_4
D) $KMnO_4$
E) Hot $KMnO_4$

22. Which monosaccharide is recovered from the hydrolysis of starch?
A) D-Galactose
* B) D-Glucose
C) D-Gulose
D) Cellobiose
E) Maltose

23. Select the reagent(s) needed to perform the following transformation.

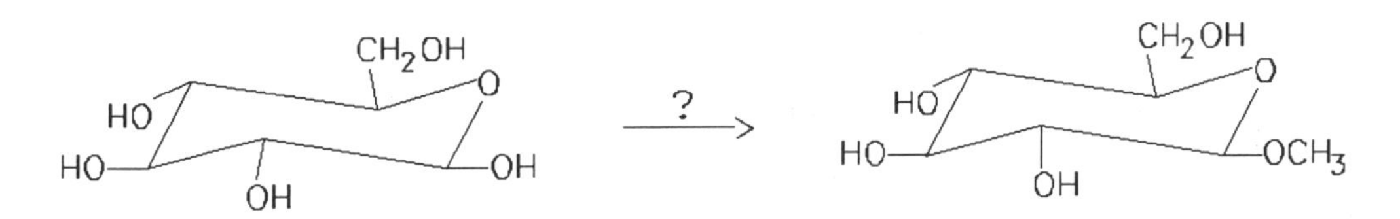

A) CH_3I, KOH

B) $(CH_3C)_2O$ (with C=O)

C) $(CH_3)_2SO_4$, NaOH

* D) CH_3OH, H^+

E) CH_3COCH_3 (with C=O)

24. Which of the following is a D-aldotetrose?

I	II	III	IV	V
CHO	CO_2H	CHO	CH_2OH	CHO
H—OH	HO—H	H—OH	H—OH	HO—H
H—OH	H—OH	HO—H	HO—H	HO—H
CH_2OH	CH_2OH	CH_2OH	CH_2OH	CH_2OH

* A) I B) II C) III D) IV E) V

25. An aldonic acid is represented by:

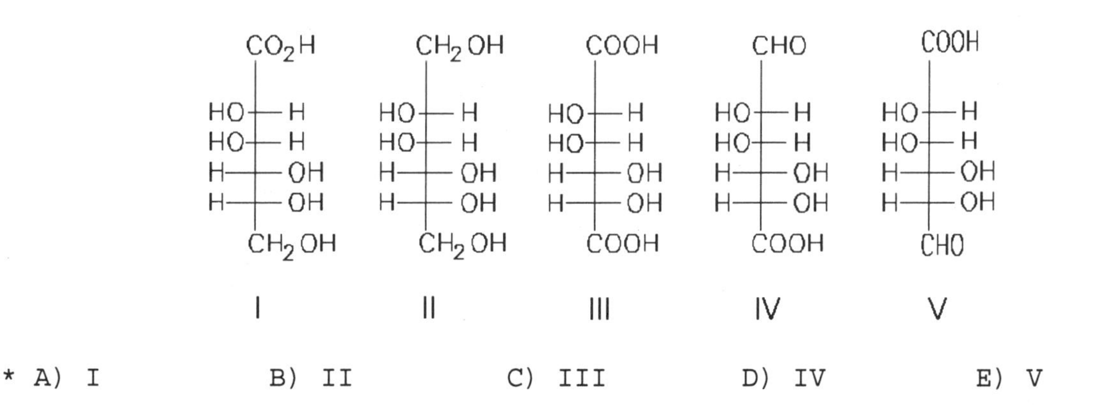

* A) I B) II C) III D) IV E) V

26. Reaction of C (below) with nitric acid would yield:

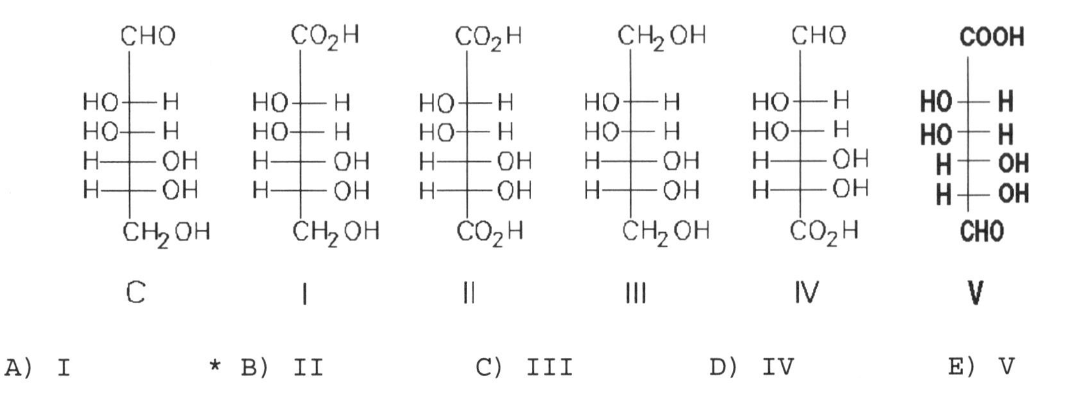

A) I * B) II C) III D) IV E) V

27. A D-aldohexose, Z, is subjected to a Ruff degradation. The degradation product is treated with nitric acid to yield an optically active aldaric acid. A possible structure for Z is:

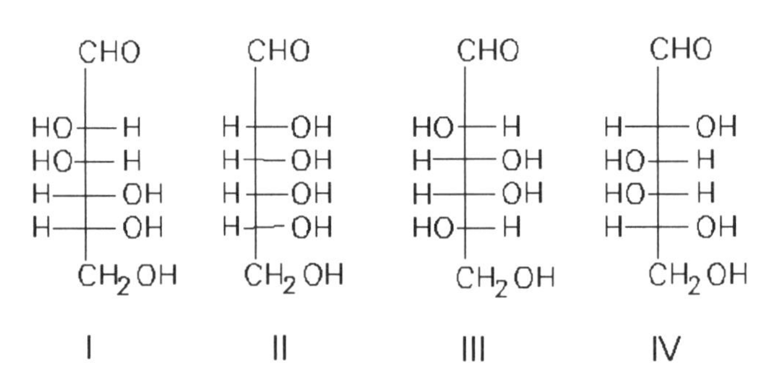

A) I
B) II
C) III
D) IV
* E) More than one of the above

28. Which of the following is an L-aldopentose?

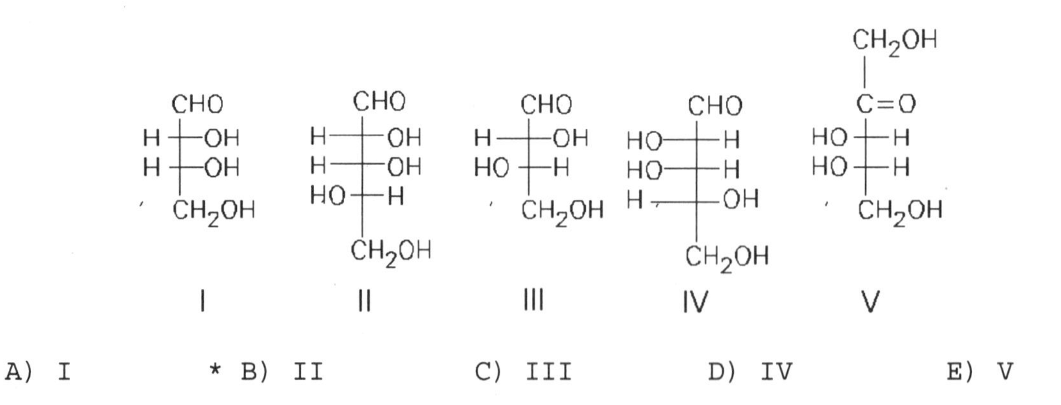

A) I * B) II C) III D) IV E) V

29. Which compound is D-galactose?

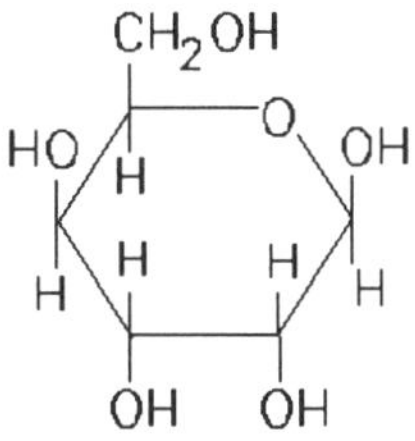

I 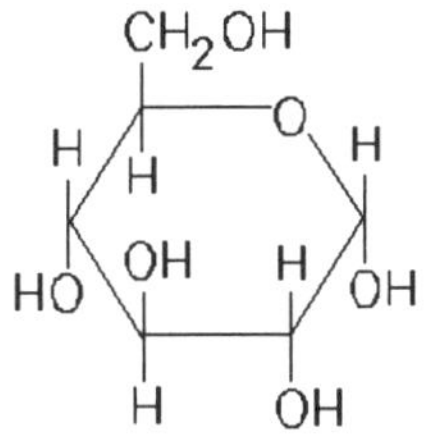 II 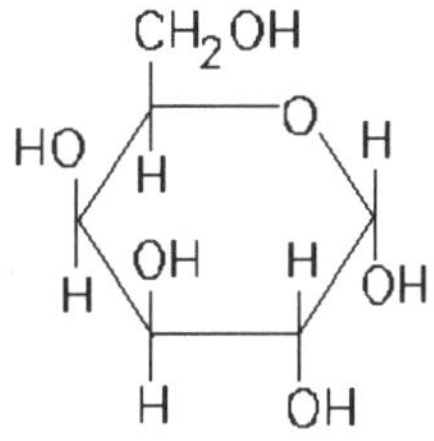 III 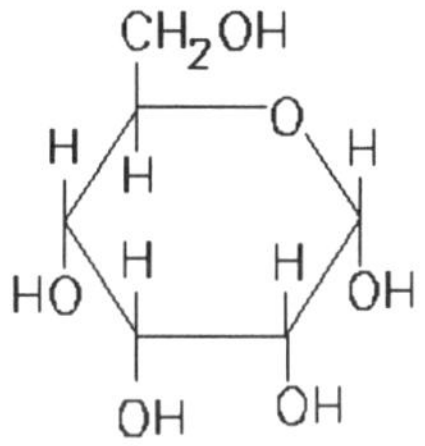 IV 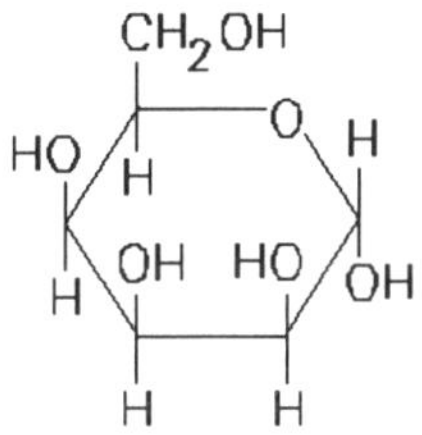 V

A) I B) II * C) III D) IV E) V

30. What is the ratio of products formed by the reaction of periodic acid with the following compound?

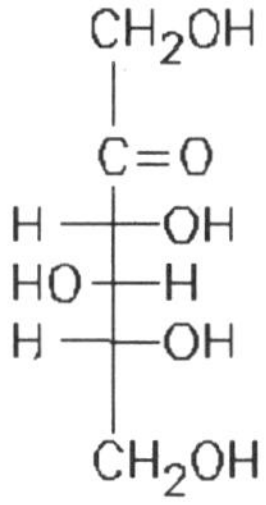

CH_2OH–C(=O)–CH(OH)–CH(OH)–CH(OH)–CH_2OH

	HCHO (O=CH–H)	HCO_2H	CO_2
A)	5	1	0
B)	3	3	0
C)	1	5	0
* D)	2	3	1
E)	0	4	2

31. Kiliani-Fischer Synthesis is the reaction of an aldose with:
A) Br_2/H_2O; then HCN; then H_3O^+; then Na-Hg, H_2O
* B) HCN; then $Ba(OH)_2$; then H_3O^+; then Na-Hg, H_2O
C) HCN; then H_3O^+; then $Ba(OH)_2$; then Na-Hg, H_2O
D) Br_2/H_2O; then H_2O_2, $Fe_2(SO_4)_3$
E) Br_2/H_2O

32. Ruff Degradation is the reaction of an aldose with:
A) Br_2/H_2O; then HCN; then H_3O^+; then Na-Hg, H_2O
B) HCN; then $Ba(OH)_2$; then H_3O^+; then Na-Hg, H_2O
C) HCN; then H_3O^+; then $Ba(OH)_2$; then Na-Hg, H_2O
* D) Br_2/H_2O; then H_2O_2, $Fe_2(SO_4)_3$
E) Br_2/H_2O

33. Which is an L-monosaccharide that would yield an optically active aldaric acid on oxidation by nitric acid?

I: CH_2OH; C=O; HO—H; H—OH; H—OH; CH_2OH

II: CHO; H—OH; HO—H; H—OH; H—OH; CH_2OH

III: CHO; HO—H; H—OH; H—OH; HO—H; CH_2OH

IV: CHO; H—OH; H—OH; HO—H; HO—H; CH_2OH

V: CHO; HO—H; HO—H; HO—H; HO—H; CH_2OH

A) I   B) II   C) III   * D) IV   E) V

34. Which reagent will distinguish between the members of the following pair?

CHO; H—OH; HO—H; H—OH; H—OH; CH_2OH

CHO; H—OCH_3; HO—H; H—OCH_3; H—OCH_3; CH_2OH

A) $Ag(NH_3)_2OH$
B) $AgNO_3/C_2H_5OH$
C) Br_2/CCl_4
D) $KMnO_4$
* E) HIO_4

35. Reaction of D-ribose with bromine water would yield an optically:

* A) active aldonic acid.
  B) inactive aldonic acid.
  C) active aldaric acid.
  D) inactive aldaric acid.
  E) active uronic acid.

36. Reaction of D-ribose with HNO_3 would yield an optically:

  A) active aldonic acid.
  B) inactive aldonic acid.
  C) active aldaric acid.
* D) inactive aldaric acid.
  E) inactive uronic acid.

37. Which are the anomers?

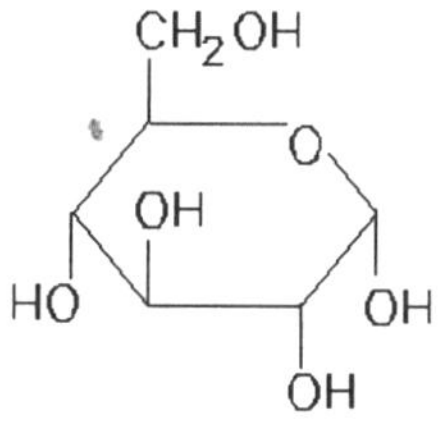

CH2OH, O, OH, HO, OH, OH — I

CH2OH, O, HO, HO, OH, OH — II

CH2OH, O, OH, OH, HO, OH — III

CH2OH, O, HO, OH, OH, OH — IV

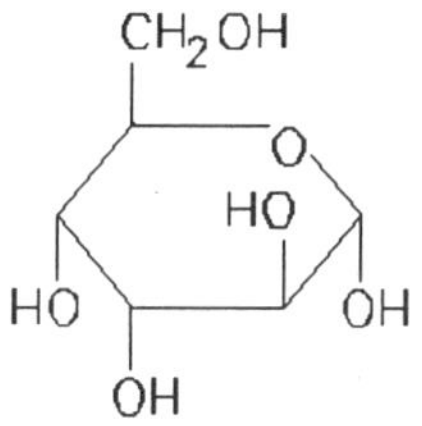

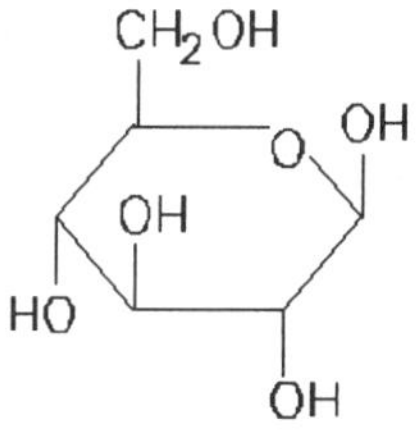

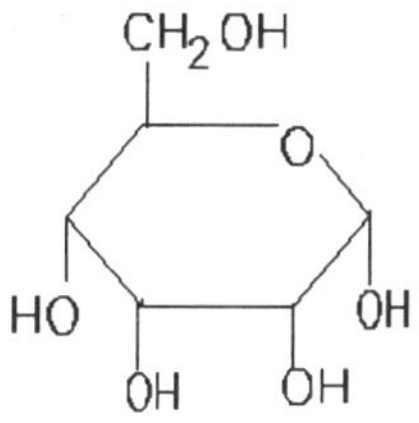

  A) I and II
* B) I and III
  C) II and III
  D) II and IV
  E) III and IV

38. Which of these is a non-reducing monosaccharide?

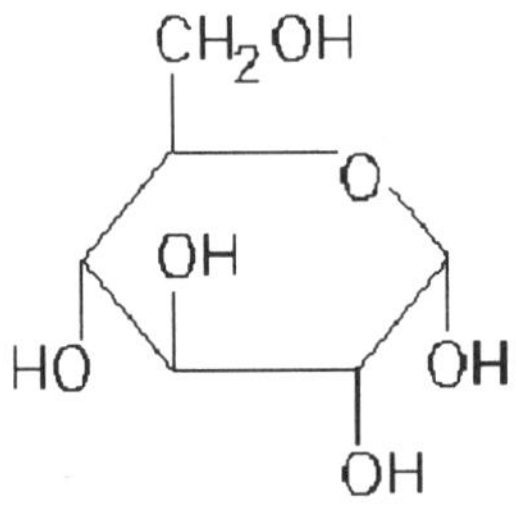

I

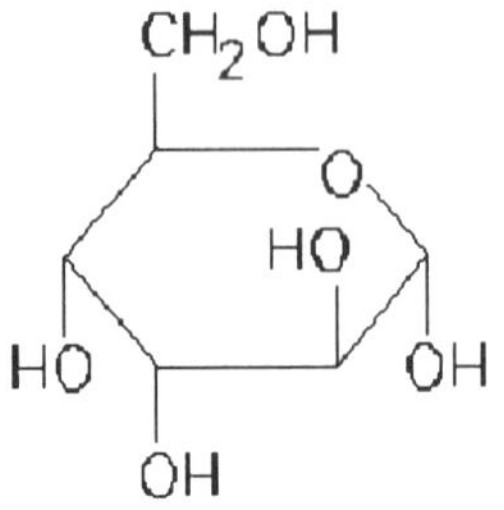

II

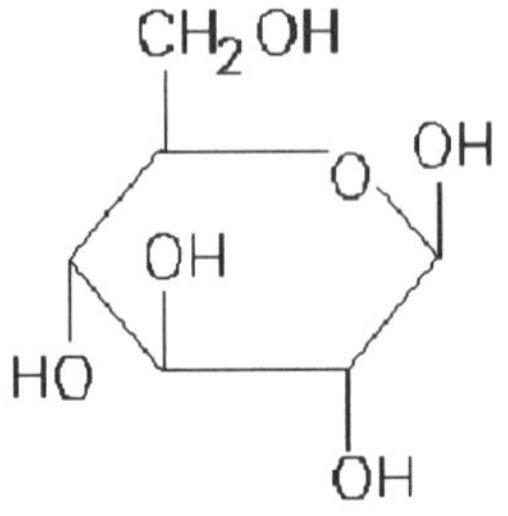

III

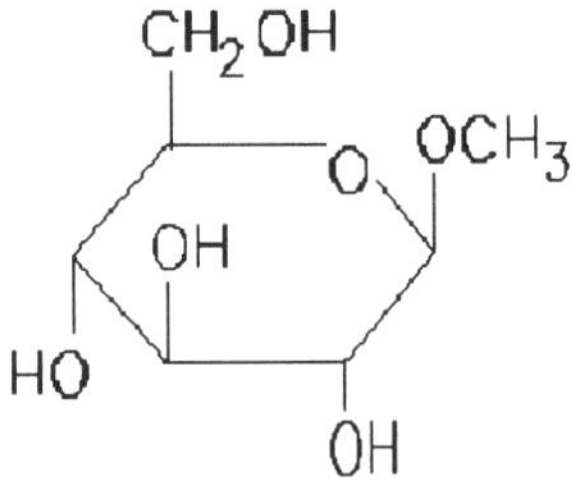

IV

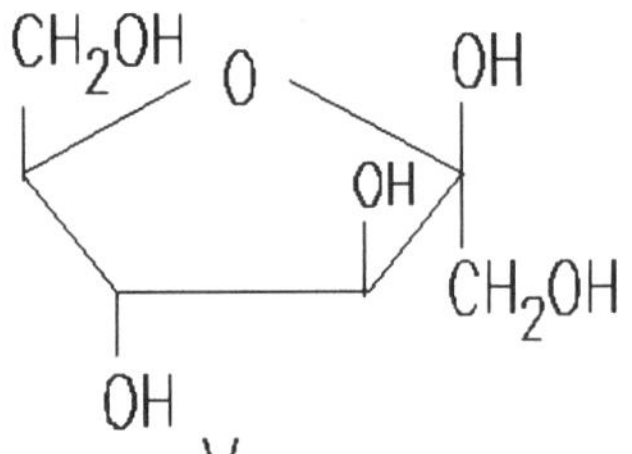

V

A) I　　B) II　　C) III　　* D) IV　　E) V

39. Which of these is ∝-D-glucopyranose?

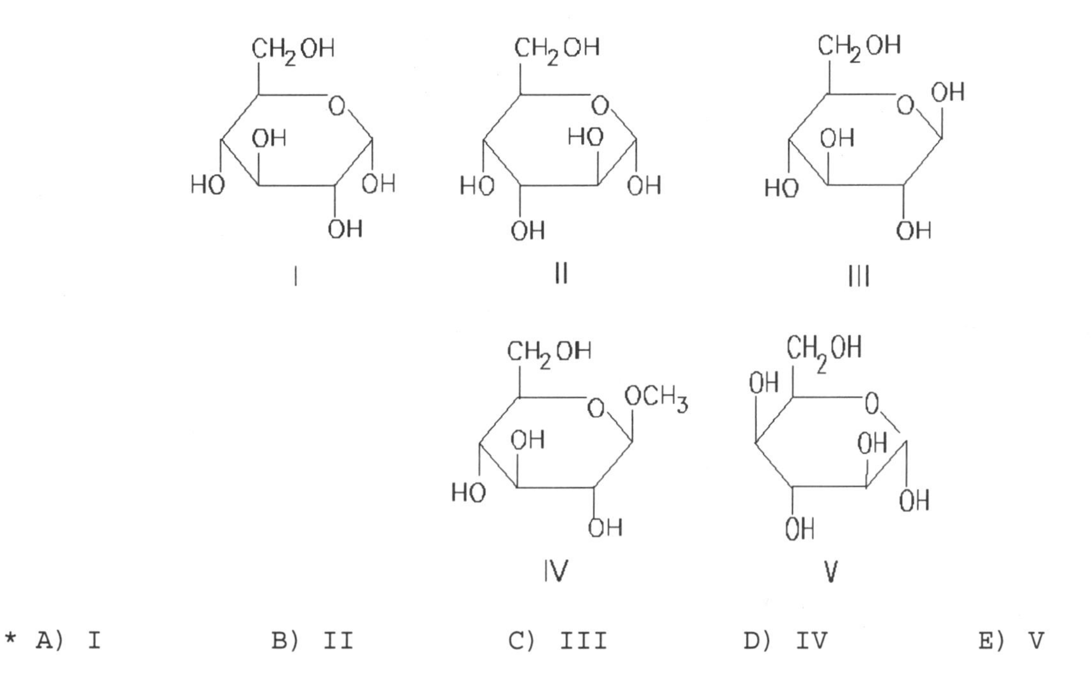

* A) I B) II C) III D) IV E) V

40. Which of these is a glycoside?

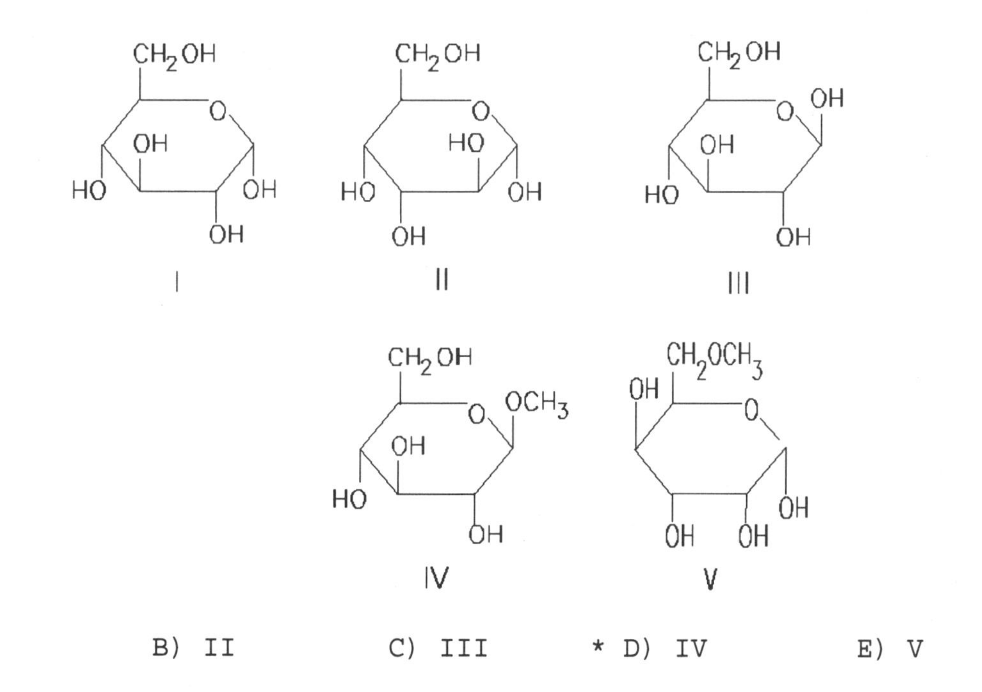

A) I B) II C) III * D) IV E) V

41. Which would undergo mutarotation in neutral aqueous solution?

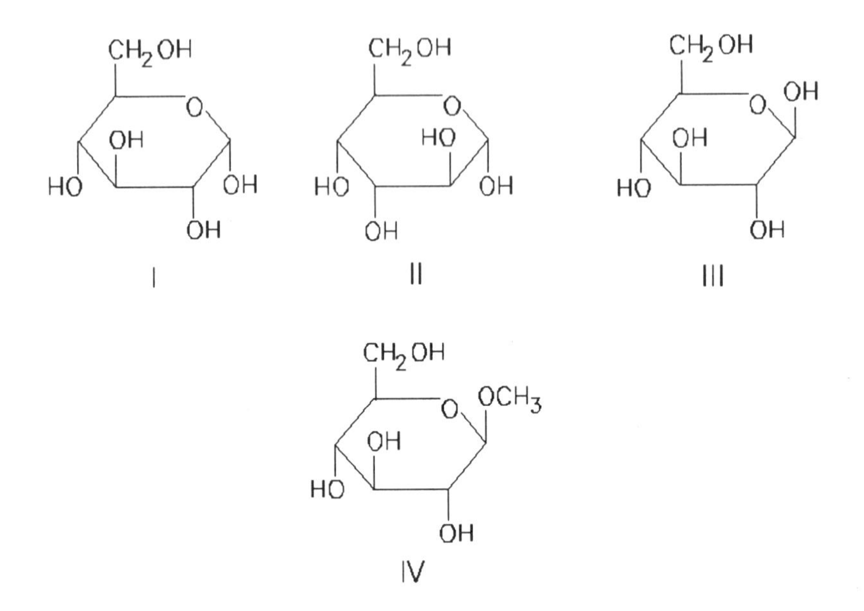

A) I
B) II
C) III
D) IV
* E) More than one of these

42. Which of the following would give a positive test with Benedict's or Fehling's solution?

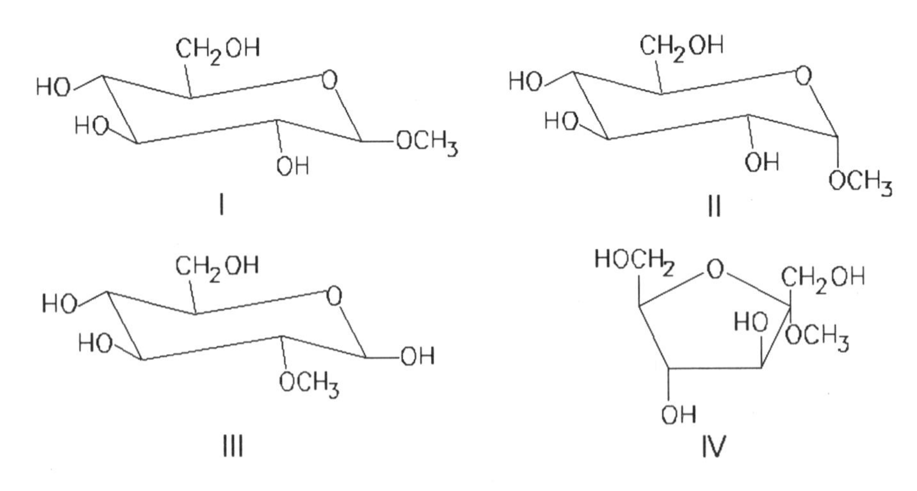

A) I
B) II
* C) III
D) IV
E) All of these

43. Which is a ketohexose?

A) D-Glucose
* B) D-Fructose
C) D-Mannose
D) D-Ribose
E) (+)-Sucrose

44. Which of these reacts with dilute HCl to produce methanol?

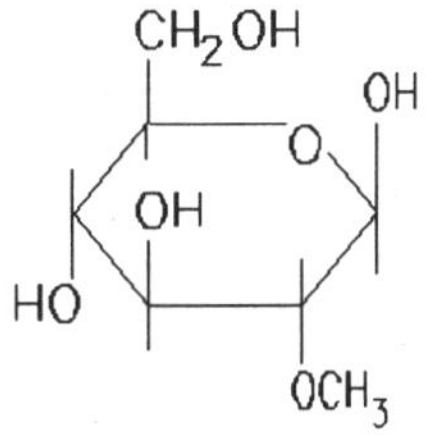

I 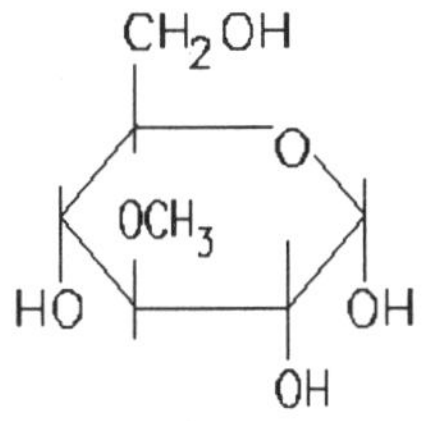 II 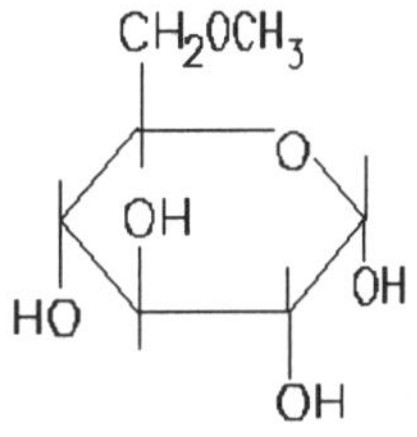 III

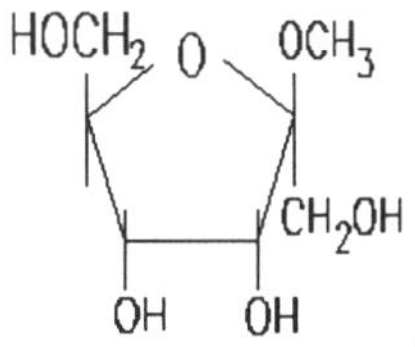

IV 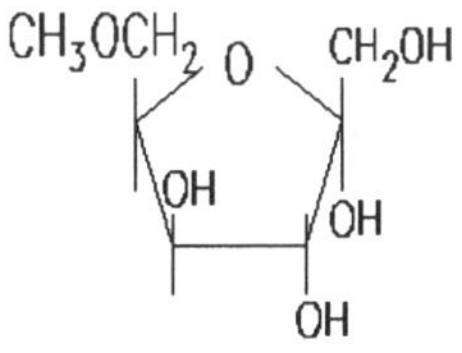 V

A) I B) II C) III * D) IV E) V

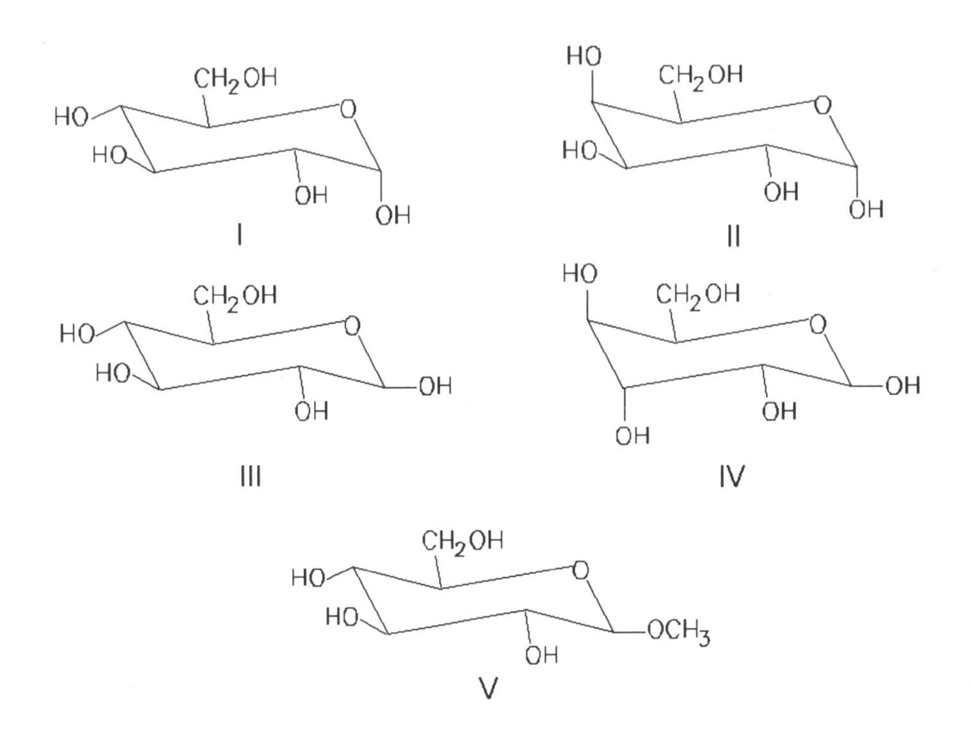

45. Which structure represents β-D-glucopyranose?
A) I B) II * C) III D) IV E) V

46. Which compound is not a reducing sugar?
A) I B) II C) III D) IV * E) V

47. Which compound or compounds would be formed when D-glucose is dissolved in methanol and then treated with anhydrous acid?

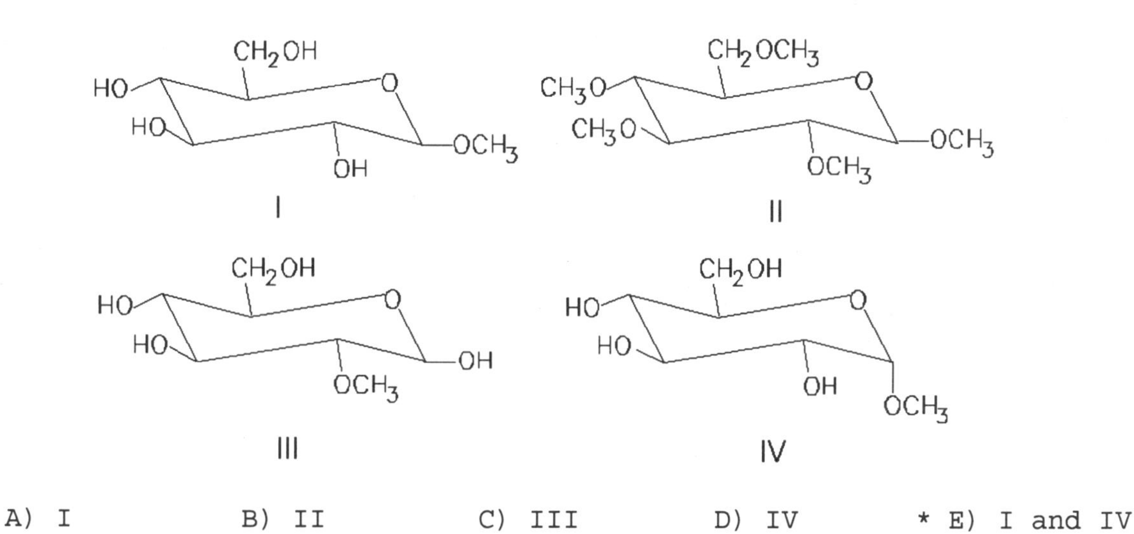

A) I B) II C) III D) IV * E) I and IV

I	II	III	IV	V
CHO	CHO	CHO	CHO	CHO
H—OH	HO—H	H—OH	H—OH	HO—H
HO—H	HO—H	H—OH	H—OH	HO—H
H—OH	H—OH	H—OH	HO—H	HO—H
H—OH	H—OH	H—OH	H—OH	HO—H
CH_2OH	CH_2OH	CH_2OH	CH_2OH	CH_2OH

48. Which structures above represent enantiomers?
A) I and II
B) II and III
C) III and IV
* D) III and V
E) IV and V

49. Which structures above represent epimers?
* A) I and II
B) II and III
C) III and IV
D) III and V
E) IV and V

50. Refer to the structures above. Which D-aldohexose would react with $NaBH_4$ to yield an optically inactive alditol?
A) I B) II * C) III D) IV E) V

51. Refer to the structures above. Which aldohexose when subjected to Fischer's end-group interchange would be converted to a compound identical with itself?
A) I * B) II C) III D) IV E) V

52. An aldopentose, X, is subjected to a Kiliani-Fischer synthesis to produce two aldohexoses, Y and Z. Both Y and Z, when oxidized with nitric acid, yield optically active aldaric acids. Which structure represents X?

I	II	III	IV	V
CHO	CHO	CHO	CHO	CHO
HO—H	HO—H	H—OH	HO—H	H—OH
HO—H	H—OH	H—OH	HO—H	H—OH
H—OH	H—OH	H—OH	HO—H	HO—H
CH_2OH	CH_2OH	CH_2OH	CH_2OH	CH_2OH

A) I * B) II C) III D) IV E) V

I	II	III	IV	V	VI
CHO	CHO	CHO	CHO	CHO	CHO
H—OH	HO—H	H—OH	HO—H	HO—H	H—OH
H—OH	H—OH	H—OH	H—OH	HO—H	H—OH
H—OH	H—OH	H—OH	HO—H	HO—H	HO—H
H—OH	HO—H	HO—H	H—OH	HO—H	HO—H
CH_2OH	CH_2OH	CH_2OH	CH_2OH	CH_2OH	CH_2OH

53. Consider the structures above. Given that structure I represents D-allose, which structure represents L-allose?
A) II B) III C) IV * D) V E) VI

54. Consider the structures above. An L-aldohexose, X, is treated with nitric acid to yield an optically **inactive** aldaric acid. The same L-aldohexose, X, is subjected to a Ruff degradation and the degradation product is oxidized with nitric acid to produce an optically **inactive** aldaric acid. Which is a possible structure for X?

A) II B) III C) IV * D) V E) VI

55. Consider the structures above. Which monosaccharides would yield an optically active aldonic acid when oxidized with bromine water?

A) I, II, and III
B) I, II, and V
C) III, IV, and VI
D) II, III, and IV
* E) All of these

56. Consider the structures above. Which monosaccharides would yield the same phenylosazone when treated with excess phenylhydrazine?

A) I and V
B) I and III
* C) II and III
D) III and VI
E) IV and V

57. Which of the following is the structure of D-galacturonic acid?

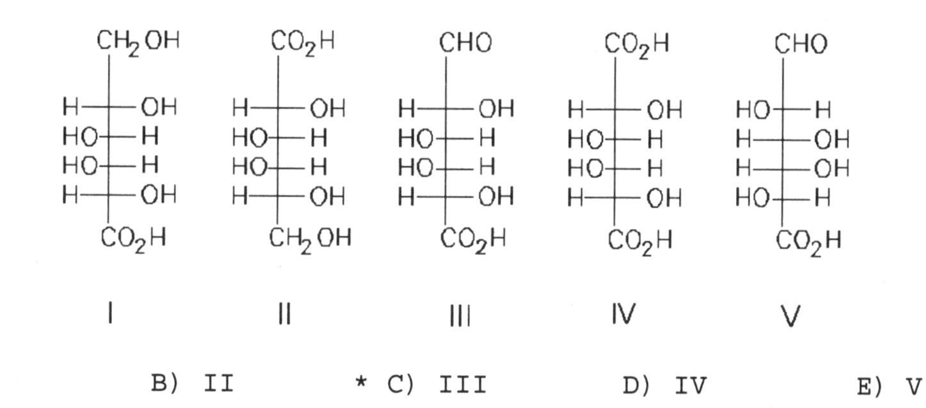

A) I B) II * C) III D) IV E) V

58. Which is a non-reducing disaccharide?

A) Maltose
B) Cellobiose
* C) Sucrose
D) Lactose
E) None of these

59. Compound X is a reducing sugar which, on hydrolysis, affords two molar equivalents of D-glucose. This hydrolysis is catalyzed by an enzyme specific for glucosides of this type,

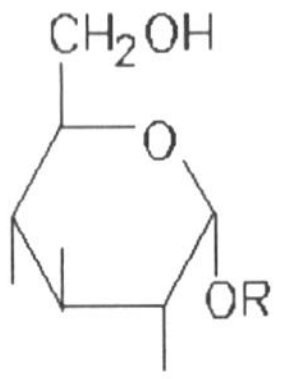

CH_2OH

O

OR

What is the identity of X?

A) Sucrose
B) Lactose
* C) Maltose
D) Cellobiose
E) None of these

60. The D-glucose unit at the branching point of amylopectin has free hydroxyl groups at which positions?

A) C-2, C-3, and C-6
* B) C-2 and C-3
C) C-3 and C-4
D) C-3, C-4, and C-6
E) C-4 and C-6

61. Cellulose differs from chitin in which way?

A) Cellulose has β-glycosidic linkages; chitin has ∝-glycosidic linkages.
* B) Cellulose contains only D-glucose units; chitin contains only N-acetyl-D-glucosamine units.
C) Cellulose cannot be hydrolyzed; chitin can be hydrolyzed.
D) Cellulose has a linear structure; chitin has a helical structure.
E) Cellulose chains are branched; chitin chains are unbranched.

62. Which is the only one of these classes of carbohydrates which can include optically inactive members?

A) Uronic acids
* B) Alditols
C) Ketoses
D) Glycosides
E) Aldonic acids

63. Cellulose lacks nutritive value for humans because:

A) the products of its digestion are excreted without utilization.
B) its conformation prevents attack by digestive enzymes.
* C) we lack the enzymes which can catalyze the hydrolysis of the glycosidic linkages.
D) it passes through the digestive tract so rapidly.
E) the molecules possess such a high molecular weight.

64. A polysaccharide, Y, undergoes hydrolysis on catalysis by an enzyme which catalyzes the hydrolysis of cellobiose, but not maltose. Complete methylation and acid hydrolysis of the permethylated Y yields better than 95% of 2,4,6-tri-O-methyl-D-glucose.

Which is a plausible structure for the repeating unit of Y?

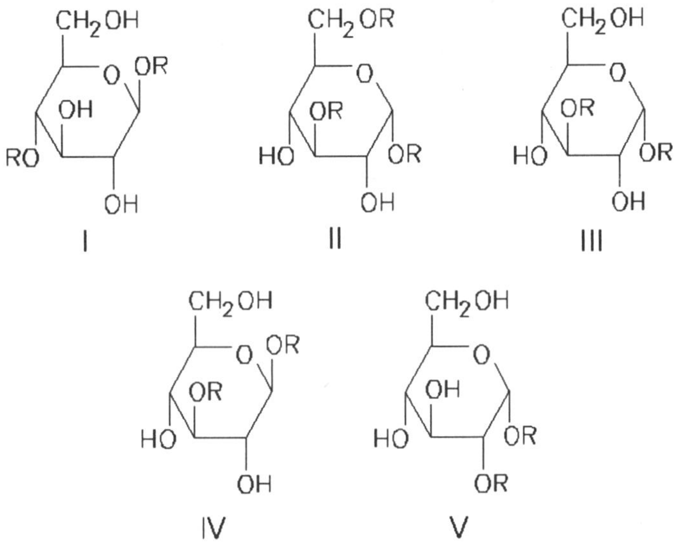

I II III

IV V

A) I B) II C) III * D) IV E) V

65. A <u>glycoside</u> is a compound which contains the structural features of these classes of organic compounds:

A) Aldehydes and alcohols
* B) Acetals and alcohols
C) Hemiacetals and alcohols
D) Ketones and alcohols
E) Alcohols and carboxylic acids

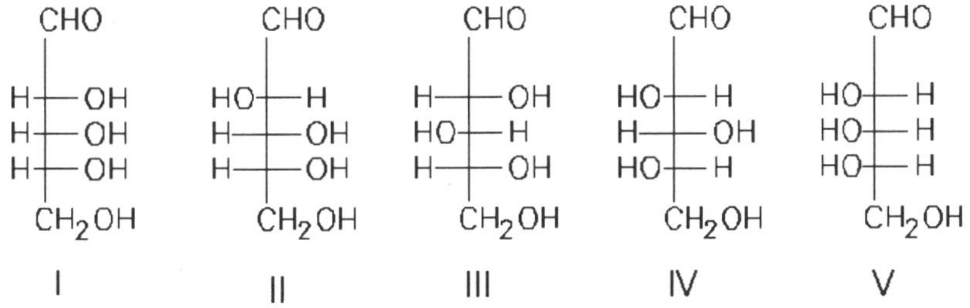

I II III IV V

66. Refer to the structures above. Which sugar(s) would yield an optically **active** aldaric acid on oxidation with nitric acid?

A) I and III
B) I, II, III, and V
* C) II
D) III and IV
E) I and V

67. Refer to the structures above. Which are L-sugars?

A) II and IV
B) I, II, and III
C) I and V
D) III, IV, and V
* E) IV and V

68. Refer to the structures above. Which sugars would react with phenylhydrazine to yield the same phenylosazone?

* A) I and II
B) III and IV
C) I and V
D) II and III
E) III and V

69. Which of these is an example of a glucan?

A) Maltose
B) Sucrose
C) Lactose
D) Cellobiose
* E) Amylose

70. What is the correct description of this disaccharide?

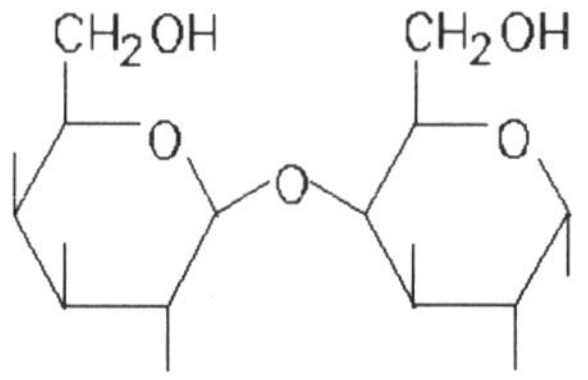

CH_2OH CH_2OH

A) The ∝-anomer of two D-glucose units joined by an ∝,1:4 linkage
B) The ∝-anomer of two D-galactose units joined by an ∝,1:4 linkage
C) The β-anomer of two D-galactose units joined by a β,1:4 linkage
D) The β-anomer of two D-glucose units joined by a β,1:4 linkage
* E) The ∝-anomer of two D-galactose units joined by a β,1:4 linkage

71. What are the correct designations for the stereocenters in this aldose:

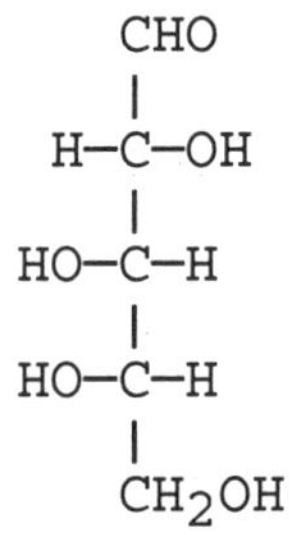

```
    CHO
     |
  H-C-OH
     |
 HO-C-H
     |
 HO-C-H
     |
    CH2OH
```

A) 2R,3S,4R * B) 2R,3S,4S C) 2S,3R,4R D) 2S,3S,4R E) 2R,3R,4S

72. What is a correct general description of the monosaccharide shown here?

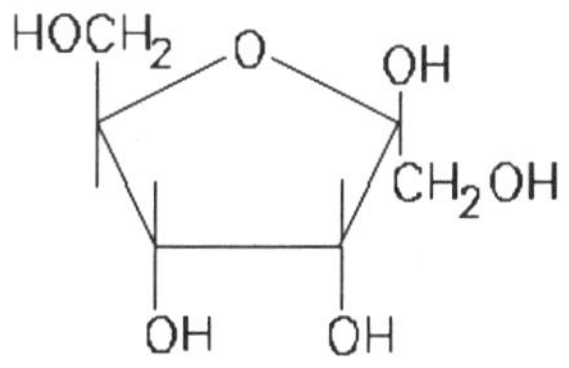

HOCH$_2$ O OH CH$_2$OH OH OH

A) The α-anomer of the pyranose form of an aldohexose
B) The β-anomer of the pyranose form of an aldohexose
C) The α-anomer of the furanose form of a ketohexose
* D) The β-anomer of the furanose form of an ketohexose
E) The β-anomer of the furanose form of an aldohexose

73. What can be said, correctly, about a monosaccharide, the name of which is preceded (only) by (+) ?
A) The compound is the α-anomer.
B) The compound exists in the pyranose form.
* C) The compound is dextrorotatory.
D) The compound has the same stereochemistry at the penultimate carbon as D-(+)-glucose.
E) The compound exists only in open-chain form.

74. If the methyl glycoside of an aldohexose is treated with HIO_4, one molar equivalent of HCHO is formed but no HCOOH. What size ring is present in the glycoside?
A) Three-membered
B) Four-membered
* C) Five-membered
D) Six-membered
E) Seven-membered

75. Which is a reducing sugar with an ∝-glycosidic linkage?

A) Sucrose
* B) Maltose
C) Lactose
D) Cellobiose
E) None of these

76. Which of these is a component of the mixture formed when D-galactose is placed in aqueous base (de Bruyn - van Ekenstein transformation)?

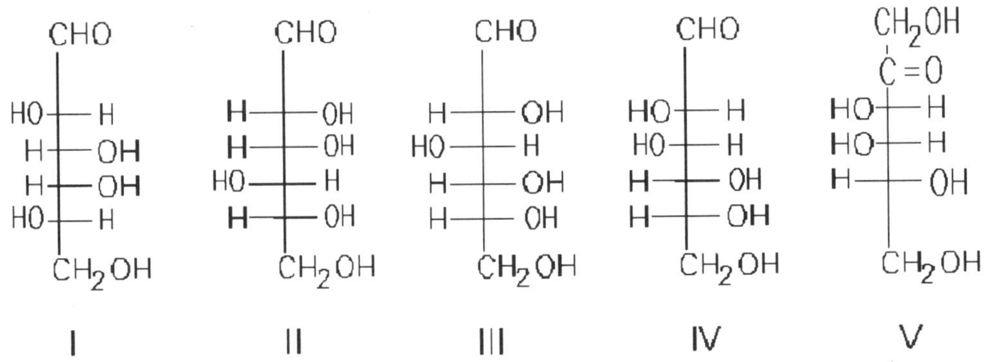

I	II	III	IV	V
CHO	CHO	CHO	CHO	CH_2OH
				C=O
HO–H	H–OH	H–OH	HO–H	HO–H
H–OH	H–OH	HO–H	HO–H	HO–H
H–OH	HO–H	H–OH	H–OH	H–OH
HO–H	H–OH	H–OH	H–OH	
CH_2OH	CH_2OH	CH_2OH	CH_2OH	CH_2OH

A) I
B) II
C) III
D) IV
* E) V

77. The pyranose form of an aldohexose which can react 1:2 with acetone in the presence of acid is which of these?

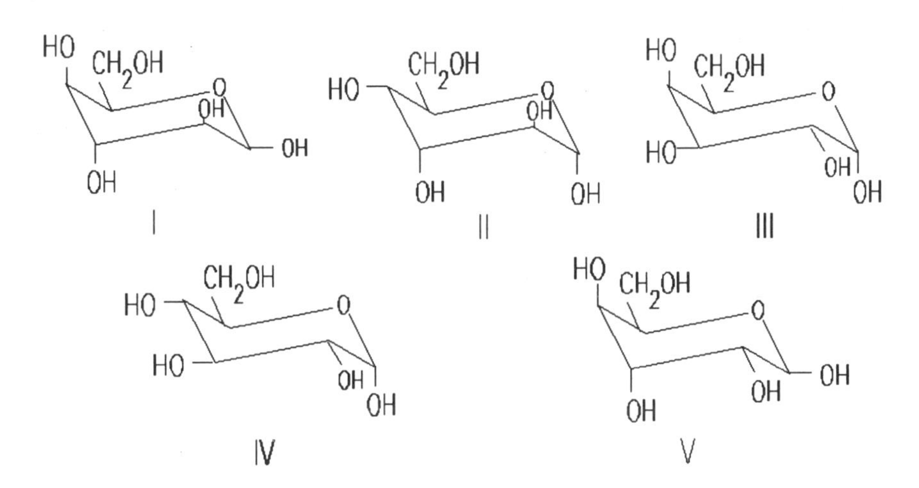

|

A) I B) II * C) III D) IV E) V

Chapter 23: Lipids

1. Which of the following would serve as the basis for a simple chemical test that would distinguish between stearic acid and oleic acid?

A) $NaOH/H_2O$
B) $NaHCO_3/H_2O$
C) HCl/H_2O
D) $Ag(NH_3)_2OH$
* E) Br_2/CCl_4

2. Which fatty acid is not likely to occur commonly in natural sources?

A) $CH_3(CH_2)_{12}COOH$

B) $CH_3(CH_2)_{14}COOH$

C)
```
CH3(CH2)5      (CH2)7COOH
         \    /
          CMC
         /   \
        H     H
```

D)
```
CH3(CH2)4      CH2       (CH2)7COOH
         \    /   \     /
          CMC      CMC
         /   \    /   \
        H     H  H     H
```

* E)
```
CH3(CH2)12CHCH2COOH
           3
           CH3
```

3. Which of the following could be used to prepare myristic acid, $CH_3(CH_2)_{12}COOH$?

A) $CH_3(CH_2)_{11}CH_2Br$, CN^-, heat; then H_3O^+, heat

B) $CH_3(CH_2)_{11}CH_2Br$, Mg, $(C_2H_5)_2O$; then CO_2; then H_3O^+

C)
```
CH3(CH2)12CH, Ag(NH3)2OH; then H3O+
           :
           O
```

D) Answers A) and B)
* E) Answers A), B), and C)

4. What product would be obtained by catalytic hydrogenation of 5-cholesten-3a-ol?

A) 5a-Cholestan-3a-ol
* B) 5!-Cholestan-3a-ol
C) 5!-Cholestan-3!-ol
D) 5a-Cholestan-3!-ol
E) 5-Cholesten-3a,6!-diol

5. How could you synthesize stearolic acid, $CH_3(CH_2)_7CpC(CH_2)_7COOH$ from oleic acid, $CH_3(CH_2)_7CHMCH(CH_2)_7COOH$?

* A) Br_2, CCl_4; then 3 $NaNH_2$, heat; then H_3O^+
B) Li, liq. NH_3; then H_3O^+
C) H_2, Pd
D) Peracid; then H_3O^+, H_2O; then H^+, H_2O, heat
E) xs HCl; then KOH, C_2H_5OH, heat

6. The product, X, of the following reaction sequence,

$SOCl_2$ NH_3 Br_2
$CH_3(CH_2)_{12}COOH$ DDDDDD4 DDDDDD4 DDDDDD4 X
OH^-

would be:

A) $CH_3(CH_2)_{11}CHCOOH$
3
Br

B) $CH_3(CH_2)_{11}CHCONH_2$
3
Br

* C) $CH_3(CH_2)_{11}CH_2NH_2$
D) $CH_3(CH_2)_{12}CONH_2$
E) $CH_3(CH_2)_{12}COBr$

7. The product, Y, of the following reaction sequence,

P OH^- H^+
$CH_3(CH_2)_{12}COOH$ DDDDDD4 DDDDDD4 DDDDDD4 Y
Br_2

would be:

A) $CH_3(CH_2)_{12}CH_2OH$

* B) $CH_3(CH_2)_{11}CHCOOH$
3
OH

C) $CH_3(CH_2)_{11}COOH$

D) $CH_3(CH_2)_{12}COBr$

E) $CH_3(CH_2)_{11}CHCOOH$
3
Br

8. The product of the following reaction sequence is nervonic acid. What is the structure of nervonic acid?

1-bromooctane $\xrightarrow{HCpCNa}$ $C_{10}H_{18}$ $\xrightarrow[liq.NH_3]{NaNH_2}$ $C_{10}H_{17}Na$ $\xrightarrow{ICH_2(CH_2)_{11}CH_2Cl}$

$C_{23}H_{43}Cl$ $\xrightarrow{NaCN}$ $C_{24}H_{43}N$ $\xrightarrow[(]{KOH}$ $C_{24}H_{43}O_2K$ $\xrightarrow{H_3O^+}$

$C_{24}H_{44}O_2$ $\xrightarrow[Ni_2B]{H_2, P-2}$ nervonic acid ($C_{24}H_{46}O_2$)

A) $CH_3(CH_2)_7$ \ / H
CMC
H / \ $(CH_2)_{13}COOH$

* B) $CH_3(CH_2)_7$ \ / $(CH_2)_{13}COOH$
CMC
H / \ H

C) $CH_3(CH_2)_9$ \ / $(CH_2)_{11}COOH$
CMC
H / \ H

D) $CH_3(CH_2)_{13}$ \ / H
CMC
H / \ $(CH_2)_7COOH$

E) $CH_3(CH_2)_{13}$ \ / $(CH_2)_7COOH$
CMC
H / \ H

9. Consider a micelle composed of phosphatidyl choline (I). Which part(s) of the molecule would form the hydrophilic surface of the micelle?

```
          CH2OOCC15H31      }  1
          3
          CHOOCC17H35       }  2
    I     3
          3   O
          3   :        +
          CH2OPOCH2CH2N(CH3)3    } 3
              3
              OD
```

* A) 3 B) 2 C) 1 D) 1 and 2 E) 2 and 3

10. Which reagent might be used to convert 5!-cholest-1-en-3-ol into 5-!-cholestan-3-ol?

A) CrO_3/pyridine
B) $KMnO_4/H_2O$
C) CH_3MgI
* D) H_2/Pt
E) $Li/C_2H_5NH_2$

11. Which reagent might serve as the basis for a simple chemical test that would distinguish between 5!-cholest-1-en-3-one and 5-!-cholestan-3-one?

A) $Ag(NH_3)_2OH$
B) CrO_3/H_2SO_4
* C) Br_2/CCl_4
D) $NaOH/H_2O$
E) $C_6H_5NHNH_2$

12. Which of the following statements regarding lipids is <u>not</u> true?

A) Lipids are soluble in non-polar organic solvents.
* B) All lipids have the same functional groups.
C) Lipids include waxes, steroids, and triacylglycerols.
D) Lipids have little in common except their solubility.
E) Many lipids have biological roles.

13. How many isoprene units are in vitamin A?

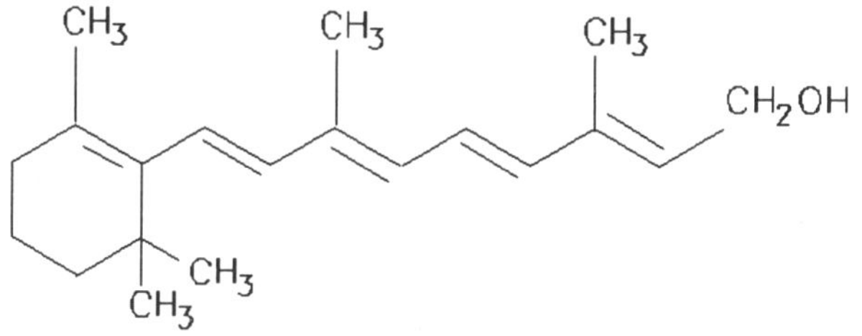

A) 1
B) 2
C) 3
* D) 4
E) More than 4

14. In a-carotene, how many tail-to-tail links are there?

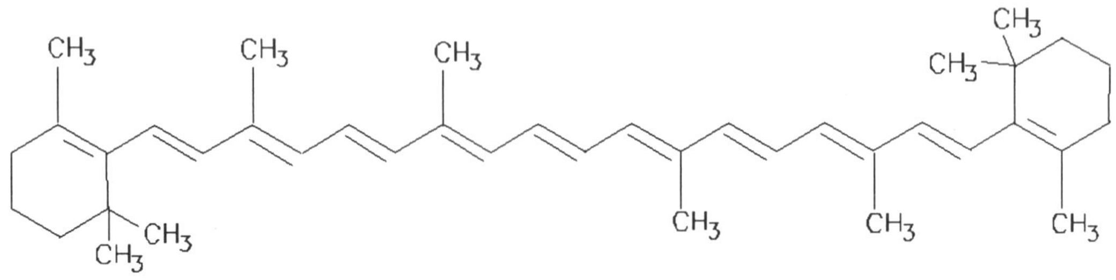

* A) 1
  B) 2
  C) 3
  D) 4
  E) More than 4

15. In the vulcanization of rubber,
  A) natural rubber is heated with sulfur.
  B) reaction occurs at allylic positions.
  C) cross-linking results in a hardening of the rubber.
  D) disulfide bridges are formed.
* E) All of the above

16. The synthesis of cortisone required placing a ketone function at the 11-position of a steroid. Where is position 11?

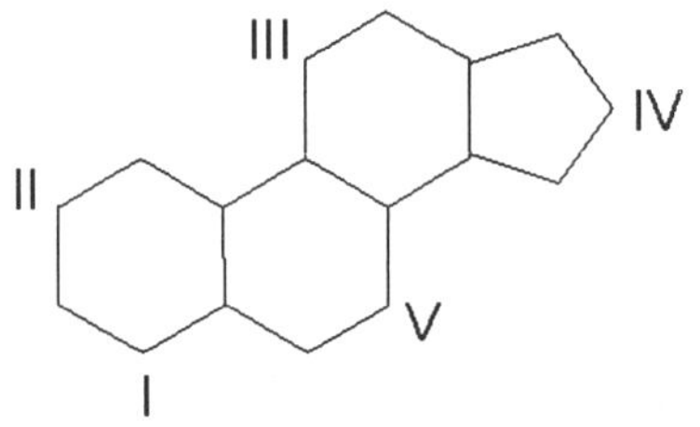

  A) I B) II * C) III D) IV E) V

17. How many stereocenters are there in cholesterol?

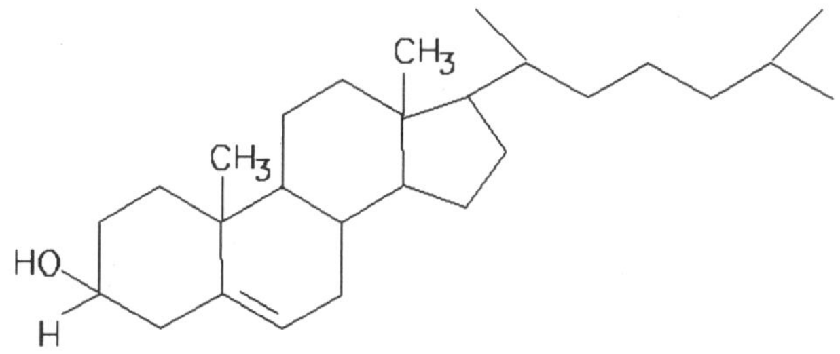

A) 2   B) 4   C) 6   * D) 8   E) 16

18. Which of the following statements regarding triacylglycerols is not true?

A) They undergo alkaline hydrolysis to yield soaps.
B) They are liquid if they have alkene bonds.
C) They are solid if they do not have alkene bonds.
D) Some can be hydrogenated.
* E) They are soluble in water.

19. In the biosynthesis of vitamin D_2, which alkane bond of ergosterol is cleaved?

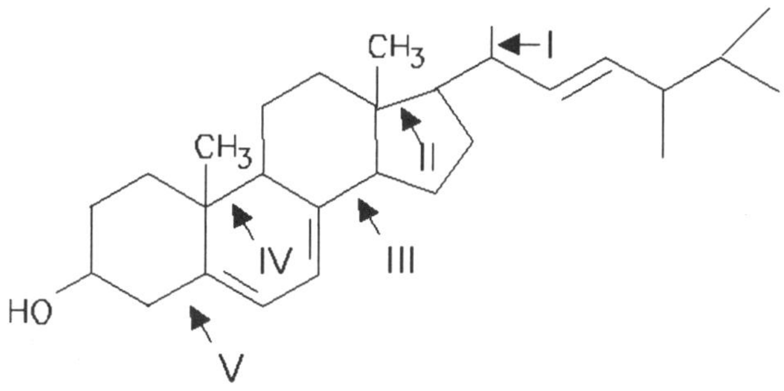

A) I   B) II   C) III   * D) IV   E) V

20. Which compound below does not obey the isoprene rule?

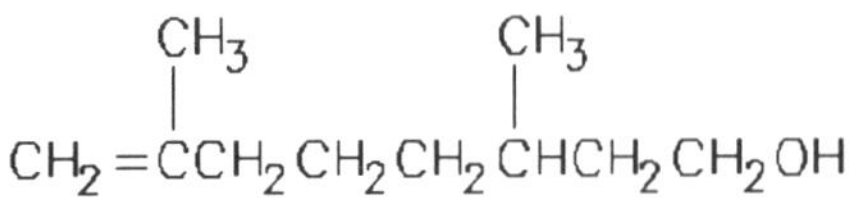

$$\begin{array}{c} \quad CH_3 \qquad\qquad\quad CH_3 \\ \quad | \qquad\qquad\qquad\quad | \\ CH_2{=}CCH_2CH_2CH_2CHCH_2CH_2OH \end{array}$$

I

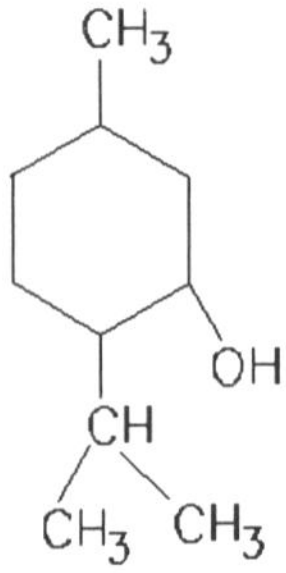

CH_3 (ring substituent), OH (ring substituent), $CH(CH_3)_2$ (ring substituent)

II

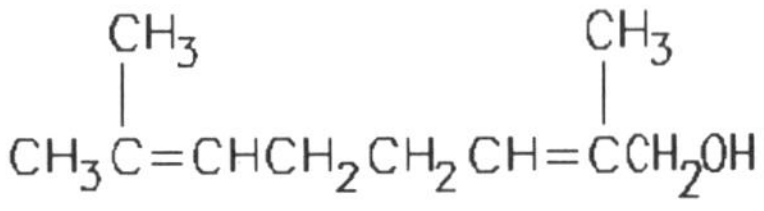

$$\begin{array}{c} \quad CH_3 \qquad\qquad\qquad\quad CH_3 \\ \quad | \qquad\qquad\qquad\qquad\quad | \\ CH_3C{=}CHCH_2CH_2CH{=}CCH_2OH \end{array}$$

III

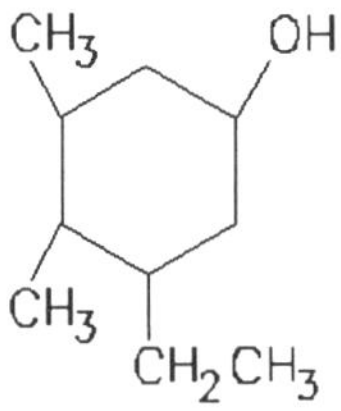

CH_3, OH, CH_3, CH_2CH_3 (ring substituents)

IV

CH_3, $C{=}CH_2$, CH_3 (ring substituents)

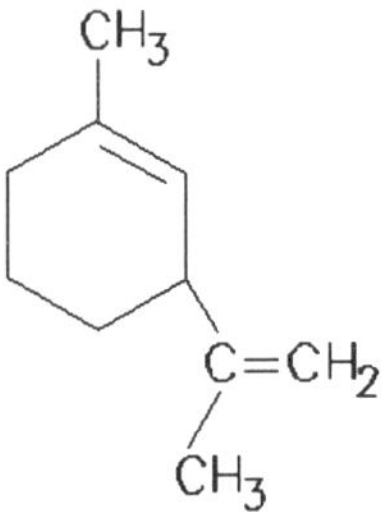

V

A) I B) II C) III * D) IV E) V

21. Which of the following compounds would be most likely to be found in nature? (Hint: recall the isoprene rule.)

A)
CH_3CH_3
3 3
$CH_2MCHCHDCHCH_2CH_2CH_2OH$

B)
CH_3 CH_3
3 3
$CH_2MCCH_2CH_2CH_2CHCH_2OH$

C)
CH_3
3
$CH_2MCCHCH_2CH_2CH_2CH_2CH_2OH$
3
CH_3

* D)
CH_3 CH_3
3 3
$CH_2MCCH_2CH_2CH_2CHCH_2CH_2OH$

E)
CH_3 CH_3
3 3
$CH_2{=}CHDCHCH_2CH_2CHCH_2OH$

22. Which compound is a sesquiterpene?

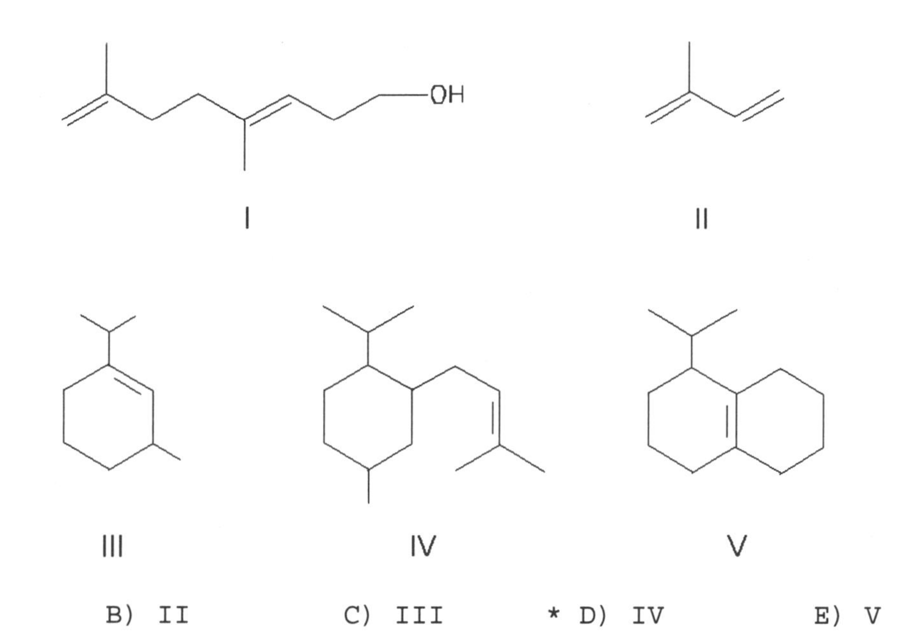

A) I B) II C) III * D) IV E) V

23. Which fatty acid is responsible for the putrid odor of rancid butter?

A) Valeric acid
B) Myristic acid
C) Stearic acid
D) Oleic acid
* E) Butyric acid

24. Which of the following is a female sex hormone?

A) Ergosterol
* B) Estradiol
C) Cortisone
D) Androsterone
E) Cholic acid

25. Which of the following is a phosphatidic acid?

A) CH_2OOCR
3
CHOOCR′
3
CH_2OOCR''

B) CH_2OH
3
CHOOCR
3
CH_2OOCR'

C) $CH_2OPO_3H_2$
3
CHOH
3
CH_2OH

* D) CH_2OOCR
3
CHOOCR′
3
$CH_2OPO_3H_2$

E) $CH_2OPO_3PO_3H_2$
3
CHOOCR
3
CH_2OOCR'

26. How could you convert an unsaturated fatty acid into a saturated fatty acid?

A) $KMnO_4$, OH^-, heat
B) OH^-, H_2O, heat; then H_3O^+
* C) H_2, Ni, pressure
D) H_3O^+, H_2O, heat
E) O_3; then Zn, H_2O

27. Which of these is a correct systematic name for progesterone?

Progesterone

A) 2-Estrene-4,20-dione
B) 5-Androstene-4,19-dione
* C) 4-Pregnene-3,20-dione
D) 5-Cholestene-5,19-dione
E) 4-Cholene-3,20-dione

28. Shown below is the formula for the antiinflammatory drug called prednisone. What is a correct systematic name for prednisone?

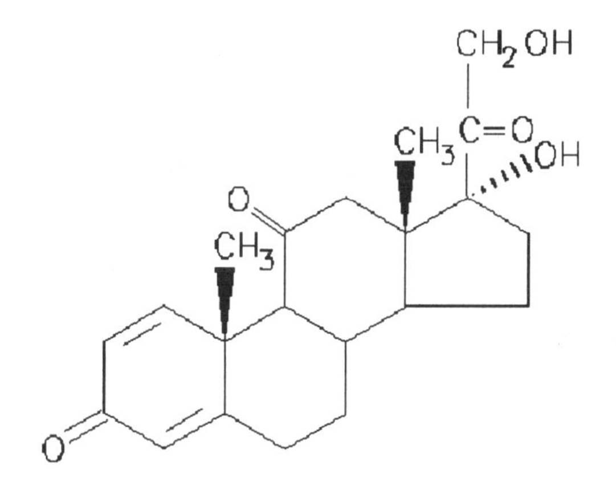

Prednisone

* A) 17!,21-Dihydroxypregna-1,4-diene-3,11,20-trione
B) 17a,21-Dihydroxypregna-1,4-diene-3,11,20-trione
C) 17!,19-Dihydroxypregna-1,4-diene-3,11,20-trione
D) 17a,19-Dihydroxypregna-1,4-diene-3,11,20-trione
E) None of the above

29. What product would you expect when progesterone is treated with one molar equivalent of hydrogen in the presence of a platinum catalyst?

Progesterone

* A) 5!-Pregnane-3,20-dione
B) 5a-Pregnane-3,20-dione
C) 5!-Estrane-3,20-dione
D) 5a-Estrane-3,20-dione
E) 5!-Androstane-3,20-dione

30. Which is the correct systematic name for the steroid shown below?

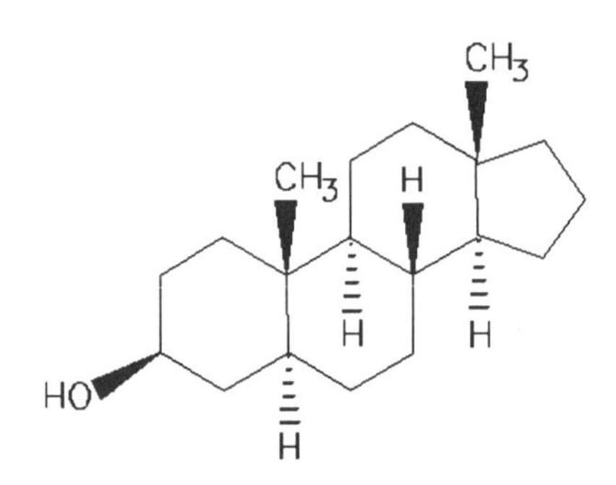

A) 5!-Androstan-3!-ol
B) 5a-Androstan-3a-ol
C) 5a-Androstan-3!-ol
* D) 5!-Androstan-3a-ol
E) 5a-Estan-3a-ol

31. Which structure represents a terpene likely to be found in nature?

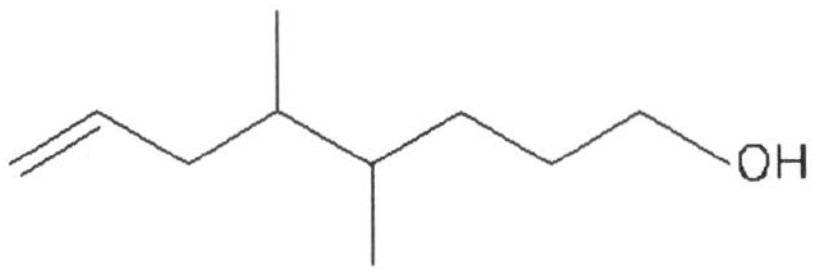

OH

I

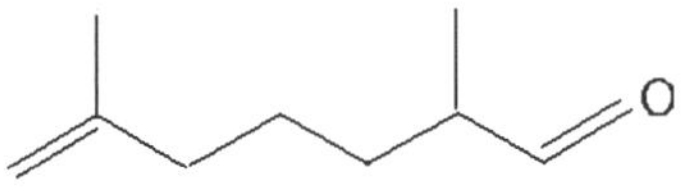

O

II

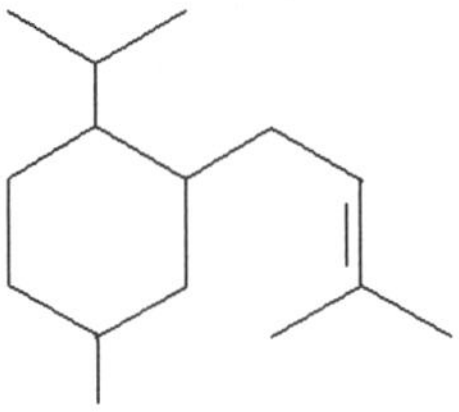

III

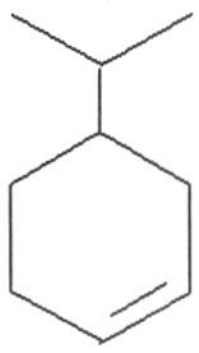

IV

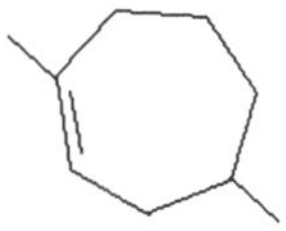

V

A) I B) II * C) III D) IV E) V

32. Which is an untrue statement concerning the fatty acid moieties of naturally-occurring triacylglycerols?

A) Generally, they possess an even number of carbon atoms.
B) Most have unbranched carbon chains.
C) The double bonds, when present, all are in the cis configuration.
* D) Where two or three double bonds are present in the same fatty acid moiety, they comprise a conjugated system.
E) The fatty acid moieties in a particular triacylglycerol usually are different.

33. Which is the repeating unit of natural rubber?

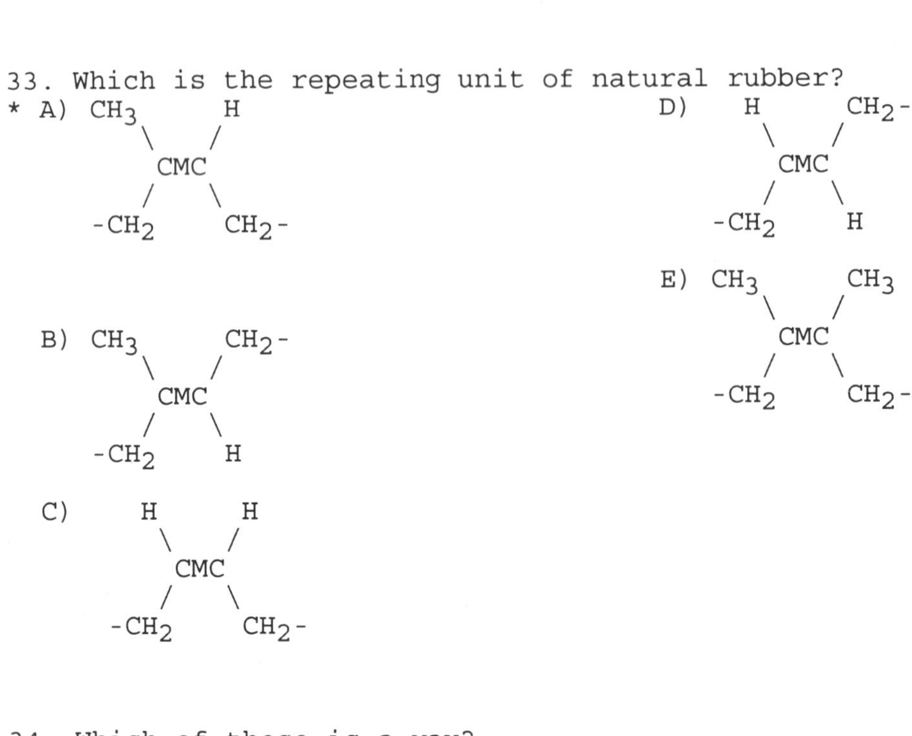

34. Which of these is a wax?

A)
```
              H
              3
CH3(CH2)20CMO
```

* B)
```
              O
              :
CH3(CH2)24COCH2(CH2)28CH3
```

C)
```
             H H       O
             3 3       :
CH3(CH2)7CMC(CH2)7COCH2CH3
```

D)
```
CH2OOC(CH2)16CH3
3
CHOOC(CH2)14CH3
3
CH2OOC(CH2)14CH3
```

E) $CH_3(CH_2)_{24}CH_2OH$

35. $-OCH_2CH_2\overset{+}{N}(CH_3)_3$ is a structural unit of which type of phosphatide?

A) Cephalins
B) Phosphatidyl serines
C) Plasmalogens
D) Lecithins
* E) Both C) and D)

36. Which of these is a male sex hormone?

A) Estrone
* B) Testosterone
C) Cholic acid
D) Cortisone
E) Estradiol

37. Which of these lipids does <u>not</u> yield glycerol upon hydrolysis?

A) A lecithin
* B) A sphingolipid
C) A cephalin
D) A triacylglycerol
E) A plasmalogen

38. Which of these reagents would not react with oleic acid?
A) H_2, Ni B) PBr_3 C) CH_3MgI D) NH_3/H_2O * E) $NaHSO_3$

39. Choline cannot be found as a product of hydrolysis of any representative of this class of lipids.
A) Sphingomyelins
B) Lecithins
C) Plasmalogens
* D) Waxes
E) Both C) and D)

40. Which of these is a detergent?
A) $CH_3(CH_2)_{16}COO^-Na^+$
B) $[CH_3(CH_2)_{14}COO^-]_2Ca^{+2}$
* C) $CH_3(CH_2)_{10}CH_2SO_3^-Na^+$
D) $HOCH_2CHOHCH_2OH$
E) $CH_3(CH_2)_{14}CH_2SH$

41. Which is not a correct statement concerning naturally-occurring triacylglycerols?
* A) The greater the degree of unsaturation, the higher the melting point.
B) Saponification yields glycerol and a mixture of carboxylic acid salts.
C) Solid examples are termed "fats."
D) Regardless of the exact nature of the R groups, such compounds are water-insoluble.
E) Such compounds frequently, but less correctly, are called "triglycerides."

42. How many isomers, including stereoisomers, exist for the triacylglycerol which, on saponification, gives 2 molar equivalents of palmitate and 1 molar equivalent of stearate?
A) 1 B) 2 * C) 3 D) 4 E) 6

43. Which is an untrue statement concerning cholesterol?
A) Cholesterol decolorizes a solution of Br_2 in CCl_4.
* B) Cholesterol reacts with 2,4-dinitrophenylhydrazine.
C) Cholesterol is optically active.
D) Cholesterol is water-insoluble.
E) All of the above are true.

44. Of the saturated fatty acids found in fats and oils, this one normally is the most abundant:
A) Capric acid
B) Lauric acid
C) Myristic acid
* D) Palmitic acid
E) Stearic acid

45. The biosynthesis of one series of prostaglandins begins with which of these fatty acids?

A) Palmitic acid
B) Stearic acid
C) Oleic acid
D) Linoleic acid
* E) Arachidonic acid

46. The ozonolysis of a fatty acid produces these fragments:

H	H	H H
3	3	3 3
$O{=}C(CH_2)_7CO_2H$	$CH_3(CH_2)_4C{=}O$	$O{=}CCH_2C{=}O$

What is the identity of the fatty acid?

A) Stearic acid
B) Palmitoleic acid
C) Oleic acid
* D) Linoleic acid
E) Linolenic acid

47. To which class of terpenes does the terpene shown below, bisabolene, belong?

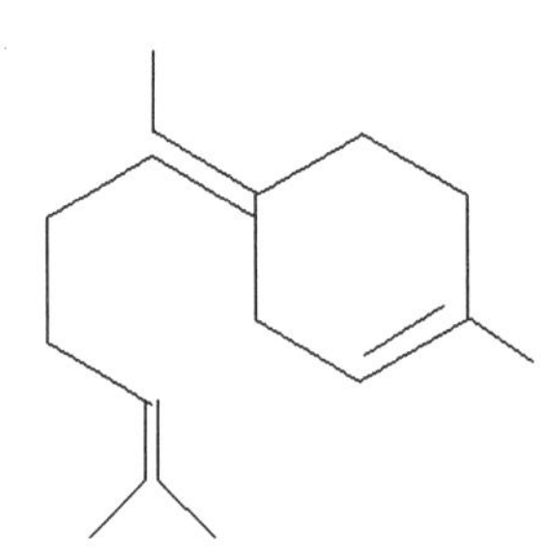

A) Monoterpenes
* B) Sesquiterpenes
C) Diterpenes
D) Triterpenes
E) Tetraterpenes

48. Which type of lipid gives these products on saponification:

$HOCH_2CHOHCH_2OH$ RCO_2^D $R'CO_2^D$ PO_4^{3D} $HOCH_2CH_2NH_2$?

A) Fat
B) Wax
C) Lecithin
* D) Cephalin
E) Plasmalogen

49. The reaction of cholesterol with dilute aqueous $KMnO_4$ at 0D5xC produces which of these compounds (A and B rings only shown)?

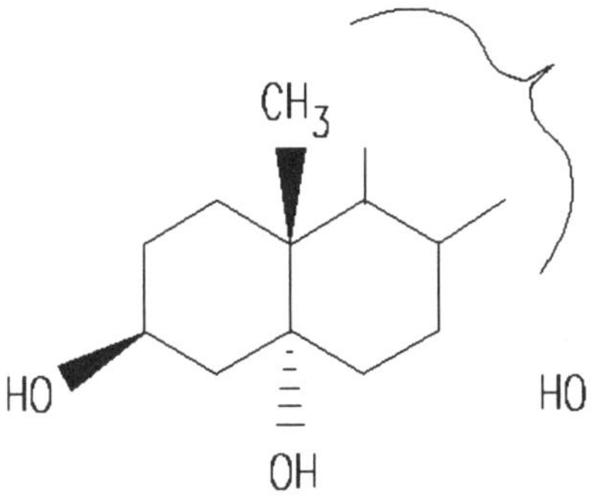

I 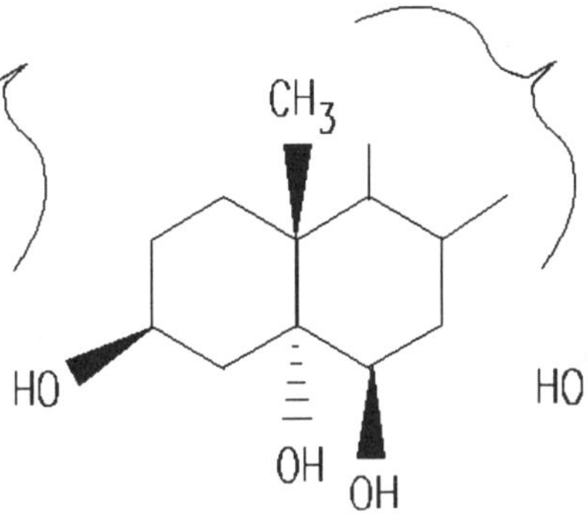 II 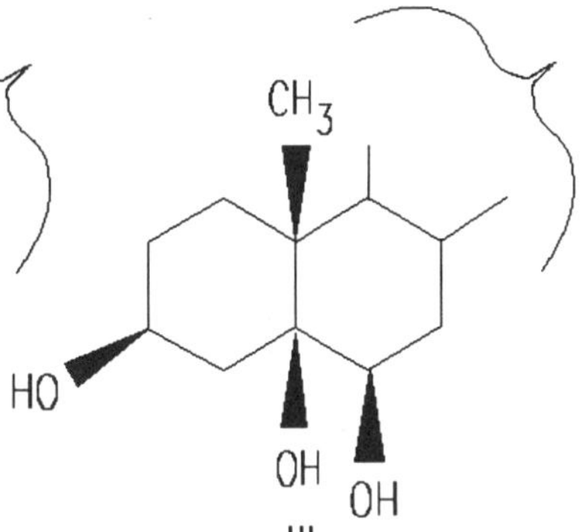 III 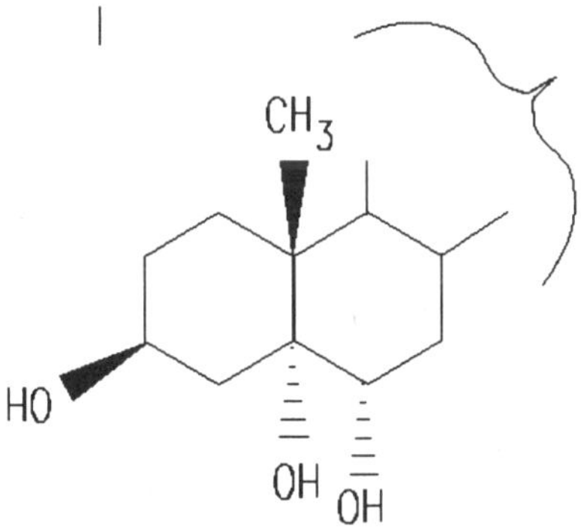 IV 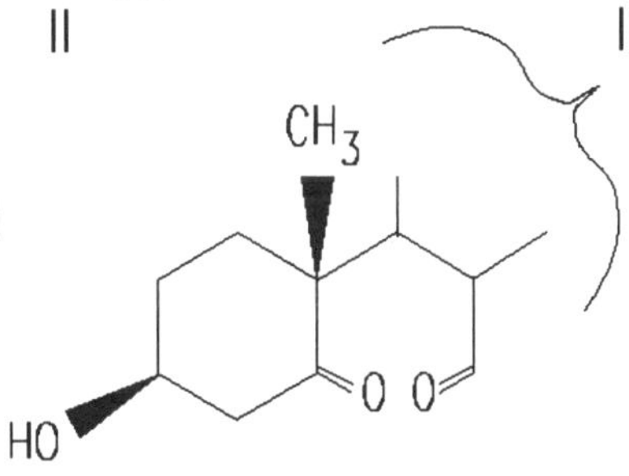 V

A) I B) II C) III * D) IV E) V

50. Which is the proper representation of three successive isoprene units in natural rubber?

A)
H_3C H H_3C H_3C H
3 3 3 3 3
$DCH_2C{=}CCH_2CH_2C{=}CCH_2CH_2C{=}CCH_2D$
3
H

* B)
H_3C H H_3C H H_3C H
3 3 3 3 3 3
$DCH_2C{=}CCH_2CH_2C{=}CCH_2CH_2C{=}CCH_2D$

C)
H_3C H_3C H_3C
3 3 3
$DCH_2C{=}CCH_2CH_2C{=}CCH_2CH_2C{=}CCH_2D$
3 3 3
H H H

D)
H_3C H_3C H_3C
3 3 3
$DCH_2C{=}CCH_2CH_2C{=}CCH_2CH_2C{=}CCH_2D$
3 3 3
H H H

E)
H_3C H H CH_3 H_3C H
3 3 3 3 3 3
$DCH_2C{=}CCH_2CH_2C{=}CCH_2CH_2C{=}CCH_2D$

1. Which amino acid would you expect to have its isoelectric point near pH 10?

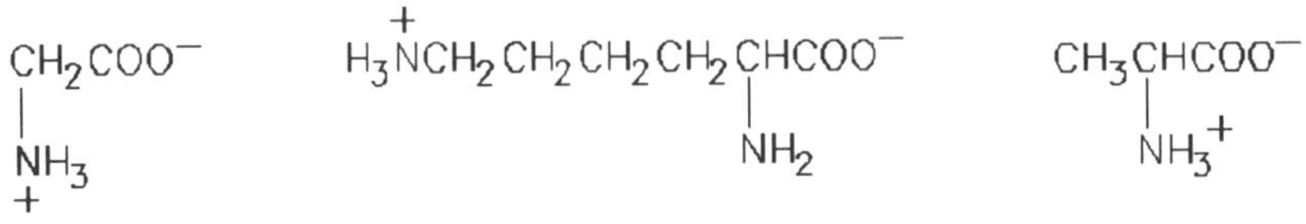

I: CH_2COO^- with $\overset{+}{N}H_3$ on the CH_2

II: $H_3\overset{+}{N}CH_2CH_2CH_2CH_2CHCOO^-$ with NH_2 on the CH

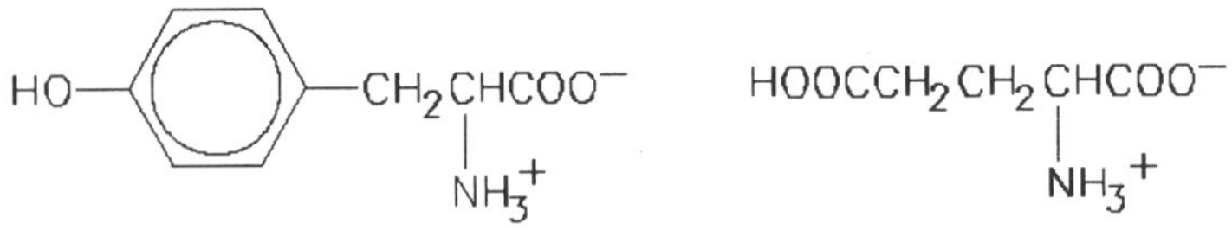

III: CH_3CHCOO^- with NH_3^+ on the CH

IV: $HO-C_6H_4-CH_2CHCOO^-$ with NH_3^+ on the CH

V: $HOOCCH_2CH_2CHCOO^-$ with NH_3^+ on the CH

A) I * B) II C) III D) IV E) V

2. Which amino acid would you expect to have its isoelectric point near pH 3?

A) CH_3CHCOO^-
3
NH_3
+

B) $C_6H_5CH_2CHCOO^-$
3
NH_3
+

* C) $HOOCCH_2CH_2CHCOO^-$
3
NH_3
+

D) +
$H_3NCH_2CH_2CH_2CH_2CHCOO^-$
3
NH_2

E) $^+NH_2$
:
$H_2NCCH_2CH_2CH_2CHCOO^D$
3
NH_2

3.

O
:
Treating alanine with $C_6H_5CH_2OCCl$ would yield:

A)
O
:
$C_6H_5CH_2OCCHCH_3$
3
NH_3
+

B) $C_6H_5CH_2CHCOO^-$
3
NH_3
+

* C)
O
:
$C_6H_5CH_2OCNHCHCOOH$
3
CH_3

D) $CH_3CHCOOH$
3
NH
3
$CH_2C_6H_5$

E) $CH_3CHCOOCH_2C_6H_5$
3
NH_2

4. The secondary structure of proteins is derived from:
A) peptide linkages.
B) disulfide linkages.
* C) hydrogen bond formation.
D) hydrophobic interactions.
E) acid-base interactions.

5. Which of the following would provide a synthesis of alanine?
* A) CH_3CH_2COOH, Br_2, P; then excess NH_3
B) Potassium phthalimide, $ClCH_2CO_2C_2H_5$; then KOH/H_2O; then HCl
C) Potassium phthalimide, $C_6H_5CH_2Br$; then KOH/H_2O; then CO_2,H^+
D) CH_3CH_2COOH, $(C_6H_5)_3CNa$; then NH_3
E) Answers A) and B)

6. What product(s) would you expect from the following reaction?

Tyrosine + Br_2(excess) $\overset{H_2O}{\text{DDDDD4}}$?

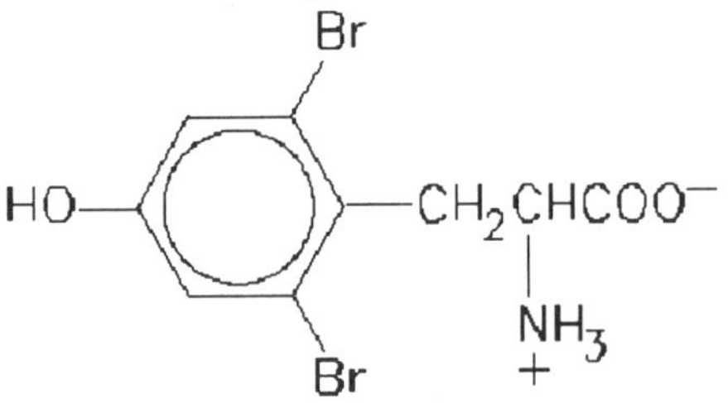

I 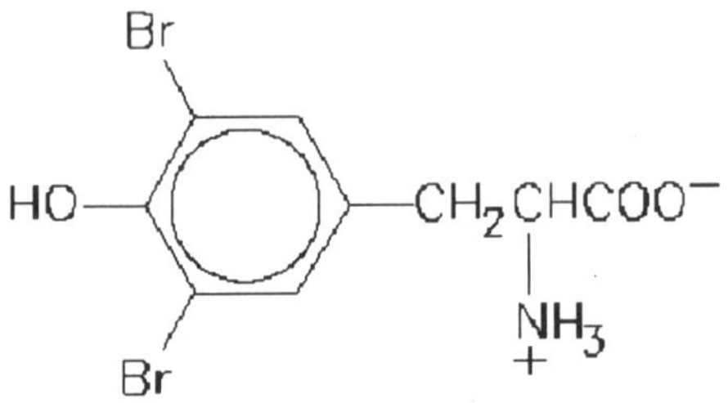 II

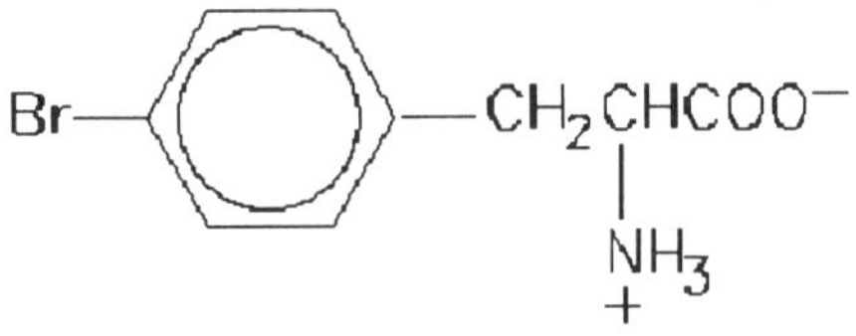

III 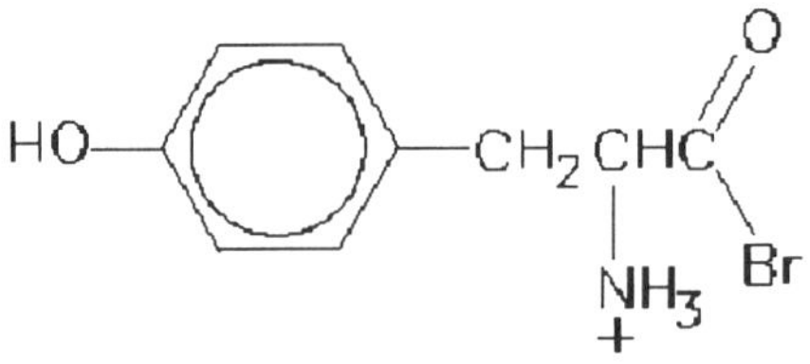 IV

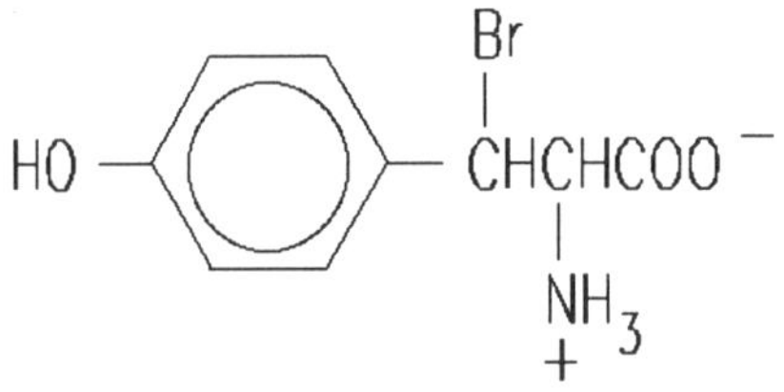

V

A) I * B) II C) III D) IV E) V

7. A pentapeptide has the molecular formula: Asp, Glu, His, Phe, Val. Partial hydrolysis of the pentapeptide gives: ValyAsp, GluyHis, PheyVal, and AspyGlu. What is the amino acid sequence of the pentapeptide?

* A) PheyValyAspyGluyHis
B) HisyGluyAspyValyPhe
C) AspyGluyHisyPheyVal
D) PheyValyGluyHisyAsp
E) GluyHisyPheyValyAsp

8. Which amino acid of a polypeptide would become labeled when the polypeptide is treated with 2,4-dinitrofluorobenzene in base, even though the amino acid is not a terminal amino acid?

* A) Lysine
B) Glycine
C) Alanine
D) Phenylalanine
E) Leucine

9. What products would you expect from the following reaction?

$$C_6H_5CH_2O\overset{O}{\overset{:}{C}}NHC_2H_5 \xrightarrow{H_2, Pd} ?$$

A) $C_6H_5CH_2OH$ + $H\overset{O}{\overset{:}{C}}NHC_2H_5$

* B) $C_6H_5CH_3$ + CO_2 + $C_2H_5NH_2$

C) $C_6H_5CH_2O\overset{O}{\overset{:}{C}}H$ + $C_2H_5NH_2$

D) $C_6H_5CH_2OCH_2NHC_2H_5$

E) $C_6H_5CH_2OH$ + CO_2 + $C_2H_5NH_2$

10. The primary structure of a protein refers to its:
* A) sequence of amino acid residues.
 B) disulfide bonds.
 C) helical structure.
 D) hydrogen bonding.
 E) All of these

11. Disulfide bonds in proteins:
 A) result from an oxidation of thiols.
 B) help to maintain the shape of proteins.
 C) can be broken by reduction.
 D) can link two cysteine amino acid residues.
* E) All of the above

12. The purple color of the anion formed in the ninhydrin test for α-amino acids is due to:
 A) the attraction of the anion to a metal in a pi-complex.
 B) intermolecular hydrogen bonding.
 C) molecular vibrations.
* D) the highly conjugated nature of the anion.
 E) the color of the ninhydrin.

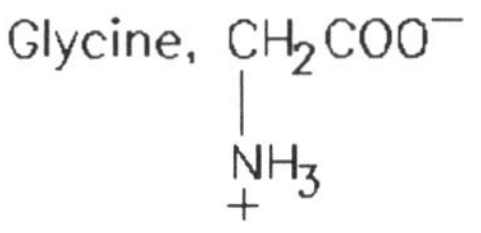

Glycine, $CH_2(NH_3^+)COO^-$ — I

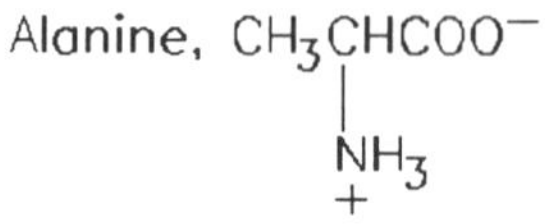

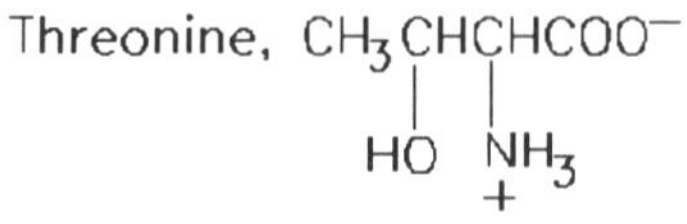

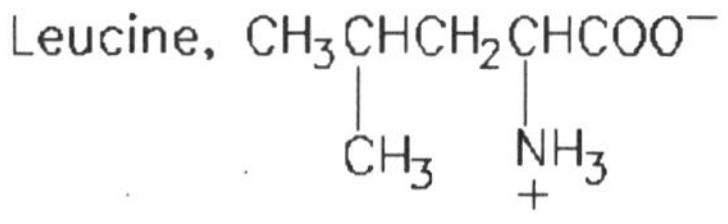

Alanine, $CH_3CH(NH_3^+)COO^-$ — II

Threonine, $CH_3CH(OH)CH(NH_3^+)COO^-$ — III

Leucine, $CH_3CH(CH_3)CH_2CH(NH_3^+)COO^-$ — IV

Proline, (pyrrolidinium ring with COO^- on C2, N^+ bearing two H) — V

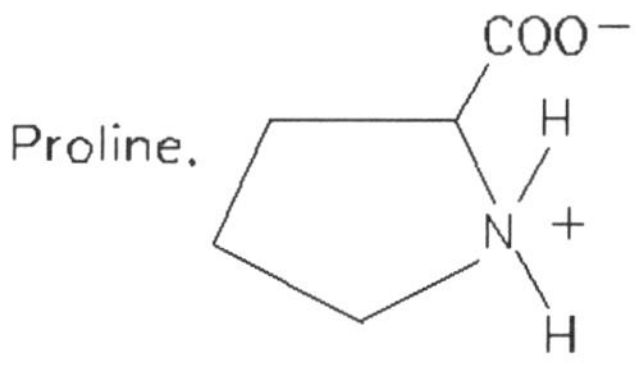

13. Which amino acid of those above is theoretically capable of existing in diastereomeric forms?

A) I B) II * C) III D) IV E) V

14. Which amino acid of those above is achiral?

* A) I B) II C) III D) IV E) V

15. What would be the predominant form of lysine in water at pH 14?

A) $\overset{+}{H_3N}(CH_2)_4CH(NH_3^+)CO_2H$

B) $\overset{+}{H_3N}(CH_2)_4CH(NH_3^+)CO_2^-$

C) $\overset{+}{H_3N}(CH_2)_4CH(NH_2)CO_2^-$

D) $H_2N(CH_2)_4CH(NH_3^+)CO_2^-$

* E) $H_2N(CH_2)_4CH(NH_2)CO_2^-$

16. The predominant form of aspartic acid in water at pH 1 would be:

* A) $HO_2CCH_2CHCO_2H$
3
NH_3^+

B) $^-O_2CCH_2CHCO_2H$
3
NH_3^+

C) $HO_2CCH_2CHCO_2^-$
3
NH_3^+

D) $^-O_2CCH_2CHCO_2H$
3
NH_2

E) $^-O_2CCH_2CHCO_2^-$
3
NH_2

17. Which amino acid is least likely to be found in a natural protein?

I: CH_2CO_2H / NH_2

II: CO_2^-, $H_3\overset{+}{N}$, C, H, CH_3

III: CO_2^-, $H_3\overset{+}{N}$, C, H, CH_2OH

IV: CO_2^-, H, C, $\overset{+}{N}H_3$, CH_3

V: CO_2^-, $H_3\overset{+}{N}$, C, H, $CH_2C_6H_5$

A) I B) II C) III * D) IV E) V

18. A heptapeptide Ala_2, Glu, Phe, Pro, Tyr, Val gives labeled alanine when heated with DNFB followed by hydrolysis. On partial hydrolysis the unlabeled heptapeptide gives the following:

AlayGlu, ProyTyr, AlayVal, TyryAla, ValyPheyPro.

What is the amino acid sequence of the heptapeptide?

A) AlayPheyProyTyryAlayGluyVal

* B) AlayValyPheyProyTyryAlayGlu

C) AlayValyPheyProyTyryGluyAla

D) AlayValyPheyTyryProyAlayGlu

E) ValyAlayPheyTyryProyAlayGlu

19. When the pentapeptide below is heated first with 2,4-dinitrofluorobenzene (and base) and then subjected to acidic hydrolysis, which amino acid will bear the dinitrophenyl group?

LeuyValyGlyyPheyIle

* A) Leucine
  B) Valine
  C) Glycine
  D) Phenylalanine
  E) Isoleucine

20. Which amino acid is unlikely to be found in a natural protein?

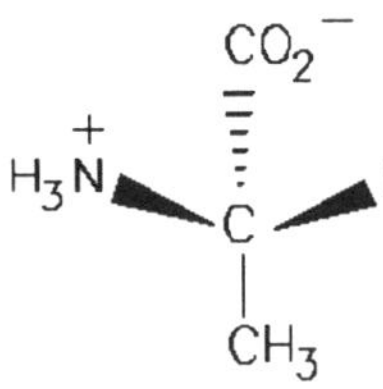

CO_2^-, $H_3\overset{+}{N}$, H, CH_3 — I

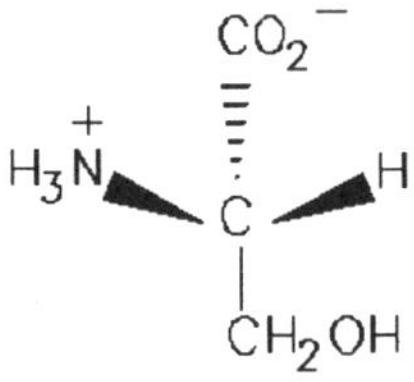

CO_2^-, $H_3\overset{+}{N}$, H, CH_2OH — II

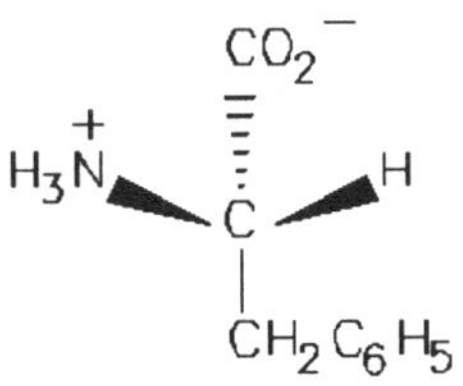

CO_2^-, $H_3\overset{+}{N}$, H, $CH_2C_6H_5$ — III

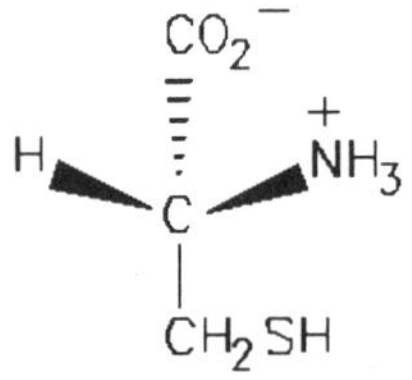

CO_2^-, H, $\overset{+}{N}H_3$, CH_2SH — IV

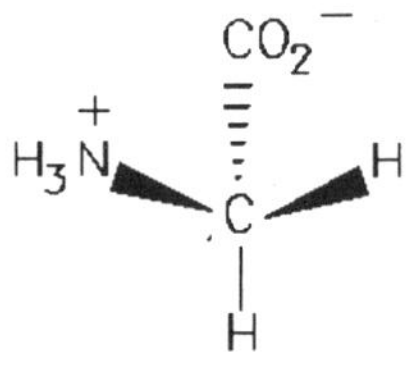

CO_2^-, $H_3\overset{+}{N}$, H, H — V

A) I B) II C) III * D) IV E) V

21. Which amino acid would have its isoelectric point near pH 10?

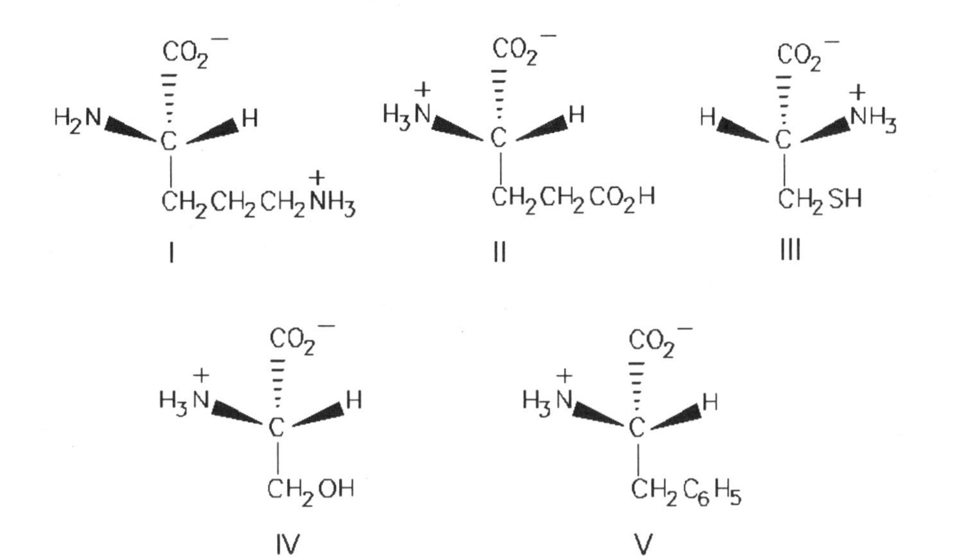

* A) I B) II C) III D) IV E) V

22. The Edman degradation uses this reagent to identify the N-terminal amino acid of a peptide or protein.

A) $C_6H_5NHNH_2$
B) $C_6H_5NH_2$
* C) C_6H_5NMCMS
D) C_6H_5NMCMO
E) Aminopeptidase

23. Which of these amino acids is formed from a precursor amino acid only after the latter has been incorporated into a polypeptide chain?

A) Serine
B) Arginine
C) Isoleucine
D) Tryptophan
* E) Hydroxyproline

24. For the accompanying fully-protonated amino acid, what is the arrangement of pK_a values in order of increasing magnitude?

I III
$HO_2CCH_2CHCO_2H$
3
$^+NH_3$
II

A) I < II < III
B) II < I < III
* C) III < I < II
D) III < II < I
E) II < III < I

25. This reagent is used to "protect" the amino group of an amino acid which is to be joined to a second amino acid by a peptide bond.

A) O
:
CH_3CCl

B) O
:
C_6H_5CCl

* C) O
:
$C_6H_5CH_2OCCl$

D) O O
: :
$C_6H_5COCC_6H_5$

E) O
:
CH_3OCCl

26. Which attractive force is responsible for maintaining the tertiary structure of a protein?

A) Disulfide linkages
B) Hydrogen bonds
C) van der Waals forces
D) Hydrophobic interactions
* E) All of these

27. The occurrence of this amino acid in a polypeptide chain disrupts an !-helix:

* A) Proline
B) Alanine
C) Methionine
D) Histidine
E) Tyrosine

28. Which of these amino acid residues is expected to prefer the interior of a protein to the exterior when the protein is in aqueous medium?

A) Threonine
* B) Valine
C) Serine
D) Aspartic acid
E) Lysine

29. Which of these is used to convert a protein into smaller, more manageable fragments for subsequent structural studies?

A) Insulin
B) Aminopeptidase
C) Carboxypeptidase
* D) Trypsin
E) 2,4-Dinitrofluorobenzene

30. Why is this two step sequence, $CH_3CH_2CO_2H$ + Br_2, P(trace); then NH_3, not a good method for the preparation of L-alanine?

A) Bromination (first step) is not regioselective.
B) The second step is an elimination reaction leading to CH_2MCHCO_2H.
C) NH_3 is not sufficiently nucleophilic to perform the second step.
D) The second step is seriously sterically hindered.
* E) Bromination produces a racemic intermediate which leads to racemic product.

31. Which of these amino acids contains a hydrophobic side chain?
A) Lysine
B) Serine
* C) Methionine
D) Arginine
E) Cysteine

32. Which one of these amino acids does not give the usual purple color with ninhydrin?
A) Histidine
* B) Proline
C) Tryptophan
D) Leucine
E) Aspartic acid

33. The pH at which the concentration of the dipolar ion (zwitterion) form of an amino acid is at a maximum and the cationic and anionic forms are at equal concentrations is termed the
A) end point.
B) equivalence point.
C) neutral point.
* D) isoelectric point.
E) dipolar point.

34. How many different tripeptides can exist, each containing one residue of glycine, one of L-threonine, and one of L-arginine?
A) 2 B) 3 * C) 6 D) 8 E) 9

35. What use is made of dicyclohexylcarbodiimide (DCC) in peptide synthesis?
A) DCC "protects" the amino group of the intended N-terminal amino acid.
* B) DCC activates the carboxyl group of one amino acid so that this amino acid reacts more readily with a second amino acid.
C) DCC cleaves the blocking groups from the final peptide.
D) DCC is the resin used in the automated synthesis of peptides.
E) DCC removes the peptide from the resin at the conclusion of the synthesis.

36. Which is an incorrect statement concerning the tetrapeptide

L-arginine-L-leucine-L-cysteine-L-phenylalanine?

A) This peptide would have an isoelectric point greater than 6.0.
B) The peptide will be modified by mild oxidizing agents.
C) Trypsin will catalyze preferentially the hydrolysis of the arginine-leucine peptide bond.
* D) Chymotrypsin will catalyze preferentially the hydrolysis of the cysteine-phenylalanine peptide bond.
E) All peptide bonds will be cleaved by refluxing with 6 M HCl for 24 hours.

37. A "conjugated protein" is one which:
A) possesses catalytic properties.
B) is a digestive enzyme.
C) exists largely as an !-helix.
D) contains unsaturated amino acids.
* E) contains a nonprotein group as part of the molecule.

38. Which is an isolable intermediate in the Strecker synthesis of an amino acid?

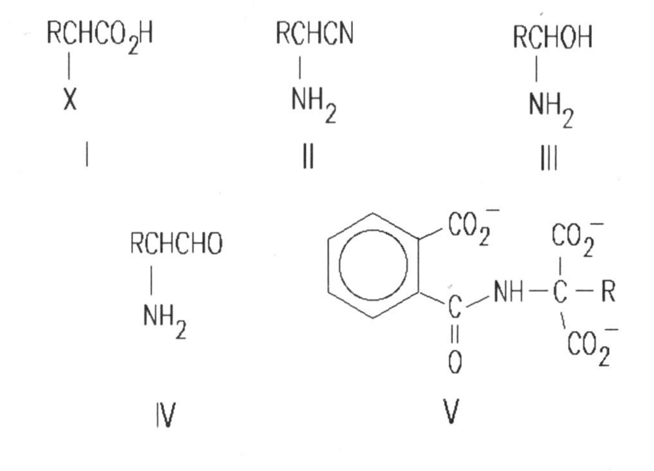

A) I   * B) II   C) III   D) IV   E) V

39. Which of these natural amino acids contains an amide function?
* A) Asparagine
B) Proline
C) Arginine
D) Histidine
E) None of these

40. For an amino acid, $HO_2CDRDCHCO_2H$,
3
NH_2

$pK_{a_1} = 2.40$, $pK_{a_2} = 10.0$, $pK_R = 4.00$ What is the pI of this amino acid?
A) 1.6   * B) 3.2   C) 5.5   D) 6.2   E) 7.0

41. Pipecolic acid logically would be substituted for which natural amino acid in the synthesis of peptide analogs?

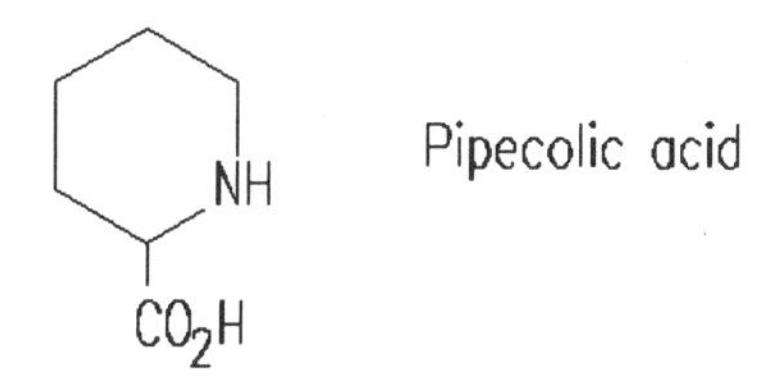

A) Histidine
* B) Proline
C) Tryptophan
D) Phenylalanine
E) Tyrosine

42. Which of these amino acids has the R configuration at the stereocenter but, nonetheless, is an L amino acid?

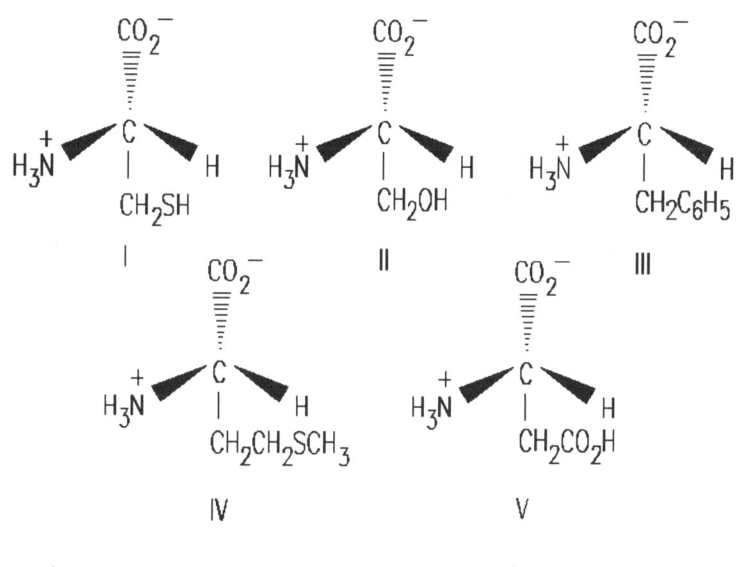

* A) I
B) II
C) III
D) IV
E) V

43. Consider this tripeptide, H_2N—leu—lys—phe—CO_2H
3
NH_2

What is the best estimate of the pI of this compound?

	$pK_{a_{COOH}}$	$pK_{a_{NH_3^+}}$	pK_R	pI
leu	2.4	9.6		6.0
lys	2.2	9.0	10.5	9.8
phe	1.8	9.1		5.5

A) 6.2 B) 7.1 C) 9.2 * D) 10.1 E) 11.3

1. The monomeric units of nucleic acids are which of these?
A) D-ribose or 2-deoxy-D-ribose
B) Phosphate ions
C) Purines
D) Nucleosides
* E) Nucleotides

2. The primary function of nucleic acids is:
A) the catalysis of biochemical reactions.
B) the regulation of reactions that occur in the body.
* C) the preservation, transcription and translation of information.
D) the acid-catalyzed hydrolysis of nucleotides.
E) the neutralization of nucleic bases.

3. Complete hydrolysis of adenylic acid would yield which of these?
A) Adenosine and a phosphate ion
* B) Adenine, D-ribose, and a phosphate ion
C) Adenine and D-ribose
D) Adenine and 2-deoxy-D-ribose
E) A pyrimidine, a pentose, and a phosphate ion

4. Supply the missing reagent in the following synthesis.

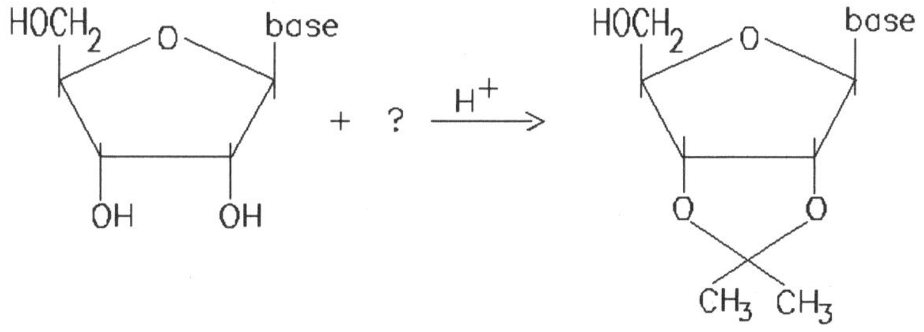

$HOCH_2$ O base, OH OH + ? $\xrightarrow{H^+}$ $HOCH_2$ O base, O O, CH_3 CH_3

* A) O
:
CH_3CCH_3

B) $HOCH_2CH_2OH$

C) $CH_3CHOHCH_3$

D) $CH_3CCl_2CH_3$

E) $CH_3CHClCH_3$

5. Supply the missing reagent(s) in the synthesis below.

$$\text{HOCH}_2\text{-(2,3-O-isopropylidene ribose)-base} + ? \xrightarrow{\text{pyridine}} (C_6H_5CH_2O)_2\overset{O}{\overset{\|}{P}}OCH_2\text{-(2,3-O-isopropylidene ribose)-base}$$

A) PO_4^{-3} and $C_6H_5CH_2OH$

B) PO_4^{-3} and $C_6H_5CH_2Cl$

* C) $(C_6H_5CH_2O)_2\overset{O}{\overset{:}{P}}Cl$

D) $C_6H_5CH_3$ + $POCl_3$

E) $C_6H_5CH_2OH$ + H_3PO_4

6. Hydrolysis of DNA from various species gives which of the following results?

A) The mole percentage of adenine is approximately equal to that of thymine.
B) The mole percentage of cytosine is approximately equal to that of guanine.
C) The total mole percentage of purines is approximately equal to that of pyrimidines.
* D) All of the above
E) None of the above

7. What is the secondary structure of DNA?

A) An alpha-helix
B) A pleated sheet
C) A flat sheet
* D) A double helix
E) A random coil

8. Which is the predominant tautomeric form of cytosine when it is present in DNA?

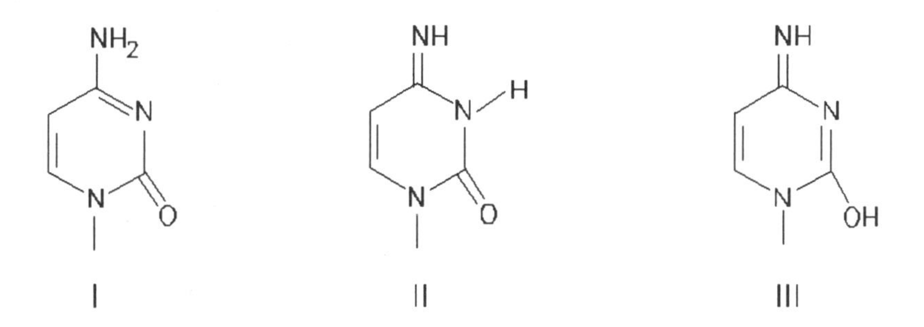

* A) I
B) II
C) III
D) None of these
E) I, II and III are present in approximately equal amounts.

9. If the base sequence along a segment of DNA were GDCDCDADT, what would be the base sequence of messenger RNA synthesized from this sequence?
A) GDCDCDADT B) CDGDGDTDA* C) CDGDGDUDA D) GDCDCDADU E) ADTDTDGDC

10. What is the transfer RNA anticodon for the messenger RNA codon, GDCDA?
A) TDADT B) GDUDT C) GDCDA D) ADCDG * E) CDGDU

11. A nucleotide unit is composed of:
A) a five carbon monosaccharide.
B) a phosphate group.
C) a heterocyclic base.
D) an amino acid.
* E) A), B) and C)

12. A purine:
A) contains four nitrogens in the ring system.
B) is bicyclic.
C) can participate in hydrogen bonding.
D) is a heterocyclic base.
* E) All of the above

13. In a nucleotide unit, the components are sequentially linked:
A) monosaccharide--phosphate--heterocyclic base.
B) amino acid--monosaccharide--phosphate.
* C) phosphate--monosaccharide--heterocyclic base.
D) monosaccharide--amino acid--phosphate.
E) heterocyclic base--phosphate--monosaccharide.

14. Which of the following is <u>not</u> a pyrimidine?
A) Cytosine
B) Thymine
* C) Guanine
D) Uracil
E) All of the above are pyrimidines.

15. The hydrogen bonding for the base pairs of DNA are between:
A) amide carbonyl and $-NH_2$.
B) amide N-H and cyclic amine nitrogens.
C) alcohols and carbonyls.
D) All of these
* E) Only A) and B)

16. Nitrous acid is a suspected mutagen because of its reaction with:
* A) amine groups.
B) monosaccharides.
C) phosphates.
D) alcohols.
E) ketones.

17. The synthesis of proteins according to "the central dogma of molecular genetics" would require:
A) replication then translation.
B) translation then replication.
C) replication then transcription.
* D) transcription then translation.
E) Any of these

18. Which heterocyclic base found in RNA does not occur in DNA?
A) Guanine B) Thymine C) Cytosine * D) Uracil E) Adenine

19. The formation of a new DNA molecule which is an exact copy of a pre-existing one is designated by what term?
A) Duplication
B) Transcription
C) Translation
* D) Replication
E) Reproduction

20. The analytical data for DNA samples which led to generalizations such as (%G + %A) w (%C + %T) came from the research of what individual?
A) James Watson
B) Francis Crick
* C) Edwin Chargaff
D) Arthur Kornberg
E) Maurice Wilkins

21. Which bases pair in DNA by hydrogen bonding?
A) Cytosine and thymine
B) Cytosine and uracil
C) Adenine and guanine
* D) Adenine and thymine
E) Adenine and uracil

22. Which is the initial N-terminal amino acid in the developing polypeptide in bacteria?
A) Cysteine
B) Methionine
* C) N-Formylmethionine
D) Glycine
E) Alanine

23. Where is found the sequence of bases termed an "anticodon"?
A) mRNA
* B) tRNA
C) rRNA
D) DNA
E) Polysomes

24. In DNA, the lactim form of guanine will pair with which base?
A) Cytosine
B) Adenine
C) Uracil
* D) Thymine
E) No pairing is possible.

25. An RNA nucleoside will undergo hydrolysis in dilute acid to yield which of the following?
A) 2-Deoxy-D-ribose and a heterocyclic base
B) 2-Deoxy-D-ribose, a heterocyclic base, and phosphate ion
* C) D-Ribose and a heterocyclic base
D) D-Ribose, a heterocyclic base, and phosphate ion
E) D-Ribose and phosphate ion

26. Which is an incorrect statement concerning the DNA double helix?
A) The sugar-phosphate backbone is on the outside of the helix and the base pairs are on the inside.
* B) The two strands are identical but proceed in opposite directions.
C) Hydrogen bonding holds together the two strands.
D) Only purine-pyrimidine base pairs can be accomodated.
E) The sugar-phosphate backbone is completely regular.

27. In DNA sequencing, what is used to convert DNA molecules into smaller, more manageable fragments?
A) DNA polymerase
B) Dilute HCl
C) Trypsin
D) Chymotrypsin
* E) Restriction endonucleases

28. Separation of the fragments produced in the chemical sequencing of a DNA segment is achieved by the use of which of these techniques?
A) Column chromatography
B) Fractional distillation
C) Thin layer chromatography
D) Gas chromatography
* E) Gel electrophoresis

29. Which of these has the smallest molecular weight?
A) rRNA
B) mRNA
* C) tRNA
D) DNA

30. If a tRNA anticodon is GUC, what was the original base sequence in DNA?
A) GUC * B) GTC C) CAG D) GAC E) CTG

31. Which is the messenger codon which calls for the initiation of protein synthesis?
* A) AUG B) GUA C) UAG D) AUC E) CAC

32. The following,

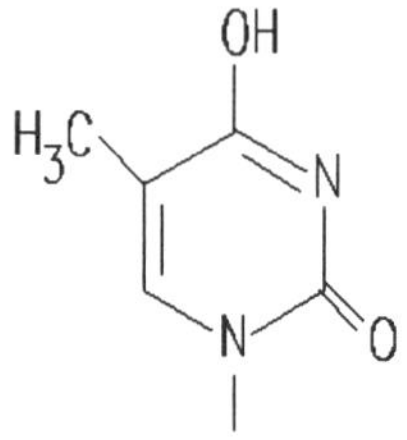

is the less common (in DNA, at least) tautomeric form of:
A) Cytosine * B) Thymine C) Adenine D) Guanine E) Uracil

33. The action of nitrous acid on 5-methylcytosine produces which nitrogen base?

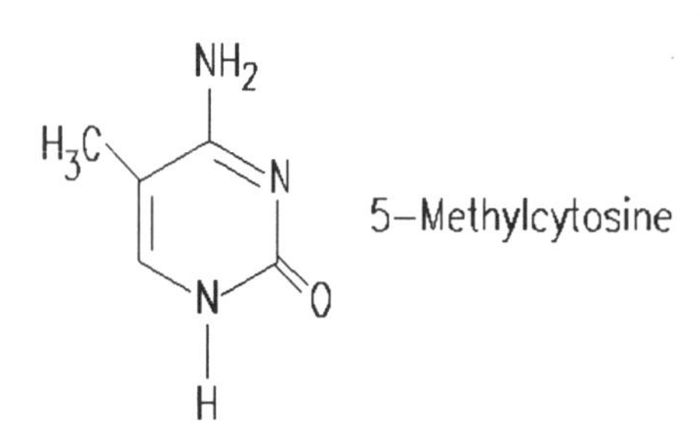

A) Adenine B) Guanine C) Cytosine * D) Thymine E) Uracil

34. Which statement concerning the monosaccharide portion of nucleotides is true in every case?
A) The monosaccharide is in the pyranose form.
B) The linkage to phosphate is at carbon #3.
* C) The heterocyclic base is attached to carbon #1 of the monosaccharide.
D) The monosaccharide - nitrogen base linkage is !.
E) The monosaccharide is 2-deoxy-D-ribose.

35. Which are the products of hydrolysis of a RNA nucleotide?
  A) D-Ribose, adenine
* B) D-Ribose, guanine, phosphate
  C) D-Ribose, thymine, phosphate
  D) 2-Deoxy-D-ribose, cytosine, phosphate
  E) 2-Deoxy-D-ribose, adenine

36. The synthesis of adenosine by the reaction shown below likely occurs by what reaction mechanism?

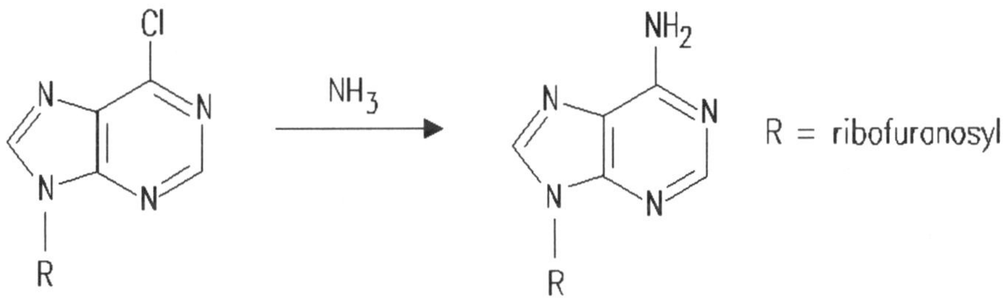

Cl
N N
N N
R
NH_3
NH_2
N N
N N
R
R = ribofuranosyl

  A) An S_N2 reaction
* B) An S_NAr reaction
  C) An elimination-addition reaction
  D) An S_N1 reaction
  E) An E_2 reaction

37. Relatively short synthetic strands of DNA complementary to certain portions of a gene are known as:
* A) antisense nucleotides.
  B) templates.
  C) palindromes.
  D) endonucleases.
  E) polymerases.

38. Concerning the genetic code, which of the following is an incorrect statement?
  A) Not all codes specify amino acids.
  B) There is a total of 64 different triplets.
* C) For each amino acid there is the same number of codes.
  D) The triplet codes are incorporated in messenger RNA.
  E) The triplet AUG is a "start" code.